International Series in Operations Research & Management Science

Volume 253

More information about this series at http://www.springer.com/series/6161

Richard W. Cottle · Mukund N. Thapa

Linear and Nonlinear Optimization

 Springer

Richard W. Cottle
Stanford University
Stanford, CA
USA

Mukund N. Thapa
Optical Fusion, Inc.
Palo Alto, CA
USA

ISSN 0884-8289   ISSN 2214-7934   (electronic)
International Series in Operations Research & Management Science
ISBN 978-1-4939-7053-7   ISBN 978-1-4939-7055-1   (eBook)
DOI 10.1007/978-1-4939-7055-1

Library of Congress Control Number: 2017940361

Printed on acid-free paper

This Springer imprint is published by Springer Nature
The registered company is Springer Science+Business Media LLC
The registered company address is: 233 Spring Street, New York, NY 10013, U.S.A.

ABOUT THE AUTHORS

Richard W. Cottle is a Professor Emeritus from the Department of Management Science and Engineering at Stanford University. He received the degrees of A.B. and A.M. from Harvard University and the Ph.D. from the University of California at Berkeley, all three in mathematics. Under the supervision of George B. Dantzig, Cottle wrote a dissertation in which the linear and nonlinear complementarity problems were introduced. Upon completion of his doctoral studies in Berkeley, Cottle became a member of the technical staff of Bell Telephone Laboratories in Holmdel, New Jersey. Two years later, he joined the operations research faculty at Stanford University where he remained until his retirement in 2005.

For nearly 40 years at Stanford, Cottle taught at the undergraduate, master's, and doctoral levels in a variety of optimization courses including linear and nonlinear programming, complementarity and equilibrium programming, and matrix theory. (The present volume is an outgrowth of one such course.) Most of Cottle's research lies within these fields.

A former Editor-in-Chief of the journals *Mathematical Programming* and *Mathematical Programming Study*, Richard Cottle is well known for *The Linear Complementarity Problem*, a Lanchester Prize winning monograph he co-authored with two of his former doctoral students, Jong-Shi Pang and Richard E. Stone. In retirement he remains active in research and writing.

Mukund N. Thapa is the President & CEO of Optical Fusion, Inc., and President of Stanford Business Software, Inc. He received a bachelor of technology degree in metallurgical engineering from the Indian Institute of Technology, Bombay. His Bachelor's thesis was on operations research techniques in iron and steel making. Later he obtained M.S. and Ph.D. degrees in operations research from Stanford University in 1981. His Ph.D. thesis was concerned with developing specialized algorithms for solving large-scale unconstrained nonlinear minimization problems. By profession he is a software developer who produces commercial software products as well as commercial-quality custom software. Since 1978, Dr. Thapa has been applying the theory of operations research, statistics, and computer science to develop efficient, practical, and usable solutions to a variety of problems.

At Optical Fusion, Inc., Dr. Thapa architected the development of a multi-point videoconferencing system for use over all IP-based networks. He holds several patents in the area. The feature-rich system focuses primarily on the needs of users and allows corporate users to seamlessly integrate conferencing in everyday business interactions. At Stanford Business Software, Dr. Thapa, ensured that the company produces high-quality turnkey software for clients. His expert knowledge of user friendly interfaces, data bases, computer science, and modular software design plays an important role in making the software practical and robust. His speciality is the application of numerical analysis methodology to solve mathematical optimization problems. An experienced modeler, he is often asked by clients to consult, prepare analyses, and to write position papers.

At the Department of Management Science and Engineering, Stanford University, Dr. Thapa has taught graduate-level courses in mathematical programming computation and numerical methods of linear programming. He is best known for his books with George B. Dantzig: *Linear Programming 1: Introduction* and *Linear Programming 2: Theory and Extensions*.

We dedicate this book to

the memory of
George B. Dantzig
our mentor, colleague, and great friend

and to

My Family

— RWC

Radhika Hegde Thapa, my wife
and best friend; Isha, my daughter;
Devi Thapa, my mother; and to
N.S. Thapa, my father, *in memoriam*

— MNT

CONTENTS

LIST OF FIGURES

LIST OF TABLES

LIST OF EXAMPLES

LIST OF ALGORITHMS

PREFACE

The subject called "optimization" is concerned with maximization and minimization. More precisely, the purpose of optimization is to find the values of *variables* that either maximize or minimize the value of a given *function*. In many cases, especially among those studied in this book, the variables are required to satisfy side conditions such as equations or inequalities in which case the term *constrained optimization* is appropriately used. When no such side conditions are imposed, the optimization problem is said to be *unconstrained*.

In an effort to convey the importance of their subject, some contemporary writers on optimization have joyfully quoted a line published in 1744 by Leonhard Euler [60], arguably the most prolific mathematician of all time. In a work on elastic curves, Euler proclaimed: *Nihil omnino in mundo contingit, in quo non maximi minimive ratio quæpiam eluceat.* That is: *Nothing at all takes place in the universe in which some rule of maximum or minimum does not appear.* In addition to being a great mathematician, Euler was a profoundly religious man. The quotation reproduced here should be construed as a pious view in line with what is called Leibnizian optimism. One sees this by considering the entire sentence from which Euler's statement is excerpted and the sentence that comes after it. As translated by Oldfather, Ellis, and Brown [153, pp. 76–77], Euler wrote:

For since the fabric of the universe is most perfect, and is the work of a most wise Creator, nothing at all takes place in the universe in which some rule of maximum or minimum does not appear. Wherefore there is absolutely no doubt that every effect in the universe can be explained as satisfactorily from final causes, by the aid of maxima and minima, as it can from the effective causes themselves.

This outlook brings to mind the character Dr. Pangloss in Voltaire's *Candide* who insisted that this is "the best of all possible worlds."

Having strong connections with human activity, both practical and intellectual, optimization has been studied since classical antiquity. Yet because of these connections, optimization is as modern as today. In their handbook on optimization [148], the editors, Nemhauser, Rinnooy Kan, and Todd, declare that

> No other class of techniques is so central to operations research and management science, both as a solution tool and as a modeling device. To translate a problem into a mathematical optimization model is characteristic of the discipline, located as it is at the crossroads of economics, mathematics and engineering.

Around the middle of the 20th century, some kinds of optimization came to be known as "mathematical programming." *Linear programming* came first; soon thereafter came *nonlinear programming, dynamic programming*, and several other types of programming. Together, these subjects and others related to them were subsumed under the title *mathematical programming*. The term "programming" was originally inspired by work on practical planning problems. Programming in this sense simply means the process of finding the levels and timing of activities. This work came to entail mathematical modeling, the development and implementation of solution techniques, and eventually the scientific study of model properties and algorithms for their solution.

The terms "mathematical programming" and "computer programming" came into existence at about the same time.[1] Whereas the latter has become a household word, the same cannot be said of the former. When "mathematical programming" turns up in everyday language, it is typically either not understood at all or confused with "computer programming." Adding to the confusion is the fact that computers and computer programming are used in the implementation of solution methods for mathematical programming problems. The contemporary use of "optimization" as a synonym for "mathematical programming" reflects the wide-spread desire among profes-

---

[1]For a book on classical and "modern" optimization problems, see Nahin [146]. The author invokes the Euler quotation given above, but it must be said that this publication gives little more of the context in which it was stated as well as a reference to Oldfather et al. [153], an annotated English translation of Euler's tract.

sionals in the field to avoid the confusion between these different senses of the word "programming." It also helps to anchor the subject historically.

In 1971, the scientific journal *Mathematical Programming* came into existence and along with it, the Mathematical Programming Society (MPS) which undertook the sponsorship of a series of triennial "International Symposiums on Mathematical Programming" that had actually begun in 1949. Today there are many journals dealing solely with mathematical programming (optimization), and there are many more that cover it along with other subjects. Likewise there are dozens of conferences every year that feature this topic. In 2010 the MPS membership voted to change the name of their organization to *Mathematical Optimization Society* (MOS). For the sake of archival continuity, however, the name of the society's journal was not changed.

Faced with this linguistic transition from "mathematical programming" to "optimization," we use both terms. But when it comes to talking about linear programming, for instance, we feel there is too much current literature and lore to abandon the name altogether in favor of "linear optimization." Furthermore, it is traditional—even natural—to refer to a linear programming problem as a "linear program." (Analogous usage is applied to nonlinear programming problems.) We are unaware of any suitable replacement for this terminology. Accordingly, we use it frequently, without reservation.

The remarkable growth of optimization in the last half century is attributable to many factors, not least of which is the effort exerted around the world in the name of national defense. This brought forth a great upsurge of activity in problem solving which, in turn, called for problem formulation, analysis, and computational methods. These are aspects of optimization, just as they are for many science and engineering disciplines.

The recognition of a "problem" is an all-important element of practical work, especially in management science. This can take many forms. For example, one is the realization that something needs to be created or put in place in order to get some job done. Today we are bombarded by advertising for products described as "solutions," for instance, "network solutions." Implicitly, these are responses to problems. Oftentimes, a problem of getting something done requires *choices* whose consequences must satisfy given *conditions*. There may even be a question of whether the given conditions can be satisfied by the allowable choices. These are called *feasibility issues*. A set of choices that result in the conditions being satisfied is called a *feasible solution,* or in everyday language, a *solution.* The problem may be to find a feasible solution that in some sense is "better" than all others. Depending on how goodness is judged, it may be necessary to find a feasible solution for which the corresponding measure is cheapest, most profitable, closest, most (or least) spacious, most (or least) numerous, etc. These goals suggest maximization or minimization of an appropriate measure of goodness based on feasible solutions. This, of course, means *optimization,* and a feasible solution that maximizes or minimizes the measure of goodness in question is called an *optimal solution.*

Modeling

After the problem is identified, there arises the question of how to represent it.[2] Some representations are graphic, others are verbal. The representations of interest in optimization are mathematical, most often algebraic. The choices mentioned above are represented by *variables* (having suitable units), the conditions imposed on the variables (either to restrict the values that are allowable or to describe what results the choices must achieve) are represented by *relations, equations,* or *inequalities* involving *mathematical functions.* These are called *constraints.* When a measure of goodness is present, it is expressed by a mathematical function called the *objective function* or simply the *objective.* This mathematical representation of the (optimization) problem is called a *mathematical model.*

Linear and nonlinear optimization models are so named according to the nature of functions used as objective and constraints. In linear optimization, *all* the functions used in the mathematical model must be linear (or strictly speaking, affine), whereas in nonlinear optimization, *at least one* of them must be nonlinear. An optimization model with a linear objective functions would be classified as nonlinear as long as it had a nonlinear function among its constraints. Some nonlinear optimization models have only linear constraints, but then of course, their objective functions are nonlinear.

For example, $c(x, y) = 3x + 4y$ is a linear function, and $c(x, y) \leq 5$, or equivalently $3x + 4y \leq 5$, is a linear inequality constraint. On the other hand the function $f(x, y) = x^2 + y^2$ is a nonlinear function, and $f(x, y) \leq 2$, or equivalently $x^2 + y^2 \leq 2$, is a nonlinear inequality constraint.

---

[2]For two books on modeling, see [4] and [191].

In general, mathematical models of problems have several advantages over mere verbal representations. Among these are *precision, clarity, and brevity.* In effect, models get down to the essence of a real situation and facilitate a precise way of discussing the problem. Some other advantages of mathematical models are their *predictive powers* and *economy.* Using mathematical models to represent real-world situations, it is possible to perform experiments (simulations) that might otherwise be expensive and time-consuming to implement in a real physical sense.

Of course, the value of a mathematical model has much to do with its accuracy. It is customary to distinguish between the *form* of a model and its *data.* In building a mathematical model, one chooses mathematical functions by which to represent some aspect of reality. In an optimization problem, there is an objective function, and there may be constraints. The representation of reality is usually achieved only by making some *approximations* or *assumptions* that affect the form of the model. This leaves room for questioning its adequacy or appropriateness. Mathematical models typically require data, also known as *parameter values.* Here the question of *accuracy* of the parameters is important as it can affect the results (the solution). If optimization is to yield useful information, the matter of acquiring sufficient, accurate data cannot be overlooked.

Computation (algorithms and software)

After the formulation of an optimization model and the collection of data, there comes a desire to "get the answer," that is, to find an optimal solution. As a rule, this cannot be done by inspection or by trial and error. It requires specialized computational methods called *algorithms.* You should have encountered maximization and minimization problems in the study of calculus. These are usually not difficult to solve: you somehow find a root of $f'(x) = 0$ or, in the multivariate case, $\nabla f(x) = 0$. Yet even this can require some work. In mathematical programming, the task is ordinarily more challenging due to the usually *much greater* number of variables and the presence of constraints.

Typically, optimization algorithms are constructed so as to generate a sequence of "trial solutions" (called *iterates*) that *converge* to an "optimal solution." In some instances (problem types), an algorithm may be guaranteed to terminate after a finite number of iterations, either with an optimal solution or with evidence that no optimal solution exists. This is true of the Simplex Algorithm for (nondegenerate) linear programming problem.

Some of the terms used in the previous paragraph, and later in this book, warrant further discussion. In particular, when speaking of a "finite algorithm" one means an algorithm that generates a finite sequence of iterates terminating in a resolution of the problem (usualy a solution or evidence that there is no solution). By contrast, when speaking of a "convergent algorithm" one means an algorithm that produces a convergent sequence of iterates. In this case it is often said that the algorithm converges (even though it is actually the sequence of iterates that converges).

A major factor in the development of optimization (and many other fields) has been the computer. To put this briefly, the steady growth of computing power since the mid-twentieth century has strongly influenced the kinds of models that can be solved. New advances in computer hardware and software have made it possible to consider new problems and methods for solving them. These, in turn, contribute to the motivation for building even more powerful computers and the relevant mathematical apparatus.

It should go without saying that algorithms are normally implemented in computer software. The best of these are eventually distributed internationally for research purposes and in many cases are made into commercially available products for use on real-world problems.

Analysis (theory)

Optimization cannot proceed without a clear and tractable definition of what is meant by an optimal solution to a problem. In everyday language (especially in advertising) we encounter the word *optimal* (and many related words such as *optimize, minimize, maximize, minimum, maximum, minimal, maximal*) used rather loosely. In management science, when we speak of optimization, we mean something stronger than "improving the status quo" or "making things better." The definition needs to be precise, and it needs to be verifiable. For example, in a minimization problem, if the objective function is $f(x)$ and $x \in S$, where S is the set of all allowable (feasible) values for x, then

$$x^* \in S \quad \text{and} \quad f(x^*) \leq f(x) \quad \text{for all } x \in S$$

is a clear definition of what we mean by saying x^* is a (global) minimizer of f on S. But in almost all instances, this definition by itself is not useful because it cannot always be checked (especially when the set S has a very large number of elements).

It is the job of analysis to define what an optimal solution is and to develop tractable tests for deciding when a proposed "solution" is "optimal." This is done by first exhibiting nontrivial properties that optimal solutions must have. As a prototype of this idea, we can point to the familiar equation $f'(x) = 0$ as a condition that must be satisfied by a local (relative) minimizer[3] $x = x^*$ of a differentiable function on an open interval of the real line. We call this a *necessary condition* of local minimality. But, as we know, a value $\bar{x}$ that satisfies $f'(x) = 0$ is not guaranteed to be a local minimizer of f. It could also be a local maximizer or a point of inflection. We need a stronger *sufficient condition* to help us distinguish between first-order stationary points, that is, solutions of $f'(x) = 0$.

We use mathematical analysis (theory) to develop conditions such as those described above. In doing so, we need to take account of the properties of the functions and sets involved in specifying the optimization model. We also need to take account of what is computable so that the conditions we propose are useful. But optimization theory is concerned with much more than just characterization of optimal solutions. It studies *properties* of the types of functions and sets encountered in optimization models. Ordinarily, optimization problems are classified according to such properties. Examples are linear programming, nonlinear programming, quadratic programmming, convex programming, ... Optimization theory also studies questions pertaining to the *existence* and *uniqueness* (or lack of it) of solutions to classes of optimization problems.

The analysis of the *behavior* of algorithms (both in practice and in the abstract) is a big part of optimization theory. Work of this kind contributes to the creation of efficient algorithms for solving optimization models.

Synergy

Modeling, computing, and theory influence each other in many ways. In practice, deeper understanding of real-world problems comes from the com-

---

[3]The same would be true for a local maximizer.

bination of mathematical modeling, computation, and analysis. It very often happens that the results of computation and analysis lead to the redefinition of a model or even an entirely new formulation of the problem. Sometimes the creation of a new formulation spurs the development of algorithms for its solution. This in turn can engender the need for the analysis of the model and the algorithmic proposals. Ultimately, the aim is to solve problems having a business, social, or scientific purpose.

About this book

This book is based on lecture notes we have used in numerous optimization courses we have taught at Stanford University. It is an introduction to linear and nonlinear optimization intended primarily for master's degree students; the book is also suitable for qualified undergraduates and doctoral students. It emphasizes modeling and numerical algorithms for optimization with continuous (not integer) variables. The discussion presents the underlying theory without always focusing on formal mathematical proofs (which can be found in cited references). Another feature of this book is its emphasis on cultural and historical matters, most often appearing among the footnotes.

Reading this book requires no prior course in optimization, but it does require some knowledge of linear algebra and multivariate differential calculus. This means that readers should be familiar with the concept of a finite-dimensional vector space, most importantly R^n (real n-space), the algebraic manipulation of vectors and matrices, the property of linear independence of vectors, elimination methods for solving systems of linear equations in many variables, the elementary handling of inequalities, and a good grasp of such analytic concepts as continuity, differentiability, the gradient vector, and the Hessian matrix. These and other topics are discussed in the Appendix.

Other sources of information

The bibliography at the end of the book contains an extensive list of textbooks and papers on optimization. Among these are Bertsimas and Freund [14], Hillier and Lieberman [96], Luenberger and Ye [125], Bradley et al. [20], Murty [144], and Nash and Sofer [147]. After consulting only a few other references, you will discover that there are many approaches to the topics we study in this book and quite a few different notational schemes in use. The same can be said of many fields of study. Some people find this downright confusing or even disturbing. But this conceptual and notational diversity is a fact of life we eventually learn to accept. It is part of the price we pay for gaining professional sophistication.

Acknowledgments

Finally, we want to express sincere thanks to Professor Frederick S. Hillier who—for many years—persistently encouraged us to complete the writing of this book for the Springer International Series in Operations Research & Management Science of which he was once the series editor. No less deserving of our gratitude are Suzanne Cottle and Radhika Thapa, our supportive wives.

1. LP MODELS AND APPLICATIONS

Overview

We begin this chapter with an example of a linear programming problem and then we go on to define linear programs in general. After that we discuss the main topic of this chapter: the classical models and applications of linear programming. In addition, we give some attention to the technique of building a linear programming model and describe various mathematically equivalent forms of its presentations.

1.1 A linear programming problem

Example 1.1: CT REFINERY. CT Gasoline has a refinery located in Richmond, California. The company has been very profitable while maintaining high quality. Carlton and Toby the founders of CT Gasoline, realized early on that, in the intensely competitive gasoline market, they would have to find a way to minimize their costs while continuing to maintain high quality. One challenge they faced was to find the most efficient way to distribute gasoline from their terminals to their gas stations. While doing their MS degrees, they had studied many management science techniques and learned a bit about linear programming, which they thought would apply to their distribution problem. To get some professional assistance, they hired a team of Operation Researchers from the local university.

When Carlton and Toby first set up the refinery, they had 2 terminals, one in Oakland (OA), and the other in San Francisco (SF). From these terminals they could ship to their 6 gas stations located in San Mateo (SM), Palo Alto (PA), San Rafael (SR), Daly City (DC), Los Gatos (LG), and Antioch (AN). They needed to distribute the gasoline in the most cost-effective way.

Carlton and Toby knew the cost to ship gasoline from each of their terminals to each of their gas stations. These are specified in Table 1.1. The station demands, which were known in advance, are shown in Table 1.2 below, together with terminals supplies. Notice that in this particular case, the total supply and total demand are equal.

At first, Carlton and Toby handled gasoline distribution by a simple,

© Springer Science+Business Media LLC 2017
R.W. Cottle and M.N. Thapa, *Linear and Nonlinear Optimization*,
International Series in Operations Research & Management Science 253,
DOI 10.1007/978-1-4939-7055-1_1

	SM	PA	SR	DC	LG	AN
SF	1.88	3.24	1.86	0.74	5.17	4.48
OA	2.81	3.59	2.20	1.68	5.01	3.66

Table 1.1: Shipping costs in cents/gallon.

Station	Demand (gallons)
SM	35,000
PA	65,000
SR	48,000
DC	53,000
LG	66,000
AN	51,000
Total	318,000

Terminal	Supply (gallons)
OA	155,000
SF	163,000
Total	318,000

Table 1.2: Station demands and terminal supplies.

intuitive scheme; they believed it would produce (near) minimum cost. For such cases, their approach amounted to supplying as much as possible to each station from the terminal having the smaller shipping cost. From their experience they knew that in such an allocation it would not always be possible to supply a station from only one terminal.

In the case shown, shipping from SF to DC has the smallest cost, namely 0.74 cents/gallon. So the full demand of 53,000 gallons at DC can be shipped from SF. Once this is done, the supply at the SF terminal is reduced to 110,000 gallons. The next smallest cost is from SF to SR, namely 1.86 cents/gallon. Once again the entire demand of 48,000 gallons at SR can be shipped from SF, reducing the supply there to 62,000 gallons. Following this, SM gets supplied its entire demand of 35,000 gallons from SF at a cost of 1.88 cents/gallon thereby reducing the supply at SF to 27,000 gallons. Among the remaining stations, the one in Palo Alto has the smallest shipping cost, 3.24 cents/gallon. Obviously only 27,000 gallons of its 65,000 gallon demand can be supplied from SF. The balance of 38,000 must be supplied from OA at the higher shipping cost of 3.59 cents/gallon. Now that supply is exhausted at SF, the other two stations, LG and AN, get their respective demands met from OA. This results in a total cost of $9,355.20.

The shipping schedule is shown in Table 1.3.

	SM	PA	SR	DC	LG	AN	Supply
SF	35,000	27,000	48,000	53,000			163,000
OA		38,000			66,000	51,000	155,000
Demand	35,000	65,000	48,000	53,000	66,000	51,000	318,000

Table 1.3: Intuitive shipping schedule (Total cost: $9,355.20).

But Carlton and Toby have decided to acquire some assets of a competitor. This would bring the number of terminals to 6 and gas stations to 47. While the acquisition was being finalized, they realized that their intuitive approach might not work so well for the larger company; they would need to consider afresh how best to do the distribution to minimize costs and thereby increase profits. They sensed that they should solve this problem by means of linear programming with the help of the consultants. They recalled being told many times that the best way to set up and validate a model is to create a small example. They already had a really small example: their two terminals supplying only six gas stations.

They remembered that this particular linear program is known as a transportation problem. The key constraints state the fact that demands must be met from the supplies on hand. The formulation starts by defining nonnegative variables as follows:

$x_{i,j} \geq 0$ is the amount (gallons) of gasoline to be shipped from terminal i to station j.

Next, the total supply at the terminals, from which shipping is to be done, needs to be taken into account. The total supply at Oakland must be the sum of shipments from Oakland to each gas station; i.e.,

$$x_{OA,SM} + x_{OA,PA} + x_{OA,SR} + x_{OA,DC} + x_{OA,LG} + x_{OA,AN} = 155,000,$$

and the total supply at San Francisco must be the sum of shipments from San Francisco to each gas station; i.e.,

$$x_{SF,SM} + x_{SF,PA} + x_{SF,SR} + x_{SF,DC} + x_{SF,LG} + x_{SF,AN} = 163,000.$$

Furthermore, the station demands for gasoline *must* be met; that is:

$$
\begin{aligned}
\text{San Mateo:} \quad & x_{OA,SM} + x_{SF,SM} = 35,000 \\
\text{Palo Alto:} \quad & x_{OA,PA} + x_{SF,PA} = 65,000 \\
\text{San Rafael:} \quad & x_{OA,SR} + x_{SF,SR} = 48,000 \\
\text{Daly City:} \quad & x_{OA,DC} + x_{SF,DC} = 53,000 \\
\text{Los Gatos:} \quad & x_{OA,LG} + x_{SF,LG} = 66,000 \\
\text{Antioch:} \quad & x_{OA,AN} + x_{SF,AN} = 51,000
\end{aligned}
$$

The objective in finding the best way to meet demand from the existing supply is to minimize the total shipping cost from the terminals to the stations:

$$
\begin{aligned}
& 1.88x_{OA,SM} + 3.24x_{OA,PA} + 1.86x_{OA,SR} + 0.74x_{OA,DC} + 5.17x_{OA,LG} + 4.48x_{OA,AN} \\
& + 2.81x_{SF,SM} + 3.59x_{SF,PA} + 2.20x_{SF,SR} + 1.68x_{SF,DC} + 5.01x_{SF,LG} + 3.66x_{SF,AN}
\end{aligned}
$$

Carlton and Toby eagerly awaited the opinion of the consultants on their "solution" using an intuitive approach. At the next meeting, one of the consultants opened her laptop and in a matter of minutes announced that there is an optimal solution with total shipping cost of $9,351.40, namely

	SM	PA	SR	DC	LG	AN	Supply
SF	35,000	65,000	10,000	53,000			163,000
OA			38,000		66,000	51,000	155,000
Demand	35,000	65,000	48,000	53,000	66,000	51,000	318,000

Table 1.4: Better shipping schedule (Total cost: $9,351.40).

Although the difference in shipping cost is very small, it did, nevertheless, convince Carlton and Toby that their intuitive approach could not be relied upon to produce the cheapest shipping schedule. They were curious about the assertion that the cost of $9,351.40 is optimal. In addition, they recognized that since their intuitive shipping schedule was for only one week, and they were planning to enlarge their distribution network, their savings with an optimal shippping schedule could be significant.

This is an instance of a *linear program*. It is an optimization problem in which the objective is a linear function and the constraints are linear inequalities and linear equations.

An important point

In general, the constraints of a linear program may possess many linear equations (and linear inequalities). Some linear programs possess only linear inequalities (and no linear equations). But a meaningful linear program must always have at least one linear inequality, for otherwise it is either trivial or unsolvable (see Exercise 1.5.)

1.2   Linear programs and linear functions

A *linear program* (LP) is an optimization problem in which a linear objective function is to be minimized (or maximized) subject to a finite number of linear constraints (linear equations and/or linear inequalities). In order to be classified as an LP it must contain *at least one* linear inequality. Example 1.1 was just such a problem. In the pages that follow, we will see several more examples of linear programming models.

The definition of a linear program makes no sense unless one knows what linear functions, linear equations, and linear inequalities are. Let's be sure we've got these concepts straight. For the time being, we will define a *linear function* to be one that is of the form

$$f(x_1, x_2, \ldots, x_n) = c_1 x_1 + c_2 x_2 + \cdots + c_n x_n$$

where $c_1, c_2, \ldots, c_n$ are *constants* and $x_1, x_2, \ldots, x_n$ are *variables* (see Exercise 1.6.) In this book we assume that all constants and variables are real-valued.

Here are some examples of linear functions:

1. $x' - x''$. This linear function is the difference of the two variables x' and x''. In this case $n = 2$; the constants c_1 and c_2 are 1 and -1, respectively.

2. $x_1 + x_2 + \cdots + x_{10}$. This is just the sum of the variables $x_1, x_2, \cdots, x_{10}$. Note that in this instance, the coefficient of each variable is the constant 1.

3. $10 \times (5.044 \, MALT + 5.217 \, YEAST + 6.503 \, HOPS)$. In this case there are 3 variables denoted $MALT, YEAST, HOPS$. (The three variables

represent ingredients used in the manufacture of beer. It is often helpful to use variable names that make it easy to interpret the model and
its solution; this is particularly imperative as the size of the problem
grows.) Note also that a constant times a linear function is still a
linear function; thus the above linear function is equivalent to

$$50.44\ MALT + 52.17\ YEAST + 65.03\ HOPS.$$

We shall see many more examples of linear functions throughout this book.

If $f(x_1, \ldots, x_n)$ is a linear function, and K is a constant, then

$$f(x_1, \ldots, x_n) = K$$

is a *linear equation*, whereas

$$f(x_1, \ldots, x_n) \geq K \quad \text{and} \quad f(x_1, \ldots, x_n) \leq K$$

are *linear inequalities*. When the right-hand side K is zero, the linear equation or linear inequality is said to be *homogeneous*. Here are some more
examples.

1. $x' - x'' = 0$ is a homogeneous linear equation. It says that the values
   of the two variables x' and x'' are equal. Note that if x' and x'' satisfy
   this equation, then for any λ we have have $\lambda x' = \lambda x''$. That is to say
   $\lambda x' - \lambda x'' = 0$.

2. $x_1 + x_2 + \cdots + x_{10} = 1$ is a linear equation. It is not a homogeneous
   linear equation since its right-hand side $(K = 1)$ is not zero.

3. $5.044\ MALT + 5.217\ YEAST + 6.503\ HOPS \leq 24$ is a linear inequality.
   Such an expression might be used to impose a budget constraint on
   what is spent for the ingredients of beer.

As stated above, in linear programming we deal with *systems* of linear
constraints that contain at least one linear inequality. Our CT Refinery
example had 8 equations in 12 variables. Furthermore, the nonnegativity
of the variables (which represent amounts of gasoline to be shipped from
terminals to service stations) is expressed by the simple homogeneous linear
inequalities

$$x_{OA,SM} \geq 0, \ x_{OA,PA} \geq 0, \ \ldots, \ x_{SF,AN} \geq 0.$$

Together the 8 equations and 12 linear inequalities constitute the *system of constraints* for that particular LP. The problem requires that all of them be satisfied *simultaneously*, as is the case for *every* constrained optimization problem.

It will typically be the case that the constraints of an LP have *many* feasible solutions. The task is to select a feasible solution which optimizes a linear objective function, that is, one which measures the "goodness" of feasible solutions in the application at hand.

1.3   LP models and applications

Our discussion here will be comprised of five parts:

(a) Some classical LP models

(b) Key assumptions of LP

(c) Building LP models

(d) Presentations of an LP

(e) Fundamental properties of an LP

(a) Some classical LP models

Here we discuss a few classes of LP models that frequently arise in practice and have been studied for a long time. These models typically have names that suggest how the models are applied.

Diet problems

A diet problem (in the sense of linear programming) is one of finding the least expensive way to meet a given set of daily nutritional goals using a particular set of foods. This is sometimes called "Stigler's diet problem" after the work of George Stigler [181]. The following is a very small example of such a problem modeled as a linear program.

Example 1.2: A SIMPLE DIET PROBLEM. Minimize the cost of achieving at least 1.00 mg of thiamine, 7.50 mg of phosphorus, and 10.00 mg of iron from two foods called FOODA and FOODB. The cost of FOODA is 10 cents per ounce (oz), and the cost of FOODB is 8 cents per ounce. The relevant nutritional contents of the foods are:

	FOODA mg/oz	FOODB mg/oz
Thiamine	0.15	0.10
Phosphorus	0.75	1.70
Iron	1.30	1.10

The LP formulation of this problem goes like this:

$$
\begin{aligned}
\text{minimize} \quad & 10\,FOODA + 8\,FOODB \\
\text{subject to} \quad & 0.15\,FOODA + 0.10\,FOODB \geq 1.00 \\
& 0.75\,FOODA + 1.70\,FOODB \geq 7.50 \\
& 1.30\,FOODA + 1.10\,FOODB \geq 10.00 \\
& FOODA \geq 0, \quad FOODB \geq 0.
\end{aligned}
$$

Note that $FOODA$ and $FOODB$ are (obviously) nonnegative variables denoting the number of ounces of FOODA and FOODB respectively used to meet the nutritional requirements (represented by in the first three linear inequalities) for thiamine, phosporus, and iron.

Definition of the diet problem

- Use a given set of food inputs to achieve a given set of nutritional requirements at least cost.

- The foods: $j = 1, \ldots, n$. In practice, these would correspond to the names of foods.

  In the example above the foods are FOODA and FOODB.

- The nutrients: $i = 1, \ldots, m$. In practice, these would correspond to the names of nutrients.

  In the example above the nutrients are thiamine, phosphorus, and iron.

Data of the diet problem

- Requirements: at least b_i units of nutrient i.

  For instance, in the example above, at least 10.00 mg of iron are required in the diet.

- Contributions: One unit of food j contributes a_{ij} units of nutrient i.

  In the example above, 1 oz of FOODB contributes 1.70 mg of phosphorus.

- Costs: c_j (monetary units) per unit of food j.

  In the example above, 1 oz of FOODA costs 10 cents.

Variables and objective of the diet problem

- Food usage levels: x_j (to be determined).

  In the example above, *FOODA* and *FOODB* denote usage levels, both measured in ounces.

- Objective: minimize total food cost.

Constraints of the diet problem

- Meet (or exceed) the nutritional requirements using the foods under consideration.

- Allow only a nonnegative level of each food to be used.

Form of the diet problem

In its simplest form, the LP model of the diet problem is

$$\text{minimize} \quad \sum_{j=1}^{n} c_j x_j$$

$$\text{subject to} \quad \sum_{j=1}^{n} a_{ij} x_j \geq b_i \quad \text{for all } i = 1, \ldots, m$$

$$x_j \geq 0 \quad \text{for all } j = 1, \ldots, n.$$

Notice that the constraints of the problem are expressed in terms of linear inequalities, rather than as linear equations. The model can be modified so as to include other linear constraints such as lower and upper bounds on the amounts of the individual foods in the diet. Such constraints might be used to enhance the palatability. One of the main applications of the diet problem is in the manufacture of various forms of animal feed, where palatability is thought not to be a great issue. For a further (possibly interactive) look at the diet problem see [6].

Product-mix/activity-analysis problems

In a *product mix*, or *activity analysis problem*, we have to use a given set of activities (production processes) and a given set of resources to maximize profit associated with the production processes subject to limitations on the resources. When the objective and constraints can be expressed linearly, we obtain an LP. Here is a tiny example of such a problem.

Example 1.3: A PRODUCT-MIX PROBLEM. This illustration is one used by R.G. Bland in an article on linear programming published in the magazine *Scientific American* (see [16]). The problem concerns the production of ale and beer from three "scarce" resources: corn, hops, and barley malt. (All other inputs to these beverages are considered to be plentiful.) The available amounts of the scarce resources are 480 pounds of corn, 160 ounces of hops, and 1,190 pounds of malt.

Producing a barrel of ale consumes 5 pounds of corn, 4 ounces of hops, and 35 pounds of malt, whereas a barrel of beer consumes 15 pounds of corn, 4 ounces of hops, and 20 pounds of malt. Assume that the profits per barrel are $13 for ale and $23 for beer.

Formulation of this product-mix problem[1]

$$
\begin{aligned}
\text{maximize} \quad & 13\,ALE + 23\,BEER \\
\text{subject to} \quad & ALE + 15\,BEER \leq 480 \\
& 4\,ALE + 4\,BEER \leq 160 \\
& 35\,ALE + 20\,BEER \leq 1{,}190 \\
& ALE \geq 0, \quad BEER \geq 0.
\end{aligned}
$$

As stated on page 6, a constant times a linear function is still a linear function. In fact, multiplication (or division) of any of the inequalities by a constant preserves the solution set of the particular linear program. For example, $4\,ALE + 4\,BEER \leq 160$ can be divided by 4 to result in the equivalent inequality $ALE + BEER \leq 40$; it is equivalent in the sense that the set of feasible solutions of the above problem is unchanged by this operation. From a numerical computation perspective, scaling (multiplication or division) plays an important role in helping provide numerically accurate solutions.

---

[1]Don't try this at home.

The product-mix/activity-analysis problem in general

As in the Example 1.2, a typical product-mix problem will have a set of "production activities" that use (consume) a set of "scarce resources." In the general case, we have the following notation:

- The activities: $j = 1, \ldots, n$. In practice, these would correspond to the *names* of activities.

  In the example above, $n = 2$. Activity $j = 1$ is the production of ale; activity $j = 2$ is the production of beer.

- The resources: $i = 1, \ldots, m$. In practice, these would correspond to the *names* of resources.

  In the example above, $m = 3$. The resources are: corn, hops, and malt.

Data for the product-mix/activity-analysis problem

- The resources: at most b_i units of resource i are available for the production process.

  In the example above, at most 480 pounds of corn can be used.

- The consumption: Operation of process j at unit level consumes a_{ij} units of resource i.

  In the example above, producing a barrel of ale consumes 5 pounds of corn.

- The profit contributions: c_j (monetary units) per unit of process j.

  In the example above, a barrel of ale generates a profit of $13.

Variables and objective of the product-mix/activity-analysis problem

- Activity levels: x_j (to be determined).

  In the example above, the activity levels are denoted by the variables *ALE* and *BEER*, both measured in barrels.

- Objective: maximize total profit.

Constraints of the product-mix/activity-analysis problem

- Produce levels of products that consume no more than the available resources.

- Allow only nonnegative production levels.

Form of the product-mix/activity-analysis problem

Assembling this information we obtain the linear program

$$\text{maximize} \quad \sum_{j=1}^{n} c_j x_j$$

$$\text{subject to} \quad \sum_{j=1}^{n} a_{ij} x_j \leq b_i \quad \text{for all } i = 1, \ldots, m$$

$$x_j \geq 0 \quad \text{for all } j = 1, \ldots, n.$$

Once again, we have an LP all of whose constraints are linear inequalities. Other linear constraints (including linear equations) could also be imposed. We'll see examples of such cases later.

Transportation and assignment problems

In a *transportation problem*[2] a single commodity is to be "shipped" from m *sources* to n *destinations*. At each source i there is a known *supply* $a_i > 0$ of the commodity. At each destination j there is a known *demand* $b_j > 0$ for the commodity. To ship a single unit of the commodity from source i to destination j costs c_{ij}, the known *unit shipping cost*. In such a problem, we let the variable x_{ij} denote amount of the commodity shipped from i to j. The objective is to minimize total shipping cost while accomplishing the shipping. Example 1.1 is an instance of a transportation problem.

The *assignment problem* can be viewed as a variant of the transportation problem in which $m = n$ and $a_i = b_j = 1$ for all i and j. We do not treat this class of problems separately in this book.

---

[2]The transportation problem is often attributed to G. Monge [137] or to F.L. Hitchcock [98], or to L.V. Kantorovich [104].

Formulation of the transportation problem (with equality constraints)

In general, the transportation problem has the following mathematical formulation:

$$\text{minimize} \quad \sum_{i=1}^{m} \sum_{j=1}^{n} c_{ij} x_{ij}$$

$$\text{subject to} \quad \sum_{j=1}^{n} x_{ij} = a_i \quad \text{for all } i = 1, \ldots, m$$

$$\sum_{i=1}^{m} x_{ij} = b_j \quad \text{for all } j = 1, \ldots, n$$

$$x_{ij} \geq 0, \quad \text{for all } i \text{ and } j.$$

Note how the problem given in Example 1.1 fits this pattern.

Properties of the transportation problem (with equality constraints)

- In a feasible transportation problem (as formulated above), *the total supply must equal the total demand.*

$$\sum_{i=1}^{m} a_i = \sum_{j=1}^{n} b_j.$$

- The constraints have *special structure.*

  For example, although $m \times n$ variables x_{ij}, each supply equation involves exactly n variables; likewise each demand equation involves exactly m variables. All the nonzero coefficients are ones, and the structure becomes even more evident when the system is written in matrix form.

- Integer supplies and demands lead to *integer solutions.*

  As to why this "integrality property" holds, see, for example [44]. This is an important characteristic of the transportation problem (but *not* of linear programs in general). Indeed, when the commodity comes in indivisible units, it is advantageous not to be faced with fractional amounts! For instance, if the commodity were a type of baseball bat, we would not be pleased with a portion of a bat.

Blending problems

A by-product of fuel refineries is residual fuel oil, as is also the case at CT refinery. When blending to create such fuel oil, rigid specifications must be met. One of these is on specific gravity. For example, residual fuel is created by blending asphalt flux (A), clarified oil (C), and kerosene (K). The specific gravity of the mix must be between 0.94 and 1.0. The specific gravity of A is 1.02, for C it is 1.10, and, for K it is 0.83. This results in two *nonlinear* inequalities:

$$\frac{1.02A + 1.10C + .83K}{A + C + K} \geq 0.94 \quad \text{and} \quad \frac{1.02A + 1.10C + .83K}{A + C + K} \leq 1.0.$$

Here we have the *ratio* of two linear functions that must be greater than or equal to 0.94 and less than or equal to 1.0. Because the denominator is positive (in this case it happens to be a positive constant because of another constraint) it is permissible to multiply both sides of the inequality by the denominator and thereby achieve ordinary linear inequalities. For instance,

$$\frac{1.02A + 1.10C + .83K}{A + C + K} \geq 0.94$$

becomes

$$1.02A + 1.10C + .83K \geq 0.94A + 0.94C + 0.94K$$

which, after collecting like terms, becomes the homogenous linear inequality

$$0.08A + 0.16C - 0.11K \geq 0.0 \,.$$

Another method to make such fractional constraints linear is to consider generating one unit of the blend. In that case the denominator $A + C + K$ is set equal to 1.0, but, in doing so, we must adjoin the equation $A + C + K = 1$ as an extra constraint.

We shall now discuss blending problems in a more general way.

The nature of blending problems

Consider a problem in which several *inputs* $i = 1, \ldots, m$ are used to produce one or more *outputs* $j = 1, \ldots, n$. Let x_{ij} be the amount of input i per unit of output j. Assume that each input i has a numerical characteristic c_i and that each output j has a numerical characteristic d_j that is given by a linearly weighted average of the characteristics of the inputs—the weights w_{ij} being unknown. For the j-th output, the unknown weights are given by $w_{ij} = x_{ij} / \sum_{i=1}^{m} x_{ij}$, which is the percentage of input i used in output j.

Thus

$$d_j := \sum_{i=1}^{m} c_i w_{ij} = \sum_{i=1}^{m} c_i \left(\frac{x_{ij}}{\sum_{i=1}^{m} x_{ij}} \right) = \frac{\sum_{i=1}^{m} c_i x_{ij}}{\sum_{i=1}^{m} x_{ij}} .$$

Bounds on characteristics of blended outputs

The characteristic of output j is often required to satisfy inequalities like $\ell_j \leq d_j \leq u_j$. By the form of d_j, this leads to

$$\ell_j \left(\sum_{i=1}^{m} x_{ij} \right) \leq \sum_{i=1}^{m} c_i x_{ij} \leq u_j \left(\sum_{i=1}^{m} x_{ij} \right)$$

and then to

$$\sum_{i=1}^{m} (\ell_j - c_i) x_{ij} \leq 0 \quad \text{and} \quad \sum_{i=1}^{m} (c_i - u_j) x_{ij} \leq 0.$$

Other applications and models

The applications of linear programming are very numerous. A few categories of them are

- Deterministic inventory problems.

- Least cost network flow problems.

- Pattern separation problems.

(b) Key assumptions of linear programming

There are three properties that must or should hold in a real optimization problem in order to make linear programming an appropriate solution tool. These are proportionality, additivity, and divisibility as described next.

- *Proportionality.* The proportionality property means that the effect on the objective function and the effect on resource utilization (or the contribution to required output) is proportional to each activity level. This means, for instance, that in a transportation problem the unit shipping cost c_{ij} does not decrease when larger quantities are shipped from i to j; i.e., there are no economies of scale.

- *Additivity.* The additivity property means that the effects of individual activities are additive (as distinct from multiplicative, for example). A simple case where such a property does *not* hold is that of finding the rectangle of largest area with a given perimeter. When the sides are of lengths x and y and the given perimeter is p, this problem is formulated as

$$\begin{aligned}
\text{maximize} \quad & xy \\
\text{subject to} \quad & 2x + 2y = p.
\end{aligned}$$

Here we see that the variables x and y appear additively (with constants of proportionality) in the constraint, but multiplicatively in the objective function.

In general, together, the properties of proportionality and additivity amount to the same thing as linearity. However, in modeling a real-world problem, it is worthwhile thinking about the *effects* of activities in terms of proportionality and additivity.

- *Divisibility.* The divisibility property refers to the variables being *continuous* rather than *integer-valued.* This is a property that is often lacking in practical optimization problems (both linear and nonlinear). For instance, there are many cases where some or all of the variables make practical sense only if they are integral. One would not build 1.7239 factories or send up 0.3826 space ships. It might be imagined that the way to deal with fractional amounts is just to round the numbers up or down. It is known however that this heuristic is not guaranteed to result in a feasible solution much less an optimal solution.

Optimization problems requiring at least some of the variables to be integer-valued are called *(mixed-)integer programs.* There is an optimization field dealing with such problems. But it is important to recognize that ordinary linear programming is an important tool used in solving them. One such application is in the technique known as *relaxation* in which the integrality requirement is temporarily ignored and the problem is solved as an ordinary LP. If the solution found lacks the required integrality, additional constraints (called *cuts*) could be introduced in an attempt to obtain a problem whose solution is feasible for the original problem and *does* possess the integrality property. Other methods for handling integer programs are also used, but it is not our purpose to cover that subject here.

(c) Steps in model building

Effective efficient model building is an art in all branches of applied science and engineering. This is especially true in linear programming where problems can be large and complex. Nevertheless, one has a rough idea of what to look for and what the overall structure will be. We cover some of the steps one takes in building an LP model. Needless to say, this has great importance in practice. More extensive development of this topic can be found in Chapter 1 in the textbook by Dantzig and Thapa [44].

Here are some steps to follow in building an LP model:

- *Identify the activities.* That is, identify all the different things that can or must be done to achieve the desired ends. These correspond to the *decision variables* whose units and values are to be determined.

- *Identify the items.* The items are the resources available to, or the products required from, the activities. This step includes identifying the amounts (in appropriate units) of the individual resources available and the requirements to be imposed.

- *Identify the input-output coefficients.* These are the constants of proportionality between the unit levels of activity and the resources or requirements. Some authors use a special sign convention for these data. For instance, a positive constant a_{ij} might mean that operating activity j at unit level *consumes* a_{ij} units of resource i. With such a convention, one might consider a negative coefficient a_{ij} to mean that activity j *contributes* to the amount of resource i. It is essential to be clear and consistent about the units of the a_{ij} and the related sign convention.

- *Write the constraints.* For a linear program, we expect the constraints to be expressed or expressible either as linear equations or linear inequalities. As we have seen in the case of the ratios appearing in blending problems, it is sometimes necessary to modify the form of certain constraints to make them linear.

- *Identify the coefficients of the objective function.* This is much the same as some of the preceding steps. Here again, it is essential to be aware of the units involved. If the objective function represents the total cost of the activities, then it needs to be reckoned in the appropriate units.

(d) Presentations of an LP

By presentations of the problem, we mean different (equivalent) ways a linear programming problem can be written down. This will span a range from Excel spreadsheets to linear algebra and matrix theory.

The LP in (verbose) standard form

A linear program expressed as

$$
\begin{array}{lrcl}
\text{minimize} & c_1 x_1 + \cdots + c_n x_n & & \\
\text{subject to} & a_{11} x_1 + \cdots + a_{1n} x_n &=& b_1 \\
& a_{21} x_1 + \cdots + a_{2n} x_n &=& b_2 \\
& \quad \vdots & & \vdots \\
& a_{m1} x_1 + \cdots + a_{mn} x_n &=& b_m \\
& x_j \geq 0, \quad j = 1, \ldots, n & &
\end{array}
\tag{1.1}
$$

is said to be in *standard form*. (The presentation used for the transportation problem Example 1.1 was in *verbose standard form*.) This form is called verbose because it takes a lot of writing or talking to express the problem this way. This form is standard in the sense that the variables are all nonnegative and the other constraints are linear equations. The latter are sometimes called *material balance equations*.

Not all linear programming models are automatically in standard form; they may have a mixture of linear inequalities and equations. Nevertheless, it is easy to convert them to standard form. For instance, the constraints of the product-mix problem, Example 1.3, are all inequalities of the form $\leq$.

Each such inequality constraint

$$\sum_{j=1}^{n} a_{ij}x_j \leq b_i$$

can be converted to a linear equation by *adding* a nonnegative variable, called a *slack* variable, to its left-hand side. It is important to use a *separate* slack variable for each such inequality. Thus the inequality above becomes a *pair* of linear constraints, one a linear equation, the other a nonnegativity constraint:

$$\sum_{j=1}^{n} a_{ij}x_j + x_{n+i} = b_i \quad \text{and} \quad x_{n+i} \geq 0.$$

When this is done the slack variable for the i-th constraint appears only in the i-th equation. Moreover the coefficient of x_{n+i} in the objective function is zero, which is to say the slack variable does not appear there.

Here is an example of how the model for the product-mix problem, Example 1.3, would be converted to standard form. It has three resource constraints pertaining to corn, hops, and malt, respectively. We introduce the nonnegative slack variables C, H, and M, adding one variable of these variables to the left-hand side of each inequality. This yields

$$
\begin{array}{lrcl}
\text{maximize} & 13\,ALE + 23\,BEER & & \\
\text{subject to} & 5\,ALE + 15\,BEER + C & = & 480 \\
& 4\,ALE + 4\,BEER + H & = & 160 \\
& 35\,ALE + 20\,BEER + M & = & 1,190 \\
& ALE \geq 0, \ BEER \geq 0, \ C \geq 0, \ H \geq 0, \ M \geq 0. &
\end{array}
$$

In the diet problem, Example 1.2, all the inequalities are of the form $\geq$. Each such inequality constraint can be converted to a linear equation by *subtracting* a nonnegative variable called a *surplus* variable from the left-hand side of the constraint. As before, each surplus variable appears in only one equation and not in the objective function either.

Row-oriented standard form

The verbose standard form shown above involves a collection of *sums*, one for the objective function and m more used in expressing the material balance equations. This observation enables us to write the problem as

$$\text{minimize} \quad \sum_{j=1}^{n} c_j x_j$$

$$\text{subject to} \quad \sum_{j=1}^{m} a_{ij} x_j = b_i, \quad i = 1, \ldots, m$$

$$x_j \geq 0, \quad j = 1, \ldots, n.$$

(1.2)

The general statement of the transportation problem was given in this row-oriented[3] standard form.

Column-oriented standard form

This approach to writing an LP emphasizes the fact that the left-hand side of the material balance equations and the objective function can be thought of as a linear combination of column vectors. Each of these column vectors is composed of coefficients of a decision variable, x_j ($j = 1, \ldots, n$). In particular, the column vector associated with x_j is

$$\begin{bmatrix} c_j \\ a_{1j} \\ \vdots \\ a_{mj} \end{bmatrix}.$$

Thus the *left-hand side* of the problem is

$$\sum_{j=1}^{n} \begin{bmatrix} c_j \\ a_{1j} \\ \vdots \\ a_{mj} \end{bmatrix} x_j.$$

Before we can use this idea to write the material balance equations, we need to express the right-hand side as a column vector as well. To facilitate this step, let us define a variable, say z, through the equation

$$z := \sum_{j=1}^{n} c_j x_j.$$

---

[3]In the next chapter, we will discuss the algebraic and geometric significance of the row-oriented presentation and column-oriented presentations.

The *right-hand side* can now be written as the vector

$$\begin{bmatrix} z \\ b_1 \\ \vdots \\ b_m \end{bmatrix}.$$

Putting all this together, we have the problem

$$\text{minimize} \qquad\qquad z \qquad\qquad\qquad (1.3)$$

$$\text{subject to} \quad \sum_{j=1}^{n} \begin{bmatrix} c_j \\ a_{1j} \\ \vdots \\ a_{mj} \end{bmatrix} x_j = \begin{bmatrix} z \\ b_1 \\ \vdots \\ b_m \end{bmatrix}$$

$$x_j \geq 0, \quad j = 1, \ldots, n.$$

Note that in this problem statement z is a (dependent) variable.

You may have noticed how much vertical space is needed in writing column vectors as above. To avoid this, we adopt the convention that the notation $x = (x_1, \ldots, x_n)$ actually means

$$x = \begin{bmatrix} x_1 \\ \vdots \\ x_n \end{bmatrix}.$$

The LP in (terse) standard form

A linear program in *standard form* can be expressed *tersely* as

$$\begin{aligned} \text{minimize} \quad & c^{\mathrm{T}}x \\ \text{subject to} \quad & Ax = b \\ & x \geq 0. \end{aligned} \qquad\qquad (1.4)$$

In this presentation, A is the coefficient matrix, b is the right-hand side vector, c is the vector of cost coefficients, and x is the vector of decision variables.

Let us consider the example of the little product-mix problem Example 1.3. After the insertion of slack variables, the problem can be put in terse standard form using

$$A = \begin{bmatrix} 5 & 15 & 1 & 0 & 0 \\ 4 & 4 & 0 & 1 & 0 \\ 35 & 20 & 0 & 0 & 1 \end{bmatrix}, \quad b = \begin{bmatrix} 480 \\ 160 \\ 1190 \end{bmatrix},$$

$$c^{\mathrm{T}} = \begin{bmatrix} 13 & 23 & 0 & 0 & 0 \end{bmatrix}, \quad \text{and} \quad x = (ALE, \ BEER, \ C, \ H, \ M).$$

Notice that by defining $z = c^{\mathrm{T}}x$, we obtain another terse version of the problem

$$\begin{aligned} \text{minimize} \quad & z & (1.5) \\ \text{subject to} \quad & \begin{bmatrix} c^{\mathrm{T}} \\ A \end{bmatrix} x = \begin{bmatrix} z \\ b \end{bmatrix} \\ & x \geq 0. \end{aligned}$$

We have illustrated four different presentations of the linear programming problem in standard form. They all have the same meaning, and each has its own advantages. For instance, the terse one is written in an abbreviated style which is often useful when brevity or physical space is an issue. These different ways of writing LPs will be used in the remainder of the book.

(e) Fundamental properties of an LP

Given a linear program, it may or may not have a feasible solution. If it does not have a feasible solution, we say its feasible region is *empty*. If the feasible region is nonempty (the problem is feasible), it may or may not be bounded. If the feasible region is nonempty and bounded, the optimal value of the objective function is always finite; if the feasible region is unbounded, the objective function may still have a finite optimal value, though this is not guaranteed. When an optimal solution exists, the optimal objective function value is unique, but there may be multiple optimal solution vectors.

1.4   Exercises

1.1   Suppose a firm produces a particular computer unit in Singapore and in Hoboken. The firm has distribution centers in Oakland, Hong Kong, and Istanbul. The supplies, demands and unit shipping costs are summarized in the following table.

	Oakland	Hong Kong	Istanbul	
Singapore	85	37	119	500
Hoboken	53	189	94	300
	350	250	200	

Formulate the corresponding transportation problem.

1.2 Represent the Diet Problem, Example 1.2, in standard form.

1.3 Identify an obvious nonzero solution to the Product-Mix Problem, Example 1.3, in standard form.

1.4 Consider the integer linear program

$$
\begin{aligned}
\text{maximize} \quad & x + 2y \\
\text{subject to} \quad & 3x + 4y \leq 12 \\
& -x + 2y \leq 4 \\
& \text{and } x \geq 0, \ y \geq 0 \\
& \text{with } x, y \text{ integers.}
\end{aligned}
$$

(a) The optimal solution to the linear program (without the integer restrictions) is $x = 1.6$, $y = 1.8$. Show that rounding this solution leads to an infeasible solution.

(b) Starting from $x = 0$, $y = 0$, systematically enumerate all possible integer solutions and pick an optimal solution. Is it unique?

(c) Replace the first inequality by $3x + 4y \leq 8$; the optimal solution to the linear program (without the integer restrictions) now changes to $x = 0.8$, $y = 1.4$. Show that rounding now leads to a feasible solution.

(d) Enumerate all possible feasible integer solutions to the problem as reformulated in (c) and show that the above integer solution is the optimal integer solution.

(e) For the integer linear program

$$
\begin{aligned}
\text{maximize} \quad & x + 2.5y \\
\text{subject to} \quad & 2x + 4.5y \leq 6 \\
& -x + 2.0y \leq 4 \\
& \text{and } x \geq 0, \ y \geq 0 \\
& \text{with } x, y \text{ integers}
\end{aligned}
$$

the optimal solution is $x = 0$, $y = 4/3$. Rounding this results in the feasible solution $x = 0$, $y = 1$. Show that this is not optimal by finding the optimal feasible integer solution.

1.5 This exercise is intended to illustrate that the minimization (or maximization) of a linear function subject to a system of linear equations is not a meaningful optimization problem.

(a) Show that the problem

$$\begin{array}{ll} \text{minimize} & x \\ \text{subject to} & x + y = 1 \end{array}$$

is feasible but has no optimal solution.

(b) Show that the problem

$$\begin{array}{ll} \text{minimize} & x + y + z \\ \text{subject to} & x + y - z = 1 \\ & -x - y + z = 1 \end{array}$$

is infeasible and hence has no optimal solution.

(c) Show that the problem

$$\begin{array}{ll} \text{minimize} & x \\ \text{subject to} & x + y = 1 \\ & x - y = 0 \end{array}$$

has only one feasible solution and hence is a trivial optimization problem.

1.6   At the beginning of Section 1.2, we gave a temporary definition of the term linear function to be one that is of the form:

$$f(x_1, x_2, \ldots, x_n) = c_1 x_1 + c_2 x_2 + \cdots + c_n x_n$$

where $c_1, c_2, \ldots, c_n$ are *constants* and $x_1, x_2, \ldots, x_n$ are *variables*. The standard definiton of a linear function f is one that satisfies

$$f(\alpha x + \beta y) = \alpha f(x) + \beta f(y)$$

for all scalars α and β and all vectors x and y.

(a) Show that a function of the form $c_1 x_1 + c_2 x_2 + \cdots + c_n x_n$ is linear in the standard definition.

(b) Let f be a linear function and let K be a nonzero constant. Show that the function $f(x) + K$ is never a linear function. (Functions of this form are said to be *affine*.)

1.7   In Section 1.3(b), we listed the key assumptions of linear programming. The first two mentioned were *proportionality* and *additivity*. We commented there that "together, the properties of proportionality and additivity amount to the same thing as linearity." Use the "standard definition of a linear function" given in Exercise 1.6 to explain this comment.

1.8   An important property of the transportation problem (with equality con-
       straints) is the feasibility requirement that the total demand must equal
       total supply. Explain why this must be so. Do the nonnegativity con-
       straints have anything to do with this requirement?

1.9   Suppose there are two warehouses from which three retail stores are sup-
       plied. The warehouses have on hand cases of chocolate candy in the
       amounts of $a_1 = 250$ and $a_2 = 375$. The demands at the retail stores are
       $b_1 = 125$, $b_2 = 300$, and $b_3 = 200$.

       (a) Using the verbose form, write out the constraints of the transporta-
           tion problem corresponding to the given data.
       (b) Write these same constraints in the terse form, that is, using matrix-
           vector notation.

1.10   A farmer purchases feed mix for his cattle from a local company which
       provides five types of mixes with different nutrient levels. The farmer has
       different requirements for his cattle feed mix than those provided by the
       company and would like to create a blend of these five different mixes to
       meet or exceed his requirements at a minimum cost. Formulate an LP to
       do this based on the data below

	Percent Nutrient					
Nutrient	Mix 1	Mix 2	Mix 3	Mix 4	Mix 5	Requirement
1	20	30	8	35	15	25
2	10	5	25	18	40	25
3	12	35	10	5	10	15
	Cost $/lb					
	.50	1.10	.55	.75	1.25	

1.11   CT Capital provides capital for home mortgages, auto loans, and personal
       loans. Their home mortgage loans are of 30 year fixed rates of three types:
       Jumbo loans (greater than $729,000), Conforming loans, and 2nd Mort-
       gages (or Home Equity Loans, abbreviated HEL). The annual interest
       rates for these loans, are as follows:

Jumbo	4.875%
Conforming	4.375%
HEL	5.750%
Auto	6.000%
Personal	8.500%

       Formulate an LP model to help CT Capital maximize the return from
       interest on these loans subject to the following:

       • The total capital available for all loans is $400 million.

- The Conforming loans, which have the lowest risk, should be at least 70% of the three home-based loans.
- HELs should be no more than 10% of the home-based loans.
- The home-based loans should be at least 80% of all loans.
- Personal loans should not be more than 30% of the auto and personal loans.

1.12   CT Coffee Manufacturers blends three types of coffee beans to produce a very popular blend. The contracts for purchasing the coffee beans that go into the blend are negotiated monthly. Aroma and taste ratings are determined subjectively by experts; the scale runs from 50 to 100 where 100 is the highest aroma or taste level. On the other hand, the degree of roast is measured quite precisely through the use of a specially modified spectrophotometer, called an Agtron. The Agtron readings range from 10 to 95, with 95 being the lightest roast and 10 being the darkest roast. The cost per pound and bean characterestics are as follows.

Coffee	Cost ($/lb)	Aroma	Taste	Roast
Bean 1	5.25	84	92	36
Bean 2	4.50	74	88	70
Bean 3	3.75	78	83	48

CT Coffee Manufacturers produce 1000 lbs of coffee a day. The blended coffee must satisfy the following: The aroma should be between 79 and 81, the taste should be between 86 and 88, and the average Agtron rating of the roast should be that of a medium roast (between 51 and 60). Formulate an LP model to determine how much of each coffee bean to buy in order to minimize the cost while satisfying the requirements for aroma, taste, and degree of roast.

1.13   A Machine Problem (Kantorovich [105]). Formulate the following machine problem to obtain the maximum number of completed items per hour. A plant manufactures an assembled item consisting of two different metal parts. For each part, the milling work can be done on different machines: milling machines, turret lathes, or automatic turret lathes according to the following

Type of Machine	Number of Machines	Maximum Output per Machine per Hour	
		First Part	Second Part
Milling Machines	3	10	20
Turret Lathes	3	20	30
Automatic Turret Lathes	1	30	80

1.14   (Based on Exercise 1.5 of Dantzig and Thapa [44].) A small refinery blends five raw gasoline types to produce two grades of motor fuel: regular and premium. The number of barrels per day of each raw gasoline type available, the performance rating, and the cost per barrel are given in the following table:

Raw Gasoline Type	Performance Rating	Barrels/day	Cost/barrel ($)
1	70	2,000	8.00
2	80	4,000	9.00
3	85	4,000	9.50
4	90	5,000	11.50
5	99	5,000	20.00

The performance rating of regular gasoline is required to be at least 85 and that of premium gasoline is required to be at least 95. (The performance rating of a blend is assumed to be a weighted average of the performance ratings of its ingredients.) The refinery can sell all of its regular and premium gasoline production at $28 and $32 per barrel, respectively, but the refinery's contract requires that it produce at least 8,000 barrels/day.

Formulate (but do not solve) the problem of maximizing the refinery's profit (net revenue based on the given data).

Be sure to define the variables you use in the formulation.

1.15 A food manufacturing company makes three products called Mix A (MA), Mix B (MB), and Mix C (MC) using four blending machines of different sizes called Mini, Small, Medium and Large. (There is one machine of each size. The machines differ only in their capacity.) Because they are used in the company's other manufacturing operations (which are not part of this problem), the scheduling people have limited the annual use of these machines for making MA, MB AND MC to 200 hours each for the two smaller machines and 100 hours each for the two larger machines. The estimated annual demands for MA, MB and MC are 190,000 pounds, 250,000 pounds and 75,000 pounds, respectively, and the company intends to meet these demands. The times required for manufacturing batches of MA, MB, and MC are 50 minutes, 30 minutes and 45 minutes, respectively, regardless of which machine is being used for the blending.

The cost of producing each of the three products on each of the four blenders as well as the loading capacity of each product on each blender is given in the following table:

	(Cost per batch / Pounds per batch)		
	Mix A	Mix B	Mix C
Mini	18 / 190	25 / 200	30 / 200
Small	72 / 795	55 / 800	50 / 820
Medium	85 / 990	100 / 1,000	135 / 1,015
Large	175 / 1975	165 / 2,000	250 / 2,025

(a) Formulate a linear program that would help the company to minimize the cost of manufacturing these three products while satisfying the given constraints.

(b) Is this truly a linear programming problem? Explain.

(c) Is this a really a blending problem? If so, in what sense?

1.16 (Based on Kantorovich, see [105].) A manufacturing firm makes a product in kit form that consists of p different parts per kit. The firm has m machines, each of which can make all the individual parts that are required for a complete kit. In fact, machine i can make $\alpha_{ik} > 0$ units of part i per day. For policy reasons it is required that the output of each part per day be the same. (There is no provision for overnight storage of extra parts.) The goal is to manufacture the greatest possible number of complete kits. Formulate the manufacturing problem as a linear program.

2. LINEAR EQUATIONS AND INEQUALITIES

Overview

As we have observed, the constraints of a meaningful linear program must include at least one linear inequality, but otherwise they may be composed of linear equations, linear inequalities, or some of each. In this chapter, we touch on two important topics within the very large subject of linear equations and linear inequalities.

The first of these topics concerns *equivalent forms of the linear programming problem*. To appreciate this subject, it is a good idea to think about what we mean by the noun "form". *Webster's New Collegiate Dictionary* defines form as "the shape and structure of something as distinguished from its material." In the present context, this has much to do with the types of mathematical structures (sets, functions, equations, inequalities, etc.) used to represent an object, such as a linear programming problem. It is equally important to have a sense of what it means for two forms of an optimization problem (or two optimization problems) to be "equivalent." This concept is a bit more subtle. Essentially, it means that there is a one-to-one correspondence between the optimal solutions of one problem and those of the other. Using the equivalence of optimization problems, one can sometimes achieve remarkable computational efficiencies. In this chapter we illustrate another use of equivalence by converting two special nonlinear optimization problems to linear programs (see Example 2.1 on page 35 and Example 2.2 on page 38).

The second important topic introduced in this chapter is about *properties of polyhedral convex sets*, by which we mean solution sets of linear inequality systems. In discussing the geometry of polyhedral sets it is customary to use words such as *hyperplane, halfspace, vertex, and edge*. In this chapter we will cover these geometric concepts and relate them to algebraic structures.

The geometric study of polyhedral convex sets goes back to classical antiquity. Much of what is now known about this subject makes use of algebraic methods and was found in more "recent" times, which is to say the 19th and 20th centuries.

In learning this material we gain valuable knowledge and useful vocabulary for practical and theoretical work alike.

© Springer Science+Business Media LLC 2017
R.W. Cottle and M.N. Thapa, *Linear and Nonlinear Optimization*,
International Series in Operations Research & Management Science 253,
DOI 10.1007/978-1-4939-7055-1_2

2.1   Equivalent forms of the LP

The material covered in this section includes: the conversion of a linear inequality to an equivalent linear equation plus a nonnegative variable; the
equivalence of a linear equation to a pair of linear inequalities; the transformation of variables that are not required to be nonnegative to variables that
are so restricted; and, finally, the conversion of a maximization problem to
an equivalent minimization problem.

In Chapter 1, we introduced the term "standard form" for a linear program: A linear programming problem will be said to be in *standard form*
if its constraints are all equations and its variables are all required to be
nonnegative. Such a problem has only constraints of the form

$$\sum_{j=1}^{n} a_{ij} x_j = b_i, \qquad i = 1, \ldots, m$$

$$x_j \geq 0, \qquad j = 1, \ldots, n.$$

More tersely, the row-oriented presentation of the standard form constraints
written above has the matrix form

$$Ax = b$$
$$x \geq 0.$$

Here A denotes an $m \times n$ matrix, and b is an m-vector.

The transportation problem discussed in Example 1.1 was presented in
standard form. Nevertheless, as we have seen in the other examples discussed in Chapter 1, some linear programs are not initially in standard form.
Granting for the moment that there is something advantageous about having a linear program in standard form, we look at the matter of converting
the constraints of a linear program to standard form when they don't happen to be that way initially. In so doing, we need to be sure that we do not
radically alter the salient properties of the problem. This statement alludes
to the notion of *equivalence* of optimization problems. We will have occasion
to use the concept of equivalence later in this and subsequent chapters.

But what is to be done with linear programs having constraints that are
not in standard form? For instance, the diet problem (Example 1.2), the
product-mix problem (Example 1.3) and the blending problem (page 13) all
have linear inequalities among the *functional constraints*, i.e., those which
are not just nonnegativity conditions or *bound constraints* as they are called.

Conversion to standard form

A single linear inequality of the form

$$\sum_{j=1}^{n} a_{ij}x_j \leq b_i$$

can be converted to a linear equation by adding a nonnegative *slack variable*. Thus, the inequality above is equivalent to

$$\sum_{j=1}^{n} a_{ij}x_j + x_{n+i} = b_i, \quad x_{n+i} \geq 0.$$

In this case, x_{n+i} is the added slack variable.

Here is a simple illustration of this process. Suppose we have the linear inequality

$$3x_1 - 5x_2 \leq 15. \tag{2.1}$$

At the moment, we are not imposing a nonnegativity constraint on the variables x_1 and x_2. This linear inequality can be converted into a linear equation by adding a nonnegative variable, x_3 to the left-hand side so as to obtain

$$3x_1 - 5x_2 + x_3 = 15, \quad x_3 \geq 0. \tag{2.2}$$

Figure 2.1 depicts how the sign of the slack variable x_3 behaves with respect to the values of x_1 and x_2 in (2.2). The shaded region corresponds to the values of x_1 and x_2 satisfying the given linear inequality.

A system of $\leq$ linear inequalities can be converted to a system of linear equations by adding a separate nonnegative slack variable in each linear inequality. Thus

$$\sum_{j=1}^{n} a_{ij}x_j \leq b_i, \quad i = 1, \ldots, m$$

becomes

$$\sum_{j=1}^{n} a_{ij}x_j + x_{n+i} = b_i, \quad x_{n+i} \geq 0, \quad i = 1, \ldots, m.$$

To see the importance of using a *separate* slack variable for each individual linear inequality in the system, consider the following system

$$\begin{aligned} 3x_1 - 5x_2 &\leq 15 \\ 2x_1 + 4x_2 &\leq 16. \end{aligned} \tag{2.3}$$

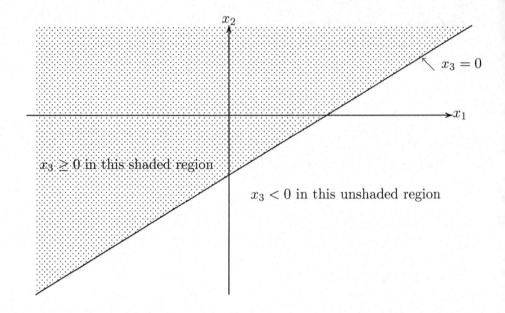

Figure 2.1: Solutions of linear inequality (2.1).

Giving each inequality a slack variable, we obtain the system

$$3x_1 - 5x_2 + x_3 \qquad = 15, \qquad x_3 \geq 0$$
$$2x_1 + 4x_2 \qquad + x_4 = 16, \qquad x_4 \geq 0. \tag{2.4}$$

Solutions of these linear inequalities are shown in the darkest region of Figure 2.2 below.

In Figure 2.2, it is plain to see that there are solutions of one inequality which are not solutions of the other. This, in turn, has a direct relationship with the *signs* of x_3 and x_4. In short, it would be incorrect to use just one slack variable for the two linear inequalities.

Suppose we now introduce nonnegativity constraints on the variables x_1 and x_2 in the inequality system (2.3). The system could then be written as

$$3x_1 - 5x_2 + x_3 \qquad = 15$$
$$2x_1 + 4x_2 \qquad + x_4 = 16 \tag{2.5}$$
$$x_j \geq 0, \quad j = 1, 2, 3, 4.$$

The feasible solutions of this system are indicated by the most darkly shaded region in Figure 2.3.

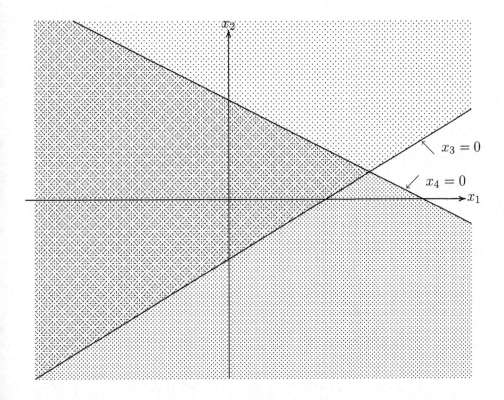

Figure 2.2: Solutions of (2.3).

The same conversion in matrix notation is as follows. Constraints

$$Ax \leq b$$

can be written as

$$Ax + Is = b, \quad s \geq 0$$

where, componentwise, $s_i = x_{n+i}$. Analogously, the constraints

$$Ax \geq b$$

can be written as

$$Ax - Is = b, \quad s \geq 0$$

where, componentwise, $s_i = x_{n+i}$. The components of the vector s are sometimes called *surplus variables*.

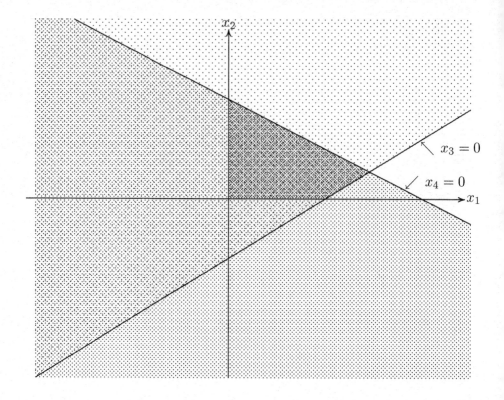

Figure 2.3: Nonnegative solutions of (2.3).

Linear programs with free variables

As you will recall, a linear programming problem is an optimization problem in which a linear function is to be minimized (or maximized) subject to a system of linear constraints (equations or inequalities) on its variables. We insist that the system contain at least one linear inequality, but we impose no further conditions on the number of equations or inequalities. In particular, the variables of a linear program need not be nonnegative. Some linear programming problems have variables that are *unrestricted in sign*. Such variables are said to be *free*. Why do we care whether variables are free or not? This has to do with the fact that the Simplex Algorithm (which we shall take up in Chapter 3) is designed to solve linear programs in standard form.

In the following example, we encounter a classic optimization problem arising in statistics that does not appear to be a linear program. The problem can, however, be converted to a linear program, albeit one that is definitely not in standard form. In particular, it will have linear inequality constraints *and* free variables. Later, we show how to bring this linear program into standard form.

Example 2.1: THE CHEBYSHEV PROBLEM[1]. Suppose we have a hypothesis that a particular variable is a linear function of certain variables (that is, a linear model). More specifically, suppose we believe that there are numbers $x_1, x_2, \ldots, x_n$ such that when the inputs[2] $a_1, a_2, \ldots, a_n$ are supplied, the output

$$b = a_1 x_1 + a_2 x_2 + \cdots + a_n x_n$$

will be observed. We would like to know what the numbers $x_1, x_2, \ldots, x_n$ are, so we run an experiment: We select values for the inputs and observe the output. In fact, we do this m times. On the i-th trial, we use inputs $a_{i1}, a_{i2}, \ldots, a_{in}$ and observe the output b_i. The question arises: Does there exist a single set of values for the numbers $x_1, x_2, \ldots, x_n$ such that all the equations

$$a_{i1} x_1 + a_{i2} x_2 + \cdots + a_{in} x_n = b_i, \quad i = 1, \ldots, m$$

hold? If not, can we choose a set of numbers $x_1, x_2, \ldots, x_n$ such that the largest deviation is minimized? By "largest deviation" we mean

$$z := \max \left\{ \left| \sum_{j=1}^{n} a_{1j} x_j - b_1 \right|, \left| \sum_{j=1}^{n} a_{2j} x_j - b_2 \right|, \ldots, \left| \sum_{j=1}^{n} a_{mj} x_j - b_m \right| \right\}.$$

We want to choose the "weights" $x_1, x_2, \ldots, x_n$ so as to minimize z.

For the sake of exposition, let us assume $m = 2$ and $n = 3$. Here is how such a Chebyshev problem can be turned into a linear program. First note that by definition

$$z \geq |a_{11} x_1 + a_{12} x_2 + a_{13} x_3 - b_1| \quad \text{and} \quad z \geq |a_{21} x_1 + a_{22} x_2 + a_{23} x_3 - b_2|.$$

Since each of these 2 absolute values is nonnegative, we have $z \geq 0$, although we shall not impose this explicitly as a constraint. Furthermore, since for any real number t, $|t| \geq \pm t$, we have

$$|a_{11} x_1 + a_{12} x_2 + a_{13} x_3 - b_1| \geq \pm (a_{11} x_1 + a_{12} x_2 + a_{13} x_3 - b_1)$$

$$|a_{21} x_1 + a_{22} x_2 + a_{23} x_3 - b_2| \geq \pm (a_{21} x_1 + a_{22} x_2 + a_{23} x_3 - b_2).$$

---

[1] For a treatment of this linear programming model and many others, see [77].

[2] Inputs are also called the independent variables and the output is called the dependent variable.

Hence we obtain the linear program

$$
\begin{aligned}
\text{minimize} \quad & z \\
\text{subject to} \quad & z - (a_{11}x_1 + a_{12}x_2 + a_{13}x_3 - b_1) \geq 0 \\
& z + (a_{11}x_1 + a_{12}x_2 + a_{13}x_3 - b_1) \geq 0 \\
& z - (a_{21}x_1 + a_{22}x_2 + a_{23}x_3 - b_2) \geq 0 \\
& z + (a_{21}x_1 + a_{22}x_2 + a_{23}x_3 - b_2) \geq 0.
\end{aligned}
$$

The linear program in general is to minimize z subject to this system of $2m$ linear inequalities in which all the variables—including z—are free. Nevertheless, by our earlier discussion, we know that for any solution of these inequalities, we must have $z \geq 0$.

Conversion of free variables to differences of nonnegative variables

A free variable can always be replaced by the *difference* of two nonnegative variables. Thus

$$\theta = \theta' - \theta'', \quad \theta' \geq 0, \ \theta'' \geq 0.$$

Notice that if $\theta = \theta' - \theta''$, it is not automatically true that the product of θ' and θ'' is zero. (For example, $-1 = 2 - 3$.) Nevertheless, θ' and θ'' can be chosen so that $\theta'\theta'' = 0$. When this is done, we have

$$\theta' = \theta^+ \quad \text{and} \quad \theta'' = \theta^-,$$

where, by definition,

$$\theta^+ = \max\{0, \theta\} \quad \text{and} \quad \theta^- = -\min\{0, \theta\}.$$

To illustrate, suppose $\theta = 5$. Then $\theta^+ = 5$ and $\theta^- = 0$. On the other hand, if $\theta = -5$, then $\theta^+ = 0$ and $\theta^- = 5$. If $\theta = 0$, then $\theta^+ = \theta^- = 0$. In each case, $\theta^+\theta^- = 0$.

Substituting for free variables

If x_j is a free variable and we use the substitution $x_j = x'_j - x''_j$, then, in the objective function, the term $c_j x_j$ becomes $c_j x'_j - c_j x''_j$, and in a constraint, the term $a_{ij}x_j$ becomes $a_{ij}x'_j - a_{ij}x''_j$.

We left Example 2.1 in the form of an inequality-constrained LP with free variables. Using the techniques described above, you can convert this linear program to one in standard form.

Important notational conventions

We often need to single out a row or column of a matrix. Accordingly, it is useful to have a clear and consistent way to do this. There are several such systems in use, but the one described in the following is preferred in this book. Suppose we have an $m \times n$ matrix A and a vector $b \in R^m$. If we wish to consider the system $Ax = b$ of m linear equations in n variables and need to emphasize the representation of b as a linear combination of the columns of A, then we let $A_{\bullet j}$ denote the j-th column of A. The equation

$$\sum_{j=1}^{n} A_{\bullet j} x_j = b$$

then gives a *column-oriented representation* of b. We say that column $A_{\bullet j}$ is *used* in the representation of b if the corresponding x_j is nonzero. In the notation $A_{\bullet j}$, the large dot "$\bullet$" represents a "wildcard". This means that all row indices i are included.

In like manner, we denote the i-th row of A by $A_{i \bullet}$. This notation is useful for singling out an individual row from the matrix A which comes up in the row-oriented presentation of the system $Ax = b$. That is, we can use

$$A_{i \bullet} x = b_i \quad \text{as an abbreviation for} \quad \sum_{j=1}^{n} a_{ij} x_j = b_i.$$

By way of illustration, let us consider the matrix

$$A = \begin{bmatrix} 5 & -1 & 0 & 8 \\ 2 & 1 & 2 & 3 \\ 0 & -1 & -1 & 4 \end{bmatrix}.$$

Then we have

$$A_{\bullet 2} = \begin{bmatrix} -1 \\ 1 \\ -1 \end{bmatrix} \quad \text{and} \quad A_{2 \bullet} = [2 \quad 1 \quad 2 \quad 3].$$

Example 2.2: LEAST-NORM SOLUTION OF A LINEAR SYSTEM. Suppose we have a linear system given by the equations

$$\sum_{j=1}^{n} a_{ij}x_j = b_i, \quad i = 1, \ldots, m.$$

Assume the system has a solution. Note that if $n > m$, the linear system could have infinitely many solutions. We seek a solution $x = (x_1, \ldots, x_n)$ having the "least 1-norm," meaning that $|x_1| + \cdots + |x_n|$ is as small as possible among all solutions of the system. So far, the problem is just

$$\begin{array}{ll} \text{minimize} & \sum_{j=1}^{n} |x_j| \\ \text{subject to} & \sum_{j=1}^{n} A_{\bullet j} x_j = b \end{array} \tag{2.6}$$

where $b = (b_1, \ldots, b_m)$. This is not quite an LP at this point. For one thing, there are no inequalities among the constraints. We can fix that by replacing each variable x_j by the difference of two nonnegative variables: $x_j = u_j - v_j$ for all $j = 1, \ldots, n$. The constraints then become

$$\sum_{j=1}^{n} A_{\bullet j} u_j - \sum_{j=1}^{n} A_{\bullet j} v_j = b$$

$$u_j \geq 0, \ v_j \geq 0, \quad j = 1, \ldots, n.$$

How do we—legitimately—get rid of the absolute value signs in (2.6)? It is easy to that if

$$x_j = u_j - v_j, \quad u_j \geq 0, \quad v_j \geq 0, \quad \text{and } u_j v_j = 0, \text{ for all } j = 1, \ldots n,$$

then

$$|x_j| = u_j + v_j, \ j = 1, \ldots, n.$$

The linear program to be solved becomes

$$\begin{array}{ll} \text{minimize} & \sum_{j=1}^{n} u_j + \sum_{j=1}^{n} v_j \\ \text{subject to} & \sum_{j=1}^{n} A_{\bullet j} u_j - \sum_{j=1}^{n} A_{\bullet j} v_j = b \end{array} \tag{2.7}$$

$$u_j \geq 0, \ v_j \geq 0, \quad j = 1, \ldots, n.$$

An appreciation of why the above formulation is valid will be gotten from reading the next section and Exercise 2.8, where you are asked to show that the condition $u_j v_j = 0$ must hold at an optimal extreme point. Based on this assertion and the notation introduced on page 36, we can also express the problem to be solved as the LP

$$\text{minimize} \qquad \sum_{j=1}^{n} x_j^+ + \sum_{j=1}^{n} x_j^-$$

$$\text{subject to} \qquad \sum_{j=1}^{n} A_{\bullet j} x_j^+ - \sum_{j=1}^{n} A_{\bullet j} x_j^- = b \qquad (2.8)$$

$$x_j^+ \geq 0, \ x_j^- \geq 0, \quad j = 1, \ldots, n.$$

Such linear programs come up in signal analysis problems in electrical engineering where estimates of *sparse solutions* of underdetermined linear systems are needed.

Converting equations into inequalities

Sometimes it is necessary to convert an equation (involving two real numbers) into a pair of inequalities. Recall that for real numbers, x and y, the equation $x = y$ is equivalent to the pair of inequalities

$$x \geq y,$$
$$x \leq y.$$

Thus, the equation $\sum_{j=1}^{n} a_{ij} x_j = b_i$ is equivalent to the pair of inequalities

$$\sum_{j=1}^{n} a_{ij} x_j \geq b_i,$$

$$\sum_{j=1}^{n} a_{ij} x_j \leq b_i.$$

Minimization versus maximization

Another general rule is that

$$\max f(x) = -\min\{-f(x)\}. \qquad (2.9)$$

Note the effects of the two minus signs in this equation. The function $-f$ is the reflection of f with respect to the horizontal axis. Suppose $\bar{x}$ minimizes $-f(x)$. Then by changing the sign of the function value $-f(\bar{x})$, we find that the maximum value of $f(x)$ is $f(\bar{x})$. The maximum of f and the minimum of $-f$ occur at the same place. The maximum and minimum *values* of these two functions sum to zero. Using this rule, we may convert a maximization problem into a minimization problem, and vice versa.

This is a handy rule when one needs to convert a maximization problem to a minimization problem. For example, if one had an optimization algorithm that solves only minimization problems, it would be possible to use (2.9) to recast any maximization problem to an equivalent minimization problem.

Figure 2.4, given below, depicts the general rule stated in (2.9) between the maxima and minima of a function (solid curve) and the minima and maxima (respectively) of the *negative* of that function (dotted curve).

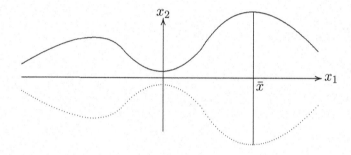

Figure 2.4: Illustration of (2.9).

Some important (LP) language

Consider an LP, say

$$\begin{array}{ll} \text{minimize} & c^{\mathrm{T}}x \\ \text{subject to} & Ax = b \\ & x \geq 0. \end{array}$$

A vector $\bar{x}$ satisfying the constraints $Ax = b$ and $x \geq 0$ is called a *feasible vector* for this LP. Even if an LP is not in standard form, the analogous definition would apply. A feasible vector for an LP is simply one that satisfies its constraints. In a minimization problem, a feasible vector $\bar{x}$ is *optimal* if $c^{\mathrm{T}}\bar{x} \leq c^{\mathrm{T}}x$ for every feasible vector x.

The feasible region of an LP

The *feasible region* of a linear program is the set of all its feasible vectors. If this set is empty, then the LP is said to be *infeasible*. The feasible region of an LP can be described as the intersection of the solution sets of a finite number of linear inequalities. As a simple illustration of this idea, consider again the linear inequalities

$$3x_1 - 5x_2 \leq 15$$
$$2x_1 + 4x_2 \leq 12$$

and Figure 2.2 which depicts their solution set. These linear inequalities could be the constraints of a linear programming problem in which case the doubly shaded region would represent (a portion of) its feasible region.

2.2 Polyhedral sets

Linear equations and linear inequalities can be interpreted geometrically. Let $a \in R^n$ be a nonzero vector and let $b \in R$. Then the solution set of a linear equation, say $a^T x = b$, is called a *hyperplane*. On the other hand, the solutions of the linear inequality $a^T x \leq b$ form what is called a *halfspace*, and the same is true for the linear inequality $a^T x \geq b$. For either of these two types of halfspaces, the boundary is a hyperplane, namely the set of x such that $a^T x = b$.

Any subset of R^n that can be represented as the intersection of a finite collection of halfspaces is called a *polyhedron*, or *polyhedral set*. This section covers some properties of polyhedral sets. In Chapter 7 we shall have more to say about polyhedral sets and their structure.

An important consequence of this definition is that the solution set of a linear inequality system is a polyhedron. And since a linear equation is equivalent to a pair of linear inequalities, we can say that the solution set of the following linear system is also polyhedral:

$$\begin{aligned} Fx &\leq f \quad &(f \in R^{m_1}) \\ Gx &= g \quad &(g \in R^{m_2}) \\ Hx &\geq h \quad &(h \in R^{m_3}). \end{aligned}$$

Within the class of linear inequality systems, we distinguish those for which the right-hand side vector is zero. A very simple example of such a system is $Ix \geq 0$. A linear inequality system of this form is said to be *homogeneous*.

Let C be the solution set of a homogeneous linear inequality system, say

$$Fx \leq 0$$
$$Gx = 0$$
$$Hx \geq 0.$$

Then C has the property that

$$x \in C \quad \Longrightarrow \quad \lambda x \in C \quad \text{for all } \lambda \geq 0. \tag{2.10}$$

Any set C for which (2.10) holds is called a *cone*; a polyhedral set for which (2.10) holds is called a *polyhedral cone*.

The feasible region of every linear programming problem is an example of a polyhedron. Indeed, a linear program whose constraints are

$$x_1 + x_2 + x_3 \leq 1$$
$$x_j \geq 0, \quad j = 1, 2, 3 \tag{2.11}$$

would have a feasible region looking like the shaded tetrahedron in Figure 2.5 below.

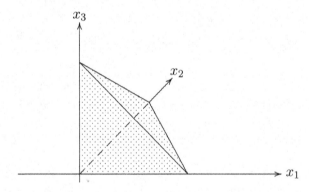

Figure 2.5: Feasible region of (2.11).

The polyhedral region specified by (2.11) is the intersection of four half-spaces. This particular set happens to be *bounded*, that is to say, it is contained within a ball of finite radius. Not every polyhedral set is a bounded,

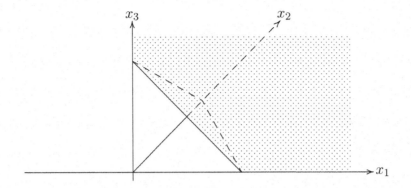

Figure 2.6: Feasible region of (2.12).

however. To illustrate this fact, we revise the example given above by making a small change in the one of the constraints. Consider the polyhedral set given by

$$x_1 + x_2 + x_3 \geq 1$$
$$x_j \geq 0, \quad j = 1, 2, 3. \tag{2.12}$$

The feasible region corresponding to (2.12) is not bounded as Figure 2.6 is intended to show.

Since a linear equation is equivalent to a pair of linear inequalities, we can regard the feasible region of linear system

$$x_1 + x_2 + x_3 = 1$$
$$x_j \geq 0, \quad j = 1, 2, 3 \tag{2.13}$$

as a polyhedral set. It is depicted in Figure 2.7.

Convexity of polyhedral sets

A set X (not necessarily polyhedral) is defined to be *convex* if it contains the line segment between any two of its point. The *line segment* between the points x^1 and x^2 in R^n is given by

$$\{x \in R^n : x = \lambda x^1 + (1 - \lambda)x^2, \ 0 \leq \lambda \leq 1\}.$$

Every element of the above set is called a *convex combination* of x^1 and x^2. More generally, if $x^1, \ldots, x^k \in R^n$ and $\lambda_1, \ldots, \lambda_k$ are nonnegative real

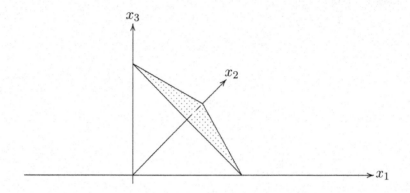

Figure 2.7: Feasible region of (2.13).

numbers satisfying $\lambda_1 + \cdots + \lambda_k = 1$, then

$$\lambda_1 x^1 + \cdots + \lambda_k x^k$$

is called a *convex combination* of $x^1, \ldots, x^k$.

The entire space R^n is clearly a convex set, and so is any halfspace. An elementary property of convex sets is that the intersection of any two of them is again a convex set.[3] It follows, then, that every polyhedron (and hence the feasible region of every linear programming problem) is convex, as is demonstrated in the next paragraph by an elementary argument. Associated with any set S (convex or not), there is another set called its *convex hull* which is defined as the intersection of all convex sets containing S. Accordingly, a set is convex if and only if it equals its convex hull.

There is an alternate definition of a convex hull of the set S. Indeed, the convex hull of S is the set of all points which can be expressed as a convex combination of finitely many points belonging to S. The two definitions of convex hull are equivalent.

The solution sets arising from (2.11), (2.12), and (2.13) are clearly convex. To see that *all* polyhedral sets are convex, consider a linear inequality system given by $Ax \le b$ where $A \in R^{m \times n}$ and $b \in R^m$. We readily verify that if x^1 and x^2 each satisfy this linear inequality system, then so does $\lambda x^1 + (1-\lambda)x^2$

---

[3]By convention, the empty set $\emptyset$ is regarded as convex.

for all $\lambda \in [0, 1]^4$. Indeed, for all $\lambda \in [0, 1]$ we have

$$A(\lambda x^1 + (1 - \lambda)x^2) = \lambda A x^1 + (1 - \lambda) A x^2 \le \lambda b + (1 - \lambda)b = b.$$

The shaded set shown on the left in Figure 2.8 is not convex because it does not contain the line segment between every pair of its elements. (It is a nonconvex polygon.) The set shown on the right in Figure 2.8 is convex, however. This set is polyhedral since it can be viewed as the intersection of a finite collection of halfspaces (in this case five of them).

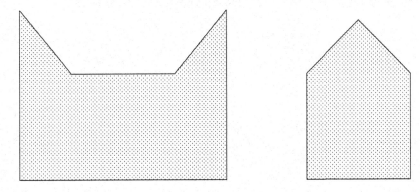

Figure 2.8: Nonconvex polygon (left) and convex polygon (right).

Furthermore, it should be noted that not all convex sets are polyhedral. For instance, disks and ellipsoids are nonpolyhedral convex sets.

Extreme points and their importance in linear programming

Let $\bar{x}$ be an element of a convex set X. We say $\bar{x}$ is an *extreme point* of X if whenever x^1 and x^2 are distinct points in X, then

$$\bar{x} = \lambda x^1 + (1 - \lambda)x^2, \quad 0 \le \lambda \le 1$$

implies that $\lambda = 0$ or $\lambda = 1$, in which case $\bar{x} = x^1$ or $\bar{x} = x^2$. Thus, $\bar{x}$ does not lie strictly between x^1 and x^2. This is the essence of the extreme point property.

It is a well-established fact (theorem) that

---

[4]The notation $[0, 1]$ means all numbers between 0 and 1 inclusive.

if a linear programming problem in standard form has an optimal solution, then it has an optimal extreme point solution.

This suggests that in trying to find an optimal solution to a linear program in standard form[5], it would be a reasonable idea to confine the search to the (finite) set of extreme points of the feasible region.

Note that this theorem does *not* say that every optimal solution of a linear program is an extreme point. Indeed, there exist linear programs having nonextreme optimal solutions. One such instance is the LP

$$\begin{aligned} \text{maximize} \quad & x_2 \\ \text{subject to} \quad & 0 \le x_1 \le 1 \\ & 0 \le x_2 \le 1 \end{aligned}$$

whose feasible region is the "unit square." Every point of the form $(x_1, 1)$ with $0 \le x_1 \le 1$ is feasible and optimal. Note that any such point lies on an edge between two extreme points.

Linear independence of vectors

Let $v^1, v^2, \ldots, v^k$ be a set of vectors in a finite-dimensional vector space V, such as R^n. The vectors $v^1, v^2, \ldots, v^k$ are *linearly independent* if and only if

$$\alpha_1 v^1 + \alpha_2 v^2 + \cdots + \alpha_k v^k = 0$$

implies

$$\alpha_1 = \alpha_2 = \cdots = \alpha_k = 0.$$

Another way to put this implication is to say that there is no *nontrivial* linear relationship among the vectors $v^1, v^2, \ldots, v^k$: none of them is a linear combination of the others.

A set of objects possessing some property is *maximal* if the set cannot be enlarged by adjoining another element while at the same time preserving the property. For instance, the set S consisting of the vectors

$$(1, 1, 1), \quad (0, 1, 1), \quad \text{and} \quad (0, 0, 1)$$

---

[5]The assumption of standard form guarantees that—if it is nonempty—the feasible region of the LP contains an extreme point.

has the property of linear independence. (Taken as an ensemble, the elements of S are linearly independent.) If we form a set of four vectors by adjoining a vector, (b_1, b_2, b_3) to the set S, we no longer have a linearly independent set of vectors. So, with respect to the property of linear independence, the set S is maximal.

A maximal linearly independent set of vectors in a finite-dimensional vector space is called a *basis* for that space. Any set of vectors that properly includes a basis must be *linearly dependent*. For example, when (b_1, b_2, b_3) is adjoined to the set S defined above, we obtain a linearly independent set.

An implication of linear independence

If w is a linear combination of the linearly independent vectors $v^1, v^2, \ldots, v^k$ in the vector space V, that is, if

$$w = \alpha_1 v^1 + \alpha_2 v^2 + \cdots + \alpha_k v^k,$$

then the scalars $\alpha_1, \alpha_2, \ldots, \alpha_k$ used in the representation of w are *unique*. To see this, suppose

$$w = \alpha_1 v^1 + \alpha_2 v^2 + \cdots + \alpha_k v^k = \beta_1 v^1 + \beta_2 v^2 + \cdots + \beta_k v^k.$$

Then we must have

$$(\alpha_1 - \beta_1)v^1 + (\alpha_2 - \beta_2)v^2 + \cdots + (\alpha_k - \beta_k)v^k = 0.$$

The linear independence of $v^1, v^2, \ldots, v^k$ implies that $\alpha_j - \beta_j = 0$, $j = 1, 2, \ldots, k$. This, in turn, is another way of saying that there is only one way to represent w as a linear combination of $v^1, v^2, \ldots, v^k$.

Matrices and systems of linear equations

When we have a system of m equations in n unknowns (or variables), say

$$\sum_{j=1}^{n} a_{ij} x_j = b_i, \quad i = 1, \ldots, m, \tag{2.14}$$

the associated matrix $A = [a_{ij}] \in R^{m \times n}$ and vector $b = (b_1, \ldots, b_m) \in R^m$ allow us to express the system as

$$Ax = b. \tag{2.15}$$

The $m \times n$ matrix A gives rise to two special sets of vectors, namely the rows $A_{1\bullet}, \ldots, A_{m\bullet}$ and the *columns* $A_{\bullet 1}, \ldots, A_{\bullet n}$.

In discussing (2.14), we are particularly interested in the columns of A and their relationship to the column vector b. Note that (2.14) has a solution, $\bar{x}$, if and only if the $n+1$ vectors $A_{\bullet 1}, \ldots, A_{\bullet n}, b$ are linearly dependent *and*, in the equation expressing linear dependence, there is a nonzero coefficient on b. Indeed, if $\bar{x}$ satisfies (2.14), we have

$$A_{\bullet 1}\bar{x}_1 + \cdots + A_{\bullet n}\bar{x}_n + b(-1) = 0,$$

and since $(\bar{x}_1, \ldots, \bar{x}_n, -1)$ is nonzero, the vectors $A_{\bullet 1}, \ldots, A_{\bullet n}, b$ are linearly dependent. Conversely, if these vectors are linearly dependent, there exist scalars $\alpha_1, \ldots, \alpha_n, \alpha_{n+1}$ not all zero such that

$$A_{\bullet 1}\alpha_1 + \cdots + A_{\bullet n}\alpha_n + b\alpha_{n+1} = 0. \tag{2.16}$$

If $\alpha_{n+1} \neq 0$, then $\bar{x} = -(\alpha_1/\alpha_{n+1}, \ldots, \alpha_n/\alpha_{n+1})$ is a solution of (2.15). On the other hand, if $\alpha_{n+1} = 0$, the columns of A are themselves linearly dependent; in such a case, it may happen that there is no solution to (2.15). For instance, if

$$A = \begin{bmatrix} 1 & 1 & -2 \\ 1 & 1 & -2 \end{bmatrix} \quad \text{and} \quad b = \begin{bmatrix} 1 \\ 0 \end{bmatrix},$$

then $(1, 1, 1, 0)$ would be a nonzero solution of the corresponding equation (2.16), yet clearly (2.15) has no solution since the system

$$x_1 + x_2 - 2x_3 = 1$$
$$x_1 + x_2 - 2x_3 = 0$$

is *inconsistent*. However, if the right-hand side vector b had been any multiple of $(1, 1)$, the system would have been *consistent* and hence solvable.

When $b = 0$ the system (2.15) is called *homogeneous*. Such systems always have at least one solution, namely $x = 0$. Solutions to such systems are called *homogeneous solutions*.

The rank of a matrix

The *column rank* of a matrix A is the maximal number of linearly independent columns of A. The *row rank* of a matrix A is the maximal number of linearly independent rows of A. (This equals the column rank of A^{T}, the transpose of A.) From matrix theory we have the following theorem: If A is an $m \times n$ matrix,

$$\text{row rank}(A) = \text{column rank}(A) = \text{rank}(A) \leq \min\{m, n\}.$$

For example, let

$$A = \begin{bmatrix} 1 & 1 & 1 \\ 1 & 1 & 1 \end{bmatrix} \quad \text{and} \quad \bar{A} = \begin{bmatrix} 1 & 1 & 1 \\ 0 & 1 & 1 \end{bmatrix}.$$

The rank of A is 1 whereas the rank of $\bar{A}$ is 2.

There is a procedure called reduction to row-echelon form for computing the rank of a nonzero $m \times n$ matrix A. The idea is to perform elementary row and column operations, including permutations, on A so as to arrive at another matrix $\tilde{A}$ with the property that for some r, $1 \leq r \leq m$,

1. $\tilde{a}_{jj} \neq 0$ for all $j = 1, \ldots, r$

2. $\tilde{a}_{ij} = 0$ if $i > j$ and $j = 1, \ldots, r$

3. $\tilde{a}_{ij} = 0$ if $i > r$ and $j > r$

The integer r turns out to be the rank of the matrix $\tilde{A}$ and the matrix A as well. Notice that the matrix

$$\bar{A} = \begin{bmatrix} 1 & 1 & 1 \\ 0 & 1 & 1 \end{bmatrix}$$

in the example above is already in row-echelon form with $r = 2$. The following matrix is also in row-echelon form:

$$\hat{A} = \begin{bmatrix} 4 & 1 & 0 & -1 \\ 0 & -3 & 1 & 2 \\ 0 & 0 & 0 & 0 \end{bmatrix}.$$

Notice that in this case the matrix $\hat{A}$ has a row of zeros. Here again $r = 2$.

A theorem of the alternative for systems of linear equations

A system of linear equations, say $Ax = b$, is either consistent or inconsistent. In the latter case, there is a consistent system of linear equations based on the same data. namely

$$\begin{bmatrix} A^\mathrm{T} \\ b^\mathrm{T} \end{bmatrix} y = \begin{bmatrix} 0 \\ 1 \end{bmatrix}. \tag{2.17}$$

The right-hand side of this linear equality system is not uniquely determined. The component 1 can be replaced by any nonzero number, θ, in which case θy would be a solution if y solves (2.17). The italicized assertion below is an example of an "alternative theorem" (also called a "transposition theorem").[6]

> *Exactly one of the following two systems of equations has a solution.*
>
> *(i)* $Ax = b,$
> *(ii)* $A^\mathrm{T}y = 0, \quad b^\mathrm{T}y = 1.$

These two systems cannot both have solutions, for otherwise

$$0 = y^\mathrm{T}Ax = y^\mathrm{T}b = 1$$

which is plainly absurd. That (ii) must have a solution if (i) does not follows easily from the form of system (i) after reduction to row echelon form.

The reduction of the system to row echelon form is based on the reduction of the coefficient matrix A to row echelon form. Indeed, the key idea, as noted by Strang [182, p. 79], is expressed by the theorem that *to any $m \times n$ matrix A, there exists a permutation matrix P, a unit lower triangular L, and an upper trapezoidal matrix U with its nonzero entries in echelon form such that $PA = LU$.* This being so, the system (i) is equivalent to $L^{-1}PAx = Ux = L^{-1}Pb$. The inconsistency of the system (i) will

---

[6]In the literature, this result has been incorrectly attributed to David Gale in whose 1960 textbook [75, p. 41] it can, indeed, be found. Actually, it goes back at least as far as the doctoral dissertation of T.S. Motzkin published in 1936 where a theorem is proved [139, p. 51] that gives the one stated here as a special case. Versions of the theorem can also be found as exercise 39 in L. Mirsky's 1955 textbook [136, p. 166] and in a 1956 paper [116, p. 222] by H.W. Kuhn discussing solvability and consistency of linear equation systems.

be revealed by the presence of a zero row, say k, in U for which the corresponding entry in $\tilde{b} = L^{-1}Pb$ is nonzero. A solution of system (ii) can then be constructed using this information.

This alternative theorem will be applied in Chapter 11.

Subspaces associated with a matrix

When the system (2.15) has a solution, we say that the columns $A_{\bullet 1}, \ldots, A_{\bullet n}$ *span* the vector b. The set of all vectors $b \in R^m$ spanned by the columns of $A \in R^{m \times n}$ is a vector space (in fact, a *subspace* of R^m). This vector space goes by various names; one of these is *column space* of A. Others are *affine hull* of A or *span* of A. Furthermore, when we think of the linear transformation $x \mapsto Ax$ associated with A, then the column space is just the *range* of this linear transformation, and is often called the *range space*. However it may be called, the object we have in mind is the vector space

$$\mathcal{R}(A) = \{b : b = Ax \text{ for some } x \in R^n\}.$$

As we have already noted, $\mathcal{R}(A)$ is a subspace of R^m. The dimension r of $\mathcal{R}(A)$ is called the *column rank* of A. This is just the cardinality of a maximal linearly independent subset of columns of A. For $A \in R^{m \times n}$, the relationship

$$r = \dim \mathcal{R}(A) \leq m$$

always holds.

Another vector space associated with A is the subspace $\mathcal{N}(A)$ of R^n consisting of all vectors that are mapped into the zero vector. Thus

$$\mathcal{N}(A) = \{x : Ax = 0\}.$$

This set is called the *null space* of A. Its elements belong to the *kernel* of the linear transformation $x \mapsto Ax$.

Just as the subspaces $\mathcal{R}(A)$ and $\mathcal{N}(A)$ are defined for the matrix A, so, too, the subspaces $\mathcal{R}(A^T)$ and $\mathcal{N}(A^T)$ are defined with respect to the matrix A^T, the transpose of A.

An important theorem of matrix algebra states that

the column rank of A equals the column rank of A^{T} (or, equivalently, the row rank of A).

This number, r, is therefore called the *rank* of A, denoted by rank(A).

Returning to the four subspaces we have mentioned, we can say that

$$
\begin{array}{lll}
\mathcal{R}(A) & \text{has dimension} & r \\
\mathcal{N}(A) & \text{has dimension} & n - r \\
\mathcal{R}(A^{\mathrm{T}}) & \text{has dimension} & r \\
\mathcal{N}(A^{\mathrm{T}}) & \text{has dimension} & m - r.
\end{array}
$$

The numbers $n-r$ and $m-r$ are called the *nullity* of A and A^{T}, respectively.

For any subspace V of R^n, there is another subspace called the *orthogonal complement*, $V^{\perp}$, defined as follows:

$$ V^{\perp} = \{u : u^{\mathrm{T}}v = 0 \text{ for all } v \in V\}. $$

The subspaces V and $V^{\perp}$ are orthogonal to each other. The dimensions of V and $V^{\perp}$ are always "complementary" in the sense that

$$ \dim V + \dim V^{\perp} = n. $$

In the special case of the subspaces associated with the $m \times n$ matrix A, we have

$$
\begin{aligned}
\mathcal{R}(A) &= (\mathcal{N}(A^{\mathrm{T}}))^{\perp} \\
\mathcal{N}(A) &= (\mathcal{R}(A^{\mathrm{T}}))^{\perp} \\
\mathcal{R}(A^{\mathrm{T}}) &= (\mathcal{N}(A))^{\perp} \\
\mathcal{N}(A^{\mathrm{T}}) &= (\mathcal{R}(A))^{\perp}.
\end{aligned}
$$

as well as

$$
\begin{aligned}
R^m &= \mathcal{R}(A) + \mathcal{N}(A^{\mathrm{T}}) \\
R^n &= \mathcal{R}(A^{\mathrm{T}}) + \mathcal{N}(A).
\end{aligned}
$$

Like any finite-dimensional vector space, the column space of A has a *basis*—a maximal linearly independent subset of vectors that span $\mathcal{R}(A)$. In fact, the basis can be taken as a set of $r = \dim \mathcal{R}(A)$ suitable columns of A itself. For further details see Strang [182, p. 102–114].

Basic solutions of systems of linear equations

In assuming that the matrix A in (2.15) is of order $m \times n$, it is reasonable to assume further that $m \le n$. (Otherwise, the equations in (2.14) are either redundant or inconsistent.) When $r = m = \min\{m, n\}$, we say that A has *full rank*. Note that when A has full rank, it follows that $\mathcal{R}(A) = R^m$ which is a way of saying that the system (2.15) has a solution for every $b \in R^m$. Moreover, A has full rank if and only if there are m columns in A, say $A_{\bullet j_1}, \ldots, A_{\bullet j_m}$, that are linearly independent. Writing

$$B = [A_{\bullet j_1} \cdots A_{\bullet j_m}],$$

we refer to B as a *basis* in A.

Here is an example to illustrate some of these concepts. Let

$$A = \begin{bmatrix} 2 & -1 & 0 & -1 \\ -1 & 2 & -1 & 1 \\ 0 & -1 & 1 & 0 \end{bmatrix} \quad \text{and} \quad b = \begin{bmatrix} 1 \\ 1 \\ 1 \end{bmatrix}. \tag{2.18}$$

We leave it as an exercise to show that the first three columns of A are linearly independent, hence

$$B = \begin{bmatrix} 2 & -1 & 0 \\ -1 & 2 & -1 \\ 0 & -1 & 1 \end{bmatrix} = [A_{\bullet 1}\ A_{\bullet 2}\ A_{\bullet 3}]$$

is a basis in A as are $[A_{\bullet 1}\ A_{\bullet 2}\ A_{\bullet 4}]$ and $[A_{\bullet 1}\ A_{\bullet 3}\ A_{\bullet 4}]$. Observe that $[A_{\bullet 2}\ A_{\bullet 3}\ A_{\bullet 4}]$ is not a basis in A since its columns are linearly dependent as shown by the fact that the equation

$$A_{\bullet 2}x_2 + A_{\bullet 3}x_3 + A_{\bullet 4}x_4 = 0$$

has the nonzero solution $(x_2, x_3, x_4) = (1, 1, -1)$.

Just as it is useful to have a notation $A_{\bullet j}$ for the j-th column of a matrix A, so it is useful to have a notation for a *set* of columns that specify a submatrix of A. A case in point occurs in the designation of a basis. To this end, let $\beta = \{j_1, \ldots, j_m\}$ and write

$$A_{\bullet \beta} = [A_{\bullet j_1} \cdots A_{\bullet j_m}].$$

Analogously, we could let $\alpha = \{i_1, \ldots, i_k\}$ and write

$$A_{\alpha\bullet} = \begin{bmatrix} A_{i_1\bullet} \\ \vdots \\ A_{i_k\bullet} \end{bmatrix}.$$

Putting these two notational devices together, we can define

$$A_{\alpha\beta} = \begin{bmatrix} a_{i_1 j_1} & \cdots & a_{i_1 j_m} \\ \vdots & & \vdots \\ a_{i_k j_1} & \cdots & a_{i_k j_m} \end{bmatrix},$$

which is a $k \times m$ *submatrix* of A.

This notational scheme for denoting a submatrix often requires that the index sets α and β be *ordered* in the sense that

$$i_1 < i_2 < \cdots < i_k \quad \text{and} \quad j_1 < j_2 < \cdots < j_m.$$

In singling out a particular set of columns of A corresponding to the index set β, we can do the same sort of thing with a vector x writing $x_\beta = (x_{j_1}, \ldots, x_{j_m})$ when β is chosen as above. If $x_j = 0$ for all $j \notin \beta$, then (2.15) becomes

$$A_{\bullet\beta} x_\beta = b, \tag{2.19}$$

and when $A_{\bullet\beta}$ is a basis (with inverse $A_{\bullet\beta}^{-1}$), this equation has the unique solution

$$x_\beta = A_{\bullet\beta}^{-1} b.$$

If we let ν denote the set of indices $j \in \{1, \ldots, n\}$ such that $j \notin \beta$, we then obtain a submatrix $A_{\bullet\nu}$ and a subvector x_ν such that (2.15) is expressible as

$$A_{\bullet\beta} x_\beta + A_{\bullet\nu} x_\nu = b.$$

When $A_{\bullet\beta}$ is a basis and we set $x_\nu = 0$, we obtain what is called a *basic solution* to (2.15), namely the vector $\bar{x}$ having $\bar{x}_\beta = A_{\bullet\beta}^{-1} b$ and $\bar{x}_\nu = 0$.

With reference to the system of linear equations given by the data in (2.18), we have

Basic index set β	Nonbasic index set ν	Basic solution $\bar{x}$	
$\{1, 2, 3\}$	4	$(3, \ \ 5, \ 6, \ 0)$	(2.20)
$\{1, 2, 4\}$	3	$(3, -1, \ 0, \ 6)$	
$\{1, 3, 4\}$	2	$(3, \ \ 0, \ 1, \ 6)$	

When $\bar{x}$ is a solution of a system (2.15), we say that the columns of A *represent* b. The representation of b in terms of A and $\bar{x}$ *uses* column $A_{\bullet j}$ if $\bar{x}_j \neq 0$. Thus, when $\bar{x}$ is a basic solution, it uses only linearly independent columns of A. This can be taken as the definition of a basic solution in cases where the rank of A is less than m (i.e., where the *rows* of A are linearly dependent).

If A is a $m \times n$ matrix of rank m, there are exactly

$$\binom{n}{m} = \frac{n!}{m!\,(n-m)!}$$

ways to choose m columns from the n columns of A, or equivalently the index set β from $\{1, 2, \ldots, n\}$. In some cases, the chosen columns maybe linearly dependent, in which case they will not form a basis. For even modest size matrices, this number is extremely large and it is impractical to generate and test every such $m \times m$ submatrix, $A_{\bullet \beta}$.

From what has already been said, it follows that a linear program in standard form can have only finitely many basic feasible solutions. By the same token, there can be only finitely many feasible bases in the matrix A.

Basic feasible solutions of linear programs

Let $\bar{x}$ be a feasible solution of a linear program with constraints $Ax = b$ and $x \geq 0$. The vector $\bar{x}$ is a *basic feasible solution* if the columns $A_{\bullet j}$ associated with all the $\bar{x}_j > 0$ are linearly independent. That is, when $\beta = \{j : \bar{x}_j > 0\}$, then the submatrix $A_{\bullet \beta}$ has linearly independent columns. When A has full rank, m, it will contain at least one $m \times m$ basis $A_{\bullet \beta}$. Relative to the system (2.15), such a matrix is a *feasible basis* if the unique solution of (2.19) is nonnegative. This amounts to saying that $A_{\bullet \beta}$ is a feasible basis if and only if

$$\bar{x}_\beta = A_{\bullet \beta}^{-1} b \geq 0,$$

and as before, the corresponding basic feasible solution (BFS) $\bar{x}$ has

$$\bar{x}_\beta = A_{\bullet \beta}^{-1} b, \qquad \bar{x}_\nu = 0.$$

Note that in some cases, the $\bar{x}$ can be a basic feasible solution without having all the components of $\bar{x}_\beta$ greater than zero; some of the components of $\bar{x}_\beta$ might be zero. A solution of this sort is said to be *degenerate*.

Basic and nonbasic variables

When $A_{\bullet\beta}$ is a basis in A and $\bar{x}$ is a basic solution of (2.15), the x_j with $j \in \beta$ are called *basic variables* (with respect to β), and those with $j \in \nu$ are known as *nonbasic variables*. To illustrate, we recall the data given in (2.18), the basic solutions of which are given in the table (2.20). Notice that only the first and third basic solutions are feasible relative the given system; the second basic solution listed is *not* feasible because it is not a nonnegative vector. It can, however, be said that each of the basic solutions is *nondegenerate* since the components corresponding to the basic columns are nonzero in each case.

Suppose we now change the right-hand side vector b to $\tilde{b} = (1, 1, -1)$. The analog of the table given in (2.20) would then be

Basic index set β	Nonbasic index set ν	Basic solution $\bar{x}$	
$\{1, 2, 3\}$	$\{4\}$	$(1, 1, \ 0, 0)$	(2.21)
$\{1, 2, 4\}$	$\{3\}$	$(1, 1, \ 0, 0)$	
$\{1, 3, 4\}$	$\{2\}$	$(1, 0, -1, 1)$	

There are a few things to notice about this case. First, because A is the same as before, there is no change in the bases. Second, we see that two different bases give rise to the same degenerate basic feasible solution. Third, we note that one of the basic solutions is infeasible but nondegenerate. With a different right-hand vector, it would be possible to have a basic solution that is infeasible and degenerate at the same time.

Basic feasible solutions and extreme points

Suppose the feasible region is given by the set X:

$$X = \{x : Ax = b, \ x \geq 0\}.$$

The following theorem is of great importance:

> *A vector $\bar{x}$ is an extreme point of X if and only if it is a basic feasible solution of the constraints $Ax = b$, $x \geq 0$.*

This theorem is a useful link between the algebraic notion of basic feasible solution and the intuitively appealing geometric notion of extreme point. Recall the theorem, stated on page 46, that if an LP in standard form has an optimal solution, then it has an optimal extreme point solution. Combining this theorem with the one stated just above, we see that to solve an LP it suffices to consider only basic feasible solutions as candidates for optimal solutions. There are only finitely many bases in A and hence $Ax = b$, $x \geq 0$, can have only finitely many basic feasible solutions. Equivalently, X can have only finitely many extreme points. But, this finite number can be extremely large, even for problems of modest size. For this reason, an algorithm that examines basic feasible solutions must be efficient in selecting them.

2.3 Exercises

2.1   (a) Graph the feasible region of the product-mix problem given in Example 1.3.

(b) On the same graph plot the set of points where the objective function value equals 750. Are any of these points optimal? Why?

(c) Do the same for the objective function value 850. Are any of these points optimal? Why?

2.2   (a) Write the coefficient matrix for the left-hand side of Exercise 1.1.

(b) Are the rows of this matrix linearly independent? Justify your answer.

(c) What is the rank of this matrix?

(d) Write the coefficient matrix for the general problem (with equality constraints, m sources and n destinations).

(e) What is the percentage of nonzero entries to total entries in the matrix given in your answer to (d)? (This is called the *density* of the matrix. When the percentage is low, the matrix is said to be *sparse*. When it is high, the matrix is said to be *dense*.)

2.3   There is a variant of the transportation problem in which the sum of the supplies is allowed to be larger than the sum of the demands, that is,

$$\sum_{i=1}^{m} a_i > \sum_{j=1}^{n} b_j.$$

This is done by adjoining an extra (fictitious) demand

$$b_{n+1} = \sum_{i=1}^{m} a_i - \sum_{j=1}^{n} b_j$$

and then allowing each supply location i to "ship" an amount $x_{i,n+1}$ to destination $n+1$ at unit cost $c_{i,n+1} = 0$. Modify Exercise 1.1 by assuming that there is a supply of 600 units at Singapore. Using the technique described above, write the corresponding transportation problem with equality constraints. (Abbreviate the fictitious destination by "FIC.")

2.4   Write down the equations of the (quadrilateral) feasible region of (2.5) depicted in Figure 2.3. Identify all the extreme points of the feasible region by their coordinates and as the intersections of the sides of the feasible region (described, as above, by their equations).

2.5   A nonempty polyhedral convex set has at most a finite number of extreme points. Why is this true?

2.6   The following table gives data for an experiment in which the linear model

$$b = a_1 x_1 + a_2 x_2 + a_3 x_3$$

is postulated.

a_1	a_2	a_3	b
14.92	−19.89	10.66	985.62
17.76	9.16	18.29	846.21
16.45	−23.07	15.77	742.67
15.93	12.04	16.85	780.32
13.99	10.47	19.55	689.62

Write down the linear program for the associated Chebyshev problem. (You need not convert the LP to standard form for this exercise.)

2.7   In general does the linear program for the Chebyshev problem always have a feasible solution? Why?

2.8   Suppose $(u_1, \ldots, u_n, v_1, \ldots, v)$ is an optimal extreme point solution of the linear program proposed in Example 2.2. Why does $u_j v_j = 0$ hold for all $j = 1, \ldots, n$? Explain why solving the LP formulated in Example 2.2 with a method that uses only extreme points of the feasible region will solve the least 1-norm problem stated there.

2.9   In (2.6) the variables x_j are free. Each of these variables can be expressed as the *difference* of two nonnegative variables x_j^+ and x_j^-. The latter variables appear in the objective function of the linear program (2.8) with a plus sign between them. Given that (2.8) will have an optimal basic feasible solution (a fact covered in Chapter 3), explain why this formulation is valid.

2.10   Suppose you are given a system of linear equations given by $Ax = b$ where
        the matrix A is of order $m \times n$, but you wish to express the solution set of
        the system as a polyhedron i.e., as the solution set of finitely many linear
        inequalities. This can be done by replacing each equation $A_i.x = b_i$ by
        two linear inequalities: $A_i.x \leq b_i$ and $A_i.x \geq b_i$, $i = 1, \ldots, m$, thereby
        leading to an equivalent system of $2m$ linear inequalities. Can this be
        done with fewer than $2m$ linear inequalities? If so, how?

2.11   We have seen that in a linear system with k free variables, each of these
        free variables can be replaced by the difference of two nonnegative vari-
        ables. This would add k more variables to the system. Would it be
        possible to do this with fewer than k more variables? Justify your an-
        swer.

2.12   In the text it is asserted that for any (b_1, b_2, b_3) the vectors

$$(1,1,1), \quad (0,1,1), \quad (0,0,1), \quad \text{and} \quad (b_1, b_2, b_3)$$

        are linearly dependent. Denote these vectors by v^1, v^2, v^3, v^4, respectively,
        and find a set of scalar coefficients $\alpha_1, \alpha_2, \alpha_3, \alpha_4$ such that

$$\alpha_1 v^1 + \alpha_2 v^2 + \alpha_3 v^3 + \alpha_4 v^4 = 0.$$

        How many such sets of coefficients are there in this case?

2.13   Verify the linear independence of columns $A_{\bullet 1}$ $A_{\bullet 2}$, $A_{\bullet 3}$, where A is the
        matrix given in (2.18).

2.14   A square matrix X of order n whose elements x_{ij} satisfy the linear con-
        ditions

$$\sum_{j=1}^{n} x_{ij} = 1, \quad i = 1, \ldots, n$$

$$\sum_{i=1}^{n} x_{ij} = 1, \quad j = 1, \ldots, n$$

$$x_{ij} \geq 0, \quad i, j = 1, \ldots, n$$

        is said to be *doubly stochastic*.

        (a) Verify that the set of all doubly stochastic matrices is convex.
        (b) What is the order (size) of the matrix of coefficients in the equations
            through which a doubly stochastic matrix of order n is defined?
        (c) Write down a verbose version of the conditions satisfied by a doubly
            stochastic matrix of order 3.
        (d) Explain why the rank of the matrix of coefficients in your answer to
            (c) is at most 5.

(e) Find a basis in the matrix of coefficients in your answer to (c).

(f) Is the basic feasible solution corresponding to your answer in (e) non-degenerate? Justify your answer.

2.15   Consider the linear program

$$\begin{array}{ll} \text{maximize} & 4x_1 + x_2 \\ \text{subject to} & x_1 - 3x_2 \leq 6 \\ & x_1 + 2x_2 \leq 4. \end{array}$$

(a) Plot the feasible region of the above linear program.

(b) List all the extreme points of the feasible region.

(c) Write the equivalent standard form as defined in (1.1).

(d) Show that the extreme points are basic feasible solutions of the LP.

(e) Evaluate the objective function at the extreme points and find the optimal extreme point solution.

(f) Modify the objective function so that the optimal solution is at a different extreme point.

3. THE SIMPLEX ALGORITHM

Overview

There are several ways to solve linear programs, but even after its invention in 1947 and the emergence of many new rivals, George B. Dantzig's Simplex Algorithm stands out as the foremost method of all. Indeed, the Simplex Algorithm was chosen as one of the top ten algorithms of the twentieth century, see [54].

This chapter covers the Simplex Algorithm beginning with some of the preliminary ideas that need to be in place before we can appreciate what it does. We have already set the stage for some of this with our consideration of standard form, basic feasible solutions, and extreme points of polyhedral convex sets. The discussion in Chapter 2 brought out the idea that in searching for an optimal solution of a linear program, we may restrict our attention to basic feasible solutions of the constraints (in standard form). To take advantage of this insight, we need to answer some key questions on feasibility and optimality:

1. How can we tell if the LP has a feasible solution?

2. How do we find a basic feasible solution (if one exists)?

3. How can we recognize whether a basic feasible solution is optimal?

4. What should we do if a basic feasible solution is known (or believed) to be nonoptimal?

These four questions are best answered in a different order. Accordingly, we will start by assuming that the given linear program is feasible and a basic feasible solution is known. After answering Questions 3 and 4, we will return to answer Questions 1 and 2. The reason for this is that we use the methodology developed to answer Questions 3 and 4 for answering Questions 1 and 2.

© Springer Science+Business Media LLC 2017
R.W. Cottle and M.N. Thapa, *Linear and Nonlinear Optimization*,
International Series in Operations Research & Management Science 253,
DOI 10.1007/978-1-4939-7055-1_3

3.1   Preliminaries

Matrices of full rank

Recall that if A is an $m \times n$ matrix with $m \leq n$, and rank$(A) = m$, then A is said to be of *full rank*. (Reduction of A to row-echelon form will reveal whether or not A has full rank.) In discussing LP algorithms, it is very common to assume that the problem is in standard form and that the coefficient matrix A has full rank. In that case, a basis in A will consist of m linearly independent columns of A which together form an $m \times m$ matrix B.

A common notational convention

If $m \leq n$ and A is an $m \times n$ matrix of full rank, it possesses at least one $m \times m$ submatrix $B = A_{\bullet\beta}$ which is a basis. The columns of B are linearly independent columns of A. Let $N = A_{\bullet\nu}$ denote the matrix formed by the rest of the columns of A. Then (by permuting the columns of A if necessary), we may assume

$$A = [B\,N].$$

The vector x has an analogous decomposition and is commonly written as (x_B, x_N), instead of (x_β, x_ν). Accordingly $x_B \in R^m$ and $x_N \in R^{n-m}$ are subvectors of x corresponding to the submatrices B and N of A. The equation $Ax = b$ is then expressed as

$$Ax = [B\,N] \begin{bmatrix} x_B \\ x_N \end{bmatrix} = Bx_B + Nx_N = b.$$

This style of representation has the advantage of being easy to read. Its disadvantage is that the use of B and N as subscripts—*when in fact they are matrices*—can lead to confusion.

Basic solutions again

If $A = [B\,N]$ and B is a basis in A, we get the *corresponding basic solution* of $Ax = b$ as follows: we set $x_N = 0$ and find the unique solution of the equation

$$Bx_B = b,$$

to obtain

$$\begin{bmatrix} x_B \\ x_N \end{bmatrix} = \begin{bmatrix} B^{-1}b \\ 0 \end{bmatrix}.$$

This can also be written as $(x_B, x_N) = (B^{-1}b, 0)$. It should be noted that *computationally* it can be a poor idea to find x_B using B^{-1}; however, we will continue to write $x_B = B^{-1}b$ for notational convenience.

Basic feasible solutions

In the LP

$$\begin{array}{ll} \text{minimize} & c^T x \\ \text{subject to} & Ax = b \\ & x \geq 0, \end{array} \qquad (3.1)$$

where A has full rank, the basic solution, $(x_B, x_N) = (B^{-1}b, 0)$ corresponding to B, is feasible if and only if $x_B = B^{-1}b \geq 0$. When this is the case, we say that B is a *feasible basis* and the corresponding solution is called a *basic feasible solution*, or BFS. Note that saying that "B is a basis in A" is independent of the vector b, whereas saying "B is a feasible basis for the LP (3.1)" does depend on b.

Example 3.1: FEASIBLE AND INFEASIBLE BASES. In the system of equations

$$3x_1 - 2x_2 + 4x_3 = 9$$
$$-3x_1 + 4x_2 - 5x_3 = 6$$

we have

$$A = \begin{bmatrix} 3 & -2 & 4 \\ -3 & 4 & -5 \end{bmatrix} \quad \text{and} \quad b = \begin{bmatrix} 9 \\ 6 \end{bmatrix}.$$

There are 3 bases in A, only two of which give nonnegative basic solutions. The bases are

$$\begin{bmatrix} 3 & -2 \\ -3 & 4 \end{bmatrix}, \quad \begin{bmatrix} 3 & 4 \\ -3 & -5 \end{bmatrix}, \quad \begin{bmatrix} -2 & 4 \\ 4 & -5 \end{bmatrix}.$$

$$\qquad\text{feasible}\qquad\qquad\text{infeasible}\qquad\qquad\text{feasible}$$

As stated above, a basis B is feasible if and only if $B^{-1}b \geq 0$. Relative to the three bases listed above, we have

$$\begin{bmatrix} 3 & -2 \\ -3 & 4 \end{bmatrix}^{-1} \begin{bmatrix} 9 \\ 6 \end{bmatrix} = \begin{bmatrix} 8.0 \\ 7.5 \end{bmatrix},$$

$$\begin{bmatrix} 3 & 4 \\ -3 & -5 \end{bmatrix}^{-1} \begin{bmatrix} 9 \\ 6 \end{bmatrix} = \begin{bmatrix} 23 \\ -15 \end{bmatrix},$$

$$\begin{bmatrix} -2 & 4 \\ 4 & -5 \end{bmatrix}^{-1} \begin{bmatrix} 9 \\ 6 \end{bmatrix} = \begin{bmatrix} 11.5 \\ 8.0 \end{bmatrix}.$$

Thus, we see that, in this example, only two of the three bases are feasible.

Dependence of z and x_B on x_N

Suppose we have the LP in standard form

$$\begin{array}{ll} \text{minimize} & c^T x \\ \text{subject to} & Ax = b \\ & x \geq 0. \end{array}$$

Suppose further that A is $m \times n$ and that $\text{rank}(A) = m$. Then there is a basis B in A. For notational simplicity, we assume[1] that the first m columns of A are the basis and write $A = [B\ N]$. Thus we have

$$x = \begin{bmatrix} x_B \\ x_N \end{bmatrix} \quad \text{and} \quad c = \begin{bmatrix} c_B \\ c_N \end{bmatrix}.$$

In terms of this partitioning, the LP becomes

$$\begin{array}{ll} \text{minimize} & c_B^T x_B + c_N^T x_N \\ \text{subject to} & Bx_B + Nx_N = b \\ & (x_B,\ x_N) \geq 0. \end{array} \tag{3.2}$$

From the constraints of LP (3.2) it follows that

$$x_B + B^{-1}Nx_N = B^{-1}b. \tag{3.3}$$

---

[1] In general the basis will not form the first m columns of A. However we can always reorder the columns (and corresponding variables) to ensure that this property holds.

We can write this in expanded form as

$$x_B + \sum_{j \in N} \bar{A}_{\bullet j} x_j = \bar{b}, \tag{3.4}$$

where

$$\bar{b} = B^{-1}b, \quad \bar{A}_{\bullet j} = B^{-1}A_{\bullet j} \text{ for all } j \in N.$$

Note that the vectors $\bar{b}$ and $\bar{A}_{\bullet j}$ depend on the specific basis B. Application of this formula with a different basis would usually lead to a different set of vectors $\bar{b}$ and $\bar{A}_{\bullet j}$.

Caution: In writing $j \in N$ as above, we are treating N as a *set*, namely the column indices of the nonbasic variables. But N is also the notation for a *matrix* made up of a set of columns in A. This practice is very common in linear programming textbooks. Nevertheless, you should be aware that it is what's called "an abuse of notation."

Solving (3.3) for the basic variables x_B, we obtain

$$x_B = B^{-1}(b - Nx_N). \tag{3.5}$$

Substituting (3.5) into the objective of the LP (3.2) yields

$$\begin{aligned}
z &= c_B^T x_B + c_N^T x_N \\
&= c_B^T B^{-1}(b - Nx_N) + c_N^T x_N \\
&= c_B^T B^{-1}b - c_B^T B^{-1}Nx_N + c_N^T x_N \\
&= c_B^T B^{-1}b - (c_B^T B^{-1}N - c_N^T)x_N \\
&= z_0 - \sum_{j \in N}(z_j - c_j)x_j
\end{aligned} \tag{3.6}$$

where

$$z_0 := c_B^T B^{-1}b, \quad z_j := c_B^T B^{-1}A_{\bullet j} \text{ for all } j \in N.$$

For any basis B (feasible or not), equation (3.6), or equivalently,

$$z = z_0 - (c_B^T B^{-1}N - c_N^T)x_N$$

expresses the value of the objective function in terms of the nonbasic variables alone. This is made possible by the fact that the basic variables are also functions of the nonbasic variables as we saw in (3.5). In the language of linear algebra, the variables in x_B would be called *dependent*, and those of x_N would be called *independent*. The following example illustrates the dependence of z and x_B upon x_N.

Example 3.2: EXPRESSIONS IN TERMS OF NONBASIC VARIABLES. In the LP

$$
\begin{aligned}
\text{minimize} \quad & 5x_1 + 3x_2 + 4x_3 + 2x_4 + x_5 \\
\text{subject to} \quad & 4x_1 - x_2 + 2x_3 - 3x_4 \quad\quad = 12 \\
& -2x_1 + 3x_2 \quad\quad + 2x_4 + 3x_5 = 9 \\
& x_1,\, x_2,\, x_3,\, x_4,\, x_5 \geq 0.
\end{aligned}
$$

We have

$$
A = \begin{bmatrix} 4 & -1 & 2 & -3 & 0 \\ -2 & 3 & 0 & 2 & 3 \end{bmatrix}, \quad b = \begin{bmatrix} 12 \\ 9 \end{bmatrix},
$$

and

$$
c^{\mathrm{T}} = \begin{bmatrix} 5 & 3 & 4 & 2 & 1 \end{bmatrix}.
$$

Suppose we select $B = [A_{\bullet 1}\ A_{\bullet 2}]$. This is certainly a basis (the columns are linearly independent) and

$$
B^{-1}b = \frac{1}{10}\begin{bmatrix} 3 & 1 \\ 2 & 4 \end{bmatrix}\begin{bmatrix} 12 \\ 9 \end{bmatrix} = \begin{bmatrix} 4.5 \\ 6.0 \end{bmatrix}
$$

is positive. Thus, B is a feasible basis. Relative to this basis, we find that the objective function can be written as

$$
z = 40.5 - (0.2x_3 - 4.9x_4 + 4.1x_5).
$$

As for the basic variables, x_1 and x_2, we obtain

$$
\begin{bmatrix} x_1 \\ x_2 \end{bmatrix} = \begin{bmatrix} 4.5 \\ 6.0 \end{bmatrix} - \begin{bmatrix} 0.6 & -0.7 & 0.3 \\ 0.4 & 0.2 & 1.2 \end{bmatrix}\begin{bmatrix} x_3 \\ x_4 \\ x_5 \end{bmatrix}
$$

$$
= \begin{bmatrix} 4.5 - 0.6x_3 + 0.7x_4 - 0.3x_5 \\ 6.0 - 0.4x_3 - 0.2x_4 - 1.2x_5 \end{bmatrix}.
$$

The canonical form of an LP relative to a basis B

Now we are going to develop the canonical form of the LP relative to the basis B. In so doing we obtain expressions for the objective function value

and the basic variables in terms of the nonbasic variables. In the next
section, this will lead to a set of conditions which, if satisfied, will assure us
that a particular BFS is optimal, thereby answering Question 3 on Page 61.

Assembling (3.4) and (3.6), we can write LP (3.2) in *canonical form*:

$$
\begin{aligned}
\text{minimize} \quad & z \\
\text{subject to} \quad & z + \sum_{j \in N} (z_j - c_j) x_j = z_0 \\
& I x_B + \sum_{j \in N} \bar{A}_{\bullet j} x_j = \bar{b} \\
& x_B \geq 0, \ x_j \geq 0 \text{ for } j \in N.
\end{aligned}
$$

Example 3.3: PUTTING A SYSTEM INTO CANONICAL FORM. Suppose
we have a linear program of the form

$$
\begin{aligned}
\text{minimize} \quad & c^{\mathrm{T}} x \\
\text{subject to} \quad & A x = b \\
& x \geq 0,
\end{aligned}
$$

where $c = (8, 4, 6, 5, 7)$, $b = (12, 15, 10)$ and

$$
A = \begin{bmatrix} 6 & -2 & 0 & 4 & 3 \\ -2 & 4 & -1 & 0 & 2 \\ -1 & 0 & 2 & -1 & 4 \end{bmatrix}.
$$

It is easy to verify that

$$
B = \begin{bmatrix} 6 & -2 & 0 \\ -2 & 4 & -1 \\ -1 & 0 & 2 \end{bmatrix}
$$

is a feasible basis for this problem. Indeed

$$
B^{-1} = \frac{1}{38} \begin{bmatrix} 8 & 4 & 2 \\ 5 & 12 & 6 \\ 4 & 2 & 20 \end{bmatrix} \quad \text{and} \quad \bar{b} = B^{-1} b = \begin{bmatrix} 88/19 \\ 150/19 \\ 139/19 \end{bmatrix}.
$$

Let us now compute

$$
\bar{A} = B^{-1} A = \begin{bmatrix} 1 & 0 & 0 & 15/19 & 20/19 \\ 0 & 1 & 0 & 7/19 & 63/38 \\ 0 & 0 & 1 & -2/19 & 48/19 \end{bmatrix}.
$$

Upon eliminating the basic variables x_1, x_2, x_3 from the objective function $z = c^T x$, we obtain the equation

$$z = \frac{2138}{19} - \frac{41}{19}x_4 - \frac{441}{19}x_5.$$

With these operations we obtain the feasible canonical form of the problem with respect to the given basis:

$$
\begin{aligned}
z \quad & + \tfrac{41}{19}x_4 + \tfrac{441}{19}x_5 = \tfrac{2138}{19} \\
x_1 \quad & + \tfrac{15}{19}x_4 + \tfrac{20}{19}x_5 = \tfrac{20}{19} \\
x_2 \quad & + \tfrac{7}{19}x_4 + \tfrac{63}{38}x_5 = \tfrac{63}{38} \\
x_3 \quad & - \tfrac{2}{19}x_4 + \tfrac{48}{19}x_5 = \tfrac{48}{19} .
\end{aligned}
$$

A sufficient condition for the optimality of a BFS

As we have seen, an equivalent form of the LP (3.1) relative to a basis B is:

$$\text{minimize} \quad z = z_0 - \sum_{j \in N}(z_j - c_j)x_j$$

$$\text{subject to} \quad x_B + \sum_{j \in N}\bar{A}_{\bullet j}x_j = \bar{b}$$

$$x_B \geq 0, \; x_j \geq 0 \text{ for } j \in N.$$

Recall that the basis B is feasible provided that $\bar{b} \geq 0$. Let us assume that this is the case. From the above representation of the LP, it is not difficult to see that the basic feasible solution is optimal if $z_j - c_j \leq 0$ for all j. The justification for this claim is that if we have a basic feasible solution $x = (x_B, x_N) = (B^{-1}b, 0)$, then in any *different* feasible (not necessarily basic) solution $\tilde{x} \geq 0$, we must $\tilde{x}_k > 0$ for at least one $k \in N$. But for each such k, we would have

$$-(z_k - c_k)\tilde{x}_k \geq 0.$$

In fact, if $(z_k - c_k) < 0$, then $-(z_k - c_k)\tilde{x}_k > 0$, so that the objective for this feasible solution $\tilde{x}$ would be greater than that for x. Hence, in these circumstances, the value of the objective function cannot be decreased by making any nonbasic variable x_j positive. This is a *sufficient* (but not necessary) condition for optimality in the minimization problem. By this we mean that the corresponding BFS is optimal if

$$z_j - c_j \leq 0 \quad \text{for all } j. \tag{3.7}$$

The quantities $z_j - c_j$ are called *reduced costs*. When B is a feasible basis and the condition (3.7) holds, we can say that B is an *optimal basis* for the LP.

Example 3.4: TESTING FOR OPTIMALITY. Let's consider again the LP used in Example 3.2:

$$
\begin{array}{rl}
\text{minimize} & 5x_1 + 3x_2 + 4x_3 + 2x_4 + \ x_5 \\
\text{subject to} & 4x_1 - \ x_2 + 2x_3 - 3x_4 \qquad\quad = 12 \\
& -2x_1 + 3x_2 \qquad\quad + 2x_4 + 3x_5 = \ 9 \\
& x_1, \, x_2, \, x_3, \, x_4, \, x_5 \geq 0
\end{array}
$$

The matrix

$$
B = \begin{bmatrix} A_{\bullet 1} & A_{\bullet 5} \end{bmatrix} = \begin{bmatrix} 4 & 0 \\ -2 & 3 \end{bmatrix}
$$

is a another *feasible basis* in A since

$$
x_B = B^{-1}b = \frac{1}{12}\begin{bmatrix} 3 & 0 \\ 2 & 4 \end{bmatrix}\begin{bmatrix} 12 \\ 9 \end{bmatrix} = \begin{bmatrix} 3 \\ 5 \end{bmatrix}.
$$

The corresponding basic feasible solution is $x = (3, 0, 0, 0, 5)$.

Is $x = (3, 0, 0, 0, 5)$ optimal? We compute the numbers

$$
z_j - c_j = c_B^{\mathrm{T}} B^{-1} A_{\bullet j} - c_j \quad \text{for } j \in \{2, 3, 4\}
$$

to see if the sufficient conditions are satisfied. In the present case

$$
c_B^{\mathrm{T}} B^{-1} = \begin{bmatrix} 5 & 1 \end{bmatrix}\frac{1}{12}\begin{bmatrix} 3 & 0 \\ 2 & 4 \end{bmatrix} = \begin{bmatrix} \frac{17}{12} & \frac{1}{3} \end{bmatrix},
$$

so

$$
z_2 - c_2 = \begin{bmatrix} \frac{17}{12} & \frac{1}{3} \end{bmatrix}\begin{bmatrix} -1 \\ 3 \end{bmatrix} - 3 = -\frac{41}{12} < 0
$$

$$
z_3 - c_3 = \begin{bmatrix} \frac{17}{12} & \frac{1}{3} \end{bmatrix}\begin{bmatrix} 2 \\ 0 \end{bmatrix} - 4 = -\frac{7}{6} < 0
$$

$$
z_4 - c_4 = \begin{bmatrix} \frac{17}{12} & \frac{1}{3} \end{bmatrix}\begin{bmatrix} -3 \\ 2 \end{bmatrix} - 2 = -\frac{67}{12} < 0,
$$

hence the BFS $x = (3, 0, 0, 0, 5)$ is an optimal solution of this LP with optimal objective value $z = 20$.

3.2   A case where the sufficient condition is also necessary

The sufficient condition for optimality of a BFS, namely (3.7), is *necessary* for its optimality in the case where $\bar{b} = B^{-1}b > 0$. Under this nondegeneracy condition, if $z_j - c_j > 0$ it is possible to obtain an improved feasible solution of the problem by increasing the variable x_j. Notice that if we merely had $\bar{b} = B^{-1}b \geq 0$ (the degenerate case), it might not be possible to increase x_j without losing feasibility. But with $\bar{b} = B^{-1}b > 0$, we are assured that at least some increase of x_j will not move the solution outside the feasible region.

Example 3.5: NECESSITY OF $z_j - c_j \leq 0$ FOR OPTIMALITY.   Consider an LP in standard form where

$$A = \begin{bmatrix} 1 & 0 & 1 \\ 0 & 1 & 1 \end{bmatrix} \quad \text{and} \quad c = (1, 1, 1).$$

Suppose $b = (1, 1)$. Writing $z = x_1 + x_2 + x_3$, we have the equations

$$\begin{aligned} z - x_1 - x_2 - x_3 &= 0 \\ x_1 \qquad\ + x_3 &= 1 \\ x_2 + x_3 &= 1 \end{aligned}$$

$$x_1, \ x_2 \text{ basic}, x_3 \text{ nonbasic}.$$

By adding the second and third equations to the first, we see that this system is equivalent to

$$\begin{aligned} z \qquad\quad + x_3 &= 2 \\ x_1 \ + x_3 &= 1 \\ x_2 + x_3 &= 1. \end{aligned}$$

In this system, we have x_1 and x_2 basic and x_3 nonbasic with $z_3 - c_3 = 1$. The basic feasible solution is $(x_1, x_2, x_3) = (1, 1, 0)$, and the corresponding objective function value is $z = 2$. Note that this BFS is not optimal since z can be decreased by increasing x_3.

Example 3.6: NONNECESSITY OF $z_j - c_j \leq 0$ FOR OPTIMALITY.   This example illustrates what can happen in the degenerate case mentioned above. Now take A and c as before but set $b = (0, 1)$. This leads to the system

$$\begin{aligned} z \qquad\quad + x_3 &= 1 \\ x_1 \ + x_3 &= 0 \\ x_2 + x_3 &= 1. \end{aligned}$$

Here we see that with x_1 and x_2 basic and x_3 nonbasic, the corresponding BFS is $(x_1, x_2, x_3) = (0, 1, 0)$. The objective function value is $z = 1$. It would seem from the objective function coefficient of x_3 that z can be decreased by increasing x_3; but doing so makes $x_1 < 0$. Suppose we now change the basis, from $[A_{\bullet 1} \ A_{\bullet 2}]$ to $[A_{\bullet 3} \ A_{\bullet 2}]$ and carry out the substitutions to put the equations in the equivalent simple form

$$
\begin{aligned}
z - x_1 & & & = 1 \\
x_1 & & + x_3 & = 0 \\
- x_1 & + x_2 & & = 1.
\end{aligned}
$$

In this representation of the equations, we see that the sufficient condition for optimality, $z_j - c_j \leq 0$, *is* satisfied. But notice that the BFS is $(x_1, x_2, x_3) = (0, 1, 0)$, just as before! This seemingly strange phenomenon illustrates what can happen when a basic variable in a BFS takes on the value zero. The same vector can be the BFS corresponding to more than one basis. In Example 3.6, the basis $[A_{\bullet 3} \ A_{\bullet 2}]$ is optimal whereas the basis $[A_{\bullet 1} \ A_{\bullet 2}]$ is not optimal, and yet they give rise to the same feasible solution.

3.3 Seeking another BFS

If $x = (x_B, x_N) = (B^{-1}b, 0)$ is a BFS and does not satisfy the sufficient conditions for optimality, there must be an index s such that $z_s - c_s > 0$. Assume that for such an s it has been decided to increase the nonbasic variable x_s while holding all other nonbasic variables at the value 0. Given that we must maintain feasibility, how much can x_s be increased? What will happen when this is done?

Example 3.7: MAINTAINING FEASIBILITY. In the linear program

$$
\begin{aligned}
\text{minimize} \quad & - 7x_2 + 4x_3 \quad\quad - 3x_5 \\
\text{subject to} \quad & 6x_2 \quad\quad + x_4 \quad\quad\quad\quad = 12 \\
& 2x_2 + 4x_3 \quad\quad + x_5 + x_6 = 10 \\
& x_1 - x_2 + 2x_3 \quad\quad - x_5 \quad\quad = 14 \\
& x_j \geq 0, \quad j = 1, \ldots, 6,
\end{aligned}
$$

we have 3 equations and 6 variables. We may regard the matrix $B = [A_{\bullet 4} \ A_{\bullet 6} \ A_{\bullet 1}]$ as a feasible basis (even though the three columns of B are not the first three columns of A). The corresponding BFS

$$(x_1, x_2, x_3, x_4, x_5, x_6) = (14, 0, 0, 12, 0, 10)$$

is clearly nonoptimal. Indeed, $c_B = 0$, so $z_j = c_B^T B^{-1} A_{\bullet j} = 0$ for all $j \in N$. In particular, we have $z_2 - c_2 = 0 + 7 > 0$ and $z_5 - c_5 = 0 + 3 > 0$. Since the basic variables are all positive, we can decrease z by increasing x_2 or by increasing x_5 without losing the feasibility of the solution generated.

Let's suppose we decide to increase x_2 on the grounds that doing so will decrease z at a rate of 7 to 1, i.e., z will decrease by 7 for every unit of increase in x_2. How much of an increase can we make? In this process, we are going to hold x_3 and x_5 at the value zero. Accordingly, we can write

$$z = 0 - 7x_2$$
$$x_4 = 12 - 6x_2$$
$$x_6 = 10 - 2x_2$$
$$x_1 = 14 + x_2.$$

To maintain feasibility, we must keep the x-variables nonnegative. From the equations above, we see that for this to hold, we must have $x_2 \leq 2$. At this limiting value, x_4 drops to zero and would become negative if x_2 were any larger than 2. As far as the other basic variables are concerned, x_6 reaches zero when $x_2 = 5$ which is larger than 2, and x_1 does not decrease as x_2 increases. Notice that this limiting value is found by computing the minimum of a set of ratios:

$$x_2 \leq \min\left\{\frac{12}{6}, \frac{10}{2}\right\} = 2.$$

The corresponding change in z would be a decrease of $7 \times 2 = 14$.

What if we had chosen to increase x_5 instead of x_2? In that case, we'd have

$$z = 0 - 3x_5$$
$$x_4 = 12$$
$$x_6 = 10 - x_5$$
$$x_1 = 14 + x_5.$$

Maintaining feasibility requires that we increase x_5 by no more than 10. The effect of that change would be to decrease z by 30. This illustrates the point that choosing the nonbasic variable to increase by the criterion of the largest (positive) $z_j - c_j$ does not always produce the greatest decrease in the objective function value. Nevertheless that is the criterion that is commonly used in the Simplex Algorithm.

Notice that in Example 3.7 there is exactly one basic variable appearing in each of the equations. In the first equation it is x_4. In the second equation it is x_6. In the third equation it is x_1. In effect, the basic variables may seem rather jumbled up. That could be changed by permuting the columns and renaming the variables, but there is no need to do so. We do, however, require a notation that helps us to keep things straight. To that end, we define j_i to be the index (subscript) of the basic variable that corresponds to equation i. Thus, in Example 3.7, $j_1 = 4$, $j_2 = 6$, and $j_3 = 1$. The basis is $[A_{\bullet j_1}\ A_{\bullet j_2}\ A_{\bullet j_3}] = [A_{\bullet 4}\ A_{\bullet 6}\ A_{\bullet 1}]$, and of course $x_B = (x_{j_1}, x_{j_2}, x_{j_3}) = (x_4, x_6, x_1)$. Understanding this will facilitate your appreciation of the more general discussion below.

Basic variables as functions of a single nonbasic variable x_s

Assume that the variables $x_{j_1}, x_{j_2} \ldots, x_{j_m}$ are basic. With $x_j = 0$ for all $j \in N$, $j \neq s$, we can write

$$x_B = B^{-1}b - B^{-1}A_{\bullet s}x_s$$
$$= \bar{b} - \bar{A}_{\bullet s}x_s$$

or, in more detail,

$$
\begin{bmatrix} x_{j_1} \\ x_{j_2} \\ \vdots \\ x_{j_m} \end{bmatrix} = \begin{bmatrix} \bar{b}_1 \\ \bar{b}_2 \\ \vdots \\ \bar{b}_m \end{bmatrix} - \begin{bmatrix} \bar{a}_{1s} \\ \bar{a}_{2s} \\ \vdots \\ \bar{a}_{ms} \end{bmatrix} x_s = \begin{bmatrix} \bar{b}_1 - \bar{a}_{1s}x_s \\ \bar{b}_2 - \bar{a}_{2s}x_s \\ \vdots \\ \bar{b}_m - \bar{a}_{ms}x_s \end{bmatrix}.
$$

The minimum ratio test

To maintain feasibility, we must have $x_s \geq 0$ and

$$x_{j_i} = \bar{b}_i - \bar{a}_{is}x_s \geq 0 \text{ for all } i = 1, \ldots, m.$$

If $\bar{a}_{is} \leq 0$ for $i = 1, \ldots, m$, then x_s can be increased indefinitely. In this case, if $z_s - c_s > 0$, the objective function is *unbounded below*; its value goes to $-\infty$. If $\bar{a}_{is} > 0$ for at least one i, then

$$x_s \leq \min\left\{\frac{\bar{b}_i}{\bar{a}_{is}} : \bar{a}_{is} > 0\right\}.$$

In fact, there will exist at least one r such that

$$\frac{\bar{b}_r}{\bar{a}_{rs}} = \min\left\{\frac{\bar{b}_i}{\bar{a}_{is}} : \bar{a}_{is} > 0\right\}.$$

This calculation is called the *minimum ratio test*. In these circumstances x_s is called the *entering variable* and x_{j_r} the *leaving variable*. Correspondingly, the column of x_s *enters* the basis, and the column of x_{j_r} *leaves* the basis.

In terms of the original data, the current basis B is

$$B = [A_{\bullet j_1}\, A_{\bullet j_2}\, \cdots A_{\bullet j_{r-1}}\, A_{\bullet j_r}\, A_{\bullet j_{r+1}} \cdots A_{\bullet j_m}].$$

The incoming column $A_{\bullet s}$ replaces the outgoing column $A_{\bullet j_r}$ resulting in the new basis

$$B^{\text{new}} = [A_{\bullet j_1}\, A_{\bullet j_2}\, \cdots A_{\bullet j_{r-1}}\, A_{\bullet s}\, A_{\bullet j_{r+1}} \cdots A_{\bullet j_m}].$$

Notice that we now define $j_r = s$.

3.4   Pivoting

We have seen how the minimum ratio test is used to preserve the feasibility of the current solution as the entering variable x_s is increased. The motivation for the increase is of course that it has the potential to decrease the value z of the objective function. When the current basic feasible solution is nondegenerate, this increase is guaranteed to make z decrease. When the increase of x_s is bounded above, say by $\bar{b}_r/\bar{a}_{rs}$, the basic variable x_{j_r} will decrease from $\bar{b}_r$ to zero as x_s increases from zero to $\bar{b}_r/\bar{a}_{rs}$. In this case, x_s will replace x_{j_r} as a basic variable.

These considerations lead us to a discussion of how the system of equations can be represented in terms of the new basis. Once we have accomplished this, we can state the Simplex Algorithm. In so doing, we will have answered the question 4 stated in the Overview of Chapter 3. Recall that question 3 has already been answered—at least in the nondegenerate case. Questions 1 and 2 will be answered by the end of the chapter.

Rerepresenting the system of equations by pivoting

Let us assume that the system of equations

$$z - c^{\mathrm{T}}x = 0$$
$$Ax = b$$

has been put into canonical form with respect to a feasible basis $B = [A_{j_1} \ A_{j_2} \ \cdots \ A_{j_m}]$. The data of the canonical form

$$z - \bar{c}^{\mathrm{T}}x = 0$$
$$\bar{A}x = \bar{b} \tag{3.8}$$

will then have the following properties:

$$\bar{A} = B^{-1}A, \quad \bar{b} = B^{-1}b, \quad \bar{c}^{\mathrm{T}} = c^{\mathrm{T}} - c_B^{\mathrm{T}}B^{-1}A.$$

These formulas imply

$$\bar{A}_{j_k} = I_{\bullet k} \quad \text{and} \quad \bar{c}_{j_k} = 0, \quad k = 1, \ldots, m.$$

Writing (3.8) in more detail, we obtain the canonical form

$$z + 0^{\mathrm{T}}x_B - \bar{c}_N^{\mathrm{T}}x_N = 0$$
$$Ix_B + \bar{A}_{\bullet N}x_N = \bar{b}. \tag{3.9}$$

Let us assume further that, having performed the minimum ratio test, we wish to change the basis from B to a new basis B^{new} whose r-th column $A_{\bullet j_r}$ is $A_{\bullet s}$, the s-th columm of A. (Notice that the columns of these bases are given in terms of the original data.) We shall want to put the system into canonical form with respect to B^{new}. From the discussion above it might appear that we need to compute $(B^{\mathrm{new}})^{-1}$, but because we already the system in canonical form with respect to B which shares $m-1$ columns with B^{new} we can achieve the same effect and save computational effort by performing a *pivot* on the positive entry $\bar{a}_{rs}$. The steps of this operation are as follows:

1. Divide each coefficient in equation r of system (3.9) by $\bar{a}_{rs}$. (Thereby, the coefficient of x_s in this equation becomes 1.)

2. Subtract $\bar{a}_{is}$ times equation r from each equation $i \neq r$. (Thereby, the coefficient of x_s in each such equation becomes 0.)

3. Subtract $\bar{c}_s$ times equation r from the objective equation. (Thereby, the coefficient of x_s in the equation corresponding to the objective function becomes 0.)

This pivot operation amounts to solving the r-th equation for x_s in terms of x_{j_r} and the other nonbasic variables x_j for $j \in N \setminus \{s\}$ and then using the result to *substitute* for x_s in each of the equations other than the r-th.

Comment: We have followed the tradition of calling this operation a pivot on the entry $\bar{a}_{rs}$. There is another way to conceptualize what is going on in pivoting. When we compare the bases B and B^{new} (as described above) we see that they have $m-1$ columns in common. The bases B and B^{new} can be likened to the parts of a hinge with the $m-1$ common columns playing the role of the piece around which the "plates" pivot. This is illustrated on Figure 3.1.

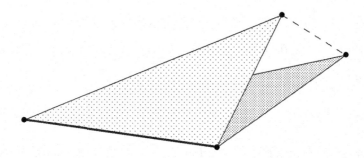

Figure 3.1: A metaphoric illustration of pivoting.

Example 3.8: A SIMPLEX ALGORITHM ITERATION (PIVOT STEP). For the LP

$$\begin{aligned}
\text{minimize} \quad & -4x_1 - 6x_2 - 3x_3 - x_4 \\
\text{subject to} \quad & x_1 + 4x_2 + 8x_3 + 6x_4 \leq 11 \\
& 4x_1 + x_2 + 2x_3 + x_4 \leq 7 \\
& 2x_1 + 3x_2 + x_3 + 2x_4 \leq 2 \\
& x_1,\, x_2,\, x_3,\, x_4 \geq 0,
\end{aligned}$$

we add slack variables x_5, x_6, and x_7 to bring the constraints to standard form. This yields

$$
\begin{aligned}
x_1 + 4x_2 + 8x_3 + 6x_4 + x_5 \qquad\qquad &= 11 \\
4x_1 + x_2 + 2x_3 + x_4 \qquad + x_6 \quad &= 7 \\
2x_1 + 3x_2 + x_3 + 2x_4 \qquad\qquad + x_7 &= 2 \\
x_1,\ x_2,\ x_3,\ x_4,\ x_5,\ x_6,\ x_7 &\geq 0.
\end{aligned}
$$

We want to minimize

$$
z = -4x_1 - 6x_2 - 3x_3 - x_4.
$$

This equation can be written as

$$
z + 4x_1 + 6x_2 + 3x_3 + x_4 = 0.
$$

Defining this to be the *0-th equation*, we append it to the system above and obtain

$$
\begin{aligned}
z + 4x_1 + 6x_2 + 3x_3 + x_4 \qquad\qquad\quad &= 0 \\
x_1 + 4x_2 + 8x_3 + 6x_4 + x_5 \qquad\qquad &= 11 \\
4x_1 + x_2 + 2x_3 + x_4 \qquad + x_6 \quad &= 7 \\
2x_1 + 3x_2 + x_3 + 2x_4 \qquad\qquad + x_7 &= 2.
\end{aligned}
$$

In this instance, $x_{j_1} = x_5$, $x_{j_2} = x_6$, $x_{j_3} = x_7$. The present BFS is clearly not optimal. Do you see why? Let the entering variable be x_2. This means we choose $s = 2$. How much can x_2 be increased? We can answer this question with the minimum ratio test.

Applying the aforementioned minimum ratio test here we find that

$$
x_2 \leq \min\left\{ \frac{b_1}{a_{21}}, \frac{b_2}{a_{22}}, \frac{b_3}{a_{23}} \right\} = \min\left\{ \frac{11}{4}, \frac{7}{1}, \frac{2}{3} \right\} = \frac{2}{3} = \frac{b_3}{a_{32}},
$$

implying $r = 3$, which means that x_7 is the leaving variable. This means that the value of the index j_3 changes from 7 to 2. The new basis is

$$
B^{\text{new}} = [A_{\bullet 5}\ A_{\bullet 6}\ A_{\bullet 2}] = \begin{bmatrix} 1 & 0 & 4 \\ 0 & 1 & 1 \\ 0 & 0 & 3 \end{bmatrix}.
$$

Pivoting on the element in row 3 and column of x_2 (i.e., on $a_{32} = 3$), we find that the system becomes

$$z \qquad + \quad x_3 - \quad 3x_4 \qquad\qquad - 2x_7 = -4$$

$$- \tfrac{5}{3}x_1 \qquad + \tfrac{20}{3}x_3 + \tfrac{10}{3}x_4 + x_5 \qquad - \tfrac{4}{3}x_7 = \tfrac{25}{3}$$

$$\tfrac{10}{3}x_1 \qquad + \tfrac{5}{3}x_3 + \tfrac{1}{3}x_4 \qquad + x_6 - \tfrac{1}{3}x_7 = \tfrac{19}{3}$$

$$\tfrac{2}{3}x_1 + x_2 + \tfrac{1}{3}x_3 + \tfrac{2}{3}x_4 \qquad\qquad + \tfrac{1}{3}x_7 = \tfrac{2}{3}.$$

Note how the value of z has decreased from 0 to -4. This completes the intended pivot step.

3.5   Steps of the Simplex Algorithm

We have seen how to test the BFS for optimality, and if not optimal, see how to improve the objective value by performing a pivot. Combining these concepts, we can write down the Simplex Algorithm for the case when a feasible basis B is known.

Algorithm 3.1 (Simplex Algorithm)   Given data A, b, and c for a linear program in standard form, assume that a starting feasible basis B is known and that the system is in canonical form with respect to this basis.

1. *Test for optimality.* Find an index s such that (relative to the current basis) $z_s - c_s = \max_j\{z_j - c_j\}$. If $z_s - c_s \leq 0$ the test for optimality is satisfied, then the basic feasible solution is optimal. Report the solution and stop.

2. *Minimum ratio test.* If $\bar{a}_{is} \leq 0$ for all i, stop. The objective is unbounded below. Otherwise, find an index r such that $\bar{a}_{rs} > 0$ and

$$\frac{\bar{b}_r}{\bar{a}_{rs}} = \min_i \left\{ \frac{\bar{b}_i}{\bar{a}_{is}} : \bar{a}_{is} > 0 \right\}.$$

3. *Pivot step.* Pivot on $\bar{a}_{rs}$. Define the new basis in which $A_{\bullet s}$ replaces the column $A_{\bullet k}$ where $k = j_r$. Define $j_r = s$. Return to Step 1.

Tabular form

Example 3.8 points up the fact that expressing a system of equations in many variables involves the writing of a lot of pluses, minuses, equal signs, and variable names. To lighten this burden somewhat, a style called *tabular form* (sometimes called *tableau form*) is used. This approach commonly appears in introductory expositions of the Simplex Algorithm and other such methods. Example 3.9 below will illustrate the use of tabular form for solving the linear program that was used in Example 3.8. But before coming to that, we present some generalities about the constituents of a linear programming tableau.

In essence, a tableau is a matrix (rectangular array of numbers) surrounded by labels. A tableau will have the following four main sections which we think of columnwise:

- The list of row numbers.

- The names of the variables and their coefficients.

- The column label 1 and the right-hand side constants of the system.

- The list of basic variables associated with the rows of the system.

min	z	x_1	x_2	$\cdots$	x_n	1	basic
row 0	1	$\bar{c}_1$	$\bar{c}_2$	$\cdots$	$\bar{c}_n$	$c_B^T B^{-1} b$	z
row 1	0	$\bar{a}_{11}$	$\bar{a}_{12}$	$\cdots$	$\bar{a}_{1m}$	$\bar{b}_1$	x_{j_1}
row 2	0	$\bar{a}_{21}$	$\bar{a}_{22}$	$\cdots$	$\bar{a}_{2m}$	$\bar{b}_2$	x_{j_2}
$\vdots$	0	$\vdots$	$\vdots$	$\cdots$	$\vdots$	$\vdots$	$\vdots$
row m	0	$\bar{a}_{m1}$	$\bar{a}_{m2}$	$\cdots$	$\bar{a}_{mn}$	$\bar{b}_m$	x_{j_m}

The labels listed in the very top row of the tableau are z (which stands for the objective function) and all the problem variables (including slack and surplus variables). In addition, the top row of the tableau includes the label 1 associated with the right-hand side. The left-most section of the tableau begins with the label "row 0" which corresponds to the objective function, and then the labels "row 1, ..., row m" which correspond to the constraint equations. In the next block of the tableau, we write the coefficients of the objective and the constraints organized columnswise underneath their labels $z, x_1, \ldots, x_n$. Then comes the one-column block of the right-hand side with its label, "1". Finally, we write the list consisting of z and the basic variables x_B. It should be noted that each basic variable corresponds to a specific row of the tableau. The entry x_{j_i}, appearing in the last column of the table, refers to the basic variable corresponding to row i. The $\bar{a}$ entries in the column corresponding to the basic variable x_{j_i} will be the i-th column of the identity matrix, that is to say: $\bar{A}_{\bullet j_i} = I_{\bullet i}$. For fixed j, the entries $\bar{c}_j$ and $\bar{a}_{kj}$ are as follows:

$$\bar{c}_j = \begin{cases} 0, & \text{if } x_j \text{ is basic,} \\ \text{arbitrary,} & \text{otherwise;} \end{cases}$$

$$\bar{a}_{kj} = \begin{cases} 0, & \text{if } x_j \text{ is basic, } j = j_i, \text{ and } k \neq i, \\ 1, & \text{if } x_j \text{ is basic, } j = j_i, \text{ and } k = i, \\ \text{arbitrary,} & \text{otherwise.} \end{cases}$$

The tableaus given in Example 3.9 illustrate the above description.

Example 3.9: SOLUTION BY THE SIMPLEX ALGORITHM. Here we solve the problem stated in Example 3.8 using the Simplex Algorithm with the system expressed in tabular form.

min	z	x_1	x_2	x_3	x_4	x_5	x_6	x_7	1	basic
row 0	1	4	6	3	1	0	0	0	0	z
row 1	0	1	4	8	6	1	0	0	11	x_5
row 2	0	4	1	2	1	0	1	0	7	x_6
row 3	0	2	3	1	2	0	0	1	2	x_7

Pivoting on the element in row 3 of the x_2 column, we obtain the following tableau

min	z	x_1	x_2	x_3	x_4	x_5	x_6	x_7	1	basic
row 0	1	0	0	1	-3	0	0	-2	-4	z
row 1	0	$-5/3$	0	$20/3$	$10/3$	1	0	$-4/3$	$25/3$	x_5
row 2	0	$10/3$	0	$5/3$	$1/3$	0	1	$-1/3$	$19/3$	x_6
row 3	0	$2/3$	1	$1/3$	$2/3$	0	0	$1/3$	$2/3$	x_2

Pivoting on the element in row 1 of the x_3 column, we obtain

min	z	x_1	x_2	x_3	x_4	x_5	x_6	x_7	1	basic
row 0	1	$1/4$	0	0	$-7/2$	$-3/20$	0	$-9/5$	$-21/4$	z
row 1	0	$-1/4$	0	1	$1/2$	$3/20$	0	$-1/5$	$5/4$	x_3
row 2	0	$15/4$	0	0	$-1/2$	$-1/4$	1	0	$17/4$	x_6
row 3	0	$3/4$	1	0	$1/2$	$-1/20$	0	$2/5$	$1/4$	x_2

Pivoting on the element in row 3 of the x_1 column, we obtain

min	z	x_1	x_2	x_3	x_4	x_5	x_6	x_7	1	basic
row 0	1	0	$-1/3$	0	$-11/3$	$-2/15$	0	$-29/15$	$-16/3$	z
row 1	0	0	$1/3$	1	$2/3$	$2/15$	0	$-1/15$	$4/3$	x_3
row 2	0	0	-5	0	-3	0	1	-2	3	x_6
row 3	0	1	$4/3$	0	$2/3$	$-1/15$	0	$8/15$	$1/3$	x_1

We now see that the sufficient condition for optimality is satisfied. The optimal solution found is

$$(x_1, x_2, x_3, x_4, x_5, x_6, x_7) = (1/3, 0, 4/3, 0, 0, 3, 0) \quad \text{and} \quad z = -16/3.$$

3.6 Exercises

3.1   Suppose you have to solve a system of m equations in n unknowns, say $Ax = b$, where $m < n$. After performing the reduction of A to row-echelon form and the corresponding operations on b, you have the equivalent system $\bar{A}\bar{x} = \bar{b}$. [Note: The reason for using a bar over the vector in the equivalent system in row-echelon form is that it may have been necessary to permute the columns of A (and hence the variables in x).] Suppose that in the process of doing this, you find the rank r of A is less than m.

  (a) How can you tell whether or not the system of equations has a solution?

  (b) Assuming the system *does* have a solution, describe how the row-echelon form would help you to find a solution.

  (c) Would this be the only solution for the system?

3.2   Let $A \in R^{m \times n}$ where $m < n$. Suppose that A possesses an $m \times m$ submatrix B with the following properties:

  (i) each column B has exactly one positive element;

  (ii) the sum of the columns of B is a positive m vector.

Show that the system $Ax = b$, $x \geq 0$ has a basic feasible solution for *every* $b \in R_+^m$.

3.3   Consider the system of equations

$$\begin{aligned} 4x_1 - x_2 + 5x_3 + 2x_4 + 2x_5 &= 20 \\ -3x_1 + x_2 + 3x_4 - 2x_5 &= 18 \\ 2x_1 - 4x_2 + 10x_3 - 6x_4 + 8x_5 &= 4. \end{aligned}$$

(a) Write the matrix A and the vector b such that the given system of equations is the verbose form of $Ax = b$.

(b) Identify the matrix $B = [A_{\bullet 3} \ A_{\bullet 2} \ A_{\bullet 4}]$ and verify that it is a basis in A.

(c) Compute the basic solution associated with the matrix B (in part (b) above) and verify that it is a nonnegative vector.

(d) What is the index set N of the nonbasic variables in this case?

[Hint: Although it is not absolutely necessary to do so, you may find it useful to compute the inverse of B in this exercise.]

3.4   Consider the LP

$$\begin{array}{ll}
\text{maximize} & x_1 + 2x_2 \\
\text{subject to} & x_1 + x_2 \le 4 \\
& -x_1 + x_2 \le 2 \\
& x_1, \ x_2 \ge 0.
\end{array}$$

(a) Plot the feasible region.

(b) Identify all the basic solutions and specify which are feasible and which are infeasible.

(c) Solve the problem using the Simplex Algorithm.

3.5   Consider the LP

$$\begin{array}{ll}
\text{maximize} & x_1 + x_2 \\
\text{subject to} & x_1 + x_2 \ge 2 \\
& -x_1 + x_2 \le 2 \\
& x_1, \ x_2 \ge 0.
\end{array}$$

(a) Plot the feasible region.

(b) Identify all the basic solutions and specify which are feasible and which are infeasible.

(c) Solve the LP by the Simplex Algorithm and show that the algorithm stops with the objective function being unbounded over the feasible region.

(d) Modify the first inequality constraint to be $x_1 + x_2 \le 2$ and show that there are infinite optimal solutions.

3.6   Suppose we wish to minimize the linear function

$$z = 3x_1 - 6x_2 + 5x_3 + 3x_4 + 7x_5$$

subject to the equations in Exercise 3.3 and of course $x_j \ge 0$, $j = 1, \ldots, 6$.

(a) Using the basis B in Exercise 3.3, compute the quantities z_j for all $j \in N$.

(b) Can you say whether the basis is optimal or not? Justify your answer.

3.7   How would the sufficient condition for optimality change if the LP were changed to a maximization problem?

3.8   Write the LP given in Example 1.3 (Chapter 1) in standard form.

(a) Is the basis associated with the slack variables optimal?

(b) Plot each of the slack variables in the standard form of the LP as a function of one of the nonbasic variables (the other nonbasic variable being held at zero). Put the varying nonbasic variable on the horizontal axis.

(c) Carry out the minimum ratio test for the column associated with ALE. How much can this be increased before one of the slack variables drops to zero?

3.9   This exercise pertains to the linear program given in Example 3.2.

(a) Put the constraints into canonical form with respect to the basis $B = [A_{\bullet 1} \, A_{\bullet 2}]$.

(b) Put the constraints into canonical form with respect to the basis $B = [A_{\bullet 3} \, A_{\bullet 4}]$.

3.10   Consider the tiny LP

$$\begin{array}{ll}
\text{maximize} & x_1 \\
\text{subject to} & x_2 \le 1 \\
& x_1, \, x_2 \ge 0.
\end{array}$$

(a) Plot the feasible region of this LP. Is the feasible region bounded, i.e, can it be contained within a circle of suitably large radius?

(b) Express the LP as a minimization problem in standard form.

(c) Apply the Simplex Algorithm to the problem stated in (b). What happens to the objective function?

(d) Change the objective function to x_2. Does the problem have a *unique* optimal solution?

3.11   Consider linear programs of the form

$$(\text{P}) \qquad \text{minimize } c^{\mathrm{T}}x \quad \text{subject to} \quad 0 \le x \le b.$$

Discuss each of the cases below from the standpoint of feasibility and optimality.

(a) b and c are real numbers.

(b) b and c are nonzero n-vectors.

That is, give necessary and sufficient conditions for the feasibility of (P) and the existence of an optimal solution to this LP. When (P) has an optimal solution, what is it?

3.12　Consider the LP solved in Example 3.8. For each iteration of the Simplex Algorithm, find the values of s, r, j_1, j_2, and j_3. Record your answer in a table like the following

Iteration	s	r	j_1	j_2	j_3
1					
2					
3					

4. THE SIMPLEX ALGORITHM CONTINUED

Overview

The discussion of the Simplex Algorithm for linear programming presented in the previous chapter included two crucial assumptions. The first was that a starting basic feasible solution was known and, moreover, that the system was in canonical form with respect to this feasible basis. If the Simplex Algorithm is to be of value, it is essential that we be able to find a basic feasible solution—if one exists. As it turns out, there are several ways to do this. Two of them are covered in the next section.

The second assumption we made in our discussion of the Simplex Algorithm was that the basic feasible solutions encountered in executing the algorithm are nondegenerate. This guaranteed improvement in every iteration and consequently the finiteness of the procedure. But degeneracy is a very commonly encountered property of linear programs, so it becomes essential to provide a way to overcome the problem of *cycling* which is a potential consequence of degeneracy. We address this issue in a later section of this chapter.

In addition to tying up the two aforementioned "loose ends" of our previous treatment of the Simplex Algorithm, we include a development of the Simplex Algorithm in matrix form; this is related to the Revised Simplex Algorithm. Here again, some effort has already been made to foreshadow the ideas involved.

4.1 Finding a feasible solution

In discussing the Simplex Algorithm of linear programming we assumed that a basic feasible solution was known. Unfortunately this is not always the case.

The Phase I Procedure

Given the system $Ax = b$, $x \geq 0$, it is easy to modify the equations so that the constants on the right-hand side (RHS) are nonnegative: if $b_i < 0$, just

© Springer Science+Business Media LLC 2017
R.W. Cottle and M.N. Thapa, *Linear and Nonlinear Optimization*,
International Series in Operations Research & Management Science 253,
DOI 10.1007/978-1-4939-7055-1_4

multiply the i-th equation through by -1. Then the right-hand side will be $-b_i > 0$. *We assume this has already been done.* Then, since $b \geq 0$ (by assumption), the system

$$Ax + Iy = b, \quad x \geq 0, \quad y \geq 0$$

has an obvious BFS: $(x, y) = (0, b)$.

A (basic) solution of the system

$$Ax + Iy = b, \quad x \geq 0, \quad y \geq 0$$

in which $y = 0$ would provide a (basic) solution of $Ax = b$, $x \geq 0$. We try to find such a solution[1] by solving the *Phase I Problem*:

$$
\begin{aligned}
\text{minimize} \quad & e^{\mathrm{T}}y \\
\text{subject to} \quad & Ax + Iy = b \\
& x \geq 0, \; y \geq 0
\end{aligned}
\tag{4.1}
$$

where $e = (1, 1, \ldots, 1)$, which is called *the vector of all ones*. The variables $y_1, \ldots, y_m$ are called *artificial variables*, or *artificials* for short. They are not to be confused with slack variables.

The reduced form of the Phase I Problem

Let $w = e^{\mathrm{T}}y = \sum_{i=1}^{m} y_i$. Then $w - e^{\mathrm{T}}y = 0$, and we can write the system as

$$
\begin{aligned}
w + 0^{\mathrm{T}}x - e^{\mathrm{T}}y &= 0 \\
Ax + Iy &= b \\
x \geq 0, \; y &\geq 0.
\end{aligned}
$$

This version is equivalent to

$$
\begin{aligned}
w + e^{\mathrm{T}}Ax + 0^{\mathrm{T}}y &= e^{\mathrm{T}}b \\
Ax + Iy &= b \\
x \geq 0, \; y &\geq 0.
\end{aligned}
$$

In this form of the problem, the basic variables y have been eliminated from the objective function. The aim now is to find w, x, and y satisfying these

---

[1] In general if $a > 0$ and $y \geq 0$, then $a^{\mathrm{T}}y \geq 0$ with equality if and only if $y = 0$.

conditions for which w is minimum. If $w = 0$, we have $y = 0$ and a BFS of $Ax = b$, $x \geq 0$. We note that the Phase I Problem is an LP in its own right, but it is one that has an obvious starting feasible basis. For that reason the Simplex Algorithm is immediately applicable to it. Moreover, if the given expressions $Ax = b$ and $x \geq 0$ are the constraints of an LP, then, as we shall see, if a BFS is found by the Phase I Procedure, it is possible to initiate the solution of the given LP from the BFS found thereby. In fact, this can be done with a "smooth transition" between Phase I and Phase II wherein the problem of interest is solved by the Simplex Algorithm.

An important remark

The Phase I Problem formulated above is in standard form: it has a system of linear equations in nonnegative variables. Since any system of linear equations and inequalities can be written as an equivalent system of linear equations, in nonnegative variables, it follows that testing the system for the existence of a solution can be carried out by formulating the corresponding Phase I Problem and solving it by the Simplex Algorithm. This approach will not only test for the existence of a solution: It will produce one if there are any. For an application of this, see [21].

Example 4.1: SOLVING AN LP IN TWO PHASES. Let us consider the LP

$$
\begin{array}{lrrrrrr}
\text{minimize} & 4x_1 + & 8x_2 + & x_3 + & 2x_4 + & 6x_5 & \\
\text{subject to} & 2x_1 - & 3x_2 - & 4x_3 + & x_4 - & 4x_5 = & -10 \\
& 5x_1 & & - 2x_3 & & - 2x_5 = & 18 \\
& -3x_1 + & 3x_2 & & + 6x_4 & - 2x_5 = & 12 \\
\end{array}
$$
$$x_j \geq 0, \ j = 1, 2, 3, 4, 5.$$

In this statement of the problem we have no obvious starting feasible basis. To find one we'll use the Phase I Procedure to accomplish that. Then we'll go on to solve the given LP with Phase II of the Simplex Algorithm.

This example will illustrate several things. The first is the formulation of the Phase I Problem in standard form. This will be easy. The second thing this example will illustrate is the idea of a "smooth transition" to Phase II. The third thing this example will illustrate (not prove) is the fact that when the problem data are all *rational numbers*, then the solution vector found by the Simplex Algorithm will have rational coordinates.

The first constraint in the given version of the LP has a negative number on the right-hand side, so we multiply through the equation by -1. Once that is done, we can insert artificial variables y_1, y_2, and y_3. Furthermore, to facilitate our smooth transition to Phase II, we include an equation for the original objective function. We express all this in tabular form as follows:

min	w	z	x_1	x_2	x_3	x_4	x_5	y_1	y_2	y_3	1	basic
row 00	1	0	0	0	0	0	0	-1	-1	-1	0	w
row 0	0	1	-4	-8	-1	-2	-6	0	0	0	0	z
row 1	0	0	-2	3	4	-1	4	1	0	0	10	y_1
row 2	0	0	5	0	-2	0	-2	0	1	0	18	y_2
row 3	0	0	-3	3	0	6	-2	0	0	1	12	y_3

This tabular form (tableau) reveals that the artificial variables must be eliminated from the equation for the artificial objective function (equivalently, row 00 of the tableau). Adding rows 1, 2, and 3 to row 00 we obtain

min	w	z	x_1	x_2	x_3	x_4	x_5	y_1	y_2	y_3	1	basic
row 00	1	0	0	6	2	5	0	0	0	0	40	w
row 0	0	1	-4	-8	-1	-2	-6	0	0	0	0	z
row 1	0	0	-2	3	4	-1	4	1	0	0	10	y_1
row 2	0	0	5	0	-2	0	-2	0	1	0	18	y_2
row 3	0	0	-3	3	0	6	-2	0	0	1	12	y_3

In an effort to minimize w, we choose x_2 as the entering variable. The leaving variable is determined by carrying out the minimum ratio test. We find that since $10/3 < 12/3$, the leaving variable is y_1. We perform a pivot on the element in row 1 and the column of x_2 obtaining

min	w	z	x_1	x_2	x_3	x_4	x_5	y_1	y_2	y_3	1	basic
row 00	1	0	4	0	-6	7	-8	-2	0	0	20	w
row 0	0	1	$-28/3$	0	$29/3$	$-14/3$	$14/3$	$8/3$	0	0	$80/3$	z
row 1	0	0	$-2/3$	1	$4/3$	$-1/3$	$4/3$	$1/3$	0	0	$10/3$	x_2
row 2	0	0	5	0	-2	0	-2	0	1	0	18	y_2
row 3	0	0	-1	0	-4	7	-6	-1	0	1	2	y_3

Next we choose x_4 as the entering variable. No minimum ratio test is needed here since there is only one positive entry in the corresponding column and rows 1, 2, and 3. We pivot on the element in row 3 and the column of x_4.

min	w	z	x_1	x_2	x_3	x_4	x_5	y_1	y_2	y_3	1	basic
row 00	1	0	5	0	-2	0	-2	-1	0	-1	18	w
row 0	0	1	-10	0	7	0	$2/3$	2	0	$2/3$	28	z
row 1	0	0	$-5/7$	1	$8/7$	0	$22/21$	$2/7$	0	$1/21$	$24/7$	x_2
row 2	0	0	5	0	-2	0	-2	0	1	0	18	y_2
row 3	0	0	$-1/7$	0	$-4/7$	1	$-6/7$	$-1/7$	0	$1/7$	$2/7$	x_4

In this tableau, we see that making x_1 basic will reduce the value of w even further. There is only one possible pivot element, namely the number 5 in row 2 and x_1 column. The outcome of this pivot operation is

min	w	z	x_1	x_2	x_3	x_4	x_5	y_1	y_2	y_3	1	basic
row 00	1	0	0	0	0	0	0	-1	-1	-1	0	w
row 0	0	1	0	0	3	0	$-10/3$	2	2	2/3	64	z
row 1	0	0	0	1	6/7	0	16/21	2/7	1/7	1/21	6	x_2
row 2	0	0	1	0	$-2/5$	0	$-2/5$	0	1/5	0	18/5	x_1
row 3	0	0	0	0	$-22/35$	1	$-32/35$	$-1/7$	1/35	1/7	4/5	x_4

This tableau reveals that w, the value of the artificial objective function, has been reduced to zero. *Now, in the computation, the row 00 and the y_1, y_2, and y_3 columns can be ignored (or deleted from the tableau).* We can also pick out the basic feasible solution $(x_1, x_2, x_3, x_4, x_5) = (18/5, 6, 0, 4/5, 0)$ of the original system that has been found by the Phase I Procedure. (As a check, try substituting these values for the variables $x_1, \ldots, x_5$ in the original problem. You should find that this is a feasible vector for which $z = 64$.) Furthermore, we see that since these variables have been eliminated from the expression for z, the true objective function value, the smooth transition has been achieved. The fact that the coefficient of x_3 is $3 > 0$ suggests that there is an opportunity to decrease z by increasing x_3. In doing so, we find there is only one possible pivot element: the element in row 1 and column x_3. The corresponding pivot operation results in the tableau

min	w	z	x_1	x_2	x_3	x_4	x_5	y_1	y_2	y_3	1	basic
row 00	1	0	0	0	0	0	0	-1	-1	-1	0	w
row 0	0	1	0	$-7/2$	0	0	-6	1	3/2	1/2	43	z
row 1	0	0	0	7/6	1	0	8/9	1/3	1/6	1/18	7	x_3
row 2	0	0	1	7/15	0	0	$-2/45$	2/15	4/15	1/45	32/5	x_1
row 3	0	0	0	11/15	0	1	$-16/45$	1/15	2/15	8/45	26/5	x_4

Remembering that we are ignoring the y_1, y_2, and y_3 columns, we declare that an optimal solution of the given LP has been found, namely

$$(x_1, x_2, x_3, x_4, x_5) = (32/5, 0, 7, 26/5, 0) \text{ and } z = 43.$$

Example 4.2: MODELING OTHER LINEAR SYSTEMS. Example 4.1 carried out the solution of a linear program with *equality constraints*. To use the Phase I Procedure we had a little bit of "preprocessing" to do in order to make the right-hand side nonnegative. Suppose the constraints of the LP had been different, for instance as in

$$\begin{aligned}
\text{minimize} \quad & 4x_1 + 8x_2 + x_3 + 2x_4 + 6x_5 \\
\text{subject to} \quad & 2x_1 - 3x_2 - 4x_3 + x_4 - 4x_5 = -10 \\
& 5x_1 \qquad - 2x_3 \qquad - 2x_5 \leq 18 \\
& -3x_1 + 3x_2 \qquad + 6x_4 - 2x_5 \geq 12 \\
& x_j \geq 0, \ j = 1, 2, 3, 4, 5.
\end{aligned}$$

A problem of this form calls for a slightly different treatment, particularly where the Phase I Procedure is concerned. In addition to the multiplication by -1 as before, we introduce slack and surplus variables x_6 and x_7 in the second and third constraints, respectively. This results in the system

$$\begin{aligned}
\text{minimize} \quad & 4x_1 + 8x_2 + x_3 + 2x_4 + 6x_5 \\
\text{subject to} \quad & -2x_1 + 3x_2 + 4x_3 - x_4 + 4x_5 \qquad\qquad = 10 \\
& 5x_1 \qquad - 2x_3 \qquad - 2x_5 + x_6 \qquad = 18 \\
& -3x_1 + 3x_2 \qquad + 6x_4 - 2x_5 \qquad - x_7 = 12 \\
& x_j \geq 0, \ j = 1, 2, 3, 4, 5, 6, 7.
\end{aligned}$$

We now observe that the column of x_6 can be used as one of the columns of a starting basis in a Phase I Problem for this LP. Instead of using a full set of artificial variables, we can use y_1, x_6, and y_3 and minimize the artificial objective function $w = y_1 + y_3$. We relegate the further discussion of this matter to Exercises 4.2 and 4.3.

Outcomes in Phase I

It can be shown that the Phase I Problem always has an optimal solution, $(\bar{x}, \bar{y})$. If $\bar{w}$, the minimum value of w, is positive, then $\bar{y} \neq 0$. In this case, the Phase I Problem has no feasible solution in which the artificial variables $y_1, \ldots, y_m$ all equal zero. This implies that there is no solution of the system $Ax = b$, $x \geq 0$. If $\bar{w} = 0$, then $Ax = b$, $x \geq 0$ does have a solution, indeed a basic solution.

It can happen that the Phase I Procedure ends with $\bar{w} = 0$ and an artificial variable—or variables—still basic. Since all such variables must have the value zero, this means we have a degenerate BFS. When this happens, there are two possible cases, depending on the existence of suitable nonzero elements for pivoting. Suppose the artificial variable y_i is basic at value zero when Phase I ends. Let the corresponding equation be

$$\sum_{j=1}^{n} \bar{a}_{ij} x_j + y_i = \bar{b}_i = 0.$$

If $\bar{a}_{ij} \neq 0$ for some j, say $j = s$, then x_s must be a nonbasic variable, and $\bar{a}_{is}$ can be used as a pivot element, *even if it is negative*. The reason for this is that when we divide through the corresponding equation by $\bar{a}_{is}$, the constant $\bar{b}_i = 0$ on the right-hand side prevents any basic variable from becoming negative after the pivot step is complete. In this situation, x_s replaces y_i as a basic variable, and of course, its value is zero. On the other hand, if $\bar{a}_{ij} = 0$ for all j, then the system is *rank deficient* and the i-th equation can be discarded.

The Big-M Method

A second approach to solving an LP when a starting BFS is not available is to solve another LP which includes the artificial basis and minimizes the original objective function *plus* a linear function that penalizes positivity of the artificial variables. Suppose the original problem is

$$\begin{array}{ll} \text{minimize} & c^{\mathrm{T}}x \\ \text{subject to} & Ax = b \quad (b \geq 0) \\ & x \geq 0. \end{array}$$

Let M be a *large* positive number. Now consider the LP

$$\begin{array}{ll} \text{minimize} & c^{\mathrm{T}}x + Me^{\mathrm{T}}y \\ \text{subject to} & Ax + \quad Iy = b \quad (b \geq 0) \\ & x \geq 0, \ y \geq 0. \end{array}$$

If M is large enough, this optimization problem will find a BFS of the given LP if one exists. This augmented LP is called the *Big-M Problem*. It raises two questions: *How big should M be?* and *Does it matter?*

Example 4.3: THE BIG-M METHOD. Consider the LP of Example 4.1 which we solved earlier by the Simplex Method.

$$\begin{array}{ll} \text{minimize} & 4x_1 + 8x_2 + \ x_3 + 2x_4 + 6x_5 \\ \text{subject to} & 2x_1 - 3x_2 - 4x_3 + \ x_4 - 4x_5 = -10 \\ & 5x_1 \qquad - 2x_3 \qquad - 2x_5 = \quad 18 \\ & -3x_1 + 3x_2 \qquad + 6x_4 - 2x_5 = \quad 12 \\ & \qquad x_j \geq 0, \ j = 1, 2, 3, 4, 5. \end{array}$$

Here we use the Big-M Method described above to find an initial feasible solution. As before, we multiply through the first equation by -1. Once that is done, we can insert artificial variables y_1, y_2, and y_3 and assign (to each of them) a cost of M, where M is a large number. In a computer implementation of the approach we would choose M to be much larger than any of the data or expected growth in the data during computations; for example, in this case we could try $M = 1000$. We express all this in tabular form below. We have kept the M-part of the objective separate from the original objective coefficients for ease of exposition.

min	z	x_1	x_2	x_3	x_4	x_5	y_1	y_2	y_3	1	basic
		0	0	0	0	0	$-M$	$-M$	$-M$	0	
row 0	1	-4	-8	-1	-2	-6	0	0	0	0	z
row 1	0	-2	3	4	-1	4	1	0	0	10	y_1
row 2	0	5	0	-2	0	-2	0	1	0	18	y_2
row 3	0	-3	3	0	6	-2	0	0	1	12	y_3

This tableau reveals that the artificial variables must be eliminated from the equation for the objective. This is done by adding a multiple, M, of each of the rows 1, 2, and 3 to row 0. The result is

min	z	x_1	x_2	x_3	x_4	x_5	y_1	y_2	y_3	1	basic
		0	$6M$	$2M$	$5M$	0	0	0	0	$40M$	
row 0	1	-4	-8	-1	-2	-6	0	0	0	0	z
row 1	0	-2	3	4	-1	4	1	0	0	10	y_1
row 2	0	5	0	-2	0	-2	0	1	0	18	y_2
row 3	0	-3	3	0	6	-2	0	0	1	12	y_3

In an effort to minimize z, we choose x_2 as the entering variable. The leaving variable is determined by carrying out the minimum ratio test. We find that since $10/3 < 12/3$, the leaving variable is y_1. We perform a pivot on the element in row 1 and the column of x_2 obtaining

min	z	x_1	x_2	x_3	x_4	x_5	y_1	y_2	y_3	1	basic
		$4M$	0	$-6M$	$7M$	$-8M$	$-2M$	0	0	$20M$	
row 0	1	$-28/3$	0	$29/3$	$-14/3$	$14/3$	$8/3$	0	0	$80/3$	z
row 1	0	$-2/3$	1	$4/3$	$-1/3$	$4/3$	$1/3$	0	0	$10/3$	x_2
row 2	0	5	0	-2	0	-2	0	1	0	18	y_2
row 3	0	-1	0	-4	7	-6	-1	0	1	2	y_3

Next we choose x_4 as the entering variable. In this case, no minimum ratio test is needed since there is only one positive entry in the corresponding column and rows 1, 2, and 3. The pivot element is in row 3 and the column of x_4. The result of this pivot step is

min	z	x_1	x_2	x_3	x_4	x_5	y_1	y_2	y_3	1	basic
		$5M$	0	$-2M$	0	$-2M$	$2M$	0	$-M$	$18M$	
row 0	1	-10	0	7	0	$2/3$	2	0	$2/3$	28	z
row 1	0	$-5/7$	1	$8/7$	0	$22/21$	$2/7$	0	$1/21$	$24/7$	x_2
row 2	0	5	0	-2	0	-2	0	1	0	18	y_2
row 3	0	$-1/7$	0	$-4/7$	1	$-6/7$	$-1/7$	0	$1/7$	$2/7$	x_4

In this tableau, we see that making x_1 basic will reduce the value of z even further. There is only one possible pivot element, namely the number 5 in row 2 and x_1 column. The outcome of this pivot operation is

min	z	x_1	x_2	x_3	x_4	x_5	y_1	y_2	y_3	1	basic
	0	0	0	0	0	0	$-M$	$-M$	$-M$	0	
row 0	1	0	0	3	0	$-10/3$	2	2	$2/3$	64	z
row 1	0	0	1	$6/7$	0	$16/21$	$2/7$	$1/7$	$1/21$	6	x_2
row 2	0	1	0	$-2/5$	0	$-2/5$	0	$1/5$	0	$18/5$	x_1
row 3	0	0	0	$-22/35$	1	$-32/35$	$-1/7$	$1/35$	$1/7$	$4/5$	x_4

This tableau reveals that the portion of the objective corresponding to M has been reduced to zero. *At this point in the computation, the y_1, y_2, and y_3 columns can be ignored (or deleted from the tableau).* The fact that the coefficient of x_3 is $3 > 0$ suggests that there is an opportunity to decrease z by increasing x_3. In doing so, we find there is only one possible pivot element: the element in row 1 and column x_3. This pivot results in

min	z	x_1	x_2	x_3	x_4	x_5	y_1	y_2	y_3	1	basic
	0	0	0	0	0	0	$-M$	$-M$	$-M$	0	
row 0	1	0	$-7/2$	0	0	-6	1	$3/2$	$1/2$	43	z
row 1	0	0	$7/6$	1	0	$8/9$	$1/3$	$1/6$	$1/18$	7	x_3
row 2	0	1	$7/15$	0	0	$-2/45$	$2/15$	$4/15$	$1/45$	$32/5$	x_1
row 3	0	0	$11/15$	0	1	$-16/45$	$1/15$	$2/15$	$8/45$	$26/5$	x_4

Remembering that we are ignoring the y_1, y_2, and y_3 columns, we declare that an optimal solution of the given LP has been found:

$$(x_1, x_2, x_3, x_4, x_5) = (32/5, 0, 7, 26/5, 0) \text{ and } z = 43.$$

It is interesting to compare the Big-M Method with Phase I of the Simplex Method. In the numerical example above we note that the same iterations were used in both methods. This will, in fact, always be true if there are no ties in choosing the incoming column. On examining the two methods we note that the Big-M Method simply adds a multiple M (which is a large number) of the Phase I objective to the true objective; this is what makes it possible to combine the Phase I and Phase II objectives into one objective. In effect it adds a penalty to including artificial variables in the system, and the idea is similar to that of penalty function methods of nonlinear programming described in Section 13.4. There is little or no advantage to using the Big-M Method instead of Phase I of the Simplex Method for linear programs. The minimization of the Phase I objective is in effect the minimization of a penalty term without the use of a large penalty parameter M (see Exercise 4.15).

It is not always clear how large M should be; choosing M too large can result in numerical errors, and choosing M not large enough can result in incorrect solutions. If the Big-M Method were to be used instead of Phase I of the Simplex Method, it would be advantageous to keep the Big-M part of the objective separate from the true objective, in order to make the calculations with the M easier to manage. However, this would make it identical to the Phase I approach. In fact there would now be no difference between the two methods because clearly we could divide the separate Big-M objective by M and obtain the Phase I objective. In the event of a tie in either a Phase I objective or a Big-M objective, the reduced cost of the original objective could be used as a tie-breaker.

In practice the Big-M Method is not used for solving linear programs. It does have its use, however, in interior-point methods where some algorithms use a Big-M approach to obtain an initial interior-point solution. This subject is discussed in Chapter 14.

4.2   Dealing with degeneracy

A BFS of the system $Ax = b$, $x \geq 0$, where the matrix A is $m \times n$ and has rank m is *degenerate* if it contains fewer than m positive variables. A linear program having one or more degenerate basic feasible solutions is said to be *degenerate*. Degeneracy is a common property among practical linear programs. In fact, as we shall see, some linear programs are inherently degenerate.

Recall that even if a BFS is nondegenerate, its successor as found by the Simplex Algorithm may be degenerate if the minimum ratio test leads to a tie in the selection of the leaving basic variable. We illustrate such a case in the next example.

Example 4.4: A DEGENERATE LP. Consider the linear program[2]

$$\text{maximize} \quad 2x_1 - 3x_2 + 5x_3$$

$$\begin{aligned}
\text{subject to} \quad 8x_1 - 5x_2 + 4x_3 + x_4 &= 24 \\
x_1 - 7x_2 \qquad\qquad\quad + x_5 \;\; &= 17 \\
5x_1 + 3x_2 + 10x_3 \qquad\qquad + x_6 &= 60
\end{aligned}$$

$$x_j \geq 0, \; j = 1, 2, 3, 4, 5, 6.$$

The BFS $(x_1, x_2, x_3, x_4, x_5, x_6) = (0, 0, 0, 24, 17, 60)$ is nondegenerate. But when we use x_3 as the incoming variable and apply the minimum ratio test, we find that

$$\frac{24}{4} = \frac{60}{10} = 6,$$

which shows that in this instance there are two choices for the index r. Either $r = 1$ or $r = 3$. Let's pick $r = 1$. The following sequence of steps and tableaus portrays the solution of this problem.

max	z	x_1	x_2	x_3	x_4	x_5	x_6	1	basic
row 0	1	-2	3	-5	0	0	0	0	z
row 1	0	8	-5	4	1	0	0	24	x_4
row 2	0	1	-7	0	0	1	0	17	x_5
row 3	0	5	3	10	0	0	1	60	x_6

Pivoting on the row 1, column x_3 element, we obtain the tableau

max	z	x_1	x_2	x_3	x_4	x_5	x_6	1	basic
row 0	1	8	$-13/4$	0	$5/4$	0	0	30	z
row 1	0	2	$-5/4$	1	$1/4$	0	0	6	x_3
row 2	0	1	-7	0	0	1	0	17	x_5
row 3	0	-15	$31/2$	0	$-5/2$	0	1	0	x_6

Here the degeneracy of the problem is explicitly revealed by the BFS shown in the tableau. We next pivot on the row 3, column x_2 element and get

max	z	x_1	x_2	x_3	x_4	x_5	x_6	1	basic
row 0	1	$301/62$	0	0	$45/62$	0	$13/62$	30	z
row 1	0	$49/62$	0	1	$3/62$	0	$5/62$	6	x_3
row 2	0	$-179/31$	0	0	$-35/31$	1	$14/31$	17	x_5
row 3	0	$-30/31$	1	0	$-5/31$	0	$2/31$	0	x_2

This tableau indicates that an optimal solution has been found.

---

[2]Instead of converting this maximization problem to a minimization problem, we simply use the appropriate changes in the sufficient conditions for optimality: $z_j - c_j \geq 0$.

Cycling

Degeneracy can cause the phenomenon called *cycling*, the infinite repetition of a finite sequence of basic feasible solutions producing no change in the value of the objective and thereby preventing the finite termination of the Simplex Algorithm. However, although the LP in Example 4.4 is degenerate, no cycling occurred in its solution by the Simplex Algorithm. This makes the point that *cycling is not bound to happen in a degenerate problem.*

Example 4.5: CYCLING. Consider the following LP:

$$
\begin{aligned}
\text{minimize} \quad & -\tfrac{3}{4}x_4 + 20x_5 - \tfrac{1}{2}x_6 + 6x_7 \\
\text{subject to} \quad x_1 \quad & + \tfrac{1}{4}x_4 - 8x_5 - x_6 + 9x_7 = 0 \\
x_2 \quad & + \tfrac{1}{2}x_4 - 12x_5 - \tfrac{1}{2}x_6 + 3x_7 = 0 \\
x_3 \quad & + x_6 = 1 \\
x_j \geq 0, \quad & j = 1, 2, 3, 4, 5, 6, 7.
\end{aligned}
$$

The starting BFS $(x_1, x_2, x_3, x_4, x_5, x_6, x_7) = (0, 0, 1, 0, 0, 0, 0)$ is degenerate. The indices of the basic variables are $(j_1, j_2, j_3) = (1, 2, 3)$.

The following seven tableaus record what happens when the Simplex Algorithm is applied to the degenerate LP above. The pivot elements $\bar{a}_{rs} > 0$ are chosen according to two rules, one for each index (subscript). The column index s is one for which $z_j - c_j$ is maximum. The row index is the one that minimizes r where

$$
\frac{\bar{b}_r}{\bar{a}_{rs}} = \min\left\{ \frac{\bar{b}_i}{\bar{a}_s} : \bar{a}_{is} > 0 \right\}.
$$

Hence at every iteration, the pivot row is one for which $\bar{b}_r$ is zero.

min	z	x_1	x_2	x_3	x_4	x_5	x_6	x_7	1	basic
row 0	1	0	0	0	$3/4$	-20	$1/2$	-6	0	z
row 1	0	1	0	0	$1/4$	-8	-1	9	0	x_1
row 2	0	0	1	0	$1/2$	-12	$-1/2$	3	0	x_2
row 3	0	0	0	1	0	0	1	0	1	x_3

min	z	x_1	x_2	x_3	x_4	x_5	x_6	x_7	1	basic
row 0	1	-3	0	0	0	4	$7/2$	-33	0	z
row 1	0	4	0	0	1	-32	-4	36	0	x_4
row 2	0	-2	1	0	0	4	$3/2$	-15	0	x_2
row 3	0	0	0	1	0	0	1	0	1	x_3

min	z	x_1	x_2	x_3	x_4	x_5	x_6	x_7	1	basic
row 0	1	-1	-1	0	0	0	2	-18	0	z
row 1	0	-12	8	0	1	0	8	-84	0	x_4
row 2	0	$-1/2$	$1/4$	0	0	1	$3/8$	$-15/4$	0	x_5
row 3	0	0	0	1	0	0	1	0	1	x_3

min	z	x_1	x_2	x_3	x_4	x_5	x_6	x_7	1	basic
row 0	1	2	-3	0	$-1/4$	0	0	3	0	z
row 1	0	$-3/2$	1	0	$1/8$	0	1	$-21/2$	0	x_6
row 2	0	$1/16$	$-1/8$	0	$-3/64$	1	0	$3/16$	0	x_5
row 3	0	$3/2$	-1	1	$-1/8$	0	0	$21/2$	1	x_3

min	z	x_1	x_2	x_3	x_4	x_5	x_6	x_7	1	basic
row 0	1	1	-1	0	$1/2$	-16	0	0	0	z
row 1	0	2	-6	0	$-5/2$	56	1	0	0	x_6
row 2	0	$1/3$	$-2/3$	0	$-1/4$	$16/3$	0	1	0	x_7
row 3	0	-2	6	1	$5/2$	-56	0	0	1	x_3

min	z	x_1	x_2	x_3	x_4	x_5	x_6	x_7	1	basic
row 0	1	0	2	0	$7/4$	-44	$-1/2$	0	0	z
row 1	0	1	-3	0	$-5/4$	28	$1/2$	0	0	x_1
row 2	0	0	$1/3$	0	$1/6$	-4	$-1/6$	1	0	x_7
row 3	0	0	0	1	0	0	1	0	1	x_3

min	z	x_1	x_2	x_3	x_4	x_5	x_6	x_7	1	basic
row 0	1	0	0	0	$3/4$	-20	$1/2$	-6	0	z
row 1	0	1	0	0	$1/4$	-8	-1	9	0	x_1
row 2	0	0	1	0	$1/2$	-12	$-1/2$	3	0	x_2
row 3	0	0	0	1	0	0	1	0	1	x_3

Notice that the tableau above is identical to the original one.

The following table summarizes the sequence of feasible bases found above while applying the Simplex Algorithm to the LP.

Entering	Leaving	Basic Variables	Objective Value
—	—	(x_1, x_2, x_3)	0
x_4	x_1	(x_4, x_2, x_3)	0
x_5	x_2	(x_4, x_5, x_3)	0
x_6	x_4	(x_6, x_5, x_3)	0
x_7	x_5	(x_6, x_7, x_3)	0
x_1	x_6	(x_1, x_7, x_3)	0
x_2	x_7	(x_1, x_2, x_3)	0

The sequence of pivot steps leads back to the starting basis. Thus we have a cycle of feasible bases. This example shows that cycling can occur in a degenerate LP with the pivot selection rules stated above.

Despite the possibility of cycling, this LP has an optimal BFS given by

$$(x_1, x_2, x_3, x_4, x_5, x_6, x_7) = (3/4, 0, 0, 1, 0, 1, 0).$$

In terms of the original data, the basis $[A_{\bullet 4} \, A_{\bullet 1} \, A_{\bullet 6}]$, i.e., $(j_1, j_2, j_3) = (4, 1, 6)$, is optimal, and the corresponding optimal value of z is $-5/4$.

Some methods of handling degeneracy

There are several methods for resolving degeneracy in the Simplex Algorithm. (One of these was used to obtain the optimal BFS exhibited above.) We give only the briefest coverage of these ideas; a full treatment is beyond the scope of this book.

1. *Random choice* [40]. The random choice rule is simply to chose the row index r randomly from among all those that yield the minimum ratio in the minimum ratio test. It is known that this works with probability 1.

2. *Perturbation of the right-hand side* [27]. Consider an LP in feasible canonical form. To each nonnegative right-hand side component b_i

add the quantity ε^i where ε is a very small positive number. Under these circumstances, the "perturbed" right-hand side vector will now be positive. It can be proved that no degenerate basic feasible solutions will be encountered when the Simplex Algorithm is applied to the modified problem. Because the perturbations are so small, the solution found can be expected to be close to the solution of the unperturbed problem.

3. *Lexicographic ordering* [43]. If x, $y \in R^m$ and $m > 1$, it need not be the case that either $x \geq y$ or $y \geq x$. This can be illustrated by taking $x = (1, 0)$ and $y = (0, 1)$. We say that R^m is *partially ordered* by the relation "$\geq$". But there is a way to achieve a *linear ordering* on R^m. We say that a nonzero vector x is *lexicographically positive*, written $x \succ 0$, if the first (least-index) nonzero component of x is positive. Thus the vector $x = (0, 1, -1)$ is lexicographically positive, whereas $y = (0, -1, 1)$ is not. (In fact, y is *lexicographically negative*.) One then says that $x \succ y$ if and only if $(x - y) \succ 0$. A fundamental fact that makes the lexicographic ordering method work is that any finite set of distinct vectors in R^m has a unique lexicographic minimum.

Now, starting with a system like $Ax = b$, where $A = [B\ N]$, $B = I \in R^{m \times m}$ (the identity matrix), and $b \geq 0$, we can (conceptually) append any $m \times m$ matrix G with linearly independent and lexicographically positive rows so as to obtain an $m \times (m+1)$ matrix $[b\ G]$. (Note that $G = I$ would do nicely.) We can then carry out pivot operations as usual and update the matrix G in the ordinary way. The difference is that instead of the original minimum ratio test, we have a *lexicographic* minimum ratio test. This means that the row index r is chosen according to the rule

$$\frac{1}{\bar{a}_{rs}} [\bar{b}_r \ \bar{G}_{r\bullet}] = \text{lexico min} \left\{ \frac{1}{\bar{a}_{is}} [\bar{b}_i \ \bar{G}_{i\bullet}] : \bar{a}_{is} > 0 \right\}.$$

It can be shown that this way of choosing the pivot element preserves the linear independence and lexicographic positivity of this more generalized right-hand side. It also produces a strict lexicographic decrease in the objective function. These features prevent cycling because the right-hand side is uniquely determined by the current basis.

4. *Bland's Rule* [16]. At any iteration, before termination of the Simplex Algorithm, there is a set of indices of variables that are eligible to become basic (these must all have $z_j - c_j > 0$) and there is a (possibly

empty) set of indices of basic variables that can be replaced by the chosen entering variable. These are the variables that are eligible to become nonbasic at this iteration. Now we have seen in Example 4.5 that choosing the leaving variable according to the least row index does not prevent cycling. Bland's Rule is a *double least-index rule*. In particular, it chooses from among the set of eligible column indices, the *least index* j such that $z_j - c_j > 0$. Let this index be denoted s. Notice that in this approach, $z_s - c_s$ need not be the largest of the $z_j - c_j$. Now for *each* row index k such that $\bar{a}_{ks} > 0$ and

$$\frac{\bar{b}_k}{\bar{a}_{ks}} = \min\left\{ \frac{\bar{b}_i}{\bar{a}_{is}} : \bar{a}_{is} > 0 \right\}, \tag{4.2}$$

there is an associated basic variable x_{j_k}. Choose r so that j_r is the smallest among all the candidate indices satisfying (4.2). This means that the row index r itself need not be the smallest among the eligible ones. Rather, it is the *basic variable* with the smallest subscript that could become nonbasic. Exercise 4.5 asks you to use Bland's Rule on the degenerate LP given in Example 4.5.

In addition to these four approaches, there are other approaches such as:

- *Wolfe's "ad hoc" rule* [194],

- *the practical anti-cycling procedure* [83], and

- *a least-distance formulation* [33].

There is some folklore to the effect that although degeneracy occurs frequently, the cycling phenomenon is not a matter for the practitioner to worry about. Commercial solvers always include anti-cycling procedures. As we have seen in Example 4.4, it can happen that in a degenerate problem, progress (a genuine decrease in the objective function) might be made. As long as this happens after a finite number of iterations, cycling does not occur.

Algorithmic efficiency, in the presence of degeneracy, is another matter, however. When the problem is degenerate, it can happen that the algorithm goes through a finite number of iterations without making an improvement in the objective function value. This phenomenon is called *stalling*, see Dantzig [41].

4.3 Revised Simplex Method

Finding a price vector

As we shall soon see, the role played by the vector $y^T := c_B^T B^{-1}$ is an important one. It is the unique solution of the equation

$$y^T B = c_B^T.$$

The vector y is called the *price vector* with respect to the basis B.

Example 4.6: USING A PRICE VECTOR. To illustrate the utility of a price vector we continue with data from Example 3.3. Relative to the given basis B in the given linear program, we have

$$y^T = c_B^T B^{-1} = [8 \ 4 \ 6] \, \frac{1}{38} \begin{bmatrix} 8 & 4 & 2 \\ 5 & 12 & 6 \\ 4 & 2 & 20 \end{bmatrix} = \begin{bmatrix} \dfrac{54}{19} & \dfrac{46}{19} & \dfrac{80}{19} \end{bmatrix}.$$

Using this price vector, you should find that for $j = 1, 2,$ and 3

$$z_j - c_j = y^T A_{\bullet j} - c_j = 0.$$

Updating the constraints

Having the equation $x_B + B^{-1} N x_N = B^{-1} b$ is useful because it facilitates the minimum ratio test:

$$x_s \le \min \left\{ \frac{\bar{b}_i}{\bar{a}_{is}} : \bar{a}_{is} > 0 \right\}.$$

But notice that

$$\bar{A}_{\bullet j} = B^{-1} A_{\bullet j} \text{ for all } j \quad \text{and} \quad \bar{b} = B^{-1} b,$$

and these are specified by (i.e., solutions of) the equations

$$B \bar{A}_{\bullet j} = A_{\bullet j} \text{ for all } j \quad \text{and} \quad B \bar{b} = b.$$

Having a known feasible basis, one can test for optimality by finding the corresponding price vector y and computing $y^T A - c^T$. Any j for which

$y^T A_{\bullet j} c_j - c_j > 0$ can be chosen as the index s of the entering variable. The minimum ratio test uses only $\bar{A}_{\bullet s}$ and $\bar{b}$. The important point here is that *no other updated columns are required at this stage.* Once the index r is found, $A_{\bullet s}$ will replace the current $A_{\bullet j_r}$ as a column of the new basis. Notice that $A_{\bullet s}$ and $A_{\bullet j_r}$ are expressed by *original data,* rather than their transformed counterparts $\bar{A}_{\bullet s}$ and $\bar{A}_{\bullet j_r}$. In a large problem, this modification of the Simplex Algorithm can save a lot of unnecessary computation. Of course, in order to take advantage of these savings, it is essential to be able to compute either an update of an appropriate factorization of the current basis or an update of the actual inverse of the current basis.

If we have a feasible basis B for a linear program

$$
\begin{aligned}
\text{minimize} \quad & c^T x \\
\text{subject to} \quad & Ax = b \\
& x \geq 0,
\end{aligned}
$$

and wish to check it for optimality, we need to find the quantities $z_j - c_j$ for all j such that x_j is nonbasic. Doing this is a matter of computing $c_B^T B^{-1} A_{\bullet j} - c_j$ for all $j \in N$. This process is called *pricing out the basis.* The reason for this terminology is that it involves the price vector y given by $y^T = c_B^T B^{-1}$. Thus, we say that B *prices out optimal* if $y^T A_{\bullet j} - c_j \leq 0$ for all $j \in N$.

The Revised Simplex Algorithm

The reader must have noticed that at each iteration of the Simplex Algorithm the whole tableau is updated as a result of pivoting. In the case of large problems, each pivot step transforms a very large amount of data as the whole tableau is updated. However, at each iteration, all one needs to do is to first determine the pivot column (to improve the objective value) and then select the pivot row (to maintain feasibility). One is inclined to question the value of the vast amount of needless computation, for once the pivot column has been selected, the minimum ratio test determines the pivot row using just *two* column vectors: the updated right-hand side and the selected pivot column. The rest of the entries play no role at all, although it must be conceded that the selection of the pivot column is based on the calculation of the quantities $z_j - c_j$, and the z_j are defined in terms of the nonbasic columns of A. But do all these columns need to be updated for this purpose? As we shall see, they do not.

In the discussion to follow, for simplicity of exposition we shall use B^{-1}; however, practical implementations do not actually use B^{-1} as we shall see later in this section itself.

Let us recall that in testing a basic feasible solution (x_B, x_N) (or equivalently, a feasible basis B) for optimality, we compute and check the sign of

$$z_j - c_j = c_B^T B^{-1} A_{\bullet j} - c_j = c_B^T \bar{A}_{\bullet j} - c_j \quad \text{for all } j \in N. \tag{4.3}$$

The column vectors $\bar{A}_{\bullet j}$ are the very ones that would be computed through the individual pivot steps.

Equation (4.3) reveals that the complete updating of the columns of A is unnecessary. Instead of performing $(c_B^T)(B^{-1} A_{\bullet j})$ for all $j \in N$, we use the price vector $y^T = c_B^T B^{-1}$ computed once for each newly encountered basis B and then compute the inner products $y^T A_{\bullet j}$. In short,

$$z_j - c_j = y^T A_{\bullet j} - c_j \text{ for all } j \in N. \tag{4.4}$$

Equation (4.4) captures one of the main ideas of the *Revised Simplex Method*. It should be clear that when n is much larger than m, this idea will result in significant computational savings. It can also be convenient when the size of A makes its internal storage in a computer an issue.

Next we need to avoid the computation of all the $\bar{A}_{\bullet j}$. Recall that once the pivot column s is identified, then to perform the minimum ratio test we need the columns $\bar{b}$ and $\bar{A}_{\bullet s}$. These are computed easily as

$$\bar{b} = B^{-1} b \quad \text{and} \quad \bar{A}_{\bullet s} = B^{-1} A_{\bullet s}.$$

The minimum ratio test

$$\alpha = \frac{\bar{b}_r}{\bar{a}_{rs}} = \min \left\{ \frac{\bar{b}_i}{\bar{a}_{is}} : \bar{a}_{is} > 0 \right\}$$

is then performed, and the basic variables are updated as follows:

$$x_{j_r} = 0$$
$$x_{j_i} = \bar{b} - \alpha \bar{a}_{rs} \quad \text{for } i = 1, 2, \ldots, m; \ i \neq r$$
$$x_s = \alpha.$$

As seen above, we need to have B^{-1} or some method of solving systems of equations using the basis matrix B. If we had to compute a new B^{-1}

for each change of basis, a significant amount of computation would be required, negating much of the advantages mentioned above, and so we need a computationally efficient technique to obtain B^{-1} from iteration to the next. Rather than going into detail on this, we simply sketch the technique. At each iteration of the Simplex Algorithm, exactly one column of the basis is replaced by a different column from A to result in the new basis $\bar{B}$. By taking advantage of the one-column change in the basis, the new basis inverse $\bar{B}^{-1}$ can be obtained with few computations. This new inverse can then be used to compute the new $\bar{A}_{\bullet s}$, the new price vector $\bar{y}$, and the new reduced costs $z_j - c_j$ for $j \in N$.

The Revised Simplex Algorithm is summarized below without taking into consideration any of the updating procedures referred to above. For a discussion on updating LU factors, see page 108.

Algorithm 4.1 (Revised Simplex Algorithm) Given data A, b, and c for a linear program in standard form, assume that a starting feasible basis B is known and consists of column $j_1, j_2, \ldots, j_m$ from the matrix A.

1. *Test for optimality.*

    (a) Solve $B^\mathrm{T} y = c_B$.

    (b) Compute $z_j = y^\mathrm{T} A_{\bullet j}$ for all nonbasic columns $A_{\bullet j}$.

    (c) Find a nonbasic index s such that (relative to the current basis) $z_s - c_s = \max_{j \in N}\{z_j - c_j\}$. If $z_s - c_s \leq 0$, stop. The current basic feasible solution is optimal.

2. *Minimum ratio test.*

    (a) Solve $B\bar{A}_{\bullet s} = A_{\bullet s}$.

    (b) If $\bar{a}_{is} \leq 0$ for all i, stop. The objective is unbounded below. Otherwise, find an index r such that $\bar{a}_{rs} > 0$ and

    $$\frac{\bar{b}_r}{\bar{a}_{rs}} = \min\left\{\frac{\bar{b}_i}{\bar{a}_{is}} : \bar{a}_{is} > 0\right\}.$$

3. *Update step.* Define the new basis in which $A_{\bullet s}$ replaces the column $B_{\bullet k}$ where $k = j_r$. Define $j_r = s$. Return to Step 1.

Example 4.7: SOLUTION BY THE REVISED SIMPLEX ALGORITHM. Here we solve the LP stated in Example 3.8

$$
\begin{aligned}
\text{minimize} \quad & - 4x_1 - 6x_2 - 3x_3 - x_4 \\
\text{subject to} \quad & x_1 + 4x_2 + 8x_3 + 6x_4 \le 11 \\
& 4x_1 + x_2 + 2x_3 + x_4 \le 7 \\
& 2x_1 + 3x_2 + x_3 + 2x_4 \le 2 \\
& x_1, x_2, x_3, x_4 \ge 0,
\end{aligned}
$$

using the Revised Simplex Algorithm. This problem was solved by the Simplex Algorithm in Example 3.9. As before, we add slack variables x_5, x_6, and x_7 to bring the constraints to standard form. This yields

$$
\begin{aligned}
x_1 + 4x_2 + 8x_3 + 6x_4 + x_5 \quad\quad\quad &= 11 \\
4x_1 + x_2 + 2x_3 + x_4 \quad + x_6 \quad\quad &= 7 \\
2x_1 + 3x_2 + x_3 + 2x_4 \quad\quad\quad + x_7 &= 2 \\
x_1, x_2, x_3, x_4, x_5, x_6, x_7 &\ge 0.
\end{aligned}
$$

An initial feasible basis B is the identity matrix corresponding to the slack variables. The columns of B are made of columns $j_1 = 5$, $j_2 = 6$, $j_3 = 7$ of the original matrix A. Corresponding to this B, we have a basic feasible solution with $x_B = (x_5, x_6, x_7) = (11, 7, 2)$; and the vector of costs associated with the basic variables is $c_B = (0, 0, 0)$.

Iteration 1:

We start by testing for optimality. To do this, we first solve $B^{\mathrm{T}}y = c_B$ obtaining $y = 0$. We next compute $z_j = y^{\mathrm{T}}A_{\bullet j}$ for nonbasic $j = 1, 2, 3, 4$ to obtain

$$(z_1, z_2, z_3, z_4) = (0, 0, 0, 0).$$

Continuing, we next find an index s such that

$$z_s - c_s = \max_{j \in N}\{z_j - c_j\} = \max\{4, 6, 4, 1\} = 6,$$

hence $s = 2$. The condition for optimality is not satisfied, and nonbasic x_2 is a candidate for improving the objective function value.

The next step is to apply the minimum ratio test to see at what level we can bring x_2 into the basic set. We first solve $B\bar{A}_{\bullet 2} = A_{\bullet 2}$ to obtain $\bar{A}_{\bullet 2} = (4, 1, 3)$. Next, we find an index r such that $\bar{a}_{r2} > 0$ and

$$
\frac{\bar{b}_r}{\bar{a}_{r2}} = \min\left\{\frac{\bar{b}_i}{\bar{a}_{i2}} : \bar{a}_{i2} > 0\right\}. = \min\left\{\frac{11}{4}, \frac{7}{1}, \frac{2}{3}\right\} = \frac{2}{3} = \frac{\bar{b}_3}{\bar{a}_{32}}.
$$

Hence we have $r = 3$. This means that the value of the index j_3 will change from 7 to 2. The new basis and associated x_B, obtained by solving $Bx_B = b$, will be

$$B = [A_{\bullet 5}\ A_{\bullet 6}\ A_{\bullet 2}] = \begin{bmatrix} 1 & 0 & 4 \\ 0 & 1 & 1 \\ 0 & 0 & 3 \end{bmatrix} \quad \text{and} \quad x_B = (x_5,\ x_6,\ x_2) = \left(\frac{25}{3},\ \frac{19}{3},\ \frac{2}{3}\right).$$

Iteration 2:

Once again we test for optimality by solving $B^T y = c_B$, where $c_B = (0, 0, 6)$ obtaining $y = (0, 0, -2)$. We next compute $z_j = y^T A_{\bullet j}$ for $j = 1, 3, 4, 7$ to obtain

$$(z_1,\ z_3,\ z_4,\ z_7) = (-4, 0, 0, 0).$$

Continuing, we next find an index s such that

$$z_s - c_s = \max_{j \in N}\{z_j - c_j\} = \max\{0, 1, -3, -2\} = 1,$$

hence $s = 3$. The condition for optimality is not satisfied and nonbasic x_3 is a candidate for improving the objective function value.

The next step is to apply the minimum ratio test to see at what level we can bring x_3 into the basic set. As before, we first solve $B\bar{A}_{\bullet 3} = A_{\bullet 3}$ to obtain $\bar{A}_{\bullet 3} = (20/3, 5/3, 1/3)$. Next, we find an index r such that $\bar{a}_{r3} > 0$ and

$$\frac{\bar{b}_r}{\bar{a}_{rs}} = \min\left\{\frac{25}{20},\ \frac{19}{5},\ \frac{2}{1}\right\} = \frac{25}{20} = \frac{\bar{b}_1}{\bar{a}_{13}}.$$

Hence we have $r = 1$. This means that the value of the index j_1 will change from 5 to 3. The new basis and associated x_B, obtained by solving $Bx_B = b$, will be

$$B = [A_{\bullet 3}\ A_{\bullet 6}\ A_{\bullet 2}] = \begin{bmatrix} 8 & 0 & 4 \\ 2 & 1 & 1 \\ 1 & 0 & 3 \end{bmatrix} \quad \text{and} \quad x_B = (x_3,\ x_6,\ x_2) = \left(\frac{5}{4},\ \frac{17}{1},\ \frac{1}{4}\right).$$

Iteration 3:

With the new basis, we obtain $y = (-3/20, 0, -9/5)$ and the corresponding $z_j = y^T A_{\bullet j}$ as

$$(z_1,\ z_4,\ z_5,\ z_7) = \left(-\frac{15}{4},\ -\frac{9}{2},\ -\frac{3}{20},\ -\frac{19}{5}\right).$$

Continuing, we next find an index s such that

$$z_s - c_s = \max_{j \in N}\{z_j - c_j\} = \max\left\{\frac{1}{4}, -\frac{7}{2}, -\frac{3}{20}, -\frac{9}{5}\right\} = \frac{1}{4},$$

hence $s = 1$. The condition for optimality is still not satisfied and nonbasic x_1 is a candidate for improving the objective function value.

We obtain $\bar{A}_{\bullet 1} = (-1/4,\ 15/4,\ 3/4)$. Applying the minimum ratio test, we find an index r such that $\bar{a}_{r1} > 0$ and

$$\frac{\bar{b}_r}{\bar{a}_{rs}} = \min\left\{\frac{17}{15}, \frac{1}{3}\right\} = \frac{1}{3} = \frac{\bar{b}_3}{\bar{a}_{31}}.$$

Hence we have $r = 3$ and the new index $j_3 = 1$. The new basis and associated x_B, obtained by solving $Bx_B = b$, will be

$$B = [A_{\bullet 3}\ A_{\bullet 6}\ A_{\bullet 2}] = \begin{bmatrix} 8 & 0 & 1 \\ 2 & 1 & 4 \\ 1 & 0 & 2 \end{bmatrix} \quad \text{and} \quad x_B = (x_3,\ x_6,\ x_1) = \left(\frac{4}{3},\ 3,\ \frac{1}{3}\right).$$

Iteration 4:

With the new basis we obtain $y = (-2/15,\ 0,\ -9/15)$ and the corresponding $z_j = y^T A_{\bullet j}$ as

$$(z_2,\ z_4,\ z_5,\ z_7) = \left(-\frac{19}{3}, -\frac{14}{3}, -\frac{2}{15}, -\frac{59}{15}\right).$$

Continuing, we next find an index s such that

$$z_s - c_s = \max_{j \in N}\{z_j - c_j\} = \max\left\{-\frac{1}{3}, -\frac{11}{3}, -\frac{2}{15}, -\frac{29}{15}\right\} = -\frac{11}{3},$$

hence $s = 4$. We now see that the sufficient condition for optimality is satisfied. The optimal solution found is

$$(x_1,\ x_2,\ x_3,\ x_4,\ x_5,\ x_6,\ x_7) = \left(\frac{1}{3},\ 0,\ \frac{4}{3},\ 0,\ 0,\ 3,\ 0\right) \quad \text{and} \quad z = -\frac{16}{3}.$$

Use of an LU factorization in the Revised Simplex Algorithm

In early implementations of the Simplex Algorithm, an explicit representation of the inverse was kept. The modifications to obtain the updated basis inverse were kept in vectors called *eta-vectors*. However, it turned out that the inverse had the potential to be numerically unstable, and in fact this was the case in many practical problems. The technique of factorizing a matrix into a product of two triangular matrices can provide better stability. One such technique is called the *LU factorization* (also called *LU decomposition or triangular decomposition*); the basis B can be written as

$$B = LU,$$

where L is unit lower triangular (i.e., lower triangular with all diagonal elements equal to 1), and U is upper triangular. This makes it easy to solve to find the price vector and also avoid some numerical instabilities associated with computation of the actual inverse of B. Furthermore, the LU factorization can be updated in a computationally efficient manner for the next basis $\bar{B}$.

This idea enables us to compute the price vector as follows. Recall that while the price vector is given by the formula $y^T = c_B^T B^{-1}$, it is also the solution of the equation

$$y^T B = y^T LU = c_B^T,$$

which we can instead write in transposed form as

$$B^T y = U^T L^T y = c_B.$$

Next, letting $w = L^T y$, we can solve the lower-triangular system $U^T w = c_B$. Once w is found, we can solve the upper-triangular system $L^T y = w$.

Recall that in order to carry out the minimum ratio test, we need the right-hand side $\bar{b}$, which we have, and the column $\bar{A}_{\bullet s}$, which is the solution to $B\bar{A}_{\bullet s} = A_{\bullet s}$. Having an LU factorization of B, we can write

$$B\bar{A}_{\bullet s} = LU\bar{A}_{\bullet s} = A_{\bullet s}.$$

Next, letting $v = U\bar{A}_{\bullet s}$, we can solve the "triangular system" $Lv = A_{\bullet s}$. Once v is found, we can solve a second triangular system $U\bar{A}_{\bullet s} = v$. The solution of the latter system is $\bar{A}_{\bullet s}$ which is what is needed for the minimum ratio test.

To be sure, finding the LU factorization of B takes some computational effort, but, as noted earlier, the LU factorization can be updated from one iteration to the next in a manner analogous to updating the basis inverse. In large-scale systems, it turns out that the majority of the coefficients in A are zero (referred to as *sparsity*). It turns out that with an LU factorization not only can we efficiently update the factorization from iteration to iteration and provide improved numerical stability over the inverse, but we can also maintain sparsity (most of the components of the LU factors remain zero). This reduces storage and calculations. As a result, the LU factorization has become an important component of commercial-grade linear programming solvers, particularly those designed for large-scale problems. For a more detailed discussion of this subject, see Golub and Van Loan, [88].

Example 4.8: USING THE LU FACTORIZATION OF A BASIS. Consider a linear program in standard form with $m = 4$ and $n = 6$, with:

$$A = \begin{bmatrix} -2 & 3 & 6 & 1 & 3 & 1 \\ 6 & -8 & -23 & -3 & -11 & 0 \\ -4 & 9 & -5 & 5 & 4 & 2 \\ 2 & 1 & -22 & -8 & -21 & 0 \end{bmatrix}, \quad \text{and} \quad b = \begin{bmatrix} 0 \\ 2 \\ 9 \\ 1 \end{bmatrix},$$

and $c = (-14,\ 34,\ -25,\ 9,\ -20,\ 10)$. Suppose the basis of interest, B, and the associated costs c_B are:

$$B = \begin{bmatrix} -2 & 3 & 6 & 1 \\ 6 & -8 & -23 & -3 \\ -4 & 9 & -5 & 5 \\ 2 & 1 & -22 & -8 \end{bmatrix} \quad \text{and} \quad c_B = \begin{bmatrix} -14 \\ 34 \\ -25 \\ 9 \end{bmatrix}.$$

As it happens, this basis has the LU factorization $B = LU$ where

$$L = \begin{bmatrix} 1 & 0 & 0 & 0 \\ -3 & 1 & 0 & 0 \\ 2 & 3 & 1 & 0 \\ -1 & 4 & -2 & 1 \end{bmatrix} \quad \text{and} \quad U = \begin{bmatrix} -2 & 3 & 6 & 1 \\ 0 & 1 & -5 & 0 \\ 0 & 0 & -2 & 3 \\ 0 & 0 & 0 & -1 \end{bmatrix}.$$

These factors transposed are the triangular matrices:

$$L^T = \begin{bmatrix} 1 & -3 & 2 & -1 \\ 0 & 1 & 3 & 4 \\ 0 & 0 & 1 & -2 \\ 0 & 0 & 0 & 1 \end{bmatrix} \quad \text{and} \quad U^T = \begin{bmatrix} -2 & 0 & 0 & 0 \\ 3 & 1 & 0 & 0 \\ 6 & -5 & -2 & 0 \\ 1 & 0 & 3 & -1 \end{bmatrix}.$$

We can compute the price vector y as the solution to $B^T y = U^T L^T y = c_B$. Recall that we set $w = L^T y$, and next solve the lower-triangular system $U^T w = c_B$ to obtain:

$$w_1 = -14/(-2) = 7$$
$$w_2 = 34 - 3w_1 = 13$$
$$w_3 = (-25 - 6w_1 + 5w_2)/(-2) = 1$$
$$w_4 = (9 - w_1 - 3w_3)/(-1) = 1.$$

Having found w, we next solve the upper-triangular system $L^T y = w$ for the price vector y:

$$y_4 = 1$$
$$y_3 = 1 + 2y_4 = 3$$
$$y_2 = 13 - 3y_3 - 4y_4 = 0$$
$$y_1 = 7 + 3y_2 - 2y_3 + y_4 = 2.$$

Next we compute the reduced costs:

$$z_5 - c_5 = y^T A_{\bullet 5} - c_5 = 18$$
$$z_6 - c_6 = y^T A_{\bullet 6} - c_6 = -2.$$

Hence $A_{\bullet 5}$ is the only candidate to enter the basis and we compute $\bar{A}_{\bullet 5}$ by solving $LU\bar{A}_{\bullet 5} = A_{\bullet 5}$. The solution process now begins by letting $v = U x_B$ and solving the equation $Lv = b$. Because of the lower-triangular form of L and the fact that the diagonal entries of L are all ones, the system is easily solved. First we find v_1; then we substitute its value into the system. The remaining components of v can be found in the same manner. It turns out that $v = (3, -11, 4, -21)$. Having computed v, we return to the equation $v = U\bar{A}_{\bullet 5}$ and solve it for $\bar{A}_{\bullet 5}$. Note that in this case we begin by finding $\bar{A}_{45}$, then $\bar{A}_{35}$, and so forth to obtain $\bar{A}_{\bullet 5} = (7, 3, 1, 2)$.

Adding more activities

Recall that we associate the variables of an LP with "activities." It sometimes happens that an LP is solved and new activities are later put into the model, and everything else stays the same, however. Suppose we let B denote an optimal basis for the original LP and then introduce a vector of

ℓ new activities $\tilde{x} = (\tilde{x}_1, \ldots, \tilde{x}_\ell)$. This yields a new larger linear program

$$\begin{aligned}
\text{minimize} \quad & c^{\mathrm{T}}x + \tilde{c}^{\mathrm{T}}\tilde{x} \\
\text{subject to} \quad & Ax + \tilde{A}\tilde{x} = b \\
& x \geq 0, \ \tilde{x} \geq 0.
\end{aligned}$$

The matrix B is a feasible basis for the enlarged problem. Let y denote the corresponding price vector. We can now say that B is an optimal basis for the enlarged problem if

$$y^{\mathrm{T}}\tilde{A}_{\bullet k} - \tilde{c}_k \leq 0, \quad \text{for all } k = 1, \ldots \ell.$$

On the other hand, if $y^{\mathrm{T}}\tilde{A}_{\bullet k} - \tilde{c}_k > 0$ for some k, we could use $\tilde{x}_k$ as an entering variable in the Simplex Algorithm applied to the enlarged LP. This approach could save some computational effort. It is, by the way, related to what are called *column generation techniques* which involve linear programs whose columns are not all explicitly known in advance. We'll encounter one of these in Chapter 7.

4.4 Exercises

4.1   (a) Use Phase I and Phase II of the Simplex Algorithm to solve the linear program

$$\begin{aligned}
\text{minimize} \quad & 3x_1 - 2x_2 - 3x_3 \\
\text{subject to} \quad & x_1 + 2x_2 - 3x_3 = 4 \\
& 2x_1 + 3x_2 + x_3 = 7 \\
& x_j \geq 0, \quad j = 1, 2, 3.
\end{aligned}$$

Do the computations "by hand."

(b) For each iteration of the two-phase Simplex Algorithm applied in part (a), find the values of s, r, j_1, and j_2. Record your answer in a table like the following

Iteration	s	r	j_1	j_2
1				
2				
⋮				

4.2   Formulate—but do not solve—the Phase I Problem for Example 4.2.

4.3   Formulate—but do not solve—the Big-M Problem for Example 4.2.

4.4   Give an example of a linear program of the form

$$
\begin{array}{llll}
\text{minimize} & & c_3 x_3 & \\
\text{subject to} & x_1 & + a_{13}x_3 & = b_1 \\
& x_2 + a_{23}x_3 & = b_2 \\
& x_j \geq 0, & j = 1, 2, 3
\end{array}
$$

with the following properties:

(i) $c_3 < 0$; (ii) $\min\{b_1, b_2\} = 0$; (iii) $a_{13}a_{23} < 0$

such that an optimal basic feasible solution will be found after one pivot.

4.5   Consider the LP given in Example 4.5. Solve this problem using the Simplex Algorithm with Bland's Rule for selecting the pivot elements. For each iteration, identify the eligible candidates for s and the eligible candidates for r. [Hint: You will find that using the first four tableaus in the discussion of Example 4.5 will reduce (but not eliminate) your work.]

4.6   Consider a linear program with m equality constraints and n variables in which z is to be minimized. Suppose it is already in feasible canonical form with respect to some nonoptimal basis $B = [A_{\bullet j_1} \cdots A_{\bullet j_m}]$. (The nonoptimality means that largest of the $z_j - c_j$ is positive.) Now suppose some other basis $\widehat{B} = [A_{\bullet k_1} \cdots A_{\bullet k_m}]$ is optimal.

(a) Describe the entries of the column vectors $A_{\bullet j_i}$ for $i = 1, \ldots, m$.
(b) Describe the entries of the column vectors $\widehat{B}^{-1} A_{\bullet j_i}$.
(c) For $i = 1, \ldots, m$ describe the objective function coefficients $z_j - c_j$ for $j = j_1, \ldots, j_m$ relative to $\widehat{B}$.
(d) Suppose the original system had been presented in tableau form. For simplicity assume (for this part only) that $j_i = i$. Relate your answers to parts (a), (b), and (c) to what you will see in the "optimal tableau", i.e., the tableau in canonical form with respect to $\widehat{B}$.

4.7   The linear programming problem given in Example 4.5 has seven decision variables $x_1, \ldots, x_7$. An optimal basis for that LP is also indicated there. Its inverse is

$$
[A_{\bullet 4}\ A_{\bullet 1}\ A_{\bullet 6}]^{-1} = \begin{bmatrix} 0 & 2 & 1 \\ 1 & -1/2 & 3/4 \\ 0 & 0 & 1 \end{bmatrix}.
$$

Now suppose three more activities, with corresponding decision variables x_8, x_9, and x_{10} are introduced. Let the unit costs for these activities be 4, 9, and 6, respectively, and let the columns of the enlarged matrix corresponding to these variables be

$$\begin{bmatrix} -3 \\ 4 \\ 5 \end{bmatrix}, \qquad \begin{bmatrix} 9 \\ -2 \\ 0 \end{bmatrix}, \quad \text{and} \quad \begin{bmatrix} 8 \\ 3 \\ -7 \end{bmatrix},$$

respectively. Determine whether the optimal basis for the original problem prices out optimal in the enlarged problem by computing $z_j - c_j$ for $j = 8, 9, 10$.

4.8   Is it possible for the linear program

$$\begin{array}{ll} \text{minimize} & c^{\mathrm{T}}x \\ \text{subject to} & Ax = b \\ & x \geq 0. \end{array}$$

to have an optimal BFS (with optimal basis B) and for the enlarged problem

$$\begin{array}{ll} \text{minimize} & c^{\mathrm{T}}x + \tilde{c}^{\mathrm{T}}\tilde{x} \\ \text{subject to} & Ax + \tilde{A}\tilde{x} = b \\ & x \geq 0, \ \tilde{x} \geq 0. \end{array}$$

to have no optimal solution? Explain your answer.

4.9   Consider the following 2-dimensional linear program

$$\begin{array}{ll} \text{minimize} & -x_1 - x_2 \\ \text{subject to} & x_1 - x_2 \geq 5 \\ & -x_1 - x_2 \geq -1 \\ & x_j \geq 0, \quad j = 1, 2. \end{array}$$

(a) Plot the feasible region.
(b) Solve the LP by the Simplex Method.
(c) Replace the second inequality by $-x_1 + x_2 \leq 1$ and plot the feasible region of the resulting LP.
(d) Solve the LP of part (c) by the Simplex Method.
(e) Replace the second inequality by $x_1 + x_2 \leq 5$ and plot the feasible region of the resulting LP.
(f) Solve the LP of part (e) by the Simplex Method.

4.10   Solve the following linear program by the Simplex Method

$$\begin{array}{ll} \text{minimize} & 2x_1 - 4x_2 + 6x_3 \\ \text{subject to} & x_1 - x_2 + 2x_3 = 10 \\ & 4x_2 - 4x_3 = -5 \\ & -4x_1 + 8x_2 + 2x_3 = 4 \\ & x_j \geq 0, \quad j = 1, 2, 3. \end{array}$$

4.11   Solve the following linear program by the Simplex Method

$$
\begin{aligned}
\text{minimize} \quad & -4x_1 - 8x_2 + 2x_3 + 4x_4 \\
\text{subject to} \quad & 2x_1 + x_2 + x_3 - 2x_4 \leq 12 \\
& 2x_2 - 2x_3 + x_4 \leq 8 \\
& 4x_1 + x_2 + 2x_3 + 3x_4 \leq 10 \\
& x_j \geq 0, \quad j = 1, 2, 3, 4.
\end{aligned}
$$

4.12   Solve the following linear program by the Simplex Method

$$
\begin{aligned}
\text{minimize} \quad & x_1 + 4x_2 + 6x_3 \\
\text{subject to} \quad & -2x_1 + 4x_2 + x_3 \geq 8 \\
& x_1 + 4x_2 + 2x_3 \geq 12 \\
& 2x_1 + 2x_3 \geq 16 \\
& x_j \geq 0, \quad j = 1, 2, 3.
\end{aligned}
$$

4.13   Solve the following linear program by the Simplex Method

$$
\begin{aligned}
\text{minimize} \quad & 3x_1 + x_2 + 6x_3 \\
\text{subject to} \quad & 3x_1 + 3x_2 + x_3 = 15 \\
& 4x_1 + 2x_2 - 5x_3 \geq -12 \\
& x_1 + 4x_3 \geq 15 \\
& x_j \geq 0, \quad j = 1, 2, 3.
\end{aligned}
$$

4.14   Solve the following linear program by the Revised Simplex Method

$$
\begin{aligned}
\text{minimize} \quad & 3x_1 + x_2 + 4x_3 + 2x_4 \\
\text{subject to} \quad & x_1 + x_2 - 2x_3 + x_4 = 10 \\
& 4x_1 + x_2 + 2x_3 + 3x_4 \geq 20 \\
& x_j \geq 0, \quad j = 1, 2, 3, 4.
\end{aligned}
$$

4.15   (a) Write down the *canonical tableau form* of the Phase I Problem for obtaining a basic feasible solution of the LP

$$
\begin{aligned}
\text{minimize} \quad & 4x_1 + 5x_2 + 3x_3 \\
\text{subject to} \quad & 6x_1 - 2x_2 + 4x_3 \geq 14 \\
& 2x_2 - 8x_3 \geq -6 \\
& -4x_1 + 2x_2 + 3x_3 \geq -3 \\
& x_j \geq 0, \quad j = 1, 2, 3.
\end{aligned}
$$

(b) Perform *one iteration* of the Simplex Algorithm on the Phase I Problem you formulated in Part (a). *You are not asked to solve the entire Phase I Problem you formulate.*

(c) Write done the canonical tableau form of the Big-M Method applied to the LP in part (a).

(d) Perform *one iteration* of the Simplex Algorithm on the Big-M Problem you formulated in part (c).

(e) Examine the objective functions and the initial canonical tableau forms of the Phase I and Big-M methods tableaus of parts (a) and (c). Compare these two tableaus and the first iteration and write down the similarities and differences, if any. Is there any advantage to using the Big-M Method instead of the Simplex Method?

4.16 On page 87 we referred to the idea of a "smooth transition" to Phase II. What steps will be required if this technique is not used?

4.17 In each of the following, specify if the statement is true or false, and justify your answer. Remember that for a statement to be considered true, it must be true in all instances of the statement.

(a) In carrying out the minimum ratio test of the Simplex Algorithm, that is, finding

$$\min\left\{\frac{\bar{b}_i}{\bar{a}_{is}} : \bar{a}_{is} > 0\right\},$$

you should choose another s if there are no positive $\bar{a}_{is}$.

(b) If two different basic feasible solutions of a linear program are optimal, they must be adjacent extreme points of the feasible region.

(c) Consider the linear programming problem

$$\begin{array}{ll} \text{minimize} & c^{\mathrm{T}}x \\ \text{subject to} & Ax = b \\ & x \geq 0 \end{array} \tag{4.5}$$

where the matrix A has m rows and n columns. Let H be a nonsingular $m \times m$ matrix, then (4.5) and the linear programming problem

$$\begin{array}{ll} \text{minimize} & c^{\mathrm{T}}x \\ \text{subject to} & HAx = Hb \\ & x \geq 0 \end{array}$$

have exactly the same optimal solutions (if any).

(d) The Phase I Method can be used to decide whether a system of linear inequalities

$$\sum_{j=1}^{n} a_{ij}x_j \geq b_i, \quad i = 1, \ldots, m$$

has a solution.

(e) If $\bar{x}$ is an optimal basic feasible solution of the linear program (4.5) and $\bar{x}_k > 0$, then we must have $z_k - c_k < 0$.

(f) If a free variable x_j in a feasible linear program is replaced by the difference of two nonnegative variables, say $x_j = x_j^+ - x_j^-$ with $x_j^+ \geq 0$ and $x_j^- \geq 0$, it is impossible for both x_j^+ and x_j^- to be positive in a basic feasible solution of the problem.

(g) An optimal basic feasible solution of the linear program (4.5) will have no more than m positive components.

(h) The linear optimization problem of the form (4.5) in which

$$A = \begin{bmatrix} 1 & 1 & 1 & 1 & 0 & 0 & 0 & 0 & 0 & 0 & 0 & 0 \\ 0 & 0 & 0 & 0 & 1 & 1 & 1 & 1 & 0 & 0 & 0 & 0 \\ 0 & 0 & 0 & 0 & 0 & 0 & 0 & 0 & 1 & 1 & 1 & 1 \\ 1 & 0 & 0 & 0 & 1 & 0 & 0 & 0 & 1 & 0 & 0 & 0 \\ 0 & 1 & 0 & 0 & 0 & 1 & 0 & 0 & 0 & 1 & 0 & 0 \\ 0 & 0 & 1 & 0 & 0 & 0 & 1 & 0 & 0 & 0 & 1 & 0 \\ 0 & 0 & 0 & 1 & 0 & 0 & 0 & 1 & 0 & 0 & 0 & 1 \end{bmatrix} \quad \text{and} \quad b = \begin{bmatrix} 300 \\ 400 \\ 300 \\ 200 \\ 300 \\ 250 \\ 250 \end{bmatrix}$$

can be recognized as a transportation problem with 3 sources and 4 destinations.

(i) When the objective function in the Phase I Problem has been reduced to zero, one can be sure that no artificial variables remain basic.

(j) If $A \in R^{m \times n}$ is the coefficient matrix of a linear program in standard form with right-hand side vector $b \geq 0$, and the rank of A is m, then an $m \times m$ invertible matrix B composed of m columns from A will be a feasible basis only if and only if every element of B^{-1} is nonnegative.

4.18 CT Gift Shop is profitable but has a cash flow problem as of the end of June. December is the busiest month, and the shop expects to make a decent profit by the end of the year. As of the end of June, the gift shop has 2,000 in cash in the bank. At the start of each month, rent and other bills are due, and the shop has to make the payments. By the end of the month the shop will receive revenues from its sales. The projected revenues and payments are as follows:

Month	Jul	Aug	Sep	Oct	Nov	Dec
Sales ($)	3,000	3,000	4,000	4,000	9,000	15,000
Payments ($)	6,000	5,000	8,000	3,000	4,000	6,000

CT Gift Shop has an excellent credit history and can take out loans to help cover the shortfalls. On July 1, the gift shop can take out a 6-month loan with a 10% interest rate for the period that is to be paid back at the end of December. The gift shop can also take out a 1-month loan at the start of each month at an interest rate of 2% per month. Formulate CT Gift Shop's cashflow problem as an LP so to minimize the amount of interest paid. Solve the LP by the Simplex Method.

5. DUALITY AND THE DUAL SIMPLEX ALGORITHM

Overview

In this chapter we present what is probably the most important theoretical aspect of linear programming: duality. This beautiful topic has more than theoretical charms. Indeed, it gives valuable insights into computational matters and economic interpretations such as the value of resources at an optimal solution. We will discuss economic interpretations of the dual variables (or multipliers) in Chapter 6 which describes postoptimality analyses.

Linear programs actually come in associated *pairs* of optimization problems. Once we start with a given LP, the other member of the pair is immediately implied. We will demonstrate how to make this association explicit and then proceed to relate their significant properties.

A second topic in this chapter is the Dual Simplex Algorithm, which is due to C.E. Lemke [124]. This is a variant of the Simplex Algorithm we have studied in Chapters 3 and 4. The usefulness of this algorithm will become apparent when we solve an LP by the (standard) Simplex Algorithm and *afterwards* add inequality constraints to the model which then requires re-solving.

5.1 Dual linear programs

For every linear programming problem, say

$$
\begin{aligned}
\text{minimize} \quad & c^{\mathrm{T}}x \\
\text{subject to} \quad & Ax = b \\
& x \geq 0,
\end{aligned}
\tag{5.1}
$$

there is an associated *dual problem* defined as

$$
\begin{aligned}
\text{maximize} \quad & y^{\mathrm{T}}b \\
\text{subject to} \quad & y^{\mathrm{T}}A \leq c^{\mathrm{T}} \\
& (y \text{ free}).
\end{aligned}
$$

The given LP is called the *primal problem*, a name suggested by George Dantzig's father, Tobias Dantzig. The primal and its dual constitute a *dual pair* of linear programs.

© Springer Science+Business Media LLC 2017
R.W. Cottle and M.N. Thapa, *Linear and Nonlinear Optimization*,
International Series in Operations Research & Management Science 253,
DOI 10.1007/978-1-4939-7055-1_5

The dual problem can also be written in the form

$$\begin{array}{ll} \text{maximize} & b^{\mathrm{T}}y \\ \text{subject to} & A^{\mathrm{T}}y \leq c \\ & (y \text{ free}). \end{array} \qquad (5.2)$$

It is very important to observe that the members of a dual pair are based on the same data, but they are used differently in the two programs. In particular, the right-hand side of the primal problem is used in the objective function of the dual problem. Likewise, the right-hand side of the dual problem is used in the objective function of the primal problem. The matrix A appearing in the primal constraints is used in the dual constraints in transposed form in (5.2), the second version of the dual. Another important feature is the difference in directions of optimization between these two problems. The primal given here is a minimization problem and its dual is a maximization problem. Had the primal been a maximization problem, its dual would have been a minimization problem. Finally, it is worthwhile remarking that when the primal constraints are equations, the dual variables are *free* (not sign restricted). In Chapter 2, we saw some linear programs with free variables. We can now put dual linear programs of the sort above in that category.

The correspondence rules relating a primal problem to its dual, and vice versa are shown in Table 5.1. It is straightforward to apply the rules in Table 5.1 to the LP (5.1) to obtain the LP (5.2).

PRIMAL		*DUAL*	
Objective	Minimization	Maximization	Objective
Objective	Coefficients	Constants	RHS
RHS	Constants	Coefficients	Objective
Constraint	$a_{i1}x_1 + \cdots + a_{in}x_n \geq b_i$	$y_i \geq 0$	Variable
Constraint	$a_{i1}x_1 + \cdots + a_{in}x_n \leq b_i$	$y_i \leq 0$	Variable
Constraint	$a_{i1}x_1 + \cdots + a_{in}x_n = b_i$	y_i free	Variable
Variable	$x_j \geq 0$	$y_1a_{1j} + \cdots + y_m a_{mj} \leq c_j$	Constraint
Variable	$x_j \leq 0$	$y_1a_{1j} + \cdots + y_m a_{mj} \geq c_j$	Constraint
Variable	x_j free	$y_1a_{1j} + \cdots + y_m a_{mj} = c_j$	Constraint

Table 5.1: Correspondence rules for primal-dual pairs.

Example 5.1: THE DUAL OF AN LP IN STANDARD FORM. Relative to the (primal) LP

$$
\begin{array}{rllllll}
\text{minimize} & -x_1 & - & 3x_2 \\
\text{subject to} & x_1 & + & x_2 & + & x_3 & & & & = 4 \\
& 2x_1 & - & x_2 & & & + & x_4 & & = 3 \\
& -x_1 & + & 2x_2 & & & & & + x_5 & = 6 \\
& & & & x_j \geq 0, & j = 1, \ldots, 5,
\end{array}
$$

the dual problem is

$$
\begin{array}{rlll}
\text{maximize} & 4y_1 + 3y_2 + 6y_3 \\
\text{subject to} & y_1 + 2y_2 - y_3 & \leq & -1 \\
& y_1 - y_2 + 2y_3 & \leq & -3 \\
& y_1 & \leq & 0 \\
& y_2 & \leq & 0 \\
& y_3 & \leq & 0.
\end{array}
$$

The dual constraints above are presented in the equivalent transposed version mentioned earlier. You should use Table 5.1 to check the usage of the data in these two problems and verify the comments made earlier about their right-hand sides and objective functions.

Another dual pair

As we know, not all linear programming models begin with equality constraints. They may instead be expressed with inequality constraints. Now when the *primal problem* is

$$
\begin{array}{rl}
\text{minimize} & c^T x \\
\text{subject to} & Ax \geq b \\
& x \geq 0,
\end{array}
$$

we have the *dual problem*

$$
\begin{array}{rl}
\text{maximize} & y^T b \\
\text{subject to} & y^T A \leq c^T \\
& y \geq 0.
\end{array}
$$

Let's see why we have this dual. So far we have *defined* dual problems in terms of primal problems in standard form (equality constraints). The primal problem given here can be put into standard form by the insertion of a vector of surplus variables, say $\bar{x}$. Doing so, we obtain

$$\begin{aligned}
\text{minimize} \quad & c^{\mathrm{T}}x + 0^{\mathrm{T}}\bar{x} \\
\text{subject to} \quad & Ax - I\bar{x} = b \\
& x \geq 0, \ \bar{x} \geq 0.
\end{aligned}$$

By our definition, it is now possible to write the dual problem

$$\begin{aligned}
\text{maximize} \quad & y^{\mathrm{T}}b \\
\text{subject to} \quad & y^{\mathrm{T}}[A \ -I] \leq [c^{\mathrm{T}} \ 0^{\mathrm{T}}] \\
& y \text{ free.}
\end{aligned}$$

Because of the dual constraints $y^{\mathrm{T}}(-I) \leq 0^{\mathrm{T}}$, we can declare that the dual requires the y vector to be nonnegative! In other words, the asserted dual problem is just another version of the one derived from the dual of the LP in standard form, and that's why we have this dual problem for the stated primal. The dual can be written down directly by applying the rules specified in Table 5.1.

In linear programming textbooks it is often said that "the dual of the dual is the primal." What does this mean? Let's consider the dual pair just presented. Taking the dual of the dual means first writing the dual as a primal problem having an objective function to be minimized and "$\geq$" constraints. Although it looks a bit silly, the dual problem can be modified to fit this requirement. It becomes

$$\begin{aligned}
\text{minimize} \quad & -b^{\mathrm{T}}y \\
\text{subject to} \quad & -A^{\mathrm{T}}y \geq -c \\
& y \geq 0.
\end{aligned}$$

According to the discussion above, this problem has a dual, namely the following maximization problem

$$\begin{aligned}
\text{maximize} \quad & -c^{\mathrm{T}}x \\
\text{subject to} \quad & -Ax \leq -b \\
& x \geq 0,
\end{aligned}$$

which is readily converted to an equivalent problem of the form we want. For this reason, *the dual of the dual is, indeed, the primal.*

Example 5.2: ALL-INEQUALITY DUAL PAIR. Relative to the (primal) LP in inequality form

$$
\begin{array}{llrcrcrcl}
\text{minimize} & & x_1 & + & x_2 & + & x_3 & & \\
\text{subject to} & & 3x_1 & - & 2x_2 & & & \geq & 3 \\
& - & x_1 & + & 3x_2 & - & x_3 & \geq & 5 \\
& & & - & x_2 & + & 2x_3 & \geq & 7 \\
& & \multicolumn{7}{l}{x_j \geq 0, \ j = 1, 2, 3,}
\end{array}
$$

we have the dual, also in inequality form,

$$
\begin{array}{llrcrcrcl}
\text{maximize} & & 3y_1 & + & 5y_2 & + & 7y_3 & & \\
\text{subject to} & & 3y_1 & - & y_2 & & & \leq & 1 \\
& - & 2y_1 & + & 3y_2 & - & y_3 & \leq & 1 \\
& & & - & y_2 & + & 2y_3 & \leq & 1 \\
& & \multicolumn{7}{l}{y_j \geq 0, \ j = 1, 2, 3.}
\end{array}
$$

An interpretation of dual feasibility

Relative to a primal problem of the form (5.1), the dual constraints $y^{\mathrm{T}}A \leq c^{\mathrm{T}}$ imply that

$$
y^{\mathrm{T}}A_{\bullet j} - c_j \leq 0 \text{ for all } j.
$$

When $y^{\mathrm{T}} = c_B^{\mathrm{T}}B^{-1}$, where B is a basis for the primal (feasible or not), *the dual feasibility condition is just the primal optimality condition*

$$
z_j - c_j \leq 0 \text{ for all } j \in N
$$

used in the Simplex Algorithm. There, you will recall, we have $z_j := y^{\mathrm{T}}A_{\bullet j}$; the definition of y gives us $z_j - c_j = 0$, when $A_{\bullet j}$ is a column of B. Thus, we can interpret the Simplex Algorithm as one that seeks concrete evidence of dual feasibility.

Example 5.3: SOLVING THE PRIMAL (IN EXAMPLE 5.1).

min	z	x_1	x_2	x_3	x_4	x_5	1	basic
row 0	1	1	3	0	0	0	0	z
row 1	0	1	1	1	0	0	4	x_3
row 2	0	2	-1	0	1	0	3	x_4
row 3	0	-1	2	0	0	1	6	x_5

Pivot to exchange x_5 and x_2 (that is, make x_2 basic in place of x_5).

min	z	x_1	x_2	x_3	x_4	x_5	1	basic
row 0	1	$5/2$	0	0	0	$-3/2$	-9	z
row 1	0	$3/2$	0	1	0	$-1/2$	1	x_3
row 2	0	$3/2$	0	0	1	$1/2$	6	x_4
row 3	0	$-1/2$	1	0	0	$1/2$	3	x_2

Pivot to exchange x_3 and x_1 (that is, make x_1 basic in place of x_3).

min	z	x_1	x_2	x_3	x_4	x_5	1	basic
row 0	1	0	0	$-5/3$	0	$-2/3$	$-32/3$	z
row 1	0	1	0	$2/3$	0	$-1/3$	$2/3$	x_1
row 2	0	0	0	-1	1	1	5	x_4
row 3	0	0	1	$1/3$	0	$1/3$	$10/3$	x_2

The optimal solution found is $(x_1, x_2, x_3, x_4, x_5) = (2/3, 10/3, 0, 5, 0)$ and $z = -32/3$.

Example 5.4: FEASIBLE (OPTIMAL) SOLUTION OF THE DUAL. The dual of the problem just solved is

$$\begin{aligned}
\text{maximize} \quad & 4y_1 + 3y_2 + 6y_3 \\
\text{subject to} \quad & y_1 + 2y_2 - y_3 \leq -1 \\
& y_1 - y_2 + 2y_3 \leq -3 \\
& y_1 \qquad\qquad \leq 0 \\
& \quad y_2 \qquad \leq 0 \\
& \qquad\quad y_3 \leq 0.
\end{aligned}$$

Note that $(y_1, y_2, y_3) = (-5/3, 0, -2/3)$ is a *feasible solution* of the dual problem and that the dual objective function has the value

$$(4)(-5/3) + (3)(0) + (6)(-2/3) = -32/3.$$

(These numbers should look familiar.) That this is an *optimal solution* of the dual problem is a consequence of the following theoretical result.

The Weak Duality Theorem

Theorem 5.1 (Weak duality)   *If $\bar{x}$ is a feasible solution for the primal problem*

$$\begin{aligned} \text{minimize} \quad & c^{\mathrm{T}}x \\ \text{subject to} \quad & Ax = b \\ & x \geq 0, \end{aligned} \qquad (5.3)$$

and $\bar{y}$ is a feasible solution for the dual problem

$$\begin{aligned} \text{maximize} \quad & b^{\mathrm{T}}y \\ \text{subject to} \quad & A^{\mathrm{T}}y \leq c \\ & (y \text{ free}), \end{aligned} \qquad (5.4)$$

then

$$\bar{y}^{\mathrm{T}}b \leq c^{\mathrm{T}}\bar{x}.$$

Furthermore, if

$$\bar{y}^{\mathrm{T}}b = c^{\mathrm{T}}\bar{x},$$

then $\bar{x}$ and $\bar{y}$ are optimal solutions of the primal problem and the dual problem, respectively.

The Weak Duality Theorem is very easily proved in a couple of lines. Its relevance is illustrated in Examples 5.3 and 5.4 where we have an optimal solution of the primal and a feasible solution of the dual for which the objective functions take on the same value. This enables us to verify the optimality of the dual feasible solution.

Example 5.5: WEAK DUALITY—OPTIMAL SOLUTION.   Consider the primal-dual pair of Example 5.1. The optimal solution of the primal and a feasible solution of dual are given in Examples 5.3 and 5.4 respectively. Application of the Weak Duality Theorem shows that $(y_1, y_2, y_3) = (-5/3, 0, -2/3)$ is an optimal solution of the dual as asserted on Page 122.

Example 5.6: WEAK DUALITY—FEASIBLE SOLUTION.   Consider the all-inequality dual pair in Example 5.2. It is easy to verify that $x_1 = 5$, $x_2 = 6$, $x_3 = 7$ is a feasible solution for the primal, and $y_1 = 0$, $y_2 = 0$, $y_3 = 0$ is a feasible solution for the dual. The respective objective values are 18 and 0. As expected, these satisfy the Weak Duality Theorem.

The next result is much deeper and, accordingly, much more powerful; its proof takes a bit of work.

The Strong Duality Theorem

Theorem 5.2 (Strong duality) *For a primal-dual pair of linear programs, exactly one of the following statements holds:*

1. *Both problems possess optimal solutions at which their objective function values are equal.*

2. *One problem is feasible but has an unbounded objective function, and the other problem is infeasible.*

3. *Both problems are infeasible.*

For a proof of the Strong Duality Theorem, see, for example, Dantzig and Thapa [45] or Schrijver [176].

One implication of the Strong Duality Theorem is that when one member of a dual pair possesses an optimal solution, then the other must as well, and their optimal values must be the same.

Example 5.7: STRONG DUALITY THEOREM CASES. The Strong Duality Theorem has three parts. These are illustrated in the following cases.

1. *Both primal and dual problems possess an optimal solution.* Examples 5.3 and 5.4 illustrate this.

2. *One problem is infeasible and the other is unbounded.*

   (a) *The primal has an unbounded objective function value and the dual is infeasible.*

$$
\begin{array}{ll}
\text{minimize} & x_1 - 2x_2 \\
\text{subject to} & -x_1 + x_2 = 10 \\
& x_1 \geq 0, x_2 \geq 0
\end{array}
\qquad
\begin{array}{ll}
\text{maximize} & 10y_1 \\
\text{subject to} & -y_1 \leq 1 \\
& y_1 \leq -2
\end{array}
$$

(b) *The primal is infeasible and the dual has an unbounded objective function value.*

$$
\begin{array}{ll}
\text{minimize} & x_1 - 2x_2 \\
\text{subject to} & x_1 + x_2 = -10 \\
& x_1 \geq 0, x_2 \geq 0
\end{array}
\qquad
\begin{array}{ll}
\text{maximize} & -10y_1 \\
\text{subject to} & y_1 \leq 1 \\
& y_1 \leq -2
\end{array}
$$

3. *Both primal and dual problems are infeasible.*

$$
\begin{array}{ll}
\text{minimize} & x_1 - 2x_2 \\
\text{subject to} & -x_1 + x_2 = 10 \\
& -x_1 + x_2 = -10 \\
& x_1 \geq 0, x_2 \geq 0
\end{array}
\qquad
\begin{array}{ll}
\text{maximize} & 10y_1 - 10y_2 \\
\text{subject to} & -y_1 - y_2 \leq 1 \\
& y_1 + y_2 \leq -2
\end{array}
$$

Farkas's Lemma

The Strong Duality Theorem can be proved by invoking a "theorem of the alternative." The most famous of these is due to Julius (Gulya) Farkas (1902). The result—which is usually called Farkas's Lemma—states that of the two linear inequality systems

$$ Ax = b, \quad x \geq 0, \tag{5.5} $$

$$ y^{\mathrm{T}}A \leq 0^{\mathrm{T}}, \quad y^{\mathrm{T}}b > 0, \quad y \text{ free}, \tag{5.6} $$

exactly one has a solution. The first of these systems, (5.5), is just the constraints of a linear program in standard form. It is clear that not both of the systems can possess solutions. Indeed, if $\bar{x}$ is a solution of (5.5) and $\bar{y}$ is a solution of (5.6), then

$$ 0 \geq \bar{y}^{\mathrm{T}}A\bar{x} = \bar{y}^{\mathrm{T}}b > 0 $$

which is impossible. The argument that the systems cannot both lack solutions is much more demanding. See [50] and the references therein.

Farkas's Lemma can be used to establish that exactly one of the two linear inequality systems
$$ Ax \geq b, \quad x \geq 0, $$
$$ y^{\mathrm{T}}A \leq 0^{\mathrm{T}}, \quad y^{\mathrm{T}}b > 0, \quad y \geq 0 $$
has a solution.

It is easy to see that if the primal linear programming problem (5.1) is infeasible, but its dual (5.2) has a feasible solution $\hat{y}$, then the dual objective function must be unbounded above, for in such a case, there exists a solution $\tilde{y}$ of (5.6), and hence for all $\lambda \geq 0$

$$(\hat{y} + \lambda\tilde{y})^{\mathrm{T}}A = \hat{y}^{\mathrm{T}}A + \lambda\tilde{y}^{\mathrm{T}}A \leq b + \lambda\tilde{y}^{\mathrm{T}}A \leq b$$

while $(\hat{y} + \lambda\tilde{y})^{\mathrm{T}}b = \hat{y}^{\mathrm{T}}b + \lambda\tilde{y}^{\mathrm{T}}b$ which goes to $+\infty$ as λ goes to ∞.

Complementary slackness

When the product of two real numbers is zero, at least one of the two factors must be zero. This fact does not, however, carry over to the inner (or dot) product of two n-vectors. For instance the vectors $u = (1, 1)$ and $v = (1, -1)$ have inner product $u^{\mathrm{T}}v = 0$, yet neither factor equals zero. Even so, there are circumstances when something of this sort does hold.

If $u, v \in R^n$ and $u_j v_j = 0$ for all $j = 1, \ldots, n$, we say that u and v have the *complementary slackness property*. A common instance of vectors with the complementary slackness property occurs when $u \geq 0$, $v \geq 0$ and $u^{\mathrm{T}}v = 0$.[1] The reason is easy to see: When the sum of nonnegative numbers is zero, the summands must all be zero. Thus, since

$$u^{\mathrm{T}}v = \sum_{j=1}^{n} u_j v_j = 0 \quad \text{and} \quad u_j v_j \geq 0, \ j = 1, 2, \ldots, n$$

it follows that $u_j v_j = 0$ for all $j = 1, \ldots, n$. This means that when $u + v > 0$ (so that for every j, at least one of u_j and v_j is positive), the conditions $u \geq 0$, $v \geq 0$, and $u^{\mathrm{T}}v = 0$ imply that the *supports*[2]

$$\mathrm{supp}(u) = \{j : u_j > 0\} \quad \text{and} \quad \mathrm{supp}(v) = \{j : v_j > 0\},$$

are *complementary* subsets of $\{1, \ldots, n\}$. In particular, no subscript j belongs to both of these sets. This much is true even when the assumption $u + v > 0$ is dropped. In this case, there may be some indices j

---

[1] Other conditions under which two orthogonal vectors u and $v \in R^n$ would have the complementary slackness property are when $u, v \leq 0$ or $u \geq 0 \geq v$, or $v \geq 0 \geq u$.

[2] The concept of "support" warrants further discussion as it is more general than how it is used here. The *support* of a numerical-valued function f is the set of all arguments x such that $f(x) \neq 0$. Regarding an n-vector u as a function defined on the numbers $\{1, 2, \ldots, n\}$, we say that $\mathrm{supp}(u) = \{j : u(j) = u_j \neq 0\}$.

such that $u_j = 0$ and $v_j = 0$. As an example of this, consider the non-negative vectors $u = (0, 2, 1)$ and $v = (1, 0, 0)$. These two vectors satisfy $u + v > 0$ and $u^T v = 0$. In this case $\text{supp}(u) = \{2, 3\}$ and $\text{supp}(v) = \{1\}$; they are complementary subsets of $\{1, 2, 3, \}$. If we define $w = (0, 2, 0)$, then $\text{supp}(w)$ and $\text{supp}(v)$ are disjoint (nonintersecting) sets, but they are not really complementary subsets of $\{1, 2, 3\}$ because $\text{supp}(w) \cup \text{supp}(v)$ is a proper subset of $\{1, 2, 3\}$.

Primal and dual feasible vectors $\bar{x}$ and $\bar{y}$ are said to satisfy the *complementary slackness condition*, if $(y^T A - c)_j \bar{x}_j = 0$ for all $j = 1, \ldots, n$. The importance of the concept of complementary slackness will be emphasized in the discussion of optimality conditions below.

Optimality conditions for LP

A vector $\bar{x}$ is optimal for the linear program

$$\begin{aligned} \text{minimize} \quad & c^T x \\ \text{subject to} \quad & Ax = b \\ & x \geq 0, \end{aligned}$$

if and only if $\bar{x}$ and some vector $\bar{y}$ satisfy the following three conditions:

$$\begin{aligned} A\bar{x} = b, \ \bar{x} \geq 0 \quad & \text{(primal feasibility)} \\ \bar{y}^T A \leq c^T \quad & \text{(dual feasibility)} \\ (\bar{y}^T A - c^T)\bar{x} = 0 \quad & \text{(complementary slackness).} \end{aligned}$$

The first two of these conditions are pretty clear. For a solution to be optimal, it must be feasible; moreover, the dual must also be feasible. As for the third condition, this is just a different way of writing $\bar{y}^T b = c^T \bar{x}$ when $\bar{x}$ is a feasible solution of the primal problem and $\bar{y}$ is a feasible solution of the dual problem.

To apply the previously discussed principle on nonnegative orthogonal vectors, we introduce the slack vector $v^T = c^T - \bar{y}^T A$ in the dual constraints, and let $u = \bar{x}$. Then, for optimality, for feasible $\bar{x}$ and $\bar{y}$, the vectors u and v are nonnegative and orthogonal. One interesting implication of this optimality criterion is that if $\bar{x}_j > 0$, then the j-th constraint of the dual problem, i.e., $\bar{y}^T A_{.j} \leq c_j$, must hold as an *equation* for any optimal solution of the dual problem. And by the same token, if $\bar{y}$ is an optimal solution of the dual problem for which $\bar{y}^T A_{\cdot j} < c_j$ holds, then $\bar{x}_j = 0$ if $\bar{x}$ is an optimal solution of the primal problem.

Notice how complementary slackness is handled in the Simplex Algorithm. In a BFS, only the basic variables are positive. And for all these basic variables (positive or zero), the coefficients $z_j - c_j$ are zero, which is just another way of saying that the corresponding dual constraint holds with equality.

Example 5.8: COMPLEMENTARY SLACKNESS. From the final tableau in Example 5.4 we note that the optimal solution x and the vector $A^T y - c$ are:

$$\begin{aligned}
x &= (2/3, \ \ 10/3, \quad\ \ 0, \ \ 5, \quad\ \ 0) \\
A^T y - c &= (\ \ 0, \quad\ \ 0, \ -5/3, \ \ 0, \ -2/3).
\end{aligned}$$

It is easy to verify that the complementary slackness conditions hold.

Optimality conditions for another pair of dual linear programs

A vector $\bar{x}$ is optimal for the linear program

$$\text{minimize } c^T x \quad \text{subject to } Ax \geq b, \quad x \geq 0$$

if and only if there exists a vector $\bar{y}$ such that

$$\begin{aligned}
&A\bar{x} \geq b, \ \bar{x} \geq 0 \quad \text{(primal feasibility)} \\
&\bar{y}^T A \leq c^T, \ \bar{y} \geq 0 \quad \text{(dual feasibility)} \\
&\bar{y}^T (A\bar{x} - b) = 0 \quad \text{(complementary slackness)} \\
&\left(\bar{y}^T A - c^T\right) \bar{x} = 0 \quad \text{(complementary slackness)}.
\end{aligned}$$

The reasoning behind these optimality conditions is much the same as for the standard form of the LP. Needless to say, we must take account of the inequality constraints in the primal and the nonnegativity of the dual variables. Under these circumstances, we get two sets of complementary slackness conditions.

Here again, the positivity of variables in one problem has consequences for the constraints of the other:

$$\bar{y}_i > 0 \Longrightarrow (A\bar{x} - b)_i = 0 \quad \text{and} \quad \bar{x}_j > 0 \Longrightarrow \left(\bar{y}^T A - c\right)_j = 0.$$

Likewise, the slackness of constraints in one problem has consequences for the variables in the other:

$$(A\bar{x} - b)_i > 0 \Longrightarrow \bar{y}_i = 0 \quad \text{and} \quad (\bar{y}^{\mathrm{T}}A - c)_j < 0 \Longrightarrow \bar{x}_j = 0.$$

The dual of the diet problem

On page 8 we stated the diet problem of linear programming. It has the form

$$\text{minimize} \quad \sum_{j=1}^{n} c_j x_j$$

$$\text{subject to} \quad \sum_{j=1}^{n} a_{ij} x_j \geq b_i \quad \text{for all } i = 1, \ldots, m$$

$$x_j \geq 0 \quad \text{for all } j = 1, \ldots, n.$$

The dual of the diet problem—or any problem of this form—is

$$\text{maximize} \quad \sum_{i=1}^{m} b_i y_i$$

$$\text{subject to} \quad \sum_{i=1}^{m} a_{ij} y_i \leq c_j \quad \text{for all } j = 1, \ldots, n$$

$$y_i \geq 0 \quad \text{for all } i = 1, \ldots, m.$$

When we think of the diet problem as one of needing to provide various amounts $b_1, \ldots, b_m$ of nutrients using foods $j = 1, \ldots, n$, there arises the question of what the nutrients themselves ought to cost. The corresponding dual problem provides an answer. Here we think of an agent offering to sell the nutrients in the required amounts. The agent needs to charge a fair price, for otherwise the foods would be a better deal. We know that a unit of food j costs c_j. One unit of food j provides the nutrients in the amounts $a_{1j}, \ldots, a_{mj}$. At prices $y_1, \ldots, y_m$, these amounts of nutrition would cost $\sum_{i=1}^{m} y_i a_{ij}$. If this exceeds c_j, then food j is more economical to the person with the (primal) diet problem. So to keep the nutrition prices competitive, we must have $\sum_{i=1}^{m} y_i a_{ij} \leq c_j$. Needless to say, these prices should be nonnegative. Taken together, these two assertions say the prices should satisfy the constraints of the dual problem. The agent proposing to sell the nutrients would want to maximize $\sum_{i=1}^{m} b_i y_i$ subject to these constraints.

5.2   Introduction to the Dual Simplex Algorithm

Sometimes it is natural to solve the dual problem instead of the primal problem. This is the case for an LP of the form

$$\text{minimize } c^{\mathrm{T}}x \quad \text{subject to } Ax \geq b, \; x \geq 0$$

in which $b \geq 0$ and $c \geq 0$. (The diet problem ordinarily has these properties.) For such a problem, the vector $y = 0$ is a feasible solution to the dual problem

$$\text{maximize } b^{\mathrm{T}}y \quad \text{subject to } A^{\mathrm{T}}y \leq c, \; y \geq 0.$$

When the dual problem is put in standard form, the slack vectors form a feasible basis for that problem. By contrast, finding a BFS for the primal problem may require a Phase I Procedure, hence more computational effort.

Another case where we might wish to solve the dual problem is for an LP in standard form (5.1) for which a basis B is known relative to which the system in canonical form has the properties

- $z_j - c_j \leq 0$ (dual feasibility);

- $z_j - c_j = 0$ if x_j is basic (complementary slackness);

- $\min_i \bar{b}_i < 0$ (primal *infeasibility*).

In plain language: the basis B prices out optimal, but is not primal feasible.

Dual problem in primal format

For ease of discussion, we assume that B is the first m columns of A; then the tableau in canonical form looks like

min	z	x_B	x_N	1	basic
row 0	1	0^{T}	$\bar{c}_N^{\mathrm{T}}$	$c_B^{\mathrm{T}}\bar{b}$	z
rows $1,\ldots,m$	0	I	$\bar{A}_{\bullet N}$	$\bar{b}$	x_B

where

$$\bar{A}_{\bullet N} = B^{-1}A_{\bullet N}, \qquad \bar{b} = B^{-1}b, \qquad \bar{c}_N^{\mathrm{T}} = c_B^{\mathrm{T}}B^{-1}A_{\bullet N} - c_N^{\mathrm{T}}$$
$$\uparrow \text{ not nonnegative} \qquad\qquad \uparrow \text{ nonpositive}$$

Steps of the Dual Simplex Algorithm

Algorithm 5.1 (Dual Simplex Algorithm)

1. *Initialization.* Input the data A, b, and c for a minimization problem. Identify a starting dual feasible basis, that is, one with $z_j - c_j \leq 0$ for all $j \in N$. Represent the problem in canonical form.

2. *Test for optimality.* Find an index r such that (relative to the current basis) $\bar{b}_r = \min_i \bar{b}_i$. If $\bar{b}_r \geq 0$, stop. The basic solution is primal feasible and hence optimal.

3. *Maximum ratio test.* If $\bar{a}_{rj} \geq 0$ for all j, stop. The primal problem is infeasible. Otherwise, find an index s such that $\bar{a}_{rs} < 0$ and

$$\frac{\bar{c}_s}{-\bar{a}_{rs}} = \max_{j \in N} \left\{ \frac{\bar{c}_j}{-\bar{a}_{rj}} : \bar{a}_{rj} < 0 \right\}.$$

4. *Pivot step.* Pivot on $\bar{a}_{rs}$. Return to Step 2.

Example 5.9: APPLICATION OF THE DUAL SIMPLEX ALGORITHM. The following tableau gives a basic solution that prices out optimal, $(z_j - c_j \leq 0)$, but lacks primal feasibility.

min	z	x_1	x_2	x_3	x_4	x_5	1	basic
row 0	1	0	-4	0	-9	0	0	z
row 1	0	0	3	1	-1	0	-6	x_3
row 2	0	0	-4	0	-2	1	-12	x_5
row 3	0	1	-2	0	-4	0	-2	x_1

We note that pivoting on a *negative* element in a row having a negative right-hand side element will produce a positive right-hand side element in that row. The pivot element cannot be chosen arbitrarily, however, because we want to *preserve the nonpositivity of the $z_j - c_j$*. Finding a suitable pivot element is accomplished with a ratio test as defined later.

Pivot to exchange x_5 and x_2 (that is, make x_2 basic in place of x_5).

min	z	x_1	x_2	x_3	x_4	x_5	1	basic
row 0	1	0	0	0	-7	-1	12	z
row 1	0	0	0	1	$-5/2$	$3/4$	-15	x_3
row 2	0	0	1	0	$1/2$	$-1/4$	3	x_2
row 3	0	1	0	0	-3	$-1/2$	4	x_1

Pivot to exchange x_3 and x_4 (that is, make x_4 basic in place of x_3).

min	z	x_1	x_2	x_3	x_4	x_5	1	basic
row 0	1	0	0	$-14/5$	0	$-31/10$	54	z
row 1	0	0	0	$-2/5$	1	$-3/10$	6	x_4
row 2	0	0	1	$1/5$	0	$-1/10$	0	x_2
row 3	0	1	0	$-6/5$	0	$-7/2$	22	x_1

The solution found is $(x_1, x_2, x_3, x_4, x_5) = (22, 0, 0, 6, 0)$ and $z = 54$.

Example 5.10: INFEASIBLE PRIMAL AND DUAL SIMPLEX ALGORITHM.

Here we take the problem presented in Example 5.9 and change just one number in the matrix A, namely a_{14}, from -1 to $+1$. The new problem will turn out to be primal infeasible. This illustrates how very important it is to get the input data right!

min	z	x_1	x_2	x_3	x_4	x_5	1	basic
row 0	1	0	-4	0	-9	0	0	z
row 1	0	0	3	1	1	0	-6	x_3
row 2	0	0	-4	0	-2	1	-12	x_5
row 3	0	1	-2	0	-4	0	-2	x_1

Pivot to exchange x_5 and x_2 (that is, make x_2 basic in place of x_5).

min	z	x_1	x_2	x_3	x_4	x_5	1	basic
row 0	1	0	0	0	-7	-1	12	z
row 1	0	0	0	1	$-1/2$	$3/4$	-15	x_3
row 2	0	0	1	0	$1/2$	$-1/4$	3	x_2
row 3	0	1	0	0	-3	$-1/2$	4	x_1

Pivot to exchange x_3 and x_4 (that is, make x_4 basic in place of x_3).

min	z	x_1	x_2	x_3	x_4	x_5	1	basic
row 0	1	0	0	-14	0	$-23/2$	222	z
row 1	0	0	0	-2	1	$-3/2$	30	x_4
row 2	0	0	1	1	0	$1/2$	-12	x_2
row 3	0	1	0	-6	0	-5	94	x_1

From this we see that the primal problem is infeasible since the right-hand side entry $\bar{b}_2 = -12$ whereas $\bar{A}_{2\bullet} = [0\ 0\ 1\ 1\ 0\ 1/2]$. The variables $x_1, \ldots, x_5$ are all required to be nonnegative, hence there is no way for the second equation to be satisfied.

In Examples 5.9 and 5.10 one can see that the value of the objective function, z, is *increasing* despite the fact that our primal problem calls for the minimization of this variable. At first this may seem strange, but in fact it is quite reasonable: we are solving the dual problem in primal format, and the dual is a maximization problem. Under the assumption that the dual objective function is not unbounded, the bases we encounter do not yield a primal feasible basic solution until the end of the run.

Finding optimal dual variables in a tableau

When the tableau for the Dual Simplex Algorithm reveals that an optimal solution for the primal problem has been found, the objective function row contains an optimal solution for the dual problem, just as in the case of the Primal Simplex Algorithm. This assumes that the primal problem is the one stated in the initial tableau.

Example 5.11: OPTIMAL DUAL VARIABLES. The linear program of Example 3.8

$$
\begin{aligned}
\text{minimize} \quad & -4x_1 - 6x_2 - 3x_3 - x_4 \\
\text{subject to} \quad & x_1 + 4x_2 + 8x_3 + 6x_4 \leq 11 \\
& 4x_1 + x_2 + 2x_3 + x_4 \leq 7 \\
& 2x_1 + 3x_2 + x_3 + 2x_4 \leq 2 \\
& x_1,\ x_2,\ x_3,\ x_4 \geq 0.
\end{aligned}
$$

has the final optimal shown below (see Example 3.9):

min	z	x_1	x_2	x_3	x_4	x_5	x_6	x_7	1	basic
row 0	1	0	$-1/3$	0	$-11/3$	$-2/15$	0	$-29/15$	$-16/3$	z
row 1	0	0	$1/3$	1	$2/3$	$2/15$	0	$-1/15$	$4/3$	x_3
row 2	0	0	-5	0	-3	0	1	-2	3	x_6
row 3	0	1	$4/3$	0	$2/3$	$-1/15$	0	$8/15$	$1/3$	x_1

The optimal primal solution found is

$$(x_1, x_2, x_3, x_4, x_5, x_6, x_7) = (1/3, 0, 4/3, 0, 0, 3, 0) \quad \text{and} \quad z = -16/3.$$

From the tableau we find the optimal dual variables are

$$(y_1, y_2, y_3) = (-2/15, 0, -29/15).$$

The optimal basis inverse is in the columns corresponding to x_5, x_6, x_7:

$$B^{-1} = \begin{bmatrix} 2/15 & 0 & -1/15 \\ 0 & 1 & -2 \\ -1/15 & 0 & 8/15 \end{bmatrix}.$$

Suppose an error was discovered, and the correct value of b_3 was initially meant to be 1 instead of 2. A change in the right-hand side does not affect the dual but does affect the primal. Hence we perform a quick check to see if the primal is still feasible, in which case, the solution values will change, but the basis will still price out optimal. We compute

$$B^{-1}b = \begin{bmatrix} 2/15 & 0 & -1/15 \\ 0 & 1 & -2 \\ -1/15 & 0 & 8/15 \end{bmatrix} \begin{bmatrix} 11 \\ 7 \\ 1 \end{bmatrix} = \begin{bmatrix} 7/5 \\ 5 \\ -1/5 \end{bmatrix},$$

hence the primal solution is infeasible. The corresponding objective function value is computed as $c_B^T x_B = -17/5$. Upon replacing the last column of the final tableau above, we obtain

min	z	x_1	x_2	x_3	x_4	x_5	x_6	x_7	1	basic
row 0	1	0	$-1/3$	0	$-11/3$	$-2/15$	0	$-29/15$	$-17/5$	z
row 1	0	0	$1/3$	1	$2/3$	$2/15$	0	$-1/15$	$7/5$	x_3
row 2	0	0	-5	0	-3	0	1	-2	5	x_6
row 3	0	1	$4/3$	0	$2/3$	$-1/15$	0	$8/15$	$-1/5$	x_1

Since the tableau is primal infeasible but dual feasible, we can apply the Dual Simplex Algorithm. Hence, we pivot to exchange x_1 and x_5 (that is, make x_5 basic in place of x_1).

min		z	x_1	x_2	x_3	x_4	x_5	x_6	x_7	1	basic
row 0		1	-2	-3	0	-5	0	0	-3	-3	z
row 1		0	2	3	1	2	0	0	3	1	x_3
row 2		0	0	-5	0	-3	0	1	-2	5	x_6
row 3		0	-15	-20	0	-10	1	0	-8	3	x_5

The solution $(1, 5, 3)$ is optimal. The new dual variables are $(0, 0, -3)$.

5.3   Exercises

5.1   State the dual of the simple diet problem given in Example 1.2.

5.2   Write the dual of the transportation problem given in Exercise 1.1 in Chapter 1.

5.3   What is the dual of the linear program given in Example 3.8?

5.4   Find the dual of the following linear program

$$
\begin{array}{ll}
\text{minimize} & c_1 x_1 + c_2 x_2 + c_3 x_3 \\
\text{subject to} & a_{11} x_1 + a_{12} x_2 + a_{13} x_3 \geq b_1 \\
& a_{21} x_1 + a_{22} x_2 + a_{23} x_3 = b_2 \\
& a_{31} x_1 + a_{32} x_2 + a_{33} x_3 \leq b_3 \\
& x_1 \geq 0, x_2 \leq 0, x_3 \text{ free.}
\end{array}
$$

Find the dual of the dual and thereby show that the dual of the dual is the primal.

5.5   Consider the linear program

$$
\text{minimize} \quad \sum_{j=1}^{k} c_j x_j + \sum_{j=k+1}^{n} c_j x_j
$$

$$
\text{subject to} \quad \sum_{j=1}^{k} a_{ij} x_j + \sum_{j=k+1}^{n} a_{ij} x_j = b_i, \quad i = 1, \ldots, \ell
$$

$$
\sum_{j=1}^{k} a_{ij} x_j + \sum_{j=k+1}^{n} a_{ij} x_j \geq b_i, \quad i = \ell + 1, \ldots, m
$$

$$
x_j \geq 0, \ j = 1, \ldots, k; \quad x_j \text{ free}, \ j = k + 1, \ldots, n.
$$

In this chapter we have examined two dual pairs of linear optimization problems. In one case the primal was in standard form. In the other the primal had linear inequality constraints and nonnegative variables. Using the principle that dual variables corresponding to equality constraints are free and those corresponding to $\geq$ constraints in a minimization problem are nonnegative, find the dual for the problem above.

5.6   Consider the linear program

$$\begin{array}{ll} \text{minimize} & c^{\mathrm{T}}x \\ \text{subject to} & Ax \geq b \\ & x \geq 0 \end{array}$$

in which

$$A = \begin{bmatrix} 3 & -1 & 2 \\ 4 & 5 & 0 \\ -6 & 9 & 5 \\ 1 & 1 & 1 \end{bmatrix}, \qquad b = \begin{bmatrix} 15 \\ 8 \\ 13 \\ 4 \end{bmatrix}, \qquad c = \begin{bmatrix} 2 \\ 3 \\ 5 \end{bmatrix}.$$

Use the optimality criteria (see page 127) to decide whether or not $\bar{x} = (2, 0, 5)$ is an optimal solution of the LP. [Hint: Assume that $\bar{x}$ is an optimal solution. Use the necessary conditions of optimality to determine what the vector $\bar{y}$ would have to be. Check to make sure that all the conditions of optimality are met.]

5.7   Sometimes it is convenient to take an LP in standard form and partition its equations into two sets of constraints:

$$\begin{array}{lll} \text{minimize} & c^{\mathrm{T}}x & \\ \text{subject to} & A_1 x = b^1 & (m_1 \text{ equations}) \\ & A_2 x = b^2 & (m_2 \text{ equations}) \\ & x \geq 0. & \end{array}$$

There is motivation to do so when m_1 is small and the constraints $A_2 x = b^2$ alone (i.e., without the "complicating constraints" $A_1 x = b^1$) lead to easily solved linear programs—even when m_2 is large. Now assume that a vector $y^1 \in R^{m_1}$ has been found such that the vector $\hat{x}$ solves the LP

$$\begin{array}{ll} \text{minimize} & (c^{\mathrm{T}} - (y^1)^{\mathrm{T}}A_1)x \\ \text{subject to} & A_2 x = b^2 \\ & x \geq 0. \end{array}$$

Assume furthermore that $\hat{x}$ satisfies the equations $A_1 x = b^1$. Would $\hat{x}$ always be an optimal solution of the entire problem (with $m_1 + m_2$ equations)? Justify your answer.

5.8   Consider the linear program

$$\begin{array}{llrcl}
\text{minimize} & 37x_1 + 29x_2 + 33x_3 & & & \\
\text{subject to} & x_2 + & x_3 & \geq & 1 \\
& x_1 + & & \geq & 1 \\
& x_1 + x_2 & & \geq & 1 \\
& x_1 + x_2 + & x_3 & \geq & 1 \\
& x_1 + & x_3 & \geq & 1 \\
& x_j \geq 0, \quad j = 1, 2, 3. & & &
\end{array}$$

(a) Use the Dual Simplex Algorithm to solve this LP.
(b) Write down the optimal primal solution and the optimal dual solution.
(c) Verify that the complementary slackness conditions hold.

5.9   Solve the following LP by the Dual Simplex Algorithm. Explain why it is advisable to solve such LPs by the Dual Simplex Algorithm.

$$\begin{array}{llrcl}
\text{minimize} & x_1 + 2x_2 & & & \\
\text{subject to} & 2x_1 + 4x_2 & \geq & 4 \\
& 4x_1 - x_2 & \geq & 2 \\
& -2x_1 + 5x_2 & \geq & 5 \\
& 5x_1 + 1x_2 & \geq & 10 \\
& 3x_1 + 6x_2 & \geq & 9 \\
& x_j \geq 0, \quad j = 1, 2. & &
\end{array}$$

Write down the optimal primal solution and the optimal dual solution.

5.10   In Chapter 1 we briefly discussed the transportation problem (with equality constraints). Here we state the problem in a slightly different way: with inequalities rather than equations. We regard the constants s_i $i = 1, \ldots, m$ as upper bounds on *supplies* of some single commodity available at locations $i = 1, \ldots, m$. On the other hand, the constants d_j $j = 1, \ldots, n$ represent lower bounds on *demands* for the commodity at locations $j = 1, \ldots, n$. The constraints c_{ij} stand for the cost to ship one unit of the commodity from source i to destination j. The form of the problem is now taken to be

$$\text{minimize} \quad \sum_{i=1}^{m} \sum_{j=1}^{n} c_{ij} x_{ij}$$

$$\text{subject to} \quad \sum_{j=1}^{n} x_{ij} \leq s_i \quad \text{for all } i = 1, \ldots, m$$

$$\sum_{i=1}^{m} x_{ij} \geq d_j \quad \text{for all } j = 1, \ldots, n$$

$$x_{ij} \geq 0 \quad \text{for all } i, \text{ for all } j.$$

(a) What condition on the s_i and the d_j must be satisfied if this LP is to have a feasible solution?

(b) Write down the dual of transportation problem stated above.

(c) Give an economic interpretation of the dual feasibility conditions.

(d) Write down the optimality conditions for this version of the transportation problem.

(e) Explain why the dual problem does or does not have a unique solution.

5.11   Gordan's Theorem dates from 1873, almost 30 years before the appearance of Farkas's Lemma. It says that for every $A \in R^{m \times n}$ exactly one of the following systems has a solution:

$$Ax = 0, \quad 0 \neq x \geq 0; \tag{5.7}$$
$$y^T A > 0. \tag{5.8}$$

The first of these systems is equivalent to the system

$$\begin{bmatrix} A \\ e^T \end{bmatrix} x = \begin{bmatrix} 0 \\ 1 \end{bmatrix}, \quad x \geq 0. \tag{5.9}$$

Use this last system and Farkas's Lemma to establish Gordan's Theorem.

5.12   Apply Farkas's Lemma to prove the theorem of the alternative presented in Chapter 2 . (Hint: Recall that $x = x' - x''$ where $x' \geq 0$ and $x'' \geq 0$.)

6. POSTOPTIMALITY ANALYSES

Overview

By *postoptimality analyses* we mean various types of analyses that can be performed after an optimal basic feasible solution for an LP has been found. In practice it is usually not enough just to find an optimal solution. Often questions are asked relating to the impact of potential data changes on the optimal solution.

- If some of the data is inexact, then how do small perturbations in the data affect the optimal solution?

- What if some data has changed? For example, suppose CT Gasoline (see Chapter 1) has determined how much gasoline to sell at one of their terminals. However, since the LP run, the spot price of oil has decreased. Will the quantity of gasoline to be sold at that location change?

- What if new activities need to be considered? For example, imagine that a new gas station has been added to CT Gasoline's distribution system. How will this affect the optimal solution to fill station demand and buy or sell gasoline?

- What if a new constraint is introduced? For example, an accident may close a section of the pipeline, making it impossible to supply some of the terminals. How would this affect the solution?

- What if the optimal solution says to use a large number of activities, but practically we would like to reduce that number without resorting to integer programming? In this case we may be willing to accept a solution that is slightly worse than the optimal but uses fewer activities. One example is when the LP model says to use Truck A to deliver unleaded premium to Station I and use Truck B to deliver to unleaded regular to the same Station I. For practical truck scheduling reasons, CT Gasoline would like to use the same truck to deliver both grades of gasoline to Station I. In this case, a modification to the model will yield the desired result. In other models, this may sometimes be achieved by examining solutions in the neighborhood of an optimal solution (including alternative optima).

© Springer Science+Business Media LLC 2017
R.W. Cottle and M.N. Thapa, *Linear and Nonlinear Optimization*,
International Series in Operations Research & Management Science 253,
DOI 10.1007/978-1-4939-7055-1_6

With modern computing power, many LP models can be re-run quickly in a matter of seconds or less; and, for instance, adding a new column or row and checking its impact can be done very quickly. However, very large problems may still require significant computing time for such changes, so there may still be some computational benefit to solving the modified problem efficiently. In many situations, a full set of sensitivity analysis runs can provide further insight into the behavior of the model.

In this chapter, we organize our discussion of postoptimality analyses according to the following purposes and techniques:

1. *Changing Model Dimensions.* One such type of analysis studies the impact of the *insertion* of an additional column (or activity) in the model. The introduction of columns comes up in the Dantzig-Wolfe Decomposition Algorithm (see Section 7.4) and, more generally, in column-generation techniques. Another is the impact of the inclusion of an additional inequality constraint in the model. As we will see, this is a situation in which the Dual Simplex Algorithm can very naturally be applied. The addition of constraints plays an important role in integer programming where new inequalities are generated to cut the feasible region so that an integer solution can be found. The Benders Decomposition Algorithm which, for instance, is used in stochastic linear programming, also makes use of cuts.

   Other changes in model dimensions can come about from the *deletion* of a column or the deletion of a constraint. The first of these changes is of interest only when the column is a part of the current optimal basis[1]. On the other hand the deletion of a constraint has the potential to enlarge the feasible region in which case the objective function value could improve (and, in some cases, improve without bound).

2. *Ranging (Sensitivity Analysis).* Within postoptimality analyses we include *sensitivity analysis* which pertains to the study of how changes in problem data affect the optimal solution of a problem. In the LP with data A, b, and c there could be changes in any or all of these. Except for the insertion of a new inequality constraint, we omit a discussion of the analysis of changes in the matrix A; this is a more difficult subject than that of changes in the right-hand side b or the cost vector c. Furthermore, we consider only changes in one data item

---

[1]Saying that B is an optimal basis means that the corresponding BFS is optimal and that the basis prices out optimal, which is to say $z_j - c_j \leq 0$ for all $j \in N$.

at a time and not simultaneous data changes in several data items. For instance, we are typically interested in the range of values of a right-hand side value b_i that leave the current basis optimal. The process of changing the right-hand side corresponds to a parallel translation of the set of points satisfying the i-th constraint.

3. *Parametric Programming.* Postoptimality analyses can also include what is called *parametric linear programming.* In contrast to standard sensitivity analysis, the goal in parametric LP is to study the optimal solution of an LP where each component of either the right-hand side vector or the cost vector (or both) changes at its own constant rate.

4. *Interpreting Shadow Prices.* We have seen that when the Simplex Algorithm solves a (primal) linear program, an optimal solution to the dual problem shows up as a by-product of the computation. (We have indicated how to find this vector in the final simplex tableau.) As we know, the dual variables are in one-to-one correspondence with the primal constraints. Economists and practitioners of linear programming often refer to the optimal dual variables as *shadow prices.* This term has to do with the marginal value of each right-hand side element of the system of constraints.

6.1 Changing model dimensions

Most of this section is concerned with modifications to a linear program in standard form, that is,

$$\text{minimize } c^{\text{T}}x \quad \text{subject to } Ax = b, \ x \geq 0,$$

and information that can be obtained by using the inverse of an optimal basis B once it has been found.

Inserting a new activity

Sometimes it happens that one wants to consider a new activity after having solved a given linear program. This amounts to adjoining an extra column to the original model. Suppose the original matrix A was of order $m \times n$. We may denote the level of the new activity by the decision variable x_{n+1}

and its associated $(m+1)$-vector of coefficients by

$$(c_{n+1}, A_{\bullet n+1}) = (c_{n+1}, a_{1,n+1}, \ldots, a_{m,n+1}).$$

The question, of course, is whether the new variable x_{n+1} will turn out to be positive in an optimal solution to the enlarged problem.

From the solution process leading to an optimal solution of the original problem, we have the inverse of an optimal basis and the corresponding price vector

$$\bar{y}^{\mathrm{T}} = c_B^{\mathrm{T}} B^{-1}.$$

The question of interest is answered by pricing out the new column, that is, examining the sign of

$$z_{n+1} - c_{n+1} = \bar{y}^{\mathrm{T}} A_{\bullet n+1} - c_{n+1}.$$

Suppose $z_{n+1} - c_{n+1} > 0$; then increasing x_{n+1} will decrease the objective value. The amount that x_{n+1} can be increased is found by applying the minimum ratio test. The critical value is

$$\min_i \left\{ \frac{\bar{b}_i}{\bar{a}_{i,n+1}} : \bar{a}_{i,n+1} > 0 \right\}.$$

If $\bar{b} > 0$ (the nondegenerate case), then x_{n+1} will attain a positive value and the objective will decrease. However, under degeneracy, it is possible that x_{n+1} becomes basic at value 0 in which case there is no change to the objective value.

Example 6.1: INSERTING A NEW ACTIVITY AFTER OPTIMIZATION.
When the linear program

$$
\begin{aligned}
\text{maximize} \quad & 2x_1 + 5x_2 + x_3 + 3x_4 \\
\text{subject to} \quad & 4x_1 \qquad\quad + 5x_3 + 2x_4 \le 6 \\
& x_2 + 2x_3 + 5x_4 \le 12 \\
& 3x_1 + 4x_2 + 4x_3 \qquad\quad \le 10 \\
& x_j \ge 0, \quad j = 1, 2, 3, 4
\end{aligned}
$$

is put into standard form as a minimization problem, we obtain the initial tableau

min	z	x_1	x_2	x_3	x_4	s_1	s_2	s_3	1	basic
row 0	1	2	5	1	3	0	0	0	0	z
row 1	0	4	0	5	2	1	0	0	6	s_1
row 2	0	0	1	2	5	0	1	0	12	s_2
row 3	0	3	4	4	0	0	0	1	10	s_3

Starting the Simplex Algorithm from the slack basis, we finally obtain the tableau

min	z	x_1	x_2	x_3	x_4	s_1	s_2	s_3	1	basic
row 0	1	$-13/10$	0	$-23/5$	0	0	$-3/5$	$-11/10$	$-91/5$	z
row 1	0	$43/10$	0	$23/5$	0	1	$-2/5$	$1/10$	$11/5$	s_1
row 2	0	$-3/10$	0	$1/5$	1	0	$1/5$	$-1/20$	$19/10$	x_4
row 3	0	$3/4$	1	1	0	0	0	$1/4$	$5/2$	x_2

From this tableau, we find that the basic variables of the optimal solution are s_1, x_4, and x_2, the inverse of the optimal basis is

$$B^{-1} = \begin{bmatrix} 1 & -2/5 & 1/10 \\ 0 & 1/5 & -1/20 \\ 0 & 0 & 1/4 \end{bmatrix},$$

and the corresponding price vector is

$$\bar{y} = (0,\ 3/5,\ 11/10).$$

Now let us sayt that—for some reason—we wish to introduce a new activity x_5 into the original model. Suppose the corresponding column is

$$(c_5,\ A_{\bullet 5}) = (10,\ 4,\ 7,\ 6).$$

Without re-optimizing, we find that

$$z_5 - c_5 = \bar{y}^T A_{\bullet 5} - c_5 = \left(0 \cdot 4 + \frac{3}{5} \cdot 7 + \frac{11}{10} \cdot 6 \right) - 10 = \frac{4}{5} > 0.$$

This implies that the basis B is still optimal after the new column is introduced into the model. Thus, even though the contribution to the objective function value per unit of activity 5 is relatively large as compared with the contributions of the other activities, it is not large enough to make a difference in the optimal objective value. For that to happen, the coefficient c_5 would have to be larger than $\bar{y}^T A_{\bullet 5} = 108/10 = 10.8$.

Inserting a new inequality (cut)

Because the solution set of a linear inequality is a halfspace—and hence divides n-space into two parts—an inequality is sometimes called a *cut*, especially when it is adjoined to another set of constraints.

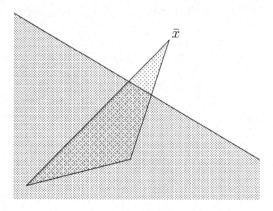

Figure 6.1: A cut slicing off part of a polyhedron.

The Dual Simplex Algorithm is frequently applied within other algorithms when a linear program has been solved and then a new inequality constraint is adjoined to the system *after* the optimization has been carried out. Let's see how this works. Suppose we have just solved the problem

$$\begin{aligned} \text{minimize} \quad & c^{\mathrm{T}}x \\ \text{subject to} \quad & Ax = b \\ & x \geq 0 \end{aligned}$$

and found an optimal basis, B. For ease of discussion, we assume that $B = [A_{\cdot 1} \ A_{\cdot 2} \ \cdots A_{\cdot m}]$ and we have the problem written in canonical form with respect to this basis as expressed in the tableau

min	z	x_B	x_N	1	basic
row 0	1	0^{T}	$\bar{c}_N^{\mathrm{T}}$	$c_B^{\mathrm{T}}\bar{b}$	z
rows $1,\ldots,m$	0	I	$\bar{A}_{\bullet N}$	$\bar{b}$	x_B

Now suppose we adjoin a linear inequality of the form

$$a_B^{\mathrm{T}}x_B + a_N^{\mathrm{T}}x_N \leq d.$$

Let x_{n+1} denote the nonnegative slack variable for this inequality. In tabular form, the enlarged system looks like

min	z	x_B	x_N	x_{n+1}	1	basic
row 0	1	0^{T}	$\bar{c}_N^{\mathrm{T}}$	0	$c_B^{\mathrm{T}}\bar{b}$	z
rows $1,\ldots,m$	0	I	$\bar{A}_{\bullet N}$	0	$\bar{b}$	x_B
row $m+1$	0	a_B^{T}	a_N^{T}	1	d	x_{n+1}

It is easy to see that, with respect to the basis

$$\begin{bmatrix} I & 0 \\ a_B^T & 1 \end{bmatrix},$$

the canonical form of the tableau is

min	z	x_B	x_N	x_{n+1}	1	basic
row 0	1	0^T	$\bar{c}_N^T$	0	$c_B^T \bar{b}$	z
rows $1,\ldots,m$	0	I	$\bar{A}_{\bullet N}$	0	$\bar{b}$	x_B
row $m+1$	0	0	$a_N^T - a_B^T \bar{A}_{\bullet N}$	1	$d - a_B^T \bar{b}$	x_{n+1}

By virtue of the previously performed optimization, $\bar{c}_N \leq 0$ and $\bar{b} \geq 0$. The question now is whether $d - a_B^T \bar{b}$ is nonnegative. If so, the corresponding BFS of the enlarged problem is optimal. If not, we are in a position to apply the Dual Simplex Algorithm.

We have assumed the adjoined inequality constraint is of the $\leq$ type. However, adding one of the $\geq$ type requires exactly the same treatment. In this case we multiply through by -1 to reverse the sense of the inequality to $\leq$.

As an aside, let B be the basis before adding the inequality constraint. Upon adding an inequality constraint of the form above, the new basis becomes

$$\widehat{B} = \begin{bmatrix} B & 0 \\ a_B^T & 1 \end{bmatrix}.$$

The basis $\widehat{B}$ is clearly nonsingular but not necessarily a feasible basis. It is easy to verify that its inverse is

$$\widehat{B}^{-1} = \begin{bmatrix} B^{-1} & 0 \\ -a_B^T B^{-1} & 1 \end{bmatrix}.$$

Example 6.2: ADDING A CUT AFTER SOLVING A PRIMAL PROBLEM.
Consider the tiny linear program

$$
\begin{array}{rrcrcl}
\text{minimize} & -x_1 & - & x_2 \\
\text{subject to} & -x_1 & + & 2x_2 & \leq & 8 \\
& 2x_1 & + & x_2 & \leq & 23/2 \\
& & & x_j & \geq & 0, \ j = 1, 2.
\end{array}
$$

In tableau form, this can be written as

min	z	x_1	x_2	x_3	x_4	1	basic
row 0	1	1	1	0	0	0	z
row 1	0	-1	2	1	0	8	x_3
row 2	0	2	1	0	1	$23/2$	x_4

Pivot to exchange x_3 and x_2 (that is, make x_2 basic in place of x_3).

min	z	x_1	x_2	x_3	x_4	1	basic
row 0	1	$3/2$	0	$-1/2$	0	-4	z
row 1	0	$-1/2$	1	$1/2$	0	4	x_2
row 2	0	$5/2$	0	$-1/2$	1	$15/2$	x_4

Pivot to exchange x_4 and x_1 (that is, make x_1 basic in place of x_4).

min	z	x_1	x_2	x_3	x_4	1	basic
row 0	1	0	0	$-1/5$	$-3/5$	$-17/2$	z
row 1	0	0	1	$2/5$	$1/5$	$11/2$	x_2
row 2	0	1	0	$-1/5$	$2/5$	3	x_1

The solution found is optimal. Now suppose we adjoin the constraint $x_2 \leq 5$, which after adding the slack x_5 becomes $x_2 + x_5 = 5$. We have

$$a_B^{\mathrm{T}} = [1 \ 0], \quad a_N^{\mathrm{T}} = [0 \ 0], \quad d = 5.$$

The vector a_B^{T} is $[1 \ 0]$ and not $[0 \ 1]$, because the basic variables are in the order x_2, x_1. We add the inequality to the tableau after performing the transformation

$$a_N^{\mathrm{T}} - a_B^{\mathrm{T}} \bar{A}_{\bullet N} = [0 \ 0] - [1 \ 0] \begin{bmatrix} 2/5 & 1/5 \\ -1/5 & 2/5 \end{bmatrix} = [-2/5 \ 1/5]$$

$$d - a_B^{\mathrm{T}} \bar{b} = 5 - [1 \ 0] \begin{bmatrix} 11/2 \\ 3 \end{bmatrix} = -1/2.$$

This results in the tableau

min	z	x_1	x_2	x_3	x_4	x_5	1	basic
row 0	1	0	0	$-1/5$	$-3/5$	0	$-17/2$	z
row 1	0	0	1	$2/5$	$1/5$	0	$11/2$	x_2
row 2	0	1	0	$-1/5$	$2/5$	0	3	x_1
row 3	0	0	0	$-2/5$	$-1/5$	1	$-1/2$	x_5

We note that the primal is infeasible but the dual is feasible. Hence we apply the Dual Simplex Algorithm. Pivot to exchange x_5 and x_3 (that is, make x_3 basic in place of x_5).

min	z	x_1	x_2	x_3	x_4	x_5	1	basic
row 0	1	0	0	0	$-1/2$	$-1/2$	$-33/4$	z
row 1	0	0	1	0	0	1	5	x_2
row 2	0	1	0	0	$1/2$	$-1/2$	$13/4$	x_1
row 3	0	0	0	1	$1/2$	$-5/2$	$5/4$	x_3

After one iteration of the Dual Simplex Algorithm, the solution is primal feasible and optimal.

Inserting an equality constraint has similarities to the above technique, but now there is no slack variable with zero cost in the objective that we can use. Instead we will need to add an artificial variable. While it may still be possible to get a dual feasible solution, in general that would be difficult. Thus, in this case we would construct an objective with one artificial variable and attempt to drive that variable to zero using the Primal Simplex Algorithm.

Suppose we adjoin a linear equation of the form

$$a_B^{\mathrm{T}} x_B + a_N^{\mathrm{T}} x_N = d.$$

Let x_{n+1} denote a nonnegative artificial variable for this equation; then

$$a_B^{\mathrm{T}} x_B + a_N^{\mathrm{T}} x_N + x_{n+1} = d.$$

Next, we introduce the Phase I objective $w = x_{n+1}$. In tabular form, the enlarged system with the Phase I objective is

min	w	z	x_B	x_N	x_{n+1}	1	basic
row 00	1	0	0^{T}	0^{T}	-1	0	w
row 0	0	1	0^{T}	$\bar{c}_N^{\mathrm{T}}$	0	$c_B^{\mathrm{T}} \bar{b}$	z
rows $1, \ldots, m$	0	0	I	$\bar{A}_{\bullet N}$	0	$\bar{b}$	x_B
row $m+1$	0	0	a_B^{T}	a_N^{T}	1	d	x_{n+1}

As before, it is easy to see that with respect to the basis

$$\begin{bmatrix} I & 0 \\ a_B^{\mathrm{T}} & 1 \end{bmatrix},$$

the canonical form of the tableau is

min	w	z	x_B	x_N	x_{n+1}	1	basic
row 00	1	0	0^{T}	$a_N^{\mathrm{T}} - a_B^{\mathrm{T}} \bar{A}_{\bullet N}$	0	$d - a_B^{\mathrm{T}} \bar{b}$	w
row 0	0	1	0^{T}	$\bar{c}_N^{\mathrm{T}}$	0	$c_B^{\mathrm{T}} \bar{b}$	z
rows $1, \ldots, m$	0	0	I	$\bar{A}_{\bullet N}$	0	$\bar{b}$	x_B
row $m+1$	0	0	0	$a_N^{\mathrm{T}} - a_B^{\mathrm{T}} \bar{A}_{\bullet N}$	1	$d - a_B^{\mathrm{T}} \bar{b}$	x_{n+1}

We then apply the Phase I Simplex Algorithm to try to eliminate the artificial variable. If this succeeds, the problem is feasible, and we proceed with Phase II.

Deleting an activity

Deleting an activity from the model after an optimal solution has been found presents two cases; these correspond to whether or not the activity to be deleted is basic. If the column of the deleted activity is nonbasic, there is no change in the optimal value because the value of the corresponding nonbasic variable is zero. It is only when the activity to be deleted corresponds to a basic variable that issues come up. The first of these is simply one of feasibility, for it may happen that the problem becomes infeasible without the deleted column. Another possibility could be that the variable corresponding to the activity is basic but has value zero (as in a degenerate basic feasible solution).

It should be noted that the elimination of an activity can be viewed as the equivalent of forcing the corresponding variable to have the value zero. Since in standard form, all the decision variables are nonnegative, we can force the variable x_k to be zero by adding the cut $x_k \leq 0$.

Example 6.3: POSTOPTIMAL DELETION OF AN ACTIVITY. The following tableau represents the Phase I Problem corresponding to an LP in which objective value z is to be minimized.

min	w	z	x_1	x_2	x_3	x_4	x_5	x_6	x_7	x_8	x_9	x_{10}	1	basic
row 00	1	0	12	15	9	12	6	3	6	0	0	0	15	w
row 0	0	1	4	6	3	2	1	6	5	0	0	0	0	z
row 1	0	0	3	6	2	5	3	0	2	1	0	0	5	x_8
row 2	0	0	5	4	6	5	1	2	4	0	1	0	7	x_9
row 3	0	0	4	5	1	2	2	1	0	0	0	1	3	x_{10}

The variables x_8, x_9, and x_{10} are artificial, and the Phase I Problem minimizes their sum. After three iterations, we arrive at the tableau

min	w	z	x_1	x_2	x_3	x_4	x_5	x_6	x_7	x_8	x_9	x_{10}	1	basic
row 00	1	0	0	0	0	0	0	0	0	-1	-1	-1	0	w
row 0	0	1	-8	0	-5	0	1	0	3	$7/3$	$-5/3$	$-8/3$	-8	z
row 1	0	0	0	0	1	1	0	0	1	$1/6$	$1/6$	$-1/3$	1	x_4
row 2	0	0	$3/2$	0	$3/2$	0	$-1/2$	1	$1/2$	$-17/36$	$13/36$	$5/18$	1	x_6
row 3	0	0	$1/2$	1	$-1/2$	0	$1/2$	0	$-1/2$	$1/36$	$-5/36$	$5/18$	0	x_2

We have now identified a feasible basis: $B = [A_{.4},\ A_{.6},\ A_{.2}]$. Phase II begins with this basis and terminates after two iterations with the tableau

min	w	z	x_1	x_2	x_3	x_4	x_5	x_6	x_7	x_8	x_9	x_{10}	1	basic
row 00	1	0	0	0	0	0	0	0	0	-1	-1	-1	0	w
row 0	0	1	-9	-2	-8	-4	0	0	0	$29/18$	$-37/18$	$-17/9$	-12	z
row 1	0	0	0	0	1	1	0	0	1	$1/6$	$1/6$	$-1/3$	1	x_7
row 2	0	0	2	1	1	0	0	1	0	$-4/9$	$2/9$	$5/9$	1	x_6
row 3	0	0	1	2	0	1	1	0	0	$2/9$	$-1/9$	$2/9$	1	x_5

We remark in passing that this tableau reveals that the dual variables for this problem are given by the vector $\bar{y} = (29/18, -37/18, -17/9)$. Because the primal problem was equality constrained, the dual variables are not restricted in sign, and indeed, the vector of dual variables turns out to have components of opposite signs. Note that the optimal basis found is still optimal if b_1 is increased from 5 to 6; and the optimal value of z increases by $29/18$, from -12 to $-187/18$. If the second component of the original vector b is increased from 7 to 8, the same basis is again optimal and the optimal value of z decreases by $37/18$, that is from -12 to $-12 - (37/18) = -253/18$.

Let us now consider what happens if we decide to delete the activity associated with basic variable x_7. This can be viewed as forcing x_7 to be zero. Since all the x-variables are nonnegative, forcing x_7 to be zero amounts to imposing the upper-bound constraint $x_7 \leq 0$.

Although our present purpose is to *delete an activity* after having solved the linear program, the technique we employ to accomplish this is the *introduce both a constraint and a new variable*, namely the constraint $x_7 \leq 0$ and its associated slack variable, x_{11}. Since we no longer have use for row 00 and the associated variable w, and to make room for the extra variable x_{11}, we revise the appearance of the tableau. Thus the updated tableau is

min	z	x_1	x_2	x_3	x_4	x_5	x_6	x_7	x_8	x_9	x_{10}	x_{11}	1	basic
row 0	1	−9	−2	−8	−4	0	0	0	29/18	−37/18	−17/9	0	−12	z
row 1	0	0	0	1	1	0	0	1	1/6	1/6	−1/3	0	1	x_7
row 2	0	2	1	1	0	0	1	0	−4/9	2/9	5/9	0	1	x_6
row 3	0	1	2	0	1	1	0	0	2/9	−1/9	2/9	0	1	x_5
row 4	0	0	0	0	0	0	0	1	0	0	0	1	0	x_{11}

which still needs to be put into canonical form. In particular, we need to eliminate x_7 from the equation represented by row 4. This leads to

min	z	x_1	x_2	x_3	x_4	x_5	x_6	x_7	x_8	x_9	x_{10}	x_{11}	1	basic
row 0	1	−9	−2	−8	−4	0	0	0	29/18	−37/18	−17/9	0	−12	z
row 1	0	0	0	1	1	0	0	1	1/6	1/6	−1/3	0	1	x_7
row 2	0	2	1	1	0	0	1	0	−4/9	2/9	5/9	0	1	x_6
row 3	0	1	2	0	1	1	0	0	2/9	−1/9	2/9	0	1	x_5
row 4	0	0	0	−1	−1	0	0	0	−1/6	−1/6	1/3	1	−1	x_{11}

This tableau represents a linear program to which the Dual Simplex Algorithm can be applied.

Note that we could have deleted the columns of the artificial variables, x_8, x_9, and x_{10}, but we have kept them to facilitate recovering the dual variables. As long as these columns remain in the tableau, we need to prevent them from becoming basic again. This means no pivoting in these columns will be allowed after the original LP is solved. With this rule in place, we begin the solution of the problem by the Dual Simplex Algorithm. The first pivot would bring in x_4 in place of the basic variable x_{11} and would change the value of z from -12 to -8 as shown in the tableau

min	z	x_1	x_2	x_3	x_4	x_5	x_6	x_7	x_8	x_9	x_{10}	x_{11}	1	basic
row 0	1	-9	-2	-4	0	0	0	0	$41/18$	$-25/18$	$-29/9$	-4	-8	z
row 1	0	0	0	0	0	0	0	1	0	0	$1/3$	1	0	x_7
row 2	0	2	1	1	0	0	1	0	$-4/9$	$2/9$	$5/9$	0	1	x_6
row 3	0	1	2	-1	0	1	0	0	$1/18$	$-5/18$	$8/9$	1	0	x_5
row 4	0	0	0	1	1	0	0	0	$1/6$	$1/6$	$-1/3$	-1	1	x_4

As usual, this tableau reveals the values of the optimal dual variables. The first three of them $(41/18, -25/18, -29/9)$ correspond to the three original constraints and—as can be verified—are exactly the values that would have been obtained if x_7 and its column had been removed from the model.

Deleting an inequality or equality constraint

The approach for deleting a constraint once an optimal solution has been obtained is the same whether it be an inequality or an equality. Accordingly, we will assume that we are deleting an equality constraint.

Suppose that after obtaining an optimal solution, say $\bar{x}$, we decide that a particular constraint $A_{k\bullet} x = b_k$ needs to be deleted. Even though $\bar{x}$ will be feasible for the modified problem, the deletion of this constraint will reduce the size of the basis and might result in dual infeasibility. To "physically" delete a constraint from the system and update the basis inverse (or factorization) could require significant computational work. It is simpler instead to make the constraint *nonbinding*, by which we mean that the left-hand side is no longer required to take on the value b_k.

We can make the k-th constraint nonbinding by introducing the free variable x_{n+1} (in the k-th constraint only) and setting its cost coefficient to zero in the objective. That is,

$$A_{k\bullet} x + x_{n+1} = b_k, \quad x_{n+1} \text{ is free.}$$

(Note: We could consider setting the right-hand side to zero; however, a constant in a nonbinding constraint does not affect the optimal objective value, so we do not set $b_k = 0$.) Next we can convert the revised constraint to fit into the standard form with the rest of the constraints by expressing x_{n+1} as the difference of two nonnegative variables, that is

$$A_{k\bullet} x + x_{n+1}^+ - x_{n+1}^- = b_k, \quad x_{n+1}^+ \geq 0, \ x_{n+1}^- \geq 0.$$

After this, we apply B^{-1} to the two new columns, and price them out. One of them will price out positive, the other one negative, or both may price out zero. We pivot to bring the positive (or zero) reduced-cost column into the basis and continue, if necessary, until the algorithm terminats with an optimal solution or an indication that none exists.

Example 6.4: POSTOPTIMAL DELETION OF A CONSTRAINT. Consider once again the tiny linear program of Example 6.2 with final tableau

min	z	x_1	x_2	x_3	x_4	x_5	1	basic
row 0	1	0	0	0	$-1/2$	$-1/2$	$-33/4$	z
row 1	0	0	1	0	0	1	5	x_2
row 2	0	1	0	0	$1/2$	$-1/2$	$13/4$	x_1
row 3	0	0	0	1	$1/2$	$-5/2$	$5/4$	x_3

The basis inverse is given by

$$B^{-1} = \begin{bmatrix} 0 & 0 & 1 \\ 0 & 1/2 & -1/2 \\ 1 & 1/2 & -5/2 \end{bmatrix}.$$

We illustrate deleting a constraint by deleting the constraint added in Example 6.2. This would not be typically done in real applications but, for our purposes, it will help to illustrate how deletion works; indeed we will obtain the solution to the original problem (that is, the problem before the addition of the constraint). So now suppose we decide to delete the inequality $x_2 \leq 5$ that was added earlier as the equation $x_2 + x_5 = 5$. We introduce the free variable x_6 as the difference of its positive and negative parts; thereby obtaining

$$x_2 + x_5 + x_6^+ - x_6^- = 5.$$

We apply B^{-1} to the two new columns and price them out. Altogether, with the previous tableau, this results in

min	z	x_1	x_2	x_3	x_4	x_5	x_6^+	x_6^-	1	basic
row 0	1	0	0	0	$-1/2$	$-1/2$	$-1/2$	$1/2$	$-33/4$	z
row 1	0	0	1	0	0	1	1	-1	5	x_2
row 2	0	1	0	0	$1/2$	$-1/2$	$-1/2$	$1/2$	$13/4$	x_1
row 3	0	0	0	1	$1/2$	$-5/2$	$-5/2$	$5/2$	$5/4$	x_3

From this tableau we see that, as expected, the BFS is not optimal (the reduced cost of x_6^- is positive). Next we pivot to exchange x_3 and x_6^- (that is, make x_6^- basic in place of x_3). This results in the optimal tableau.

min	z	x_1	x_2	x_3	x_4	x_5	x_6^+	x_6^-	1	basic
row 0	1	0	0	$-1/5$	$-3/5$	0	0	0	$-17/2$	z
row 1	0	0	1	$2/5$	$1/5$	0	0	0	$11/2$	x_2
row 2	0	1	0	$-1/5$	$2/5$	0	0	0	3	x_1
row 3	0	0	0	$2/5$	$1/5$	-1	-1	1	$1/2$	x_6^-

The optimal solution is the same as that in Example 6.2 prior to adjoining the constraint $x_2 \leq 5$. To see this, remove the last row and columns x_5, x_6^+, and x_6^- from the above tableau.

6.2 Ranging (Sensitivity analysis)

As mentioned earlier, ranging, or sensitivity analysis, has to do with the impact of small changes to the optimal solution. With ranging we examine the extent to which *changes in a single element do not affect the optimal basis*. For example, we would be interested in how much a right-hand side element b_i could change without causing the optimal basis to change. This would be important for key resources where we would be concerned about the impact of market variations, or simply concerned that we were off in our estimate of its value. In a similar vein, it would be comforting to know that minor variations in cost (or elements of the data matrix) do not impact our solution much. Such analysis has its limitations since we are only discussing modifications of single elements; later, in Section 6.3, we shall consider changing multiple elements through the use of parametric programming.

Changing the right-hand side vector

Let $B = [A_{\bullet j_1} \cdots A_{\bullet j_m}]$ be an optimal basis for the LP

$$\text{minimize } c^T x \quad \text{subject to } Ax = b, \ x \geq 0$$

where A is $m \times n$. Let $I_{\bullet k}$ denote the k-th column of the $m \times m$ identity matrix. Changing only the k-th component of b, from b_k to $\hat{b}_k = b_k + \delta$, amounts to creating a new right-hand side vector, say

$$\hat{b} = b + \delta I_{\bullet k}.$$

The raises the question: Does B remain a feasible basis under this change, and, if so, for what values of δ, or equivalently for what values of $\hat{b}_k$? Fortunately, under this change, B will remain *dual feasible*. The dual feasibility of B really means that $y^T = c_B^T B^{-1}$ is a dual feasible vector. So, in the present circumstances, B is an optimal basis for the LP

$$\text{minimize } c^T x \quad \text{subject to } Ax = \hat{b}, \ x \geq 0$$

if and only if $x_B = B^{-1}\hat{b} \geq 0$. Now

$$B^{-1}\hat{b} = B^{-1}(b + \delta I_{\bullet k}) = \bar{b} + \delta (B^{-1})_{\bullet k},$$

so the basis B will be feasible (and hence optimal) provided $\bar{b} + \delta(B^{-1})_{\bullet k} \geq 0$. Since $x_B = (x_{j_1}, \ldots, x_{j_m})$, the feasibility criterion just means

$$x_{j_i} = (\bar{b} + \delta(B^{-1})_{\bullet k})_i = \bar{b}_i + \delta(B^{-1})_{ik} \geq 0, \ i = 1, 2, \ldots m.$$

The key question is: When does the variation of δ cause $\bar{b}_i + \delta(B^{-1})_{ik}$ to become zero? This question is answered by the bounds given in Table 6.1 which displays the restrictions on δ that ensure the nonnegativity of x_{j_i}.

Let's now summarize this discussion of changing the right-hand side. If B is an optimal basis for

$$\text{minimize } c^T x \quad \text{subject to } Ax = b, \ x \geq 0,$$

then B is still optimal when b_k is changed to $\tilde{b}_k = b_k + \delta$ if and only if

$$\sup_i \left\{ \frac{-\bar{b}_i}{(B^{-1})_{ik}} : (B^{-1})_{ik} > 0 \right\} \leq \delta \leq \inf_i \left\{ \frac{-\bar{b}_i}{(B^{-1})_{ik}} : (B^{-1})_{ik} < 0 \right\}. \quad (6.1)$$

	$B_{ik}^{-1} < 0$	$B_{ik}^{-1} = 0$	$B_{ik}^{-1} > 0$
Increasing δ	$\delta \leq \dfrac{-\bar{b}_i}{(B^{-1})_{ik}}$	none	$\delta < +\infty$
Decreasing δ	$\delta > -\infty$	none	$\delta \geq \dfrac{-\bar{b}_i}{(B^{-1})_{ik}}$

Table 6.1: Bounds on δ for keeping $x_{j_i} \geq 0$.

In (6.1), the operator sup changes to max when the set is (finite but) nonempty. When the set is empty, the sup is $-\infty$. Analogously, the operator inf changes to min when the set is (finite but) nonempty. When the set is empty, the inf is $+\infty$.

The above pair of inequalities gives a range (interval) of values of δ over which the change of b_k to $b_k + \delta$ does not affect the feasibility, and hence optimality, of the basis B. Thus the range of values of b_k is:

$$b_k + \sup_i \left\{ \frac{-\bar{b}_i}{(B^{-1})_{ik}} : (B^{-1})_{ik} > 0 \right\} \leq \hat{b}_k \leq b_k + \inf_i \left\{ \frac{-\bar{b}_i}{(B^{-1})_{ik}} : (B^{-1})_{ik} < 0 \right\}.$$

The process of determining these values (one component at a time) is called *ranging* the right-hand side vector.

Example 6.5: RIGHT-HAND SIDE RANGING ON A TINY LP. Consider the tiny linear program:

$$\begin{aligned} \text{minimize} \quad & - 6x_1 - 3x_2 - 9x_3 \\ \text{subject to} \quad & - 3x_1 - 2x_2 + 8x_3 \leq 10 \\ & 4x_1 + 10x_2 + 5x_3 \leq 12 \\ & x_j \geq 0, \ j = 1, 2, 3. \end{aligned}$$

In tableau form, this is

min	z	x_1	x_2	x_3	x_4	x_5	1	basic
row 0	1	6	3	9	0	0	0	z
row 1	0	-3	-2	8	1	0	10	x_4
row 2	0	4	10	5	0	1	12	x_5

After two pivots, we obtain

min	z	x_1	x_2	x_3	x_4	x_5	1	basic
row 0	1	0	$-597/47$	0	$-6/47$	$-75/47$	$-960/47$	z
row 1	0	0	$22/47$	1	$4/47$	$3/47$	$76/47$	x_3
row 2	0	1	$90/47$	0	$-5/47$	$8/47$	$46/47$	x_1

From the initial and final tableaus, we see that the optimal basis and its inverse are

$$B = \begin{bmatrix} 8 & -3 \\ 5 & 4 \end{bmatrix} \quad \text{and} \quad B^{-1} = \begin{bmatrix} 4/47 & 3/47 \\ -5/47 & 8/47 \end{bmatrix}.$$

We can also find the optimal solutions of the primal and dual problems.

Ranging the right-hand side element b_1. The formulas for the limits on δ when modifying b_k to $b_k + \delta$ are given by (6.1). At this moment, $k = 1$. Looking back at the final tableau and applying (6.1), we find that

$$\frac{-76/47}{4/47} = -19 \leq \delta \leq \frac{-46/47}{-5/47} = \frac{46}{5}.$$

Ranging the right-hand side element b_2. Next set $k = 2$ in formula (6.1). From the final tableau and (6.1), we find that

$$\frac{-23}{4} \leq \delta < +\infty.$$

Summary of the right-hand side ranging results.

RHS Parameter	Minimum	Maximum
b_1	-9	$96/5$
b_2	$25/4$	$+\infty$

This sort of analysis does not allow for simultaneous changes of two or more right-hand side constants. Figure 6.2 illustrates this fact with respect to the cone spanned by the basis $B = [A_{.3} \ A_{.1}]$.

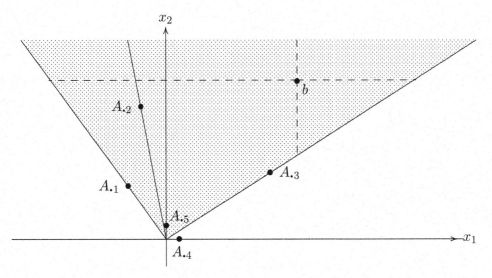

Figure 6.2: Ranging the parameters of the right-hand side.

Changing the cost vector

Let B be an optimal basis for the LP

$$\text{minimize } c^{\mathrm{T}}x \quad \text{subject to } Ax = b, \ x \geq 0$$

where A is $m \times n$. Let $I_{\bullet k}$ denote the k-th column of the $m \times m$ identity matrix. Changing only the k-th component of c, from c_k to $\hat{c}_k = c_k + \delta$, amounts to creating a new cost vector

$$\hat{c} = c + \delta I_{\bullet k}.$$

This change will not affect the *feasibility* of B since this depends only on B and b, neither of which has been changed at this stage.

Case 1: x_k is nonbasic. In this case $\hat{c}_B = c_B$, so

$$\hat{y}^{\mathrm{T}} = \hat{c}_B^{\mathrm{T}} B^{-1} = c_B^{\mathrm{T}} B^{-1} = y^{\mathrm{T}}$$

and

$$\hat{z}_j = z_j = y^{\mathrm{T}} A_{\bullet j} \text{ for all } j.$$

Hence for all $j \neq k$ we have $\hat{z}_j - \hat{c}_j = z_j - c_j \leq 0$. If $\hat{z}_k - \hat{c}_k = z_k - c_k - \delta \leq 0$, then B is still an optimal basis. Otherwise, the Simplex Algorithm applied to the revised problem will make x_k basic on the next iteration.

Case 2: x_k is basic. Assume $k = j_t$, that is, $x_k = x_{j_t}$. Since

$$\hat{c}_B = c_B + \delta I_{\bullet t} \quad \text{and} \quad \hat{c}_N = c_N,$$

we have

$$\hat{y}^{\mathrm{T}} = \hat{c}_B^{\mathrm{T}} B^{-1} = c_B^{\mathrm{T}} B^{-1} + \delta I_{t\bullet} B^{-1} = y^{\mathrm{T}} + \delta (B^{-1})_{t\bullet}.$$

This just says that the new price vector $\hat{y}^{\mathrm{T}}$ is the old price vector y^{T} plus δ (the amount by which c_k is changed) times the t-th row of the *inverse* of B.

For x_j nonbasic, we have

$$\hat{z}_j - \hat{c}_j = \hat{y}^{\mathrm{T}} A_{\bullet j} - \hat{c}_j$$

$$= c_B^{\mathrm{T}} B^{-1} A_{\bullet j} + \delta (B^{-1})_{t\bullet} A_{\bullet j} - c_j$$

$$= z_j - c_j + \delta \bar{a}_{tj}.$$

Note how the current t-th row of the tableau affects the new cost coefficients. The basis will be optimal if

$$\hat{z}_j - \hat{c}_j = z_j - c_j + \delta \bar{a}_{tj} \leq 0 \text{ for all } j \in N.$$

In summary, this discussion says that the basis B will remain an optimal basis under a change of the k-th cost coefficient c_k to $c_k + \delta$ provided

$$\sup_{j \in N} \left\{ \frac{-(z_j - c_j)}{\bar{a}_{tj}} : \bar{a}_{tj} < 0 \right\} \leq \delta \leq \inf_{j \in N} \left\{ \frac{-(z_j - c_j)}{\bar{a}_{tj}} : \bar{a}_{tj} > 0 \right\} \quad \text{if } k \in B$$

$$z_k - c_k \leq \delta \qquad\qquad\qquad\qquad\qquad \text{if } k \in N.$$

This process of analyzing changes in cost vector elements is also called *ranging*. It differs from right-hand side ranging in that here it matters whether the variable corresponding to the cost coefficient being modified is basic or nonbasic.

Example 6.6: COST RANGING ON A TINY LP. Once again consider the tiny linear program of Example 6.5. How much can the individual components of $(c_1, c_2, c_3) = (-6, -3, -9)$ be changed without affecting the optimality of the BFS we have found?

Cost ranging: c_1. Note that x_1 is basic and $x_1 = x_{j_2}$. In the final tableau we find the $z_j - c_j$. We want $z_j - c_j + \delta \bar{a}_{2j} \le 0$ for $j = 2, 4, 5$.

$$
\begin{aligned}
j = 2: && -597/47 + \delta(90/47) \le 0 && \implies && \delta \le 597/90 \\
j = 4: && -6/47 + \delta(-5/47) \le 0 && \implies && \delta \ge -6/5 \\
j = 5: && -75/47 + \delta(8/47) \le 0 && \implies && \delta \le 75/8
\end{aligned}
$$

Thus, $-6/5 \le \delta \le 597/90$.

Cost ranging: c_2. Ranging c_2: Note that x_2 is nonbasic. We need δ to satisfy

$$z_2 - c_2 - \delta \le 0.$$

Thus, we must have $-597/47 \le \delta$.

Cost ranging: c_3. Since $x_3 = x_{j_1}$ is basic, we want $z_j - c_j + \delta \bar{a}_{1j} \le 0$ for $j = 2, 4, 5$.

$$
\begin{aligned}
j = 2: && -597/47 + \delta(22/47) \le 0 && \implies && \delta \le 597/22 \\
j = 4: && -6/47 + \delta(4/47) \le 0 && \implies && \delta \le 3/2 \\
j = 5: && -75/47 + \delta(3/47) \le 0 && \implies && \delta \le 25
\end{aligned}
$$

Thus, $\delta \le 3/2$.

Summary of the cost ranging results. The basis $[A_{\bullet 3}, A_{\bullet 1}]$ is optimal for the following ranges of the cost coefficients:

Cost Coefficient	Minimum	Maximum
c_1	$-36/5$	$-19/30$
c_2	$-738/47$	$-\infty$
c_3	$-\infty$	$-15/2$

6.3 Parametric linear programming

In the preceding section on ranging, we began with an optimal basic solution $\bar{x}$ to an LP and considered the effects of changing either individual right-hand side (RHS) constants or individual cost coefficients. The goal in that sort of postoptimality analysis was to determine the range of values of the parameter being varied that would preserve the optimality of the basis B from which the optimal basic solution $\bar{x}$ was obtained.

In this section we turn to a more general sort of problem: parametric linear programming. The purpose of parametric linear programming is to obtain an optimal solution of an LP and the corresponding objective function value where each component of either the right-hand side vector b or the cost vector c changes at its own constant rate. In the case of an LP with a parametric right-hand side vector, the right-hand side has the form $b + \lambda p$ where p is nonzero and λ belongs to a closed interval. Analogously, for LPs with a parametric objective function, the vector c is replaced by one of the form $c + \lambda d$ where d is nonzero and again λ belongs to a closed interval of the real line. Although it is possible to handle both types of parametric problems simultaneously, we present them one at a time.

Before going further with this discussion, we pause to point out why this class of problems is more general than ranging. In the first place, it does not proceed from a *given* optimal solution; rather, it requires that we find optimal solutions (if they exist) for each member of an infinite family of linear programs. Second, the linear programs within this infinite family may differ from one another in more than one data element. For example, if the right-hand side is parameterized, it could happen that *all* the right-hand side constants are simultaneously varying with the parameter. An analogous statement applies to the case where the objective function is parameterized, in which case all the cost coefficients could vary simultaneously. This is not to say that *all* such constants or coefficients must actually change under the parametrization. Indeed, in parametric LP problems, there could be as few as one data element that is changing. In such a case, the vector p or the vector d would have only one nonzero component. The ranging problem is of this sort, but it proceeds from a known optimal solution and, in effect, asks us to find the interval that maintains the optimality of the known basis for such variation.

Preliminaries

Since both kinds of parametric LPs involve a closed *interval* of the real line as the set to which the special parameter belongs, it is important to understand what such a set can be. Closed intervals are essentially sets of the form

$$\Lambda = \{\lambda : \underline{\lambda} \leq \lambda \leq \bar{\lambda}\}$$

where $\underline{\lambda}$ and $\bar{\lambda}$ are given lower and upper bounds, respectively. This description is precise when $\underline{\lambda}$ and $\bar{\lambda}$ denote real numbers. But, for any infinite bound, the corresponding weak inequality needs to be taken informally. That is, if the lower bound $\underline{\lambda} = -\infty$, we really mean $\underline{\lambda} < \lambda$, and if the upper bound $\bar{\lambda} = \infty$, we really mean $\lambda < \bar{\lambda}$. Thus, an interval is one of four types of subset of the real line R:

1. $\Lambda = [\underline{\lambda}, \bar{\lambda}]$ $= \{\lambda : -\infty < \underline{\lambda} \leq \lambda \leq \bar{\lambda} < \infty\}$;

2. $\Lambda = [\underline{\lambda}, \infty)$ $= \{\lambda : -\infty < \underline{\lambda} \leq \lambda < \infty\}$;

3. $\Lambda = (-\infty, \bar{\lambda}]$ $= \{\lambda : -\infty < \lambda \leq \bar{\lambda} < \infty\}$;

4. $\Lambda = (-\infty, \infty) = \{\lambda : -\infty < \lambda < \infty\}$.

In the last of these cases, $\Lambda = R$.

For both kinds of parametric linear programs considered here, it will be convenient to "normalize" these intervals by a redefinition of the parameter for the problem. By this we mean that each of these interval types (and the parametric problem at hand) will be stated as an equivalent problem in a new parameter μ belonging to a simpler interval: either $[0, \bar{\mu}]$ or $[0, \infty)$. We now describe these reformulations.

1. $\Lambda = [\underline{\lambda}, \bar{\lambda}]$.   Let $\mu = \lambda - \underline{\lambda}$ and $\bar{\mu} = \bar{\lambda} - \underline{\lambda}$. Then $\lambda \in [\underline{\lambda}, \bar{\lambda}] \Longleftrightarrow$ $\mu \in [0, \bar{\lambda} - \underline{\lambda}] = [0, \bar{\mu}]$. When this is done, $b + \lambda p = (b + \underline{\lambda}p) + \mu p$, and $c + \lambda d = (c + \underline{\lambda}d) + \mu d$.

2. $\Lambda = [\underline{\lambda}, \infty)$.   This is handled exactly as the previous case except for the fact that instead of the new interval $[0, \bar{\mu}] = [0, \bar{\lambda} - \underline{\lambda}]$ we have the interval $[0, \infty)$.

3. $(-\infty, \bar{\lambda}]$.   In this case, define $\mu = \bar{\lambda} - \lambda$. Then $\mu \geq 0$, and the new interval becomes $[0, \infty)$. Note that since we now have $\lambda = \bar{\lambda} - \mu$, it follows that $b + \lambda p = (b + \bar{\lambda}p) - \mu p = (b + \bar{\lambda}p) + \mu(-p)$. Similarly, $c + \lambda d = (c + \bar{\lambda}d) - \mu d = (c + \bar{\lambda}d) + \mu(-d)$.

4. $\Lambda = (-\infty, +\infty)$.   In this situation, we can consider the set $\Lambda = R$ to be the union of two intervals: $(-\infty, 0]$ and $[0, \infty)$. (It does not matter that these two sets both contain zero.) The objective functions are reformulated in the manner of cases 3 and 2, respectively.

When one of these normalizations is used in a parametric linear program, $P(\lambda)$, the resulting solution $(\bar{x}(\lambda), \bar{z}(\lambda))$ can be restated in terms of the original formulation.

In parametric linear programming, we face the following three questions:

1. Are there values of $\lambda \in \Lambda$ for which $P(\lambda)$ has an optimal solution? If so, how do we find them?

2. Is there an algorithm for solving problems of this sort?

3. How do $\bar{x}(\lambda)$ and $\bar{z}(\lambda)$ behave as functions of λ?

We are now ready to take up our two parametric linear programming models.

Linear programs with parametric right-hand side

Let p be a nonzero vector and let λ be a scalar parameter that belongs to an interval Λ of the real line; for example, we might have $0 \leq \lambda \leq \bar{\lambda}$. For a given $\lambda \in \Lambda$, let $P(\lambda)$ denote the parametric linear program

$$\text{minimize } c^{\mathrm{T}}x \quad \text{subject to } Ax = b + \lambda p, \ x \geq 0. \tag{6.2}$$

The family of these problems is a *linear program with parametric right-hand side*.

Problems of the form (6.2) arise when there is uncertainty in the precise values of right-hand side elements or when one wants to examine the impact on the optimal solution and optimal value when one or more right-hand side values are varied. To give an example of the latter sort, a linear program could have a constraint that measures the rate of return of a set of investments, or it might have a constraint that measures the level of service delivered by a company. Constraints like these might be expressed in the form $a^{\mathrm{T}}x \geq r$.

In the present case, we temporarily assume that the constraints of the dual problem are feasible.[2] We begin by addressing the first of the three questions stated above.

---

[2]This sort of assumption can be checked with a Phase I type of procedure.

Question 1

Given that the dual problem has feasible constraints, we know that problem $P(\lambda)$ will have an optimal solution if and only if it is feasible. We begin by finding the smallest and largest values of λ for which $P(\lambda)$ is feasible. This is done by solving the linear programs

$$\lambda_{\min} = \inf \{\lambda : Ax - \lambda p = b, \ x \geq 0\}]$$

and

$$\lambda_{\max} = \sup \{\lambda : Ax - \lambda p = b, \ x \geq 0\}].$$

It follows that $P(\lambda)$ has an optimal solution for each value of λ in the interval

$$\lambda_{\min} \leq \lambda \leq \lambda_{\max}.$$

It is, of course, possible that there is no value of λ for which the problem $P(\lambda)$ is feasible. In that case we take $\lambda_{\min} = \infty$ or $\lambda_{\max} = -\infty$.

We take up Questions 2 and 3 after the following illustrative example.

Example 6.7: LP WITH PARAMETRIC RIGHT-HAND SIDE. For all $\lambda \in R$ solve

$$P(\lambda) \quad \begin{array}{ll} \text{minimize} & -3x_1 - 2x_2 \\ \text{subject to} & 2x_1 + 3x_2 + x_3 \qquad = 6 - \lambda \\ & 2x_1 + x_2 \qquad + x_4 = 4 + \lambda \\ & x_j \geq 0, \quad j = 1, 2, 3, 4. \end{array}$$

In this example, $p = (-1, 1)$. As it happens, the problem $P(0)$ has an optimal solution which we now obtain. For this purpose, we set up the tableau

min	z	x_1	x_2	x_3	x_4	1	λ	basic
row 0	1	3	2	0	0	0	0	z
row 1	0	2	3	1	0	6	-1	x_3
row 2	0	2	1	0	1	4	1	x_4

After two pivots we find the optimal solution of $P(0)$ from the tableau

min	z	x_1	x_2	x_3	x_4	1	λ	basic
row 0	1	0	0	$-1/4$	$-5/4$	$-13/2$	-1	z
row 1	0	0	1	$1/2$	$-1/2$	1	-1	x_2
row 2	0	1	0	$-1/4$	$3/4$	$3/2$	1	x_1

Now to solve this parametric linear programming problem, we break the set $\lambda = R$ into two parts: $\Lambda_+ = R_+$ and $\Lambda_- = R_- = -R_+$.

Increasing λ from 0. The tableau reveals that

$$z = -13/2 - \lambda$$
$$x_2 = 1 - \lambda \geq 0 \qquad \text{which implies} \quad \lambda \leq 1$$
$$x_1 = \tfrac{3}{2} + \lambda \geq 0 \qquad \text{for all} \qquad \lambda \geq 0.$$

This means that for $0 \leq \lambda \leq 1$, the basic variables will have nonnegative values, and the current basis will be optimal for this range of values.

When $\lambda = 1$, the basic variable x_2 decreases to zero and must be made nonbasic; it should be replaced by a nonbasic variable whose coefficient (in the row of x_2) is *negative*, as in the Dual Simplex Algorithm. In this case, there is only one choice: x_4 must replace x_2 as a basic variable. (Had there been more choices, the pivot column would have been chosen as in the Dual Simplex Algorithm.) Thus, we obtain

min	z	x_1	x_2	x_3	x_4	1	λ	basic
row 0	1	0	$-5/2$	$-3/2$	0	-9	$3/2$	z
row 1	0	0	-2	-1	1	-2	2	x_4
row 2	0	1	$3/2$	$1/2$	0	3	$-1/2$	x_1

We now find that

$$z = -9 + 3\lambda/2$$
$$x_4 = -2 + 2\lambda \geq 0 \qquad \text{for all} \qquad \lambda \geq 1$$
$$x_1 = 3 - \lambda/2 \geq 0 \qquad \text{which implies} \quad \lambda \leq 6.$$

The basic variables x_4 and x_1 are nonnegative for all values of λ such that $1 \leq \lambda \leq 6$. For $\lambda > 6$, we have

$$0 \leq x_1 + \tfrac{3}{2}x_2 + \tfrac{1}{2}x_3 = 3 - \tfrac{1}{2}\lambda < 0,$$

which is a contradiction. In other words, the feasible region is empty when $\lambda > 6$, a fact that could have discovered by first computing $\lambda_{\max}$.

Decreasing λ from 0. Returning to the tableau showing the optimal solution of $P(0)$, we begin to decrease λ from 0. The tableau is

min	z	x_1	x_2	x_3	x_4	1	λ	basic
row 0	1	0	0	$-1/4$	$-5/4$	$-13/2$	-1	z
row 1	0	0	1	$1/2$	$-1/2$	1	-1	x_2
row 2	0	1	0	$-1/4$	$3/4$	$3/2$	1	x_1

It is clear that

$$z = -\tfrac{13}{2} - \lambda$$
$$x_2 = 1 - \lambda \geq 0 \qquad \text{for all} \qquad \lambda \leq 0$$
$$x_1 = \tfrac{3}{2} + \lambda \geq 0 \quad \text{which implies} \quad \lambda \geq -\tfrac{3}{2}.$$

The basic variables are nonnegative for $-\tfrac{3}{2} \leq \lambda \leq 0$. When $\lambda = -\tfrac{3}{2}$, the basic variable x_1 decreases to 0 and must be made nonbasic. A negative pivot element is needed and again, there is only one choice: x_3 must become basic. After pivoting, we obtain the tableau

min	z	x_1	x_2	x_3	x_4	1	λ	basic
row 0	1	-1	0	0	-2	-8	-2	z
row 1	0	2	1	0	1	4	1	x_2
row 2	0	-4	0	1	-3	-6	-4	x_3

We now want to decrease λ even further. We find that

$$z = -8 - 2\lambda$$
$$x_2 = 4 + \lambda \geq 0 \qquad \text{which implies} \quad \lambda \geq -4$$
$$x_3 = -6 - 4\lambda \geq 0 \quad \text{for all} \qquad \lambda \leq -3/2.$$

This shows that the current basis is optimal for all λ in the subinterval $[-4, -3/2]$. It is also possible to see from the tableau that the problem $P(\lambda)$ is infeasible for $\lambda < -4$.

In solving this problem for $\lambda \leq 0$ we did not use the technique of defining $\mu = -\lambda$ and $q = -p$ so as to work only with nonnegative parameter values. The solution of this problem is summarized in the table below. Notice that we have combined the intervals $[-3/2, 0]$ and $[0, 1]$ since they do not involve a change of basis.

Interval	Optimal Solution	Optimal Value
$-4 \leq \lambda \leq -3/2$	$(0, 4+\lambda, -6-4\lambda, 0)$	$-8-2\lambda$
$-3/2 \leq \lambda \leq 1$	$(3/2+\lambda, 1-\lambda, 0, 0)$	$-13/2-\lambda$
$1 \leq \lambda \leq 6$	$(3-\lambda/2, 0, 0, -2+2\lambda)$	$-9+3\lambda/2$

The values of the optimal objective function values for this example are indicated in the figure below. Over the interval $-4 \leq \lambda \leq 6$, the function $\bar{z}(\lambda)$ is continuous, piecewise linear, and convex with *breakpoints* at $-3/2$ and 1.

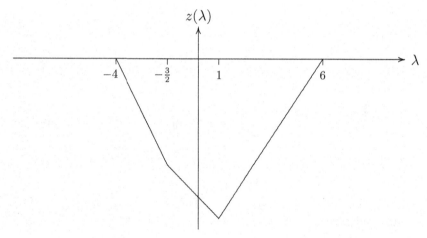

Figure 6.3: Optimal-value function for LP with parametric RHS.

Question 2

To solve a linear program with a parametric right-hand side vector, we assume the problem has the form shown in (6.2), but drop the assumption of dual feasibility.

Algorithm 6.1 (LP with parametric right-hand side)

1. *Initialization.* Check the feasibility of the dual problem. If none exists, stop and report the dual problem is infeasible.

2. *Standardization.* Check the range of feasibility of the primal problem. Obtain $\lambda_{\min}$ and $\lambda_{\max}$. Standardize the parameter to $[0, \bar{\mu}]$ or $[0, \infty)$. Set $k = 0$ and let $\mu_k = 0$.

3. *Solution of an LP.* Solve $P(\mu_k)$ by the Simplex Algorithm.

4. *Continuation.* Determine

$$\mu_{k+1} = \max\{\mu : \mu_k \leq \mu \leq \bar{\mu}, \quad \bar{b} + \mu\bar{p} \geq 0\}$$

where $\bar{b}$ and $\bar{p}$ come from the right-hand side columns of the tableau. If $\mu_{k+1} = \bar{\mu}$ (or if $\mu_{k+1} = \infty$), stop. Otherwise, repeat Step 3 with k replaced by $k + 1$.

Note that the conditions on μ imposed in Step 3 involve a minimum ratio test in addition to the restriction $\mu_k \leq \mu \leq \bar{\mu}$. That is, we must compute

$$\min_i\left\{-\frac{\bar{b}_i}{\bar{p}_i} : \bar{p}_i < 0\right\}.$$

The statement of the algorithm above does not mention how to avoid cycling due to degeneracy. This issue can be handled by the usual methods.

Figure 6.4 below also pertains to Example 6.7. It depicts the set of point $b + \lambda p$ with $\lambda \in \Lambda$ and the basic cones containing the right-hand side vector $b(\lambda) = b + \lambda p$. The rays emanating from the origin in this figure are (reading clockwise) the sets of nonnegative multiples of the columns $A_{\bullet 4}$, $A_{\bullet 1}$, $A_{\bullet 2}$, $A_{\bullet 3}$. Pairs of these vectors span basic cones, and the basic cones containing b correspond to feasible bases. This figure gives an interesting perspective on pivoting.

Question 3

In the case of a linear program with parametric right-hand side, the interval Λ is composed of subintervals. The subintervals are defined by the endpoints of the interval Λ and the interior values of λ (called breakpoints) at which basis changes occur.

The optimal-value function in (6.2) is linear between successive breakpoints. Indeed, the optimal-value function is continuous, piecewise linear, and convex as illustrated in Figure 6.3. Corresponding to each subinterval there is an associated basis.

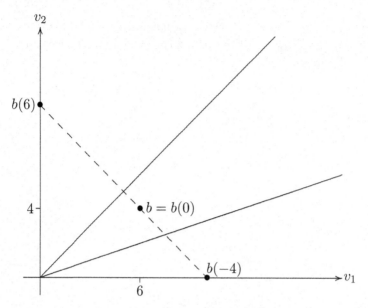

Figure 6.4: Right-hand side vector $b(\lambda)$ moving through basic cones.

Linear programs with parametric objective function

For each $\lambda \in \Lambda$ let $P(\lambda)$ denote the linear program

$$\text{minimize } (c + \lambda d)^{\mathrm{T}}x \quad \text{subject to } Ax = b, \ x \geq 0, \tag{6.3}$$

The family of these problems is a *linear program with parametric objective function*.

Problems of the form (6.3) can arise in different ways, usually connected with some uncertainty as to the precise values of the cost coefficients. In one setting, the cost coefficients are each believed to lie within a prescribed interval or to be varying at a constant rate over time (as represented by λ). In another setting, there might be two distinct objective functions under consideration; one might wish to determine the optimal values of all convex combinations of the two functions as a way of examining the tradeoffs between the two particular measures of effectiveness.

At this time we assume that the constraints $Ax = b$, $x \geq 0$ are feasible.[3]

---

[3]The question of whether this property holds can be checked with the Phase I Procedure. If the constraints are not feasible, there is no point in trying to solve this parametric linear programming problem.

Question 1

Since each problem $P(\lambda)$ is feasible, the existence of an optimal solution of $P(\lambda)$ depends on (is equivalent to) the feasibility of its dual. Define

$$\lambda_{\min} = \inf \left\{ \lambda : w^{\mathrm{T}}A - \lambda d^{\mathrm{T}} \leq c^{\mathrm{T}}, \ w \ \text{free} \right\}$$

and

$$\lambda_{\max} = \sup \left\{ \lambda : w^{\mathrm{T}}A - \lambda d^{\mathrm{T}} \leq c^{\mathrm{T}}, \ w \ \text{free} \right\}.$$

Hence $P(\lambda)$ has an optimal solution for each value of λ in the interval $[\lambda_{\min}, \lambda_{\max}]$.

The definitions $\lambda_{\min}$ and $\lambda_{\max}$ need to be clarified just a bit. It could happen for example that $\lambda_{\min} = -\infty$ or $\lambda_{\max} = \infty$. If either of these holds, then $[\lambda_{\min}, \lambda_{\max}]$ is not a bounded interval, but this is not a serious problem. It could also be the case that $\lambda_{\min} = \infty$. This would mean that there is no vector w and value of λ such that $w^{\mathrm{T}}A - \lambda d^{\mathrm{T}} \leq c$, and hence there is no value of λ such that $P(\lambda)$ has an optimal solution. An analogous statement can be made if $\lambda_{\max} = -\infty$. Whatever the case may be, it should be clear from the definitions that $\lambda_{\min}$ and $\lambda_{\max}$ are the optimal values of linear programs—albeit LPs that are not yet in standard form.

Example 6.8: LP WITH PARAMETRIC COST FUNCTION. Suppose we have a parametric LP of the form

$$\begin{array}{rl} \text{minimize} & (c + \lambda d)^{\mathrm{T}}x \\ P(\lambda) \qquad \text{subject to} & Ax = b \\ & x \geq 0, \end{array}$$

where $\lambda \in [0, 1]$ and

$$A = \begin{bmatrix} 2 & 3 & 4 & 6 \\ 1 & 2 & 9 & 6 \end{bmatrix}, \quad b = \begin{bmatrix} 12 \\ 18 \end{bmatrix}, \quad c = (5, 4, 6, 7), \quad \text{and} \quad d = (-3, 0, 2, 2).$$

It is easily demonstrated that columns $A_{\bullet 3}$ and $A_{\bullet 4}$ yield a feasible basis for the constraints of this problem, so the solvability of the problem depends on the feasibility of the dual constraints

$$A^{\mathrm{T}}w \leq c + \lambda d, \quad \text{and} \quad 0 \leq \lambda \leq 1.$$

To settle this issue, we can consider the problems

$$\text{maximize} \quad \lambda$$
$$\text{subject to} \quad A^T w - \lambda d \leq c$$

and

$$\text{minimize} \quad \lambda$$
$$\text{subject to} \quad A^T w - \lambda d \leq c.$$

In solving these problems for the data given above, we find that the objective function is unbounded in each case, that is: $\lambda_{\min} = -\infty$ and $\lambda_{\max} = +\infty$. Accordingly, the dual constraints are feasible for all $\lambda \in R$ and *a fortiori* for all $\lambda \in [0, 1]$.

These observations assure us that our parametric problem has an optimal solution for every $\lambda \in \Lambda = [0, 1]$. Thus, we may begin by solving the ordinary linear programming problem $P(0)$.

Because we shall need to initiate the parametrization of the objective function from the optimal tableau of $P(0)$, we devise a special tableau of the following form for problems of this type.

min	z_c	z_d	x	1	basic
row c	1	0	$-c^T$	0	z_c
row d	0	1	$-d^T$	0	z_d
rows $1, \ldots, m$	0	0	A	b	x_B

In our example we have

min	z_c	z_d	x_1	x_2	x_3	x_4	1	basic
row c	1	0	-5	-4	-6	-7	0	z_c
row d	0	1	3	0	-2	-2	0	z_d
row 1	0	0	2	3	4	6	12	?
row 2	0	0	1	2	9	6	18	?

The question marks in the "basic" column signify that we have not yet identified a feasible basis.

Solving $P(0)$ leads to the tableau

min	z_c	z_d	x_1	x_2	x_3	x_4	1	basic
row c	1	0	$-44/15$	$-23/30$	0	0	$78/5$	z_c
row d	0	1	$53/15$	$13/15$	0	0	$24/5$	z_d
row 1	0	0	$-1/5$	$-1/5$	1	0	$6/5$	x_3
row 2	0	0	$7/15$	$19/30$	0	1	$6/5$	x_4

From this tableau, we can see that the BFS $\bar{x}(0) = (0, 0, 6/5, 6/5)$ is optimal for all problems $P(\lambda)$ with $\lambda \leq 0$. But we are interested in making $\lambda > 0$.

The current basis will be optimal provided

$$-44/15 + \lambda\, 53/15 \leq 0 \quad \text{and} \quad -23/30 + \lambda\, 13/15 \leq 0.$$

For both of these inequalities to hold, we must have

$$\lambda \leq \min\{44/53,\ 23/26\} = 44/53.$$

If $\lambda > 44/53$, the coefficient of x_1 will be positive. (If $44/53 < \lambda < 23/26$, we can be sure that the coefficient of x_2 remains negative.)

On these grounds, we make x_1 basic in place of x_4. Doing this results in the tableau

min	z_c	z_d	x_1	x_2	x_3	x_4	1	basic
row c	1	0	0	$45/14$	0	$44/7$	$162/7$	z_c
row d	0	1	0	$-55/14$	0	$-53/7$	$-30/7$	z_d
row 1	0	0	0	$1/14$	1	$3/7$	$12/7$	x_3
row 2	0	0	1	$19/14$	0	$15/7$	$18/7$	x_1

Again we seek the largest value of $\lambda \geq 44/53$ that will preserve the nonpositivity of objective function coefficients. Since $45/55 < 44/53$ (the current value of λ), it clear that λ we can increase λ indefinitely and still not make any cost coefficient to become positive. Be that as it may, we are only interested in having λ increase to 1.

The following table summarizes our findings on this problem:

Interval	Optimal Solution	Optimal Value
$0 \leq \lambda \leq 44/53$	$(0,\, 0,\, 6/5,\, 6/5)$	$78 + 24\lambda$
$44/53 \leq \lambda \leq 1$	$(18/7,\, 0,\, 12/7,\, 0)$	$162/7 - 30\lambda/7$

The graph of the optimal objective function value is shown in Figure 6.5.

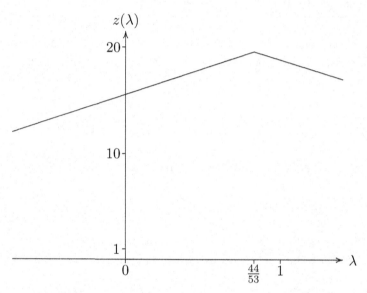

Figure 6.5: Optimal value of the parametric objective function.

Question 2

The solution of a linear program with parametric objective function can be handled by the following procedure. Let the problem be of the form (6.3).

Algorithm 6.2 (LP with parametric objective function)

1. *Initialization.* Obtain a basic feasible solution of the constraints of (6.3) if one exists. If none exists, stop.

2. *Standardization.* Check the feasibility of the dual constraints. Obtain $\lambda_{\min}$ and $\lambda_{\max}$. Standardize the parameter to $[0, \bar{\mu}]$ or $[0, \infty)$. Set $k = 0$ and let $\mu_k = 0$.

3. *Solution of an LP.* Solve $P(\mu_k)$ by the Simplex Algorithm.

4. *Continuation.* Determine

$$\mu_{k+1} = \max\{\mu \geq \mu_k : -\bar{c} - \mu\bar{d} \leq 0, \ \mu \leq \bar{\mu}\}$$

where $-\bar{c}^{\mathrm{T}}$ and $-\bar{d}^{\mathrm{T}}$ come from the top two rows of the current tableau. If $\mu_{k+1} = \bar{\mu}$ (or if $\mu_{k+1} = \infty$), stop. Otherwise, repeat Step 3 with k replaced by $k + 1$.

Note that the condition on μ imposed in Step 3 involves a minimum ratio test in addition to the restriction that $\mu \leq \bar{\mu}$. Thus

$$\mu_{k+1} = \min\left\{\bar{\mu}, \ \min_j\left\{-\frac{\bar{c}_j}{\bar{d}_j} : -\bar{d}_j > 0\right\}\right\}.$$

Just as in the case of Algorithm 6.1, the statement of Algorithm 6.2 assumes that degeneracy is handled in the usual way.

Question 3

Although the example above is almost too small to illustrate the point, it is a fact that the interval Λ is—by the process outlined above—partitioned into subintervals. The subintervals are defined by the endpoints of the interval Λ and the interior values of λ (called breakpoints) at which basis changes occur.

The optimal-value function in (6.3) is linear between successive breakpoints. Indeed, it is a general fact that the optimal value function is continuous, piecewise linear, and concave as illustrated in Figure 6.5 above. Corresponding to each subinterval there is an associated basis.

6.4   Shadow prices and their interpretation

Recall that if B denotes the optimal basis found by solving the primal problem

$$\begin{aligned} \text{minimize} \quad & c^{\mathrm{T}}x \\ \text{subject to} \quad & Ax = b \\ & x \geq 0, \end{aligned} \tag{6.4}$$

given here in standard form, then the price vector $\bar{y}$ corresponding to B is given by

$$\bar{y}^{\mathrm{T}} = c_B^{\mathrm{T}} B^{-1}. \tag{6.5}$$

This vector satisfies $\bar{y}^{\mathrm{T}} A \le c^{\mathrm{T}}$, the dual feasibility condition, for every right-hand side vector b.

Given $A \in R^{m \times n}$ and $c \in R^n$, let us now define the function of $b \in R^m$ given by

$$z(b) = \min\{c^{\mathrm{T}} x : Ax = b, \ x \ge 0\}.$$

with the extra convention that

$$z(b) = \begin{cases} +\infty & \text{if the primal is infeasible,} \\ -\infty & \text{if the primal is feasible and the dual is infeasible.} \end{cases}$$

We are interested in the behavior of z as a function of b. Clearly $z(b)$ is finite-valued if and only if the linear program (6.4) has an optimal solution. (This would be equivalent to both the primal and the dual being feasible.)

Now let $b \in R^m$ satisfy $-\infty < z(b) < +\infty$. Then the LP (6.4) has an optimal basis, say B. It then follows that with $\bar{y}^{\mathrm{T}} = c_B^{\mathrm{T}} B^{-1}$, we have

$$z(b) = \bar{y}^{\mathrm{T}} b = c_B^{\mathrm{T}} B^{-1} b.$$

Next suppose the basic feasible solution corresponding to B is *nondegenerate*, that is,

$$B^{-1} b > 0.$$

One implication of this assumption is that small variations of the components of b will not affect the feasibility of the basis B with respect to the modified problem. In addition, as we have already noted, B will yield a feasible solution to the dual constraints. Let $\hat{b}$ denote a modified right-hand side vector. All we require of $\hat{b}$ is that

$$x_B = B^{-1} \hat{b} \ge 0.$$

When this holds, B^{-1} is an optimal basis for the primal LP

$$\begin{aligned} \text{minimize} \quad & c^{\mathrm{T}} x \\ \text{subject to} \quad & Ax = \hat{b} \\ & x \ge 0, \end{aligned}$$

and moreover the vector $\bar{y}^{\mathrm{T}}$ given in (6.5) satisfies

$$z(\hat{b}) = \bar{y}^{\mathrm{T}}\hat{b}.$$

Let us now restrict our attention to the following type of variation of b:

$$\hat{b} = b + \theta I_{\bullet k}$$

where $I_{\bullet k}$, the k-th column of the identity matrix of order m, and θ belongs to an ε-neighborhood of zero. (This resembles what is done in ranging the right-hand side vector.)

The nondegeneracy assumption implies that for sufficiently small values of θ

$$B^{-1}(b + \theta I_{\bullet k}) \geq 0. \tag{6.6}$$

This permits us to write

$$z(b + \theta I_{\bullet k}) - z(b) = \bar{y}^{\mathrm{T}}(b + \theta I_{\bullet k}) - \bar{y}^{\mathrm{T}}b = \theta \bar{y}^{\mathrm{T}} I_{\bullet k} = \theta \bar{y}_k.$$

Thus,

$$\lim_{\theta \to 0} \frac{\bar{y}^{\mathrm{T}}(b + \theta I_{\bullet k}) - \bar{y}^{\mathrm{T}}b}{\theta} = \lim_{\theta \to 0} \frac{\theta \bar{y}_k}{\theta} = \bar{y}_k.$$

In other words,

$$\bar{y}_k = \frac{\partial z(b)}{\partial b_k}.$$

This tells us that when we have a nondegenerate basic feasible solution, the signs of the entries of the price vector indicate the effect that a slight change in a right-hand side constant will have on the optimal value of the objective function. If $\bar{y}_k > 0$, a slight increase in the value of b_k will increase the minimum value of the objective function, and if $\bar{y}_k < 0$, a slight increase in the value of b_k will decrease the minimum value of the objective function. If $c^{\mathrm{T}}x$ is considered to be the cost of the vector of activities x subject to the constraints, then $\bar{y}_k$ is the *marginal cost* relative to b_k.

The interpretation of the price vector corresponding to a *degenerate* basic feasible solution is just a little more complicated than above. The key question is whether the matrix B remains a feasible basis under the variation of b. In symbols, this amounts to having the condition (6.6) hold for all θ in a neighborhood of zero. If it does, the interpretation of $\bar{y}$ is the same as in the nondegenerate case.

Changing the right-hand side vector in a more general way, from b to $b+\theta d$, where now $d \in R^m$ is *arbitrary* and $B^{-1}(b+\theta d) \geq 0$ for all sufficiently small θ, leads to the *directional derivative*

$$z'(b; d) = \bar{y}^{\mathrm{T}} d.$$

In the economics (and operations research) literature, the multipliers or dual variables $\bar{y}_i$ are called *shadow prices*.[4] One might imagine that shadow prices can be interpreted as one-sided partial derivatives, but this is not always true. Nevertheless there are circumstances when it is. We take this up in Exercise 6.16.

An economic interpretation of optimal dual variables

Let B denote a basis for a *nondegenerate* optimal basic solution for a (primal) LP and let $\bar{x}$ be the corresponding BFS. Let $\bar{y}$ be the corresponding optimal solution of the dual problem (given by $\bar{y}^{\mathrm{T}} = c_B^{\mathrm{T}} B^{-1}$). Then z is a differentiable function of b. Indeed

$$z = c^{\mathrm{T}} \bar{x} = \bar{y}^{\mathrm{T}} b = c_B^{\mathrm{T}} B^{-1} b.$$

Since the solution is nondegenerate, the basis and the hence the simplex multipliers will not change for small changes in b_i. By taking the partial derivative of z with respect to b_i, we obtain

$$\frac{\partial z}{\partial b_i} = \left(c_B^{\mathrm{T}} B^{-1}\right)_i = \bar{y}_i.$$

The above implies that the multiplier, or dual variable, $\bar{y}_i$ is the amount by which the objective function changes per unit change in the value of the

---

[4] Although the term "shadow price" is widely used, it seems as though its provenance is rarely given. Nering and Tucker [149, p. 102] call it "a term due to the economist Paul A. Samuelson" while Dantzig and Thapa [44, p. 173] merely say it is "a term attributed to Paul Samuelson." Samuelson's writings [174], [175], [56] reveal his use of the term in connection with linear programming as early as 1949, yet it can be found elsewhere at least ten years earlier. Indeed, in his monograph *Value and Capital* [95, pp. 110–111] published in 1939, the economist J.R. Hicks wrote "There is a real price, which is fixed as a datum, and there is a 'shadow price', which is determined by equilibrium conditions." Also in 1939, L.V. Kantorovich [105] referred to dual variables as "resolving multipliers" and hinted at the economic significance of the optimal dual variables $\bar{y}_i$ which he called "objective valuations." As an article by Samuelson [175] points out, F.H. Knight [113] expressed the *idea* (though not the term shadow price) in 1925.

right-hand side b_i; that is, $\bar{y}_i$ is the "price" per unit change in the right-hand side b_i. If $\bar{y}_i > 0$, then the optimal value of z will increase with increasing b_i; whereas if $\bar{y}_i < 0$, then the optimal value of z will decrease with increasing b_i.

Example 6.9: SHADOW PRICES UNDER NONDEGENERACY. The shadow prices at an optimal solution of the LP in Example 3.9 are $(\bar{y}_1, \bar{y}_2, \bar{y}_3) = (-2/15, 0, -29/15)$ as can be seen in its final tableau quoted below.

min	z	x_1	x_2	x_3	x_4	x_5	x_6	x_7	1	basic
row 0	1	0	$-1/3$	0	$-11/3$	$-2/15$	0	$-29/15$	$-16/3$	z
row 1	0	0	$1/3$	1	$2/3$	$2/15$	0	$-1/15$	$4/3$	x_3
row 2	0	0	-5	0	-3	0	1	-2	3	x_6
row 3	0	1	$4/3$	0	$2/3$	$-1/15$	0	$8/15$	$1/3$	x_1

This says that if b_1 is increased by 1, so that $b_1 = 12$, then the objective function will decrease by $-2/15$. On the other if b_1 is decreased by 1, so that $b_1 = 10$, then the objective function will increase by $2/15$. While this can be easily verified for a change to b_1, the shadow prices reflect instantaneous changes, and it may well be that the basis changes if a modification by 1 is made to the right-hand side b_1. For example, if b_3 is decreased by 1 to $b_3 = 1$, the basis changes, and the objective changes by a larger amount than that indicated by the shadow prices. Notice further that the shadow price on constraint 2 is zero. This implies that small changes to b_2 do not affect the objective function. This makes sense because the constraint has slack in it and so there is no benefit from having additional resources. Likewise, a small reduction in b_2 is covered by the slack, so there is no loss in value.

If, on the other hand, B is a basis for a *degenerate* optimal solution for a (primal) LP, then the interpretation can be complicated. The reason for this is that even a small change in one direction may cause a change in basis and hence a change in the multipliers (shadow prices), as illustrated in the following example. However, a complete discussion of this is outside the scope of this book.

Example 6.10: SHADOW PRICES UNDER DEGENERACY. The shadow prices at the degenerate optimal solution of the LP in Example 4.4 are $(\bar{y}_1, \bar{y}_2, \bar{y}_3) = (45/62, 0, 13/62)$ and the optimal solution has $x_2 = 0$ as can be seen in its final tableau quoted below.

max	z	x_1	x_2	x_3	x_4	x_5	x_6	1	basic
row 0	1	301/62	0	0	45/62	0	13/62	30	z
row 1	0	49/62	0	1	3/62	0	5/62	6	x_3
row 2	0	−179/31	0	0	−35/31	1	14/31	17	x_5
row 3	0	−30/31	1	0	−5/31	0	2/31	0	x_2

If we were to decrease the right-hand side $b_3 = 60$ of the LP in Example 4.4 by even an infinitesimal amount the basis would change, and it would not be possible to predict the change to the objective value. However, if we were to increase the value of b_3 by 1, that is, set $b_3 = 61$, the basis would remain the same and the objective would in fact increase by the shadow price $\bar{y}_3 = 13/62$.

6.5 Exercises

6.1   Under the heading of "Cuts" we looked at the matter of adding a constraint of the form $a^T x \leq d$ after an optimal solution has been found. Discuss how you would handle a model like the one on page 143 where the adjoined constraint has the form

$$a_B^T x_B + a_N^T x_N \geq d.$$

6.2   Consider the linear programming problem

$$
\begin{array}{rl}
\text{minimize} & 4x_1 + 3x_2 + 2x_3 + 6x_4 + x_5 \\
\text{subject to} & 3x_1 + 4x_2 - x_3 + 2x_4 - x_5 = 14 \\
& -x_1 + 4x_3 + x_4 - 2x_5 = 15 \\
& -2x_1 + x_2 - 2x_3 + 5x_5 = 10 \\
& x_j \geq 0, \ j = 1, \ldots, 5.
\end{array}
$$

(a) Verify that the matrix $B = [A_{.1}\ A_{.3}\ A_{.5}]$ is a feasible but not optimal basis for the above LP.

(b) Compute the inverse of $\tilde{B} = [A_{.2}\ A_{.3}\ A_{.5}]$ and verify that $\tilde{B}$ is an optimal basis for the given LP.

(c) Perform ranging on the right-hand side of the given LP.

(d) Perform ranging on the cost coefficients of the given LP.

6.3   The following LP was used in Example 6.5:

$$\begin{array}{llrcl}
\text{minimize} & -6x_1 & - & 3x_2 & - & 9x_3 \\
\text{subject to} & -3x_1 & - & 2x_2 & + & 8x_3 & \leq 10 \\
& 4x_1 & + & 10x_2 & + & 5x_3 & \leq 12 \\
& & & x_j \geq 0, & j = 1, 2, 3.
\end{array}$$

Regarding the ranging procedure, it was said (see page 156) that

> "this sort of analysis does not allow for simultaneous changes
> of two or more right-hand side constants. Figure 6.2 illustrates
> this fact with respect to the cone spanned by the basis $B =$
> $[A_{\cdot 3} \ A_{\cdot 1}]$."

Explain why this is so, and prove your point by showing what goes wrong
in Figure 6.2.

6.4   Let $d = (1, 0, -1, 1, -1)$ and solve the parametric linear program

$$P(\lambda) \qquad \begin{array}{ll}
\text{minimize} & z(\lambda) = (c + \lambda d)^{\mathrm{T}} x \\
\text{subject to} & Ax = b \\
& x \geq 0,
\end{array}$$

where A, b, and c are as in Exercise 6.2 and $\lambda \in R$. Organize your answer
in a table and graph the optimal objective value as a function of λ.

[Suggestion: Start from $P(0)$ and the optimal BFS corresponding to $\tilde{B}$
in Exercise 6.2(b). Then consider only nonnegative λ. After this is done,
replace d by its negative and go back to solve the modified $P(0)$ with
$\lambda \geq 0$.]

6.5   Solve the parametric linear program

$$(P_\lambda) \qquad \begin{array}{ll}
\text{minimize} & 8x_1 + 3x_2 + 5x_3 \\
\text{subject to} & 2x_1 - x_2 + 5x_3 \leq 12 + \lambda \\
& -2x_1 + 4x_2 + 8x_3 \geq 6 + \lambda \\
& x_1, x_2, x_3 \geq 0.
\end{array}$$

where λ is a parameter belonging to the interval $[-15, 15]$. Organize your
answer in a table which gives the optimal solution (including the objective
function value) as a function of the parameter λ.

6.6   Consider the minimization problem LP in standard form with data

$$A = \begin{bmatrix} 1 & 3 & -1 & 5 \\ 0 & 5 & -5 & 10 \\ 2 & 1 & 3 & 0 \end{bmatrix}, \qquad b = \begin{bmatrix} 8 \\ 7 \\ 6 \end{bmatrix}, \qquad c = (1, 1, 1, 1).$$

(a) Does this problem have a feasible solution?

(b) Let $d = (7, 10, 2)$. Do the constraints

$$Ax = b + \lambda d, \quad x \geq 0, \quad \lambda \in R$$

have a solution?

(c) Relate the answer given in part (b) to the symbols $\lambda_{\min}$ and $\lambda_{\max}$ for the parametric linear program

$$
\begin{array}{rl}
& \text{minimize} \quad z(\lambda) = c^T x \\
P(\lambda) & \text{subject to} \quad Ax = b + \lambda d \\
& \qquad\qquad\quad x \geq 0, \ \lambda \in R.
\end{array}
$$

6.7 For all $\lambda \geq 0$, solve

$$
\begin{array}{rl}
& \text{minimize} \quad -(1 - 2\lambda)x_1 - (4 - \lambda)x_2 \\
P(\lambda) & \text{subject to} \qquad\qquad x_1 + \qquad 2x_2 \leq 6 \\
& \qquad\qquad\qquad\qquad -x_1 + \qquad 3x_2 \leq 6 \\
& \qquad\qquad\qquad x_j \geq 0, \ j = 1, 2.
\end{array}
$$

6.8 The LP of Example 3.8 is of the form commonly encountered in optimal resource allocation problems where the right-hand side represents amounts of resources and the coefficient matrix represents the how much the different activities consume of the different resources when each activity is operated at unit level. The objective function coefficients are then the contributions of the activities when operated at unit level. With this as an interpretation of the model, how much would an additional unit of each resource add to the optimal value of the objective?

6.9 Use the Simplex Method to solve the LP product-mix problem described in Example 1.3 in Chapter 1.

(a) What are the shadow prices on the scarce resources: corn, hops, and malt?

(b) Interpret the shadow prices in economic terms. That is, estimate the financial impact per unit change of each of the scarce resources. Verify your answer by re-solving the linear program. (Hint: B^{-1} is available so it is a simple check to determine if the new solutions are optimal and feasible, etc.)

6.10 Verify the discussion on shadow prices in Example 6.9.

6.11 Verify the discussion on shadow prices in Example 6.10.

6.12   Solve the parametric linear programming problem

$$\begin{array}{rl}
\text{maximize} & x_1 + x_2 \\
\text{subject to} & 2x_1 + x_2 \leq 8 + 2\lambda \\
& x_1 + 2x_2 \leq 7 + 7\lambda \\
& x_2 \leq 3 + 2\lambda \\
& x_1 \geq 0, \ x_2 \geq 0.
\end{array}$$

for all values of λ such that $-1 \leq \lambda \leq 3/4$. In your answer, express the optimal values of the decision variables, slack variables and optimal objective value as functions of λ over the given interval.

[Suggestion: For a given value of λ, let $P(\lambda)$ denote the linear program

$$\begin{array}{rl}
\text{maximize} & x_1 + x_2 \\
\text{subject to} & 2x_1 + x_2 \leq 8 + 2\lambda \\
& x_1 + 2x_2 \leq 7 + 7\lambda \\
& x_2 \leq 3 + 2\lambda \\
& x_1 \geq 0, \ x_2 \geq 0.
\end{array}$$

In this case, it is obvious that $P(0)$ has an optimal solution; begin by finding it.]

6.13   Consider the linear program $P(\lambda)$

$$\begin{array}{rl}
\text{minimize} & 3x_1 + (4 + 2\lambda)x_2 + 2x_3 - x_4 + (1 - 3\lambda)x_5 \\
\text{subject to} & 5x_1 + 3x_2 + 2x_3 - 2x_4 + x_5 = 20 \\
& x_1 + x_2 + x_3 - x_4 + x_5 = 8 \\
& x_j \geq 0, \quad j = 1, \ldots, 5.
\end{array}$$

where $\lambda \leq -1/4$.

(a) Show that $B = [A_{\bullet 2}, A_{\bullet 3}]$ is a feasible basis for this problem.

(b) For what range of λ values is B an optimal basis for $P(\lambda)$?

(c) Solve this parametric linear programming problem for all $\lambda \leq -1/4$. That is, find the set of values of the parameter λ for which $P(\lambda)$ has an optimal solution, and for all such values, find an optimal point $\bar{x}(\lambda)$ and the corresponding optimal objective value $\bar{z}(\lambda)$.

6.14   Imagine you are a decisionmaker for a firm having a product mix problem of the form

$$\begin{array}{rl}
\text{maximize} & 2x_1 + 4x_2 + x_3 + x_4 \\
\text{subject to} & x_1 + 3x_2 + x_4 \leq 8 \\
& 2x_1 + x_2 \leq 6 \\
& x_2 + 4x_3 + x_4 \leq 6 \\
& x_j \geq 0, \quad j = 1, 2, 3, 4.
\end{array}$$

The three numbers on the right-hand side of this linear program represent available amounts of raw materials measured in the same units.

When this LP is solved by the Simplex Algorithm, the optimal basis index set is $\{2, 1, 3\}$ and the inverse of the corresponding optimal basis is

$$B^{-1} = \frac{1}{20} \begin{bmatrix} 8 & -4 & 0 \\ -4 & 12 & 0 \\ -2 & 1 & 5 \end{bmatrix}.$$

(a) Suppose you have the opportunity to arrange for the marginal increase of *one* of the three raw materials at no additional cost. Which one would you choose and why?

(b) Over what range of values of the available amount of raw material 1 is the aforementioned basis B optimal?

(c) What is the most you would be willing to pay for 7 more units of raw material 1?

(d) Assume your firm has an option to manufacture a new product. Let the level of this production activity be x_8. The corresponding vector of coefficients is

$$(4, \ 2, \ 6 - 3\theta, \ -2 + \theta)$$

where θ is a parameter that can be set anywhere between 0 and 6. What is the *minimum* value of θ at which it *becomes* profitable to manufacture the new product?

6.15   In solving the resource allocation problem

$$\begin{aligned} \text{maximize} \quad & 45x_1 + 80x_2 \\ \text{subject to} \quad & 5x_1 + 20x_2 \leq 400 \\ & 10x_1 + 15x_2 \leq 450 \\ & x_1, \ x_2 \geq 0. \end{aligned}$$

one obtains the tableau

min	z	x_1	x_2	x_3	x_4	1	basic
row 0	1	0	0	1	4	2200	z
row 1	0	0	1	$\frac{2}{25}$	$-\frac{1}{25}$	14	x_2
row 2	0	1	0	$-\frac{3}{25}$	$\frac{4}{25}$	24	x_1

(a) Identify the optimal basis in this problem.

(b) Identify the optimal solution of the primal and the dual problems in this case.

(c) Determine the range of values of the first resource for which the current basis remains optimal.

(d) Determine the range of values of the coefficient of x_2 in the objective function for which the current basis remains optimal.

(e) Determine whether the basis would remain optimal if a third activity x_3 with associated column $(90, 15, 20)^{\mathrm{T}}$ were introduced.

6.16   Let

$$A = \begin{bmatrix} 3 & 2 & 1 & 0 & 4 & -1 \\ 2 & 4 & 2 & -4 & 3 & 2 \\ 1 & 2 & 3 & 4 & 5 & 6 \\ 0 & -4 & 4 & 5 & -2 & 1 \end{bmatrix}$$

$$b = (11, 22, 21, 8) \quad \text{and} \quad c = (11, 14, 17, 1, 26, 20).$$

The first four columns of the matrix A constitute an optimal basis B in the linear program

$$\begin{aligned} \text{minimize} \quad & c^{\mathrm{T}}x \\ \text{subject to} \quad & Ax = b \\ & x \geq 0. \end{aligned}$$

(a) Find the inverse of B.

(b) Find the basic solution corresponding to B and verify that it is a degenerate optimal solution of the LP.

(c) Find the optimal solution of the dual problem.

(d) For each of the components of the right-hand side vector b determine whether (positive or negative) perturbations of that component yield a feasible solution of the LP. (Hint: See Section 6.2.)

(e) What does all this say about giving an economic interpretation to the shadow prices in this LP?

(f) Investigate what can be said about shadow prices in other LPs having degenerate optimal basic solutions such that only one basic variable has value zero?

6.17   In this exercise we investigate a method for solving integer programs, that is, LPs which require the final solution to be integral. Consider the integer program

$$\begin{aligned} \text{minimize} \quad & -x_1 - x_2 \\ \text{subject to} \quad & -5x_1 + 3x_2 \leq 0 \\ & 3x_1 + x_2 \leq 15 \\ & x_1, x_2 \geq 0 \\ & x_1, x_2 \text{ integers.} \end{aligned}$$

(a) Plot the feasible region on a grid assuming nonintegral values.

(b) Mark all integral solutions.

(c) Add slack variables to put it in standard form.

(d) Solve the problem using the Simplex Algorithm ignoring the integer constraints.

(e) The optimal solution $x_1 = 45/14$ and $x_2 = 75/14$ is not integral. Create a new constraint, called a *cut*, as follows:

   (a) Pick a row in the final tableau where the basic variable is not an integer. The second row of the tableau, which has x_1 as a basic variable is, $x_1 + 0x_2 - (1/14)s_1 + 3/14s_2 = 45/14$.

   (b) From this row create a new i-th constraint of the form

$$\sum_j \underline{a}_{ij} x_j \geq \underline{b}_i,$$

   where the underline symbol is used to denote the positive fractional part. In this case it would be $(13/14)s_1 + (3/14)s_2 \geq 3/14$. (Note: In this case, a further simplification is possible resulting in the inequality $x_1 \leq 3$.)

(c) Add the cut $x_1 \leq 3$ to the LP and plot the new region.

(d) Use the Dual-Simplex Algorithm as discussed on page 143 to solve this problem. In this case, you will find that the optimal solution is integral.

(e) Instead of adding $x_1 \leq 3$, add the cut $(13/14)s_1 + (3/14)s_2 \geq 3/14$ and re-solve the problem.

7. SOME COMPUTATIONAL CONSIDERATIONS

Overview

In this chapter we take up some additional techniques of a mostly practical nature: the handling of linear programs with explicitly bounded variables; the construction of a starting (feasible) basis; structured linear programs; the steepest-edge rule for column selection; the rare (but possible) exponential behavior of the Simplex Algorithm. Each of these is a large subject in its own right, so we shall limit our discussion to the most important fundamental concepts.

In practice, linear programs tend to be *large*.[1] They also tend to be *structured*, that is, they possess block-matrix patterns of zeros and nonzeros that arise for various modeling reasons, usually spatial or temporal. In an effort to achieve computational efficiency, investigators have devised algorithms to take advantage of matrix structure. Generally speaking, one important aim is to reduce the size of the problems that need to be solved. In this chapter, we will briefly discuss the best-known examples of these algorithms: the Dantzig-Wolfe Decomposition Procedure and the Benders Decomposition Method.

The remainder of the chapter is devoted to a class of linear programs (found by Klee and Minty) on which the conventional Simplex Algorithm can require an exponential number of iterations. Following its publication in 1972, this discovery sparked an intense burst of activity in linear programming methodology, particularly in the area called *interior-point algorithms*. Because the study of polynomial-time algorithms for solving linear programs employs concepts and methods of nonlinear optimization (described in Part II), we postpone the detailed discussion of interior-point algorithms to Chapter 14.

7.1 Problems with explicitly bounded variables

There are several considerations that go into an implementation of the Simplex Algorithm on a computer. For example, the constraints in a given model

---

[1] The numerical examples exhibited in books such as this are necessarily very small. For one thing, they need to fit on the page!

© Springer Science+Business Media LLC 2017

R.W. Cottle and M.N. Thapa, *Linear and Nonlinear Optimization*,
International Series in Operations Research & Management Science 253,
DOI 10.1007/978-1-4939-7055-1_7

will not necessarily all be equations, or all be inequalities. Further, many activity levels may have a lower bound different from 0 and an upper bound that is finite. It is a routine exercise to convert such a linear program into standard form. However, from a computational standpoint, it is inefficient to consider the explicit representation of upper-bounded variables as separate inequality constraints. We shall see that with suitable modification of the Simplex Algorithm, it is unnecessary to carry out such transformations.

To appreciate the distinction between explicit and implicit representations of upper-bounded variables consider the following two systems

$$x_1 + x_2 + \cdots + x_n = 1$$
$$0 \le x_j \le 1, \quad j = 1, \ldots, n, \tag{7.1}$$

and

$$x_1 + x_2 + \cdots + x_n = 1$$
$$x_j \ge 0, \quad j = 1, \ldots, n. \tag{7.2}$$

In (7.1) the variables x_j all have *explicit* upper-bound constraints. In (7.2) the variables are bounded above, but the upper-bound constraints $x_j \le 1$ are *implicit*; they are a consequence of the other constraints. Needless to say, there are LP formulations in which some of the variables are explicitly bounded while others are implicitly bounded.

Often when practitioners develop linear programming models, one or more of the constraints may turn out to be redundant. This poses a problem since the Simplex Algorithm requires that the basis be a nonsingular $m \times m$ matrix. Accordingly, techniques have been devised to handle such cases. See page 197.

While many practical linear programs can be solved very quickly on computers available today, there are real-world cases where computational efficiency is important. For example, there are very large linear programs having hundreds of thousands of constraints and millions of variables; in addition there are also cases that require the solution of linear programs in "real time." Hence from a computational efficiency perspective, the implementation needs careful thought. Techniques for efficiency include taking advantage of sparsity, using different incoming column selection rules, pricing out only some of the columns, etc. Often very large problems have a structure that can be exploited by using decomposition techniques; see [120].

It is not always convenient to start with an identity matrix as the initial basis. Often for practical problems an initial basis is known. If such a basis is not known, then one can be constructed by choosing columns from A. In yet other cases, the known "basis" may have dependencies in its columns, in which case, it is not, strictly speaking, a basis; other columns (possibly columns associated with artificial variables) may need to be added to construct a basis. In each of these cases, the basis formed might not be feasible; in which case a Phase I Procedure would then need to be invoked.

Lower and upper bounds on variables

Consider an LP with m equality constraints in n variables having explicit lower and upper bounds that are not necessarily 0 and ∞ respectively.

$$
\begin{aligned}
\text{minimize} \quad & c^{\mathrm{T}}x \\
\text{subject to} \quad & Ax = b, \\
& l_j \le x_j \le u_j \quad \text{for } j = 1, 2, \ldots, n.
\end{aligned}
\tag{7.3}
$$

We will assume that each variable has at least one finite bound. Any variable, not satisfying this assumption must be a free variable and hence can be replaced by the difference of two nonnegative variables, thereby making the assumption hold. Furthermore, any linear program can be put in this form.

For the moment, let us assume that for $j = 1, 2, \ldots, n$ we have $l_j = 0$, and every upper bound u_j finite and positive. Such a problem can be put in standard form by adding slack variables $x'_j \ge 0$ to the constraints $x_j \le u_j$ as follows:

$$
\begin{aligned}
\text{minimize} \quad & c^{\mathrm{T}}x \\
\text{subject to} \quad & Ax \qquad = b \\
& Ix + Ix' = u \\
& x \ge 0, \ x' \ge 0.
\end{aligned}
$$

This formulation would lead to bases that are of size $(m+n) \times (m+n)$, which is plainly undesirable. As the following discussion demonstrates, we need not make any such assumptions or undesirable transformations.

Problem (7.3) can be solved by the Simplex Algorithm with fairly minor modifications but with a little more bookkeeping. The following is a brief description of the changes to the Simplex Algorithm designed to facilitate the solution of problems of the form (7.3).

Modifications of the Simplex Algorithm for bounded-variable LPs

In this discussion we assume that at least one of the explicit bounds on each variable is finite. The following is a brief description of such modifications.

1. Earlier we defined a basic solution of a linear program in standard form as one obtained by setting the nonbasic variables to 0 (i.e., at their lower bounds). In the bounded-variable case above, we define a *basic solution* to be one obtained by setting each nonbasic variable to a finite (lower or upper) bound.

2. Given a basis B and the subvector $\bar{x}_N$, where each nonbasic variable is at its lower bound or upper bound, we can compute the values of the basic variables by solving

$$Bx_B = b - N\bar{x}_N.$$

3. A basic feasible solution (BFS) in a bounded-variable problem $x = \bar{x}$ is one that satisfies

$$B\bar{x}_B = b - N\bar{x}_N, \quad l_B \le \bar{x}_B \le u_B, \quad \text{and} \quad \bar{x}_j = l_j \text{ (or } u_j) \text{ for all } j \in N.$$

A *nondegenerate* basic feasible solution is one that satisfies the above and further satisfies

$$l_B < \bar{x}_B < u_B.$$

4. Recall that the optimality condition for solving a linear program in standard form is that $z_j - c_j \le 0$ for $j \in N$, where the nonbasic variables were at their lower bound of 0. For (7.3), we compute $z_j - c_j$ in exactly the same way, but here we require a slight modification to the optimality criterion:

   - $z_j - c_j \le 0$ for the nonbasic variables x_j that are at their *lower bound*. The argument to show this is identical to that discussed for the standard form.

   - $z_j - c_j \ge 0$ for the nonbasic variables x_j that are at their *upper bound*. The argument to show this is analogous to that discussed for the standard form for nonbasic variables at their lower bound.

5. When applying the Simplex Algorithm to a linear program in standard form, a nonbasic variable chosen to become basic does so at the end

of the iteration. However, in the case of (7.3), a nonbasic variable—
selected by the minimum ratio test—may possibly remain nonbasic at
the end of an iteration of the Simplex Algorithm. The reason for this
is that as it is increased (decreased) from its lower (upper) bound, it
may reach the other bound before a current basic variable reaches one
of its bounds. In this case there is a change in the basic solution but
not of the basis. This phenomenon is illustrated in Example 7.1 below.

Modifications of the Simplex Algorithm for LPs with free variables

Earlier we made the assumption that at least one bound on each variable
was finite. Any variable not satisfying this assumption must be free, and we
have remarked that a free variable can be written as the difference of two
nonnegative variables which are its positive and negative parts. In practice,
this is not necessary; simple modifications to the Simplex Algorithm for the
bounded-variable case allow us to handle this quite easily. The following is
a brief description of such modifications.

1. From the earlier definition of nonbasic variables, it appears that a free
   variable (one which does not have a lower or upper bound) will always
   be basic. However, it may turn out that the column corresponding to
   the free variable is a linear combination of the columns in the basis.
   In this case, the nonbasic free variable will be maintained at the fixed
   value of zero.

2. A basic feasible solution (BFS) $x = \bar{x}$ is one that satisfies

$$B\bar{x}_B = b - N\bar{x}_N, \quad l_B \leq \bar{x}_B \leq u_B,$$

   and each nonbasic $\bar{x}_j$ is at one of its bounds, or, if nonbasic x_j is a
   free variable, then its value is 0.

3. For optimality we require $z_j - c_j = 0$ for the nonbasic free variables
   x_j that have been set to zero. A free variable can take on any value.
   Hence, depending on the sign of $z_j - c_j$ it can be adjusted to decrease
   the objective function value. The only time it will not affect the objective function value is when $z_j - c_j = 0$.

The brief discussion above shows the modifications needed to apply the
Simplex Algorithm to the bounded-variable problem. We now state the

algorithm under the assumption that a feasible basis B is known.

Algorithm 7.1 (Simplex Algorithm for bounded-variable LPs)

1. *Initialization.* Input the data A, b, c, l, u for a minimization problem with bounds on the variables. Assume the system is in canonical form with respect to a known feasible basis B. Let $x = \bar{x}$ be the current basic feasible solution.

2. *Test for optimality.* Find an index s such that (relative to the current basis)

$$z_s - c_s = \max \begin{cases} (z_j - c_j) & \text{if } \bar{x}_j = l_j \quad \text{and } l_j \neq u_j, \\ -(z_j - c_j) & \text{if } \bar{x}_j = u_j \quad \text{and } l_j \neq u_j, \\ |z_j - c_j| & \text{if } x_j \text{ is free.} \end{cases}$$

If $z_s - c_s \leq 0$, stop. The basic feasible solution is optimal.

3. *Minimum ratio test.* Find an index r that results in the largest amount α_r by which x_s can change while retaining feasibility. (Note that in the bounded-variable case, if every bound is finite, then the linear program can never have an unbounded objective value.) If x_s is to be increased, i.e., $x_s = l_s$ or x_s is a free variable and $z_s - l_s > 0$, then

$$\alpha_r = \min \begin{cases} \dfrac{\bar{x}_{j_i} - l_{j_i}}{\bar{a}_{is}} & \text{for all } i \text{ such that } \bar{a}_{is} > 0, \\ \dfrac{u_{j_i} - \bar{x}_{j_i}}{-\bar{a}_{is}} & \text{for all } i \text{ such that } \bar{a}_{is} < 0, \\ u_s - l_s. \end{cases}$$

If x_s is to be decreased; i.e., $x_s = u_s$ or x_s is a free variable and $z_s - l_s < 0$, then

$$\alpha_r = \min \begin{cases} \dfrac{\bar{x}_{j_i} - l_{j_i}}{-\bar{a}_{is}} & \text{for all } i \text{ such that } \bar{a}_{is} < 0, \\ \dfrac{u_{j_i} - \bar{x}_{j_i}}{\bar{a}_{is}} & \text{for all } i \text{ such that } \bar{a}_{is} > 0, \\ u_s - l_s. \end{cases}$$

In either of these two situations (increasing or decreasing x_s) it could happen that $\alpha_r = +\infty$, and if this occurs, terminate the algorithm and report an unbounded objective function.

4. *Pivot step.* There are two cases to consider here.

- If $r \neq s$, pivot on $\bar{a}_{rs}$. Define the new basis in which $A_{.s}$ replaces the column $A_{.k}$ where $k = j_r$. Define $j_r = s$.

- If $r = s$, recompute $\bar{x}_B$, using the updated value for the nonbasic variable x_s, which switches bounds but remains nonbasic.

Return to Step 1.

Example 7.1: SOLVING A BOUNDED-VARIABLE LP. Consider the linear program

$$
\begin{aligned}
\text{minimize} \quad & - 2x_1 - 4x_2 - 2x_3 - 3x_4 \\
\text{subject to} \quad & x_1 + 3x_2 - 2x_3 + 4x_4 \leq 10 \\
& \quad\quad -2x_2 + 4x_3 + 2x_4 \leq 20 \\
& 2x_1 + 3x_2 \quad\quad - 5x_4 \leq 30 \\
& x_1 \geq 0, \ 0 \leq x_2 \leq 10, \ 0 \leq x_3 \leq 8, \ x_4 \geq 0.
\end{aligned}
$$

After adding slacks and converting the problem to tableau form we obtain

min	z	x_1	x_2	x_3	x_4	x_5	x_6	x_7	1	basic
row 0	1	2	4	2	3	0	0	0	0	z
row 1	0	1	3	-2	4	1	0	0	10	x_5
row 2	0	0	-2	4	2	0	1	0	20	x_6
row 3	0	2	3	0	-5	0	0	1	30	x_7
lb	$-\infty$	0	0	0	0	0	0	0		
ub	∞	∞	10	8	∞	∞	∞	∞		
nonbasic		0	0	0	0					

Let $\bar{c}_j = z_j - c_j$ for all j. Next we try to find an index s for the incoming nonbasic column.

$$
\bar{c}_s = \max \begin{cases} \{\bar{c}_1, \bar{c}_2, \bar{c}_3, \bar{c}_4\} & \text{for nonbasic } \bar{x}_j = l_j, \\ \{\} & \text{for nonbasic } \bar{x}_j = u_j, \\ \{\} & \text{for nonbasic } x_j \text{ free}, \end{cases}
$$

$$
= \max \begin{cases} \{2, 4, 2, 3\} & \text{for nonbasic } \bar{x}_j = l_j, \\ \{\} & \text{for nonbasic } \bar{x}_j = u_j, \\ \{\} & \text{for nonbasic } x_j \text{ free}, \end{cases}
$$

$$
= 4 = \bar{c}_2.
$$

Thus $s = 2$ and the incoming column is chosen to be the one corresponding to x_2. The variable corresponding to the outgoing column is determined by

$$
\alpha_r = \min \left\{
\begin{array}{l}
\left\{ \dfrac{\bar{b}_1 - l_5}{\bar{a}_{12}}, \dfrac{\bar{b}_3 - l_7}{\bar{a}_{32}} \right\} \\[2ex]
\dfrac{u_6 - \bar{b}_2}{-\bar{a}_{22}} \\[2ex]
u_2 - l_2,
\end{array}
\right.
$$

$$
= \min \left\{
\begin{array}{l}
\left\{ \dfrac{10}{3}, \dfrac{30}{3} \right\} \\[2ex]
\infty \\[2ex]
10,
\end{array}
\right.
$$

$$
= \frac{10}{3} = \alpha_1.
$$

Pivot to make x_2 basic in place of x_5.

min	z	x_1	x_2	x_3	x_4	x_5	x_6	x_7	1	basic
row 0	1	2/3	0	14/3	−7/3	−4/3	0	0	−40/3	z
row 1	0	1/3	1	−2/3	4/3	1/3	0	0	10/3	x_2
row 2	0	2/3	0	8/3	14/3	2/3	1	0	80/3	x_6
row 3	0	1	0	2	−9	−1	0	1	20	x_7
lb	−∞	0	0	0	0	0	0	0		
ub	∞	∞	10	8	∞	∞	∞	∞		
nonbasic		0		0	0	0				

The basic variable values above are computed using

$$
x_B = B^{-1}(b - N\bar{x}_N) = \bar{b}
$$

$$
= \begin{bmatrix} 1/3 & 0 & 0 \\ 2/3 & 1 & 0 \\ -1 & 0 & 1 \end{bmatrix} \begin{bmatrix} 10 \\ 20 \\ 30 \end{bmatrix} = \begin{bmatrix} 10/3 \\ 80/3 \\ 20 \end{bmatrix}
$$

because all the nonbasic variables are at their lower bound of 0, the vector $N\bar{x}_N$ is zero. The objective function value is computed using $c^T x$. However, in this case, because the nonbasic variables are all at their lower bound of 0, ordinary pivoting would have yielded the same values.

Next we examine the updated reduced costs again,

$$\bar{c}_s = \max \begin{cases} \{\bar{c}_1, \bar{c}_3, \bar{c}_4, \bar{c}_5\} = \{2/3, 14/3, -7/3, -4/3\} & \text{for nonbasic } \bar{x}_j = l_j \\ \{\} & \text{for nonbasic } \bar{x}_j = u_j \\ \{\} & \text{for nonbasic } x_j \text{ free} \end{cases}$$

$$= 14/3 = \bar{c}_3.$$

Hence the incoming column is chosen to be the one corresponding to x_3. The outgoing column, if any, is determined by:

$$\alpha_r = \min \begin{cases} \left\{ \dfrac{\bar{b}_2 - l_6}{\bar{a}_{23}}, \dfrac{\bar{b}_3 - l_7}{\bar{a}_{33}} \right\} = \left\{ \dfrac{80/3}{8/3}, \dfrac{20}{2} \right\} \\ \dfrac{u_5 - \bar{b}_1}{-\bar{a}_{13}} = \infty \\ u_3 - l_3 = 8 \end{cases}$$

$$= 8 = \alpha_3.$$

That is, the nonbasic variable x_3 increases from its lower bound to its upper bound of 8 *without becoming basic*. The tableau does not change except for the values of the basic variables and objective function. The basic variables are computed by

$$x_B = B^{-1}(b - N\bar{x}_N) = B^{-1}b - \bar{A}_{\bullet 3}\bar{x}_3$$

$$= \begin{bmatrix} 1/3 & 0 & 0 \\ 2/3 & 1 & 0 \\ -1 & 0 & 1 \end{bmatrix} \begin{bmatrix} 10 \\ 20 \\ 30 \end{bmatrix} - \begin{bmatrix} -2/3 \\ 8/3 \\ 2 \end{bmatrix} 8 = \begin{bmatrix} 26/3 \\ 16/3 \\ 4 \end{bmatrix}.$$

The objective function value is

$$z = c^T \bar{x} = -2\bar{x}_1 - 4\bar{x}_2 - 2\bar{x}_3 - 3\bar{x}_4 + 0\bar{x}_5 + 0\bar{x}_6 + 0\bar{x}_7 = -152/3.$$

The tableau at the end of this iteration is

min	z	x_1	x_2	x_3	x_4	x_5	x_6	x_7	1	basic
row 0	1	2/3	0	14/3	-7/3	-4/3	0	0	-152/3	z
row 1	0	1/3	1	-2/3	4/3	1/3	0	0	26/3	x_2
row 2	0	2/3	0	8/3	14/3	2/3	1	0	16/3	x_6
row 3	0	1	0	2	-9	-1	0	1	4	x_7
lb	$-\infty$	0	0	0	0	0	0	0		
ub	∞	∞	10	8	∞	∞	∞	∞		
nonbasic		0		8	0	0				

Because there was no change to the basis at the previous iteration, the reduced costs are the same, except that $\bar{c}_3$ is now the reduced cost of a variable at its upper bound. We determine the incoming column by

$$\bar{c}_s = \max \begin{cases} \{\bar{c}_1, \bar{c}_4, \bar{c}_5\} = \{2/3, -7/3, -4/3\} & \text{for nonbasic } \bar{x}_j = l_j \\ \{-\bar{c}_3,\} = -14/3 & \text{for nonbasic } \bar{x}_j = u_j \\ \{\} & \text{for nonbasic } x_j \text{ free} \end{cases}$$

$$= 2/3 = \bar{c}_1.$$

Thus the incoming column is chosen to be the one corresponding to x_1. The outgoing column is determined by

$$\alpha_r = \min \begin{cases} \left\{\dfrac{\bar{b}_1 - l_2}{\bar{a}_{11}}, \dfrac{\bar{b}_2 - l_6}{\bar{a}_{21}}, \dfrac{\bar{b}_3 - l_7}{\bar{a}_{31}}\right\} = \left\{\dfrac{26/3}{1/3}, \dfrac{16/3}{2/3}, \dfrac{4}{1}\right\} \\ \{\} = \infty \\ u_1 - l_1 = \infty \end{cases}$$

$$= 4 = \alpha_3.$$

Thus the incoming column is the one corresponding to x_1, and we pivot to make x_1 basic in place of x_7. Relative to the new basis (the inverse of which appears in the next tableau), the basic variable values are given by

$$x_B = B^{-1}(b - N\bar{x}_N) = B^{-1}(b - \bar{A}_{\bullet 3}\bar{x}_3)$$

$$= \begin{bmatrix} 2/3 & 0 & -1/3 \\ 4/3 & 1 & -2/3 \\ -1 & 0 & 1 \end{bmatrix} \begin{bmatrix} 10 \\ 20 \\ 30 \end{bmatrix} - \begin{bmatrix} -4/3 \\ 4/3 \\ 2 \end{bmatrix} 8 = \begin{bmatrix} 22/3 \\ 8/3 \\ 4 \end{bmatrix}.$$

The objective function value is computed by

$$z = c^T\bar{x} = -160/3.$$

The tableau at the end of this iteration is

min	z	x_1	x_2	x_3	x_4	x_5	x_6	x_7	1	basic
row 0	1	0	0	10/3	11/3	-2/3	0	-2/3	-160/3	z
row 1	0	0	1	-4/3	13/3	2/3	0	-1/3	22/3	x_2
row 2	0	0	0	4/3	32/3	4/3	1	-2/3	8/3	x_6
row 3	0	1	0	2	-9	-1	0	1	4	x_1
lb	$-\infty$	0	0	0	0	0	0	0		
ub	∞	∞	10	8	∞	∞	∞	∞		
nonbasic				8	0	0		0		

We determine the incoming column by:

$$\bar{c}_s = \max \begin{cases} \{\bar{c}_4, \bar{c}_5\,\bar{c}_7\} = \{11/3, -2/3, -2/3\} & \text{for nonbasic } \bar{x}_j = l_j \\ \{-\bar{c}_3,\} = -10/3 & \text{for nonbasic } \bar{x}_j = u_j \\ \{\} & \text{for nonbasic } x_j \text{ free} \end{cases}$$

$$= 11/3 = \bar{c}_4.$$

Thus the incoming column is chosen to be the one corresponding to x_4. The outgoing column is determined by

$$\alpha_r = \min \begin{cases} \left\{ \dfrac{\bar{b}_1 - l_2}{\bar{a}_{14}}, \dfrac{\bar{b}_2 - l_6}{\bar{a}_{24}} \right\} = \left\{ \dfrac{22/3}{13/3}, \dfrac{8/3}{32/3} \right\} \\ \left\{ \dfrac{u_7 - \bar{b}_3}{-\bar{a}_{34}} \right\} = \infty \\ u_4 - l_4 \qquad\qquad\qquad = \infty \end{cases}$$

$$= 8/32 = \alpha_2.$$

Thus the outgoing column is the one corresponding to x_6, and we pivot to make x_4 basic in place of x_6. The basic variable values are computed using

$$x_B = B^{-1}(b - N\bar{x}_N) = B^{-1}(b - \bar{A}_{\bullet 3}\bar{x}_3)$$

$$= \begin{bmatrix} 1/8 & -13/32 & -1/16 \\ 1/8 & 3/32 & -1/16 \\ 1/8 & 27/32 & 7/16 \end{bmatrix} \begin{bmatrix} 10 \\ 20 \\ 30 \end{bmatrix} - \begin{bmatrix} -15/8 \\ 1/8 \\ 25/8 \end{bmatrix} 8 = \begin{bmatrix} 25/4 \\ 1/4 \\ 25/4 \end{bmatrix}.$$

The objective function value is

$$z = c^T \bar{x} = -217/4.$$

These details are revealed in the final tableau which is

min	z	x_1	x_2	x_3	x_4	x_5	x_6	x_7	1	basic
row 0	1	0	0	23/8	0	−9/8	−11/32	−7/16	−217/4	z
row 1	0	0	1	−15/8	0	1/8	−13/32	−1/16	25/4	x_2
row 2	0	0	0	1/8	1	1/8	3/32	−1/16	1/4	x_4
row 3	0	1	0	25/8	0	1/8	27/32	7/16	25/4	x_1
lb	−∞	0	0	0	0	0	0	0		
ub	∞	∞	10	8	∞	∞	∞	∞		
nonbasic				8		8	0	0		

Example 7.2: AN LP WITH AN UNBOUNDED OBJECTIVE. Consider the LP

$$\begin{aligned}
\text{minimize} \quad & -2x_1 - 4x_2 - 2x_3 - 3x_4 \\
\text{subject to} \quad & x_1 + 3x_2 - 2x_3 - 1x_4 \le 10 \\
& \qquad\quad -2x_2 + 4x_3 - 9x_4 \le 20 \\
& 2x_1 + 3x_2 \qquad\quad - 5x_4 \le 30 \\
& x_1 \ge 0,\ 0 \le x_2 \le 10,\ 0 \le x_3 \le 8,\ x_4 \ge 0.
\end{aligned}$$

After adding slacks and converting to tableau form we obtain

min	z	x_1	x_2	x_3	x_4	x_5	x_6	x_7	1	basic
row 0	1	2	4	2	3	0	0	0	0	z
row 1	0	1	3	-2	-1	1	0	0	10	x_5
row 2	0	0	-2	4	-9	0	1	0	20	x_6
row 3	0	2	3	0	-5	0	0	1	30	x_7
lb	$-\infty$	0	0	0	0	0	0	0		
ub	∞	∞	10	8	∞	∞	∞	∞		
nonbasic		0	0	0	0					

As in Example 7.1 we determine the incoming column to be the one corresponding to x_2 and the outgoing one as that corresponding to x_5. Pivot to make x_2 basic in place of x_5.

min	z	x_1	x_2	x_3	x_4	x_5	x_6	x_7	1	basic
row 0	1	2/3	0	14/3	13/3	$-4/3$	0	0	$-40/3$	z
row 1	0	1/3	1	$-2/3$	$-1/3$	1/3	0	0	10/3	x_2
row 2	0	2/3	0	8/3	$-29/3$	2/3	1	0	80/3	x_6
row 3	0	1	0	2	-4	-1	0	1	20	x_7
lb	$-\infty$	0	0	0	0	0	0	0		
ub	∞	∞	10	8	∞	∞	∞	∞		
nonbasic		0		0	0	0				

Next, the incoming column is chosen to be the one corresponding x_3. However, in this case, x_3 reaches its upper bound before any column can leave the basis. The resulting tableau is

min	z	x_1	x_2	x_3	x_4	x_5	x_6	x_7	1	basic
row 0	1	2/3	0	14/3	13/3	$-4/3$	0	0	$-152/3$	z
row 1	0	1/3	1	$-2/3$	$-1/3$	1/3	0	0	26/3	x_2
row 2	0	2/3	0	8/3	$-29/3$	2/3	1	0	16/3	x_6
row 3	0	1	0	2	-4	-1	0	1	4	x_7
lb	$-\infty$	0	0	0	0	0	0	0		
ub	∞	∞	10	8	∞	∞	∞	∞		
nonbasic		0		8	0	0				

Next, the incoming column is chosen to be the one corresponding x_4. Now it is easy to see that the objective value is unbounded below.

7.2   Constructing a starting (feasible) basis

Practitioners often provide a set of activities—or equivalently, an index set—that corresponds to the columns of a starting basis. Sometimes, the new problem to be solved differs from a previously solved problem only in some of the objective function coefficients; in such a case, any feasible basis of the previous problem is also a feasible basis for the new problem. This situation is straightforward. However, if the new problem differs because of changes in the constraints, then some work is usually required to find a basis from the supplied index set. In other situations, the practitioner may "know" that certain columns must be basic and may provide the corresponding index set. The provided index set need not result in a "real" basis because there may be unforeseen dependencies among the corresponding columns. Alternatively, the practitioner may provide fewer than m columns as required for a basis. If so, additional columns must be chosen either from the matrix A (which possibly includes columns corresponding to slack and/or surplus variables) or columns associated with artificial variables.

When choosing columns from A, we can select them so as to obtain a basis that is sparse. In this case we can construct a sparse LU factorization of the basis, see [88]. Of course, in all these approaches, the basis so constructed may not be feasible, in which case we invoke a Phase I Procedure, such as "minimizing the sum of infeasibilities" (discussed in the next subsection).

A question that needs to be answered is: How do we determine whether a proposed basis column is linearly dependent on those already selected? We can do this as a part of the LU factorization process. The computation of the LU factorization is organized so that it works on one column at a time as it is added into the basis. This procedure easily identifies dependent columns. If a dependent column is identified, then it may be possible to choose another column that is independent, but it may also be the case that there is no independent column due to one or more redundant constraints being present. In commercial LP software (often called *solvers*), whenever a dependent column is found, an artificial variable (one with upper and lower bounds of 0) is added to the basic set. Some commercial solvers go further and perform a step known as "pre-solve." In this step, among other things, redundant constraints are identified to the extent possible, and removed from the problem data before an LP solver is invoked. See, for example, [3].

When solving a linear program in standard form, once the LU factorization has been obtained, an initial x_B (or $\bar{b}$) is computed in a manner analogous to that of obtaining $\bar{A}_{\bullet s}$, i.e., by solving

$$Bx_B = LUx_B = b.$$

For a linear program with explicitly bounded variables, we solve

$$Bx_B = LUx_B = b - Nx_N.$$

Of course, it is possible that such a basis may not be feasible, that is, at least one of the corresponding basic variables fails to satisfy its bounds. For example, $x_j = \bar{x}_j$ may be below its lower bound. Note that artificial variables must be zero for a feasible solution and thus we assume that each artificial variable, if any, has a lower bound of 0 and an upper bound of 0. If the chosen initial basis is infeasible, a Phase I Procedure that minimizes the sum of infeasibilities is used.

Minimizing the sum of infeasibilities

A basic solution to a bounded-variable linear program (7.3) is said to be *infeasible* if a basic variable violates one of its bounds. We define *the infeasibility value* (of a basic variable) to be the positive distance from its violated bound. That is, if $x = \bar{x}$ is a basic solution, then the infeasibility value of a basic variable x_j is

$$
\begin{aligned}
l_j - \bar{x}_j \quad &\text{if } \bar{x}_j < l_j, \\
\bar{x}_j - u_j \quad &\text{if } \bar{x}_j > u_j.
\end{aligned}
$$

Corresponding to the above, the *infeasibility term* of a basic variable x_j will be

$$
\begin{aligned}
l_j - x_j \quad &\text{if } \bar{x}_j < l_j, \\
x_j - u_j \quad &\text{if } \bar{x}_j > u_j.
\end{aligned}
$$

The approach for creating a Phase I Procedure that minimizes the sum of infeasibilities is as follows:

1. Let $x = \bar{x}$ be the current solution. Construct a Phase I objective by adding all the infeasibility terms. Start with an objective function w with no terms in it. For each basic variable x_j, if $\bar{x}_j < l_j$, add

the infeasibility term $l_j - x_j$ to the objective; if $\bar{x}_j > u_j$, add the infeasibility term $x_j - u_j$ to the objective. Thus the Phase I objective function to be minimized is

$$
w = \sum_{j:\bar{x}_j > u_j} (x_j - u_j) + \sum_{j:\bar{x}_j < l_j} (l_j - x_j)
$$

$$
= \sum_{j:\bar{x}_j > u_j} x_j - \sum_{j:\bar{x}_j < l_j} x_j + \sum_{j:\bar{x}_j < l_j} l_j - \sum_{j:\bar{x}_j > u_j} u_j.
$$

Since additive constants do not affect the location of an optimal solution $x = x^*$, we can write the Phase I objective function as

$$
w = \sum_{j:\bar{x}_j > u_j} x_j - \sum_{j:\bar{x}_j < l_j} x_j.
$$

2. Modify the bounds on the variables that have been included in the Phase I objective so that they are feasible with respect to their new bounds. That is, define the new bounds as

$$
\begin{aligned}
-\infty \leq x_j \leq u_j \quad &\text{if } \bar{x}_j < l_j, \\
l_j \leq x_j \leq \infty \quad &\text{if } \bar{x}_j > u_j.
\end{aligned}
$$

3. At the end of each iteration of the Simplex Algorithm, examine the values of the basic variables with respect to their *original bounds*. If, for any infeasible x_j, the original bounds are satisfied, then modify the Phase I objective function by dropping the infeasibility contribution of that x_j, and reinstate its original bounds. If the Phase I objective function is empty, then a basic feasible solution has been found.

4. If the conditions of optimality are satisfied but some terms remain in the objective function, then the problem is infeasible.

Note 1: When using a Phase I approach that minimizes the sum of infeasibilities, more than one variable can become feasible at a given iteration.

Note 2: In practical computations, with the Simplex Algorithm applied to a linear program with explicitly bounded variables (which includes the standard form), one or more basic variables can violate a bound as a result of numerical error buildup. In this case also, the method of minimizing the sum of infeasibilities can be used to recover a feasible basis.

Example 7.3: MINIMIZING THE SUM OF INFEASIBILITIES. We illustrate this procedure on a slight modification of the problem that we solved in Example 4.1, i.e.,

$$
\begin{array}{ll}
\text{minimize} & 4x_1 + 8x_2 + x_3 + 2x_4 + 6x_5 \\
\text{subject to} & 2x_1 - 3x_2 - 4x_3 + x_4 + 4x_5 = 10 \\
& 5x_1 \qquad\quad - 2x_3 \qquad\quad - 2x_5 = 5 \\
& -3x_1 + 3x_2 \qquad\qquad + 6x_4 - 2x_5 = 10 \\
& x_j \geq 0, \ j = 1, 2, 3, 4, 5.
\end{array}
$$

We start by writing this in tabular form as follows:

min	z	x_1	x_2	x_3	x_4	x_5	1	basic
row 0	1	-4	-8	-1	-2	-6	0	z
row 1	0	2	-3	-4	1	4	10	$-$
row 2	0	5	0	-2	0	-2	5	$-$
row 3	0	-3	3	0	6	-2	10	$-$

A starting basis is guessed at by choosing the columns corresponding to x_1, x_2, x_3. Fortunately this is nonsingular, and we obtain the tabular form

min	z	x_1	x_2	x_3	x_4	x_5	1	basic
row 0	1	0	0	0	$-105/22$	$-602/33$	$725/66$	z
row 1	0	1	0	0	$-7/11$	$-6/11$	$-10/11$	x_1
row 2	0	0	1	0	$15/11$	$-40/33$	$80/33$	x_2
row 3	0	0	0	1	$-35/22$	$-4/11$	$-105/22$	x_3

The basic solution is infeasible because x_1 and x_3 violate their lower bound of 0, so we modify the problem and create a new objective to minimize the sum of infeasibilities. The Phase I objective to minimize the sum of infeasibilities is $w = -x_1 - x_3$. Furthermore, the modified problem has x_1 and x_3 as free variables because the lower bounds of 0 were violated and the upper bounds were ∞ to start with. Thus the modified bounds on x_1 and x_3 are set to $-\infty \leq x_1 \leq +\infty$ and $-\infty \leq x_3 \leq +\infty$. The tableau is

min	w	z	x_1	x_2	x_3	x_4	x_5	1	basic
row 00	1	0	1	0	1	0	0	0	w
row 0	0	1	0	0	0	$-105/22$	$-602/33$	$725/66$	z
row 1	0	0	1	0	0	$-7/11$	$-6/11$	$-10/11$	x_1
row 2	0	0	0	1	0	$15/11$	$-40/33$	$80/33$	x_2
row 3	0	0	0	0	1	$-35/22$	$-4/11$	$-105/22$	x_3

Pricing out row 00 and inserting in rows to keep track of the bounds on the variables results in

min	w	z	x_1	x_2	x_3	x_4	x_5	1	basic
row 00	1	0	0	0	0	$49/22$	$10/11$	$14/11$	w
row 0	0	1	0	0	0	$-105/22$	$-602/33$	$725/66$	z
row 1	0	0	1	0	0	$-7/11$	$-6/11$	$-10/11$	x_1
row 2	0	0	0	1	0	$15/11$	$-40/33$	$80/33$	x_2
row 3	0	0	0	0	1	$-35/22$	$-4/11$	$-105/22$	x_3
lb	$-\infty$	$-\infty$	$-\infty$	0	$-\infty$	0	0		
ub	∞	∞	∞	∞	∞	∞	∞		
nonbasic						0	0		

From the Phase I objective, the incoming column is chosen to be the one corresponding to x_4. The outgoing column is determined by

$$\alpha_r = \min \begin{cases} \left\{ \dfrac{\bar{b}_2 - l_2}{\bar{a}_{24}} \right\} = \left\{ \dfrac{80/33}{15/11} \right\} = 16/9 \\ \left\{ \dfrac{u_3 - \bar{b}_3}{-\bar{a}_{34}} \right\} = \infty \\ u_4 - l_4 \quad = \infty \end{cases}$$

$$= 16/9 = \alpha_2.$$

Thus the outgoing column is the one corresponding to x_2, and we pivot to make x_4 basic in place of x_2.

min	w	z	x_1	x_2	x_3	x_4	x_5	1	basic
row 00	1	0	0	$-49/30$	0	0	$26/9$	$31/18$	w
row 0	0	1	0	$-7/2$	0	0	-14	$5/2$	z
row 1	0	0	1	$7/15$	0	0	$-10/9$	$2/9$	x_1
row 2	0	0	0	$11/15$	0	1	$-8/9$	$16/9$	x_4
row 3	0	0	0	$7/6$	1	0	$-16/9$	$-35/18$	x_3
lb	$-\infty$	$-\infty$	$-\infty$	0	$-\infty$	0	0		
ub	∞	∞	∞	∞	∞	∞	∞		
nonbasic						0	0		

Basic variable x_3 still does not satisfy its lower bound, but x_1 now does. The new Phase I objective to minimize the sum of infeasibilities is $w = -x_3$. The modified problem has only x_3 as a free variable because its lower bound of 0 is violated and the upper bound, which was ∞ to start with, is clearly not violated. Furthermore the bounds on x_1 are restored to the original ones. After pricing out x_3 from row 00 (Phase I objective) the modified tableau is

min	w	z	x_1	x_2	x_3	x_4	x_5	1	basic
row 00	1	0	0	$-7/6$	0	0	$16/9$	$35/18$	w
row 0	0	1	0	$-7/2$	0	0	-14	$5/2$	z
row 1	0	0	1	$7/15$	0	0	$-10/9$	$2/9$	x_1
row 2	0	0	0	$11/15$	0	1	$-8/9$	$16/9$	x_4
row 3	0	0	0	$7/6$	1	0	$-16/9$	$-35/18$	x_3
lb	$-\infty$	$-\infty$	0	0	$-\infty$	0	0		
ub	∞	∞	∞	∞	∞	∞	∞		
nonbasic						0	0		

From the Phase I objective, the incoming column is chosen to be the one corresponding to x_5. As we now try to determine the outgoing column we find that the objective is unbounded! However, it is unbounded because of the manner in which we constructed the Phase I problem. At this point all we are trying to do is find a feasible solution. Hence, we now reset the upper bound on x_3 be 0, the original lower bound. Accordingly, the outgoing column is determined by

$$\alpha_r = \min\left\{\infty, \ \frac{u_3 - \bar{b}_3}{-\bar{a}_{35}} = \frac{35/18}{16/9} = 35/32, \ u_5 - l_5 = \infty\right\} = 35/32 = \alpha_3.$$

Thus the outgoing column is the one corresponding to x_3, and we pivot to make x_5 basic in place of x_3.

min	w	z	x_1	x_2	x_3	x_4	x_5	1	basic
row 00	1	0	0	0	1	0	0	0	w
row 0	0	1	0	$-203/16$	$-63/8$	0	0	$285/16$	z
row 1	0	0	1	$-21/80$	$-5/8$	0	0	$23/16$	x_1
row 2	0	0	0	$3/20$	$-1/2$	1	0	$11/4$	x_4
row 3	0	0	0	$-21/32$	$-9/16$	0	1	$35/32$	x_5
lb	$-\infty$	$-\infty$	0	0	$-\infty$	0	0		
ub	∞	∞	∞	∞	∞	∞	∞		
nonbasic						0	0		

All basic variables now satisfy the nonnegativity bounds of the original problem, and so a basic feasible solution has been found.

Example 7.4: MINIMIZING THE SUM OF INFEASIBILITIES AGAIN. We modify slightly the first element on the right-hand side of Example 4.1 to be 15 instead of -10; thus, the problem becomes

$$
\begin{aligned}
\text{minimize} \quad & 4x_1 + 8x_2 + \ x_3 + 2x_4 + 6x_5 \\
\text{subject to} \quad & 2x_1 - 3x_2 - 4x_3 + \ x_4 - 4x_5 = 15 \\
& 5x_1 \qquad\quad - 2x_3 \qquad\quad - 2x_5 = 18 \\
& -3x_1 + 3x_2 \qquad\quad + 6x_4 - 2x_5 = 12 \\
& x_j \geq 0, \ j = 1, 2, 3, 4, 5.
\end{aligned}
$$

Choosing the columns corrsponding to x_1, x_2, x_3 as a starting basis, we obtain

min	z	x_1	x_2	x_3	x_4	x_5	1	basic
row 0	1	0	0	0	$-105/22$	$254/33$	$-746/11$	z
row 1	0	1	0	0	$-7/11$	$2/11$	$34/11$	x_1
row 2	0	0	1	0	$15/11$	$-16/33$	$78/11$	x_2
row 3	0	0	0	1	$-35/22$	$16/11$	$-14/11$	x_3

The solution is infeasible because x_3 violates its lower bound. We create the Phase I problem as shown earlier:

min	w	z	x_1	x_2	x_3	x_4	x_5	1	basic
row 00	1	0	0	0	0	$35/22$	$-16/11$	$153/22$	w
row 0	0	1	0	0	0	$105/22$	$-254/33$	$767/22$	z
row 1	0	0	1	0	0	$-7/11$	$2/11$	$9/11$	x_1
row 2	0	0	0	1	0	$15/11$	$-16/33$	$53/11$	x_2
row 3	0	0	0	0	1	$-35/22$	$16/11$	$-153/22$	x_3
lb	$-\infty$	$-\infty$	0	0	$-\infty$	0	0		
ub	∞	∞	∞	∞	∞	∞	∞		
nonbasic						0	0		

From the Phase I objective, the incoming column is chosen to be the one corresponding to x_4. The outgoing column is determined by

$$
\alpha_r = \min \left\{ \begin{aligned}
& \left\{ \frac{\bar{b}_2 - l_2}{\bar{a}_{24}} \right\} = \left\{ \frac{53/11}{15/11} \right\} \\
& \left\{ \frac{u_3 - \bar{b}_3}{-\bar{a}_{34}} \right\} = \infty \\
& u_4 - l_4 \qquad = \infty
\end{aligned} \right.
$$

$$
= 53/15 = \alpha_2.
$$

Thus the outgoing column is the one corresponding to x_2, and we pivot to make x_4 basic in place of x_2.

min	w	z	x_1	x_2	x_3	x_4	x_5	1	basic
row 00	1	0	0	$-7/6$	0	0	$-8/9$	$4/3$	w
row 0	0	1	0	$7/2$	0	0	6	18	z
row 1	0	0	1	$7/15$	0	0	$-2/45$	$46/15$	x_1
row 2	0	0	0	$11/15$	0	0	$-16/45$	$53/15$	x_4
row 3	0	0	0	$7/6$	1	1	$8/9$	$-4/3$	x_3
lb	$-\infty$	$-\infty$	0	0	$-\infty$	0	0		
ub	∞	∞	∞	∞	∞	∞	∞		
nonbasic						0	0		

The variable x_3 still violates its lower bound; so the Phase I objective remains the same. However, it is not possible to improve the solution any further because all the reduced costs in the Phase I objective function are nonpositive. This reveals that the problem is infeasible.

7.3   Steepest-edge rule for incoming column selection

At the expense of some additional computations per iteration, it is possible to reduce the number of iterations resulting from the most negative reduced cost rule by using a modified incoming selection rule, called the *steepest-edge rule*, see [71]. We now describe how this works.

For simplicity of exposition, we shall consider a linear program in standard form

$$\text{minimize} \quad c^\mathrm{T} x$$
$$\text{subject to} \quad Ax = b$$
$$x \geq 0.$$

Recall, that in the Simplex Algorithm (Section 3.5), we choose the incoming column s as the one with the most positive reduced cost, i.e, $z_s - c_s = \max_{j \in N}\{z_j - c_j\}$. Equivalently, we can change the signs and choose s by $c_s - z_s = \min_{j \in N}\{c_j - z_j\}$, the most negative reduced cost rule. That is,

$$z_s - c_s = \max_{j \in N}\{z_j - c_j\} \quad \Longleftrightarrow \quad c_s - z_s = \min_{j \in N}\{c_j - z_j\}.$$

The reason for this reversal will be apparent shortly.

To avoid the use of complicated notation, we assume that at the current iteration the first m columns constitute a feasible basis. A basic feasible solution has the form

$$x = \begin{bmatrix} B^{-1}b \\ 0 \end{bmatrix}.$$

It is straightforward to see that this is the solution to the system of equations

$$\begin{bmatrix} B & N \\ 0 & I_{n-m} \end{bmatrix} \begin{bmatrix} x_B \\ x_N \end{bmatrix} = \begin{bmatrix} b \\ 0 \end{bmatrix},$$

or, equivalently,

$$\begin{bmatrix} I_m & B^{-1}N \\ 0 & I_{n-m} \end{bmatrix} \begin{bmatrix} x_B \\ x_N \end{bmatrix} = \begin{bmatrix} B^{-1}b \\ 0 \end{bmatrix},$$

where I_{n-m} is an $(n-m) \times (n-m)$ identity matrix and I_m is an $m \times m$ identity matrix.

We let $B^{-1}A = \bar{A}$. The incoming column is chosen from the $n \times (n-m)$ matrix

$$\begin{bmatrix} B^{-1}N \\ I_{n-m} \end{bmatrix},$$

which corresponds to the nonbasic columns of A. Let $s > m$ be the index of the incoming nonbasic column. The adjustment made to the vector of basic variables x_B, while ensuring $x_B \geq 0$, is given by the minimum ratio test. This is the same as finding the smallest number α, in the equation below, that causes one of the basic variables to become 0; thus

$$x_B = B^{-1}b - \bar{A}_{\bullet s}\alpha$$
$$x_s = \alpha.$$

In matrix notation, this can be written as

$$x = \begin{bmatrix} x_B \\ x_N \end{bmatrix} = \begin{bmatrix} B^{-1}b \\ 0 \end{bmatrix} + \alpha \begin{bmatrix} -\bar{A}_{\bullet s} \\ e^s \end{bmatrix},$$

where e^s is an $(n-m)$-vector with 1 in the $(s-m)$ position and zeros elsewhere. The column $p^s = (-\bar{A}_{\bullet s}, e^s)$ can be thought of as the *direction* along which x changes in accordance with the scalar multiplier α. That is, letting $\bar{x} = (B^{-1}b, 0)$, we have

$$x = x(\alpha) = \bar{x} + \alpha p^s.$$

Now we note that for nonbasic column index $j > m$, and column-vector p^j defined as $(-\bar{A}_{\bullet j}, e^j)$, the reduced cost is the dot product

$$c^T p^j = [c_B^T \ c_N^T]\begin{bmatrix} -\bar{A}_{\bullet j} \\ e^j \end{bmatrix} = c_j - c_B^T \bar{A}_{\bullet j} = c_j - c_B^T B^{-1} A_{\bullet j} = c_j - z_j,$$

where e^j is an $(n-m)$-vector with 1 in the $(j-m)$ position and zeros elsewhere. In the most negative reduced cost rule, the direction p^s is chosen so as to yield the most negative inner product of c and p^j, i.e.,

$$c^T p^s = \min_{j \in N} c^T p^j.$$

To use a steepest-edge rule we require that the direction p^s be the "most downhill" relative to the gradient (cost) vector; i.e.,

$$\frac{c^T p^s}{\|p^s\|} = \min_{j \in N} \frac{c^T p^j}{\|p^j\|} = \min_{j \in N} \frac{c_j - z_j}{\|p^j\|}.$$

In practice using the steepest-edge rule results in fewer iterations but requires the computation of the $\|p^j\|$, for $j = m+1, m+2, \ldots, n$, which is computationally expensive. Techniques outside the scope of this textbook have been developed to do this efficiently through recurrence relations. For further details, see [71].

As we noted earlier, in the Simplex Algorithm (Section 3.5), we choose the incoming column s as the one with the most-positive reduced cost, i.e, $z_s - c_s = \max_{j \in N}\{z_j - c_j\}$. In this case, the steepest-edge rule equivalent is to choose the incoming column s by:

$$\frac{c^T p^s}{\|p^s\|} = \max_{j \in N} \frac{z_j - c_j}{\|p^j\|}.$$

Example 7.5: APPLYING THE STEEPEST-EDGE RULE. Consider the linear program (due to Chvátal [28])

$$
\begin{aligned}
\text{minimize} \quad & -9x_1 - 3x_2 - x_3 \\
\text{subject to} \quad & x_1 \leq 1 \\
& 6x_1 + x_2 \leq 9 \\
& 18x_1 + 6x_2 + x_3 \leq 81 \\
& x_j \geq 0, \quad j = 1, 2, 3.
\end{aligned}
$$

After adding the slack variables x_4, x_5, and x_6 we obtain the tableau form

min	z	x_1	x_2	x_3	x_4	x_5	x_6	1	basic
row 0	1	9	3	1	0	0	0	0	z
row 1	0	1	0	0	1	0	0	1	x_4
row 2	0	6	1	0	0	1	0	9	x_5
row 3	0	18	6	1	0	0	1	81	x_6

To determine the incoming column by the steepest-edge rule, we start by first computing $\|p^j\|_2 = \|(-\bar{A}_{\bullet j}, e^j)\|_2$ for $j \in N$, i.e., for $j = 1, 2, 3$. Thus,

$$
\begin{aligned}
\|p^1\|_2 &= \sqrt{1 + 6^2 + 18^2 + 1} = \sqrt{362} \\
\|p^2\|_2 &= \sqrt{0 + 1 + 6^2 + 1} = \sqrt{38} \\
\|p^3\|_2 &= \sqrt{0 + 0 + 1 + 1} = \sqrt{2}.
\end{aligned}
$$

The index of the incoming column is then $s = 3$ by the steepest edge rule:

$$
\frac{c^{\mathsf{T}} p^s}{\|p^s\|} = \max_{j=1,2,3} \left\{ \frac{9}{\sqrt{362}}, \frac{3}{\sqrt{38}}, \frac{1}{\sqrt{2}} \right\} = \frac{1}{\sqrt{2}} = \frac{c^{\mathsf{T}} p^3}{\|p^3\|}.
$$

After pivoting to bring in column 3 into the basis, we obtain the optimal tableau

min	z	x_1	x_2	x_3	x_4	x_5	x_6	1	basic
row 0	1	-9	-3	0	0	-1	0	-81	z
row 1	0	1	0	0	1	0	0	1	x_4
row 2	0	6	1	0	0	1	0	9	x_5
row 3	0	18	6	1	0	0	1	81	x_3

in one iteration. If we were to use the usual rule to select an incoming column, we would have chosen column 1. The optimal tableau would then have been obtained in 7 iterations. We will encounter problems of this nature again in Section 7.5.

7.4   Structured linear programs

Typically linear programs, especially those of large size, exhibit structure that can be used to advantage. What do we mean by "structure" in a linear program? Perhaps the best way to describe it is in terms of the coefficient matrix A. Structure can occur in several different ways; for example, large linear programs are sparse and advantage can be taken of sparsity, or, there can be embedded network models within the linear program and specialized algorithms can be used for this. Here we will focus on specific patterns whereby the coefficient matrix can be partitioned into blocks. Let's say the LP is given in standard form

$$\begin{aligned}
\text{minimize} \quad & c^T x \\
\text{subject to} \quad & Ax = b \\
& x \geq 0,
\end{aligned}$$

where A is an $m \times n$ matrix. Suppose A can be partitioned as a *block matrix*[2]

$$A = \begin{bmatrix}
A_{11} & A_{12} & \cdots & A_{1\ell} \\
A_{21} & A_{22} & \cdots & A_{2\ell} \\
\vdots & \vdots & & \vdots \\
A_{k1} & A_{k2} & \cdots & A_{k\ell}
\end{bmatrix}$$

with the property that many of the blocks are zero matrices. If the nonzero blocks have a pattern, we say that the matrix has *structure*. There are several different structures that come up in linear programming problems. (See, for example, [20, Section 12.1, especially p. 506].) One of these, called *block-angular structure*, nicely illustrates the idea.

Figure 7.1 depicts a block matrix A with block-angular structure. In this case $k = 5$ and $\ell = 4$. The matrix has the property that blocks $A_{1\bullet}$, A_{21}, A_{32}, A_{43}, and A_{54} are nonzero, whereas all other blocks are zero.[3]

---

[2]The blocks A_{ij} in a block matrix are themselves matrices.

[3]When we say that a block is nonzero, we do *not* mean that every element of the block (submatrix) is nonzero but rather that the block is not a zero matrix.

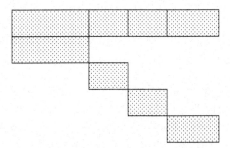

Figure 7.1: Block-angular structure.

The general definition of block angularity is easily inferred from Figure 7.1. We simply require that $A_1.$ be nonzero and all blocks, except for A_{ij} for $j = 1, 2, \ldots, l$, and $i = j + 1$, be zero.

When an LP (like the one above) has a block matrix A, the vectors b and c have a corresponding structure; they are each composed of subvectors. Thus

$$b = (b^1, \ldots, b^k) \quad \text{and} \quad c = (c^1, \ldots, c^\ell).$$

Linear programs with block-angular structure arise naturally. A typical case is one in which an organization has several units that, collectively, must satisfy a certain set of constraints and, individually, must observe resource (or production) constraints on their own activities. From the standpoint of the organization's highest level, there is an objective that embraces all the activities of all the individual units.

In such an LP, the constraints $A_1.x = b^1$ that express the sharing of resources are called *coupling constraints*. (This terminology is used in many situations, not necessarily in the context of block-angular LPs.)

Another setting in which block-angular structure comes up is that of planning the activities of a firm over a finite sequence of time periods. In this situation, there may be resources (for instance, financial) that are to be expended and not replenished over the planning horizon. These, of course, lead to coupling constraints. When single-period activities period are independent of each other (which would preclude carrying specific resources forward for use in a later period) and the objective is the sum of the individual period objectives, the model's coefficient matrix has block-angular structure (see Figure 7.1).

Notice that if it were not for the coupling constraints, a block-angular linear program would break into a set of ℓ independent linear programs. The total computational effort for the ℓ subproblems would be less than it is for the one large problem.

We will describe two types of approaches to handle such types of linear programs: the Dantzig-Wolfe Decomposition Algorithm and the Benders Decomposition Algorithm. These two algorithms are based on structure of a different sort, that of polyhedral sets which we describe next.

On the structure of polyhedral sets

As we discussed in Section 2.2, the feasible region of any feasible linear program is a nonempty polyhedral convex set. It is known that any nonempty polyhedron X in can be "resolved" into the sum of a bounded polyhedron and a polyhedral cone. This means that for every $x \in X$, there exist points $p \in P$ (a bounded polyhedron), $q \in C$ (a polyhedral cone) such that $x = p + q$. Accordingly, we write

$$X = P + C.$$

There is an even sharper resolution of a polyhedron X based on the fact that a polyhedral cone C can be expressed as the sum of a linear subspace S (called the *lineality space* of C) and another cone C_0 with the property that it does not contain a line. A cone with the latter property is said to be *pointed*. Thus,

$$X = P + C_0 + S.$$

When the cone C does not contain a line, we put $C = C_0$ and $S = \{0\}$. For a bounded polyhedron X, we put $X = P$ and $C = \{0\}$ which of course means $S = \{0\}$. (This is in contrast to the case of an unbounded polyhedron of the form $X = P + S$ for which the pointed cone C_0 is just the singleton $\{0\}$.)

It is well worth observing that if $X \neq \emptyset$ is the feasible region of a linear program in standard form, it cannot contain a line. This follows from the fact that X is a subset of the nonnegative orthant which cannot contain an entire line. For this reason, the lineality space in the resolution of X must be $\{0\}$, which is to say the cone $C = C_0$.

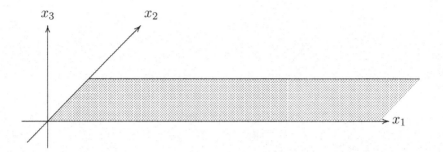

Figure 7.2: A "slice" of an unbounded polyhedron.

Example 7.6: RESOLUTION OF AN UNBOUNDED POLYHEDRON. Let X be the set of all points $x \in R^3$ such that

$$x_1 \geq 0, \quad 0 \leq x_2 \leq 1, \quad x_3 \text{ free.} \tag{7.4}$$

As Figure 7.2 shows, X is the sum of the bounded polyhedron

$$P = \{(x_1, x_2, x_3) : x_1 = x_3 = 0, \ 0 \leq x_2 \leq 1\},$$

the pointed polyhedral cone

$$C_0 = \{(x_1, x_2, x_3) : x_1 \geq 0\},$$

and the lineality space

$$S = \{(x_1, x_2, x_3) : x_1 = x_2 = 0, \ x_3 \in R\}.$$

The set X in Figure 7.2 extends indefinitely in the positive x_1 direction. The shaded region here is actually just the "slice" of the set X corresponding to $x_3 = 0$. A better mental image of X is obtained by sliding the shaded region up and down the x_3 axis.

The Dantzig-Wolfe Decomposition Algorithm rests on the fact that polyhedral sets can be represented in two ways.

1. The *external representation*, which we have been using all along, specifies a polyhedron as the intersections of halfspaces. The unbounded polyhedron X defined in Example 7.6 is given by an external representation (7.4).

2. The *internal representation,* specifies each element of the polyhedron as a certain kind of linear combination of "generators" chosen from a finite set. (The same finite set of generators is used in the linear combination for every element of the polyhedron, but the coefficients used in such expressions vary from one point to another.)

Our aim here is to elaborate the notion of the internal representation of polyhedra and then show how it is used in the Dantzig-Wolfe Algorithm.

Let X be a nonempty polyhedral set that does not contain a line (as in the case of a linear program in standard form). Then X is the sum of a bounded polyhedron P and a pointed cone C. Now to describe the internal representation of X, we begin with that of the bounded polyhedron P. The prevailing assumptions on X imply that P has a finite number $K \geq 1$ of extreme points, and that, indeed, P is the set of all convex combinations of its K extreme points $x^1, \ldots, x^K$. Thus, for every $x \in P$, there exist nonnegative scalars $\lambda_1, \ldots, \lambda_K$ which sum to 1 such that $x = \lambda_1 x^1 + \cdots + \lambda_K x^K$. This internal representation of P can be put as follows:

$$x \in P \iff \begin{bmatrix} x \\ 1 \end{bmatrix} = \sum_{i=1}^{K} \lambda_i \begin{bmatrix} x^i \\ 1 \end{bmatrix}, \quad \lambda_i \geq 0, \ i = 1, \ldots, K. \qquad (7.5)$$

The name given to a set which is the collection of all convex combinations of finitely many points is *polytope.* In fact, polytopes and bounded polyhedra amount to the same thing. The term "polytope" emphasizes the *internal* representation of the elements of the set as a convex combination of extreme points; the term "bounded polyhedron" stems from its boundedness and *external* representation as the intersection of halfspaces.

To deal with the internal representation of a cone C we need to introduce a little terminology. Let y be a nonzero vector in R^n. The set of all nonnegative multiples of y is called the *ray* generated by y. We denote this set by $\langle y \rangle$. Thus we have $\langle y \rangle = \{q : q = \lambda y, \ \lambda \geq 0\}$. Note that all nonzero elements of $\langle y \rangle$ generate the same set. By definition, a ray is a cone. When y belongs to a cone C, the ray $\langle y \rangle$ is a subset of C; it is called an *extreme ray* of C if and only if

$$y = \lambda y^1 + (1 - \lambda)y^2, \text{ where } y^1, \ y^2 \in C \implies y^1, y^2 \in \langle y \rangle.$$

Now let $y^1, \ldots, y^L$ be vectors in R^n. The set

$$C = \langle y^1 \rangle + \cdots + \langle y^L \rangle$$

can be seen to be a convex cone. Having finitely many *generators* $y^1, \ldots, y^L$, C is said to be a *finite cone*. Another way to express the definition of such a cone C is to say

$$C = \left\{ x \in R^n : x = \sum_{j=1}^{L} \mu_j y^j, \quad \mu_j \geq 0, \ j = 1, \ldots, L \right\}.$$

To highlight the similarities and differences between membership in P and membership in C, we can even write

$$y \in C \Longleftrightarrow \begin{bmatrix} y \\ 0 \end{bmatrix} = \sum_{j=1}^{L} \mu_j \begin{bmatrix} y^j \\ 0 \end{bmatrix}, \quad \mu_j \geq 0, \ j = 1, \ldots, L. \qquad (7.6)$$

A theorem due to H. Minkowski [135] states that *every polyhedral cone is finitely generated*. The converse of this theorem, namely that *every finite cone is polyhedral*, is due to H. Weyl. See [176, p. 87]. These two theorems express the equivalence for cones of the two different characterizations; one of them being by hyperplanes and the other being by generators.

When it comes to finitely generated cones, it is customary to single out the generators that produce extreme rays. Thus, every pointed polyhedral cone is the sum of its extreme rays.[4] The internal representation (7.6) is the analogue of (7.5) used for bounded polyhedra.

Accordingly, there exist two finite sets of vectors $x^1, \ldots, x^K$ (extreme points of X) and $y^1, \ldots, y^L$ (generators of extreme rays) such that

$$x \in X \implies x = \lambda_1 x^1 + \cdots + \lambda_K x^K + \mu_1 y^1 + \cdots + \mu_L y^L \qquad (7.7)$$

where

$$\begin{aligned} \lambda_1 + \cdots + \lambda_K = 1, \quad \lambda_i \geq 0, \quad i = 1, \ldots, K, \\ \mu_j \geq 0, \quad j = 1, \ldots, L. \end{aligned} \qquad (7.8)$$

This is sometimes called the *Goldman Resolution Theorem* [87].

Figure 7.3 conveys the central idea of this theorem. The polyhedron X on the left-hand side of the equation can be seen as the sum of the polytope P (convex hull of the extreme points x^1 and x^2) and the pointed finite cone C (the sum of the extreme rays $\langle y^1 \rangle + \langle y^2 \rangle$).

---

[4]As the example $C = \{x \in R^2 : x_1 + x_2 \geq 0\}$ shows, the assumption of pointedness is essential to the truth of the statement about every finite cone being the sum of its extreme rays. This cone is not pointed and has two extreme rays which generate only its lineality space, rather than the entire cone.

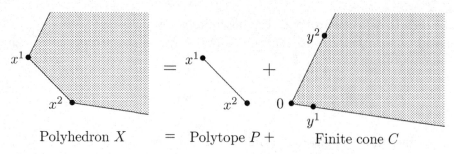

$$\text{Polyhedron } X \quad = \quad \text{Polytope } P + \quad \text{Finite cone } C$$

Figure 7.3: Goldman resolution of polyhedron X.

The Dantzig-Wolfe Decomposition Principle/Algorithm

The main idea in a decomposition algorithm for a linear program is to break the problem into smaller, more manageable parts, solve each part separately, and then coordinate the results into a solution of the grand problem. It is beneficial to apply such an algorithm to structured linear programs. This "philosophy" is easy enough to appreciate, but the devil, as the saying goes, is in the details. In this section, we sketch out some of the ideas that go into the Dantzig-Wolfe Decomposition Algorithm [46], [47], [40]. You may find it helpful to recall Exercise 5.7 in Chapter 5.

For the sake of concreteness, we will study the case of an LP with block-angular structure and $\ell = 2$. Although this is a rather unrealistic assumption[5], it imposes no serious loss of generality, for if ℓ were greater than 2, there would just be more work of a similar nature to do. So our LP has the form

$$
\begin{aligned}
\text{minimize} \quad & (c^1)^{\mathrm{T}}x^1 + (c^2)^{\mathrm{T}}x^2 \\
\text{subject to} \quad & A_{11}x^1 + A_{12}x^2 = b^0 && (b^0 \in R^{m_0}) \\
& A_{21}x^1 \qquad\qquad = b^1 && (b^1 \in R^{m_1}) && (7.9) \\
& \qquad\qquad A_{32}x^2 = b^2 && (b^2 \in R^{m_2}) \\
& x^1 \geq 0, \quad x^2 \geq 0.
\end{aligned}
$$

To facilitate reference to this LP, we call it the *two-division problem* with the thought in mind that we are solving a problem that pertains to an enterprise having two "nearly independent" divisions. The divisions are not totally

---

[5] Nevertheless, there are circumstances in which $\ell = 1$ for which the decomposition algorithm would make sense and be computationally beneficial. For example, if the equations $Ax = b$ of the full problem have a large but easily solved subset of constraints, it can be advantageous to separate the constraints into two groups.

independent since their activities are affected by the coupling constraints.

In order to develop an understanding of the Dantzig-Wolfe Decomposition Algorithm, we apply the Goldman Resolution Theorem to represent feasible solutions of each of the two blocks in terms of its extreme points and extreme rays. In doing this, we unfortunately have to resort to a more detailed notational scheme. In particular, for feasible solutions of

$$A_{21}x^1 = b^1, \ x^1 \geq 0 \tag{7.10}$$

we write

$$\begin{bmatrix} x^1 \\ 1 \end{bmatrix} = \begin{bmatrix} x^{11} \\ 1 \end{bmatrix} \lambda_{11} + \cdots + \begin{bmatrix} x^{1K_1} \\ 1 \end{bmatrix} \lambda_{1K_1} + \begin{bmatrix} y^{11} \\ 0 \end{bmatrix} \mu_{11} + \cdots + \begin{bmatrix} y^{1L_1} \\ 0 \end{bmatrix} \mu_{1L_1},$$

$$\lambda_{1i} \geq 0, \ i = 1, \ldots, K_1, \qquad \mu_{1j} \geq 0, \ j = 1, \ldots, L_1.$$

In like manner, for feasible solutions of

$$A_{32}x^2 = b^2, \ x^2 \geq 0 \tag{7.11}$$

we write

$$\begin{bmatrix} x^2 \\ 1 \end{bmatrix} = \begin{bmatrix} x^{21} \\ 1 \end{bmatrix} \lambda_{21} + \cdots + \begin{bmatrix} x^{2K_2} \\ 1 \end{bmatrix} \lambda_{2K_2} + \begin{bmatrix} y^{21} \\ 0 \end{bmatrix} \mu_{21} + \cdots + \begin{bmatrix} y^{2L_2} \\ 0 \end{bmatrix} \mu_{2L_2},$$

$$\lambda_{2i} \geq 0, \ i = 1, \ldots, K_2, \qquad \mu_{2j} \geq 0, \ j = 1, \ldots, L_2.$$

Notice how the equations above capture the conditions

$$\lambda_{11} + \cdots + \lambda_{1K_1} = 1 \quad \text{and} \quad \lambda_{21} + \cdots + \lambda_{2K_2} = 1.$$

Such equations in nonnegative variables are called *convexity constraints*.

In this way the set of feasible solutions to the two blocks are *generated* by the vectors

$$x^{11}, \ldots, x^{1K_1}, \ y^{11}, \ldots, y^{1L_1} \quad \text{and} \quad x^{21}, \ldots, x^{2K_2}, \ y^{21}, \ldots, y^{2L_2}.$$

It has to be acknowledged that these generators *are not usually known in advance* partly because they may be too numerous. But this is not a problem. We know that they exist, and the algorithm finds them as needed.[6] For the time being, we grant their existence and proceed with this assumption. Substituting the above representations of x^1 and x^2 into the coupling

---

[6]The Dantzig-Wolfe Algorithm belongs to the category of *column-generation methods*.

constraints, we arrive at a linear program with $m_0 + 2$ equations in the nonnegative variables $\lambda_{11}, \ldots, \lambda_{1K_1}, \mu_{11}, \ldots, \mu_{1L_1}, \lambda_{21}, \ldots, \lambda_{2K_2}, \mu_{21}, \ldots, \mu_{2L_2}$. We get m equations from the coupling constraints and one equation from each of the "convexity constraints" associated with the divisions. The resulting linear program is named the *full master problem*.[7]

$$
\text{minimize} \quad (c^1)^{\mathrm{T}} \left(\sum_{i=1}^{K_1} x^{1i} \lambda_{1i} + \sum_{j=1}^{L_1} y^{1j} \mu_{1j} \right) + (c^2)^{\mathrm{T}} \left(\sum_{i=1}^{K_2} x^{2i} \lambda_{2i} + \sum_{j=1}^{L_2} y^{2j} \mu_{2j} \right)
$$

$$
\text{subject to} \quad A_{11} \left(\sum_{i=1}^{K_1} x^{1i} \lambda_{1i} + \sum_{j=1}^{L_1} y^{1j} \mu_{1j} \right) + A_{12} \left(\sum_{i=1}^{K_2} x^{2i} \lambda_{2i} + \sum_{j=1}^{L_2} y^{2j} \mu_{2j} \right) = b^0
$$

$$
\sum_{i=1}^{K_1} \lambda_{1i} = 1
$$

$$
\sum_{i=1}^{K_2} \lambda_{2i} = 1
$$

$$
\lambda_{11}, \ldots, \lambda_{1K_1}, \mu_{11}, \ldots, \mu_{1L_1}, \lambda_{21}, \ldots, \lambda_{2K_2}, \mu_{21}, \ldots, \mu_{2L_2} \geq 0.
$$

We observe that the variables in the full master problem are the nonnegative weights

$$
\lambda_{11}, \ldots, \lambda_{1K_1}, \mu_{11}, \ldots, \mu_{1L_1}, \lambda_{21}, \ldots, \lambda_{2K_2}, \mu_{21}, \ldots, \mu_{2L_2}.
$$

As noted earlier, the full set of generators (vectors) is not known in advance. Furthermore, it would be impractical to generate all the extreme points and rays of the two blocks. Instead, we generate just the column that would be selected as the incoming column for the master problem. A problem formed by using only a subset of the generators (vectors)

$$
x^{11}, \ldots, x^{1K_1}, y^{11}, \ldots, y^{1L_1} \quad \text{and} \quad x^{21}, \ldots, x^{2K_2}, y^{21}, \ldots, y^{2L_2}
$$

is called the *restricted master problem*. A feasible solution of a restricted master problem can be viewed as a feasible solution of the full master problem in which the variables λ and μ corresponding to undetermined generators are zero.

We are now ready to describe how the Dantzig-Wolfe Decomposition Algorithm works. We assume that we have $m_0 + 2$ extreme points and

---

[7]It is often the case that the number m_0 of coupling constraints is very small relative to the overall number of constraints, namely $m_0 + m_1 + m_2$. This would make it relatively easy to solve the full master problem.

extreme rays that form a feasible solution to the restricted master problem. That is,

minimize

$$
(c^1)^{\mathrm{T}} \left(\sum_{i=1}^{\bar{K}_1} x^{1i} \lambda_{1i} + \sum_{j=1}^{\bar{L}_1} y^{1j} \mu_{1j} \right) + (c^2)^{\mathrm{T}} \left(\sum_{i=1}^{\bar{K}_2} x^{2i} \lambda_{2i} + \sum_{j=1}^{\bar{L}_2} y^{2j} \mu_{2j} \right)
$$

subject to

$$
A_{11} \left(\sum_{i=1}^{\bar{K}_1} x^{1i} \lambda_{1i} + \sum_{j=1}^{\bar{L}_1} y^{1j} \mu_{1j} \right) + A_{12} \left(\sum_{i=1}^{\bar{K}_2} x^{2i} \lambda_{2i} + \sum_{j=1}^{\bar{L}_2} y^{2j} \mu_{2j} \right) = b^0
$$

$$
\sum_{i=1}^{\bar{K}_1} \lambda_{1i} \qquad\qquad\qquad = 1
$$

$$
\sum_{i=1}^{\bar{K}_2} \lambda_{2i} \qquad = 1
$$

$$
\lambda_{11}, \ldots, \lambda_{1\bar{K}_1},\ \mu_{11}, \ldots, \mu_{1\bar{L}_1},\ \lambda_{21}, \ldots, \lambda_{2\bar{K}_2},\ \mu_{21}, \ldots, \mu_{2\bar{L}_2} \geq 0.
$$

The number of variables at this stage is $\bar{K}_1 + \bar{L}_1 + \bar{K}_2 + \bar{L}_2 = m_0 + 2$.

At the next iteration of the Simplex Algorithm, we need to price out the nonbasic columns to see if the objective function value can be improved. In order to do this we have to price out all the other columns of the full master; that is, compute the reduced costs of the other columns. Unfortunately, though, we do not really know *all* the columns of the full master problem. So how are we to do the pricing? The answer is the ingenious part. For the restricted master problem there is a *price vector* having $m_0 + 2$ components which we will denote by $w_1, \ldots, w_{m_0}, \omega_1, \omega_2$.

For the time being, we set aside the extreme rays and consider only the extreme point solutions of the subproblems that make up the full master problem. First consider subproblem 1. We would need price out each column and choose a column with the largest reduced cost, that is, find

$$
\omega_1 + \max_{i=1,\ldots,K_1} \left\{ w^{\mathrm{T}} A_{11} x^{1i} - (c^1)^{\mathrm{T}} x^{1i} \right\}. \tag{7.12}
$$

The x^{1i} are the extreme point solutions of (7.10). So, we note that we are essentially using the prices (multipliers) $w_1, \ldots, w_{m_0}, \omega_1$ as cost coefficients with each extreme point. This is equivalent to solving the following linear program

$$\begin{aligned} \text{maximize} \quad & \rho^1 x^1 + \omega_1 \\ \text{subject to} \quad & A_{21} x^1 = b^1 \\ & x^1 \geq 0 \end{aligned} \tag{7.13}$$

where $\rho^1 = w^{\mathrm{T}} A_{11} - (c^1)^{\mathrm{T}}$. Similarly we solve the following linear program for subproblem 2

$$\begin{aligned} \text{maximize} \quad & \rho^2 x^2 + \omega_2 \\ \text{subject to} \quad & A_{32} x^2 = b^2 \\ & x^2 \geq 0 \end{aligned} \tag{7.14}$$

where $\rho^2 = w^{\mathrm{T}} A_{12} - (c^2)^{\mathrm{T}}$.

So now there is the question of handling the extreme rays. To start with, we price them out as we did for the extreme point solutions, and choose one with the largest reduced cost; that is

$$\max_{i=1,\dots,L_1} \left\{ w^{\mathrm{T}} A_{11} y^{1i} - (c^1)^{\mathrm{T}} y^{1i} \right\}. \tag{7.15}$$

This is equivalent to solving

$$\begin{aligned} \text{maximize} \quad & \rho^1 x^1 \\ \text{subject to} \quad & A_{21} x^1 = b^1 \\ & x^1 \geq 0 \end{aligned} \tag{7.16}$$

where $\rho^1 = w^{\mathrm{T}} A_{11} - (c^1)^{\mathrm{T}}$. Notice that the objective function for (7.16) is the same as that for (7.13) except for an additive constant, ω_1. Hence we could solve either problem. If the LP is unbounded, then we obtain an extreme homogeneous solution. However, every multiple of such a solution is also a solution. In the event that the LP (7.13) is unbounded, we could solve the following LP to obtain a normalized extreme homogeneous solution:

$$\begin{aligned} \text{maximize} \quad & \rho^1 x^1 \\ \text{subject to} \quad & A_{21} x^1 = 0 \\ & e^{\mathrm{T}} x^1 = 1 \\ & x^1 \geq 0 \end{aligned} \tag{7.17}$$

where $\rho^1 = w^{\mathrm{T}} A_{11} - (c^1)^{\mathrm{T}}$. However, it is not necessary to solve (7.17). We accept any homogeneous solution to either (or both) of the subproblems and add it to the restricted master problem.

Strictly speaking, it is not necessary to solve the two subproblems to optimality if a positive value is obtained for the objective function $\rho^1 x^1 + \omega_1$ in (7.13) or $\rho^2 x^2 + \omega_2$ in (7.14). In either case, the corresponding basic feasible solution (an extreme point of one of the divisional subproblem polyhedra) yielding this positive value would be a candidate for improving the objective function value of the restricted master problem. It could then be *adjoined* to the set of known (subproblem) extreme points being used to define the restricted master problem. This redefinition of the restricted master problem would initiate a new iteration of the Dantzig-Wolfe Decomposition Algorithm.

The above account gives some meaning to a frequently made statement about the Dantzig-Wolfe Decomposition Algorithm:

> The restricted master problem passes prices to the subproblems, and the subproblems pass proposals (basic feasible solutions) to the restricted master problem.

We can now write down the steps of the Dantzig-Wolfe Decomposition Algorithm.

Algorithm 7.2 (Dantzig-Wolfe Decomposition Applied to (7.9)) Assume that a starting basic feasible solution is known for the restricted master problem.

1. *Apply the Simplex Algorithm to the restricted master problem.* If an optimal solution is obtained, go to the next step. Otherwise, if the Simplex Algorithm reports an unbounded solution, stop and report the original problem as having no optimal solution.

2. *Solve the subproblems.* Using the multipliers from the solution of the restricted master problem, set up and solve the subproblems (7.13) and (7.14) using the Simplex Algorithm. One of the following occurs.

   (a) *Optimal solution obtained to one or both subproblems with positive objective values.* Add one or both extreme point solutions to the restricted master problem.

   (b) *One or both subproblems are unbounded.* Add one or both extreme homogeneous solutions to the restricted master problem.

(c) *Optimal solutions obtained to both subproblems with optimal objective function values equal to* 0. The solution is optimal for the original problem. Construct the solution to the original problem by combining the extreme points and extreme homogeneous solution and stop.

3. *Iterate.* Go to Step 1.

For finding an initial feasible solution of the master problem there is a technique analogous to the Phase I Procedure of the standard Simplex Algorithm. For the Decomposition Algorithm, one can start by taking a basic feasible solution from each divisional polyhedron. A set of suitable artificial columns with corresponding artificial variables will provide a basic feasible solution to this artificial restricted master problem. The objective is to reduce the sum of the artificial variables to zero, using, of course, the Decomposition Algorithm outlined above to do this. All the while, the subproblems will be passing extreme points of their feasible regions to the master problem. Assuming feasibility, eventually all the artificial variables will be eliminated from the master problem (or fixed at zero), and thereby a BFS for the master problem will have been found.

The Benders Decomposition Algorithm

In [11], Benders devised a decomposition method to handle "mixed-variables programming problems" of the form

$$\max\{c^{\mathrm{T}}x + f(y) : Ax + F(y) \le b, \ x \in R^n, \ y \in S\}$$

where $S \subseteq R^\ell$ is arbitrary. The symbol $f(y)$ denotes a numerical function of y whereas $F(y)$ denotes a mapping from S to R^m. The principal application Benders had in mind was mixed-integer linear programming, but the formulation above is more general than that. As it turned out, Benders's approach has been widely used in the field of *two-stage stochastic linear programming*. The development given here reflects that sort of application but without the stochastic element.

While the Benders Decomposition Algorithm was developed independently after the Dantzig-Wolfe Decomposition Algorithm, it turns out that the Benders Decomposition Algorithm is the same as the Dantzig-Wolfe Algorithm applied to the dual problem. See [45, p. 299] or [15, p. 259].

We consider a linear program of the form

$$
\begin{aligned}
\text{minimize} \quad & c^{\mathrm{T}}x + \sum_{\omega=1}^{\Omega}(f^{\omega})^{\mathrm{T}}y^{\omega} \\
\text{subject to} \quad & Ax = b \\
& B_{\omega}x + D_{\omega}y^{\omega} = d^{\omega}, \quad \omega = 1,\ldots,\Omega \\
& x, y^{\omega} \geq 0, \quad \omega = 1,\ldots,\Omega.
\end{aligned}
\tag{7.18}
$$

The ωs correspond to "scenarios" in the stochastic programming setting.

For ease of discussion, we shall ignore the details of "scenarios" in stochastic programming applications and focus on problem (7.18) in the case where $\Omega = 2$. The form of the problem is then

$$
\begin{aligned}
\text{minimize} \quad & c^{\mathrm{T}}x + (f^1)^{\mathrm{T}}y^1 + (f^2)^{\mathrm{T}}y^2 \\
\text{subject to} \quad & Ax = b \\
& B_1x + D_1y^1 = d^1 \\
& B_2x + D_2y^2 = d^2 \\
& x, y^1, y^2 \geq 0.
\end{aligned}
\tag{7.19}
$$

To simplify the discussion, *we assume that the above problem is feasible and has an optimal solution.*

Rewriting (7.19) by fixing $x \geq 0$ such that it satisfies $Ax = b$, and, moving the x terms in the last two sets of equations to the right, we obtain[8]

$$
\begin{aligned}
\text{minimize} \quad & c^{\mathrm{T}}x + \min_{y^1|x}(f^1)^{\mathrm{T}}y^1 + \min_{y^2|x}(f^2)^{\mathrm{T}}y^2 \\
\text{subject to} \quad & Ax = b \\
& D_1y^1 = d^1 - B_1x \\
& D_2y^2 = d^2 - B_2x \\
& x, y^1, y^2 \geq 0.
\end{aligned}
\tag{7.20}
$$

By fixing such a feasible $x \in R^n$ for the *first-stage problem*, we can solve (7.19) by solving the two subproblems

$$
\begin{aligned}
\text{minimize} \quad & f^{\mathrm{T}}y^{\omega} = z_{\omega}(x) \\
\text{subject to} \quad & Dy^{\omega} = d^{\omega} - B_{\omega}x \\
& y^{\omega} \geq 0.
\end{aligned}
\tag{7.21}
$$

---

[8]The notation $\min_{y^i|x}$ means find the minimum over y^i given a fixed x.

for $\omega = 1, 2$. Solving each of these subproblems separately decomposes the problem (7.18) into $\Omega = 2$ smaller ones (each depending on the previous choice of x). If the true solution $x = x^*$ for (7.18) were known in advance, then the solution of the subproblems (7.21) would result in the solution of the whole problem (7.18). The method devised by Benders estimates an x, uses this to solve the subproblems, and continues in this fashion until a solution to (7.18) is obtained.

We now start examining for each separate ω the dual of (7.21)

$$\text{maximize} \quad (d^\omega - B_\omega x)^\mathrm{T} p^\omega = z_\omega(x)$$

$$\text{subject to} \quad D_\omega^\mathrm{T} p^\omega \leq f^\omega \qquad (7.22)$$

$$p^\omega \quad \text{free.}$$

For each ω, let P_ω denote the feasible region of the corresponding dual problem (7.22). We now assume that the polyhedral sets P_ω are nonempty and that each of them contains at least one extreme point[9].

We next apply the Goldman Resolution Theorem (see page 213) to P_ω. Denote the extreme points of P_ω by $p^{\omega,1} \ldots, p^{\omega,K_\omega}$, and let $q^{\omega,1}, \ldots, q^{\omega,L_\omega}$ denote the complete set of extreme rays of P_ω. Then any feasible solution of (7.22) can be written as

$$p^\omega = \sum_{i=1}^{K_\omega} \lambda_{\omega,i} p^{\omega,i} + \sum_{j=1}^{L_\omega} \mu_{\omega,j} q^{\omega,j}, \qquad (7.23)$$

for some choice of $\lambda_{\omega,i} \geq 0$, $i = 1, \ldots, K_\omega$, where

$$\sum_{i=1}^{K_\omega} \lambda_{\omega,i} = 1, \text{ and } \mu_{\omega,j} \geq 0, \ j = 1, \ldots, L_\omega.$$

Then, substituting (7.23) in (7.22), we obtain

maximize

$$\sum_{i=1}^{K_\omega} \left[(d^\omega - B_\omega x)^\mathrm{T} p^{\omega,i} \right] \lambda_{\omega,i} + \sum_{j=1}^{L_\omega} \left[(d^\omega - B_\omega x)^\mathrm{T} q^{\omega,j} \right] \mu_{\omega,j} = z_\omega(x)$$

subject to   (7.24)

$$\sum_{i=1}^{K_\omega} \lambda_{\omega,i} \qquad\qquad = \quad 1$$

$$\lambda_{\omega,i} \geq 0, \ i = 1, \ldots, K_\omega, \text{ and } \mu_{\omega,j} \geq 0, \ j = 1, \ldots, L_\omega.$$

---

[9]In practice this can be tested computationally.

The assumptions on P_ω imply that either (7.22) has an optimal solution with finite objective function value $z_\omega(x)$ or else the objective function is unbounded above (i.e., $z_\omega(x) = \infty$) thereby implying that its dual (7.21) is infeasible. However, we have assumed that its dual (7.21) is feasible, hence we must have $z_\omega(x) < \infty$. Therefore from (7.24) it follows that x satisfies

$$(d^\omega - B_\omega x)^{\mathrm{T}} q^{\omega,j} \leq 0, \quad j = 1, \ldots L_\omega,$$

otherwise it would be easy to construct a solution that makes $z_\omega(x) \to \infty$. Since $z_\omega(x) < \infty$, it follows that the optimal solution of (7.22) must be attained at an extreme point of P_ω. Thus

$$z_\omega(x) = \max_{i=1,\ldots,K_\omega} (d^\omega - B_\omega x)^{\mathrm{T}} p^{\omega,i}.$$

This is equivalent to saying $z_\omega(x)$ is the smallest number z_ω such that

$$(d^\omega - B_\omega x)^{\mathrm{T}} p^{\omega,i} \leq z_\omega, \quad i = 1, \ldots, K_\omega.$$

From the above development we can write the Benders full master problem as

$$
\begin{aligned}
\text{minimize} \quad & c^{\mathrm{T}} x + z_1 + z_2 \\
\text{subject to} \quad & Ax = b \\
& (d^1 - B_1 x)^{\mathrm{T}} p^{1,i} \leq z_1 \quad \text{for all } i = 1, \ldots, K_1 \\
& (d^1 - B_1 x)^{\mathrm{T}} q^{1,j} \leq 0 \quad \text{for all } j = 1, \ldots, L_1 \qquad (7.25)\\
& (d^2 - B_2 x)^{\mathrm{T}} p^{2,i} \leq z_2 \quad \text{for all } i = 1, \ldots, K_2 \\
& (d^2 - B_2 x)^{\mathrm{T}} q^{2,j} \leq 0 \quad \text{for all } j = 1, \ldots, L_2 \\
& x \geq 0, \ z_1, z_2 \ \text{free.}
\end{aligned}
$$

In this problem, the variables are $x_1, \ldots, x_n, z_1$, and z_2. As compared with the original problem, this form of the problem has a greatly reduced number of variables. However, this is achieved at the expense of introducing a potentially huge number of initially unspecified constraints.

In general it is impractical to write out all the constraints to create a Benders full master problem. Instead a *relaxed master problem* is formed by using the objective in (7.25) with only a *subset* of the constraints stated therein. Then additional constraints (or *cuts* as they are called) are generated as needed. Let us see how this is done.

We can initiate the process by solving the first-stage problem

$$
\begin{aligned}
\text{minimize} \quad & c^{\mathrm{T}}x \\
\text{subject to} \quad & Ax = b \\
& x \geq 0
\end{aligned}
\tag{7.26}
$$

to obtain a feasible $x = \bar{x}$ and then proceed to create a relaxed master problem by using the solutions obtained by solving the subproblems (7.22) for $\omega = 1, 2$. Assume that we have a Benders restricted master problem with $\bar{K}_1, \bar{K}_2, \bar{L}_1$, and $\bar{L}_2$ constraints. Let $x = \bar{x}, z_1 = \bar{z}_1, z_2 = \bar{z}_2$ be a solution to this restricted master and, using $x = \bar{x}$, create and solve the subproblems (7.22) for $\omega = 1, 2$. Then for each ω, one of the following two cases arises:

1. *Optimal solution found.* In this case the following new inequality is generated and adjoined to the restricted master

$$
(d^{\omega} - B_{\omega}\bar{x})^{\mathrm{T}}p^{\omega, \bar{L}+1} \leq z_{\omega}.
$$

   This is called an *optimality cut*.

2. *Problem is unbounded.* In this case a homogeneous solution is obtained and the following inequality is generated and adjoined to the restricted master

$$
(d^{\omega} - B_{\omega}\bar{x})^{\mathrm{T}}q^{\omega, \bar{K}+1} \leq 0.
$$

   This is called an *infeasibility cut*.

Once these cuts are adjoined to the restricted master, the new restricted master is solved and the process is repeated until it is impossible to make an improvement in any of the current subproblem objective values.

7.5 Computational complexity of the Simplex Algorithm

Everyday experience with the Simplex Algorithm suggests that the number of pivot steps is usually a small multiple of the number of equations in the system. Furthermore, use of the Revised Simplex Method, basis handling techniques, and *partial pricing* strategies[10] assure that each iteration is efficient. However, in 1972, Victor Klee and George Minty [112] published an

---

[10] A partial pricing strategy is one where, at the next iteration, the incoming nonbasic column is selected from a *subset* of all the nonbasic columns; see, for example, Orchard-Hays [154].

important paper in which they exhibited a class of linear programs on each of which the Simplex Algorithm with its customary rule for choosing the pivot column as that with the most negative reduced cost (i.e., the "greedy" rule) takes an exponential number of iterations to reach an optimal solution. The Klee-Minty example shows that the worst-case complexity of the Simplex Algorithm is *exponential time*.

Example 7.7: A SMALL EXPONENTIAL EXAMPLE. Consider the linear program (due to Chvátal [28]) previously discussed in Example 7.5.

$$\begin{aligned} \text{maximize} \quad & 9x_1 + 3x_2 + x_3 \\ \text{subject to} \quad & x_1 \leq 1 \\ & 6x_1 + x_2 \leq 9 \\ & 18x_1 + 6x_2 + x_3 \leq 81 \\ & x_j \geq 0, \quad j = 1,2,3. \end{aligned}$$

We shall add slack variables x_4, x_5, x_6 and treat this as a maximization problem in standard form. In tableau form this becomes

max	z	x_1	x_2	x_3	x_4	x_5	x_6	1	basic
row 0	1	-9	-3	-1	0	0	0	0	z
row 1	0	1	0	0	1	0	0	1	x_4
row 2	0	6	1	0	0	1	0	9	x_5
row 3	0	18	6	1	0	0	1	81	x_6

We are *maximizing* z, so we want to make the coefficients in the objective function row become *nonnegative*.

In a maximization problem, choosing the pivot column as the one with "the most negative" $z_j - c_j$ is called the *greedy rule*. Under this pivot column selection criterion, the Simplex Algorithm applied to the problem above does seven pivot steps as shown in Table 7.1. In this table ← means "replaces as a basic variable." There are three things to notice about this sequence of pivots.

1. x_4 and x_1 are interchanged on the odd-numbered iterations.

2. x_5 and x_2 are interchanged every other even-numbered iteration.

3. The number of iterations is of the form $7 = 2^n - 1 = 2^3 - 1$.

Iteration
1	$x_4 \leftarrow x_1$		
2		$x_5 \leftarrow x_2$	
3	$x_1 \leftarrow x_4$		
4			$x_6 \leftarrow x_3$
5	$x_4 \leftarrow x_1$		
6		$x_2 \leftarrow x_5$	
7	$x_1 \leftarrow x_4$		
8		Done!	

Table 7.1: Exponential pivot sequence.

General form of the LPs in this class of problems

To generate an n-variable problem in this class of linear programs, let C be a constant such that $C > 2$. Consider the LP problem

$$
\begin{aligned}
\text{maximize} \quad & \sum_{j=1}^{n} C^{n-j} x_j \\
\text{subject to} \quad & 2 \sum_{j=1}^{i-1} C^{i-j} x_j + x_i \leq (C^2)^{i-1}, \quad i = 1, \ldots, n \\
& x_j \geq 0, \quad j = 1, \ldots, n.
\end{aligned}
\tag{7.27}
$$

It is known from the work of Klee and Minty [112] that this type of problem will take $2^n - 1$ pivot steps if the greedy pivot selection rule is used in the Simplex Algorithm.[11] Indeed the algorithm visits each extreme point of the feasible region.

Just how large is $2^n - 1$?

When $n = 50$, the required number of iterations to solve a problem of the form (7.27) is $2^{50} - 1$ which is approximately 10^{15}. In a year with 365 days, there are

$$
365 \times 24 \times 60 \times 60 = 31,536,000 \approx 3 \times 10^7
$$

---

[11]Actually, Klee and Minty used a different, but equivalent, form of the problem. Chvátal [28, pp 47-49 and 255-258] analyzes a form of the problem like the one discussed here.

seconds. At the rate of T iterations per second, it would then take approximately $10^{15}/(3T \times 10^7) = 10^8/3T$ *years* to solve the problem (7.27) by the Simplex Algorithm with the greedy pivot selection rule!

The discovery of linear programs that require an exponential number of iterations of the Simplex Algorithm intensified the search for *polynomial-time algorithms*, which is to say, algorithms that could be guaranteed to find a solution in a number of iterations that is bounded by a polynomial in a suitable measure of the size of the problem. Some methods for doing are discussed in Chapter 14.

Basic code vectors

Let us consider an n-variable member of the family of linear programs (7.27). In its original form, there will be n constraints in n variables plus n slack variables when the problem is expressed in standard form. Denote the original variables by $x_1, \ldots, x_n$ and the slack variables by $x_{n+1}, \ldots, x_{2n}$. In any BFS of the standard form version, each of the original variables x_i is either basic or nonbasic.[12] We can record this by defining a binary vector $v = (v_1, \ldots, v_n)$ according to the following rule: For each $i = 1, \ldots, n$ set $v_i = 1$ if x_i is basic and $v_i = 0$ otherwise. We shall call these "code vectors."

Let's now have a look at the basic feasible solutions in Example 7.7. There we have $n = 3$ and the following code vectors:

Iteration	Basic Variables	Code Vector v
1	x_4, x_5, x_6	$(0, 0, 0)$
2	x_1, x_5, x_6	$(1, 0, 0)$
3	x_1, x_2, x_6	$(1, 1, 0)$
4	x_4, x_2, x_6	$(0, 1, 0)$
5	x_4, x_2, x_3	$(0, 1, 1)$
6	x_1, x_2, x_3	$(1, 1, 1)$
7	x_1, x_5, x_3	$(1, 0, 1)$
8	x_4, x_5, x_3	$(0, 0, 1)$

Thus, to each set of 3 basic variables there corresponds a code vector v which

---

[12] The same is true of the slack variables, but at the moment we are concentrating on the original variables.

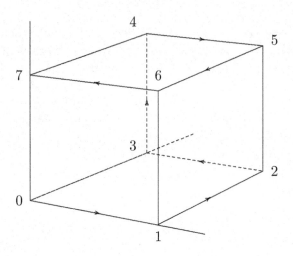

Figure 7.4: A Hamiltonian path on a 3-cube.

can be interpreted as the coordinates of a vertex of the 3-cube.

Figure 7.4 indicates what is called a *Hamiltonian path* on a 3-cube. This path visits each vertex on the cube exactly once. Notice that this Hamiltonian path sweeps out a Hamiltonian path on the bottom face of the cube and then moves to the top face of the cube where is sweeps out a Hamiltonian path in the opposite direction to what it did on the bottom face.

The 8 vertices on the cube in Figure 7.4 are labeled from 0 to 7 to bring out the connection between these code vectors (which are precisely the coordinates of these vertices). These 8 vectors also happen to be the *Gray code* (see Gray [90] and Gardner [76]) representations of the integers from 0 to 7.

7.6   Exercises

7.1   In Section 7.5 we discussed the Klee-Minty example which demonstrated that the Simplex Algorithm takes an exponential number of steps when using the greedy rule. Solve the LP of Example 7.7 using Bland's Rule. How does this compare to the use of the greedy rule?

7.2   In an LP with bounded variables, show that, at an optimal solution, $z_j - c_j \leq 0$ for the nonbasic variables x_j that are at their *upper bound*.

7.3   Solve the following LP using the Simplex Algorithm for bounded variables.

$$
\begin{array}{lrrrrrr}
\text{minimize} & 4x_1 + 3x_2 + 5x_3 + & 2x_4 + & 3x_5 - & x_6 \\
\text{subject to} & 3x_1 + 2x_2 - 2x_3 & & + 3x_5 - 2x_6 & = 9 \\
& -x_1 + 5x_2 & & + 11x_5 + 4x_6 & = 6 \\
& 4x_2 + 6x_3 + 10x_4 & & & = 18
\end{array}
$$

$$0 \le x_1 \le 2, \quad 0 \le x_2 \le 2, \quad 0 \le x_3 \le 6$$
$$0 \le x_4 \le 3, \quad 0 \le x_5 \le 3, \quad 0 \le x_6 \le 2.$$

7.4   Consider the linear program in standard form where

$$
A = \begin{bmatrix} 1 & 0.4 & 0 & 3 & 0 & 5 \\ 0 & 1 & 0.4 & -1 & 2 & 2 \\ 0 & 0 & 1 & 2 & 3 & 3 \end{bmatrix}, \quad b = \begin{bmatrix} 20 \\ 15 \\ 10 \end{bmatrix}
$$

and
$$c = (2, 3, 5, 8, 2, 6).$$

(a) Verify that $B = [A_{\bullet 1}, A_{\bullet 2}, A_{\bullet 3}]$ is a feasible basis.
(b) Put the problem into canonical tableau form with respect to the basis B.
(c) Price out the basis B in the usual way and then by the steepest-edge rule. Compare the results and the effect on the corresponding steps.

7.5   (a) Sketch an algorithm for determining which variables in a given linear program are implicitly bounded.
(b) Apply the algorithm in part (a) of this exercise to the polyhedral set given by the following linear inequalities:

$$
\begin{array}{rrrrr}
3x_1 + & x_2 - & x_3 & \le 8 \\
-x_1 + & 2x_2 + & x_3 + x_4 & \le 5 \\
& x_2 + & 2x_3 + x_4 & \ge 1 \\
& & x_j \ge 0, & j = 2, 3, 4
\end{array}
$$

7.6   Let P denote the polyhedral set defined by the linear inequality system

$$
\begin{array}{rrr}
-4x_1 + & x_2 & \le 12 \\
-2x_1 + & 3x_2 & \le 6 \\
2x_1 + & 5x_2 & \le 10 \\
3x_1 + & x_2 & \le 15.
\end{array}
$$

(a) Plot the inequalities and thereby graph the set P.
(b) Determine the extreme points of P.

(c) Indicate (draw) what the Goldman Resolution Theorem has to say about P.

(d) Write down the extreme rays of the cone in (c).

7.7   (After Bertsimas & Tsitsiklis [15, p. 261].) A limiting case of the block-angular structure is that in which $k = 2$ and $\ell = 1$ thereby giving constraints of the form

$$A_1 x = b^1$$
$$A_2 x = b^2.$$

This may be of interest in an LP situation where one or both of the subsystems $A_i x = b^i$ are easy to solve, but together they are more difficult. The Dantzig-Wolfe Decomposition Principle applies to this situation just as it does to problems with the more general block-angular structure.

Consider an LP model of a distribution problem in which there are 2 warehouses ($i = 1, 2$) each with supply capacity 20 of some commodity and three retail outlets ($j = 1, 2, 3$) having demands of 20, 10, and 10 for that commodity. The decision variables are the amounts x_{ij} of the commodity to ship from warehouse i to retail outlet j. So far, this is just like a set of transportation problem constraints. In this case it is of particular interest to *maximize* the sum of x_{12}, x_{22}, and x_{23}. There is one further constraint. The sum of x_{11} and x_{23} may not exceed 15.

(a) Formulate this problem as a linear program with 2 blocks in which the last constraint is treated as the only "coupling" or complicating constraint and the remaining constraints correspond to the single "subproblem."

(b) The vectors

$$x^1 = (x_{11}, x_{12}, x_{13}, x_{21}, x_{22}, x_{23}) = (20, 0, 0, 0, 10, 10)$$

and

$$x^2 = (x_{11}, x_{12}, x_{13}, x_{21}, x_{22}, x_{23}) = (0, 10, 10, 20, 0, 0)$$

are feasible for the subproblem. Construct a restricted master problem in which the vector x is required to be a convex combination of these two vectors. Solve the restricted problem and obtain the price vector corresponding to its optimal feasible basis.

7.8   *Benders Decomposition.*

(a) A manufacturing company makes p different products for which it has firm orders, say $d_1, \ldots, d_p$. To meet these demands, the company uses its n manufacturing processes (some of which may be alternatives for each other), a set of resources in given amounts, say $r_1, \ldots, r_m$, and its linear technology. The unit cost of process j is $\$ c_j$. Formulate the manufacturer's problem as an equality constrained linear program.

(b) Now assume that—in addition to the p products—the same manufacturing company also produces 3 independent groups of pollutants. The pollutant groups are identified by what they affect: Air, Earth (i.e., Soil), and Water. (There are no Fire pollutants to speak of.) As it happens, each manufacturing process produces an amount of each pollutant that is proportional to the level of the process (activity). Assume that the total amount of any pollutant generated is additive across the production processes. Assume further that there are given limits to the amount of each pollutant the firm may produce. However, the firm may buy or sell pollution credits according to whether its production of a particular pollutant is over or under the prescribed limit. The price of a single credit for (or of) pollutant k is \$$\varphi_k$ per unit. Taking these new conditions into account, formulate the manufacturer's problem as an equality constrained linear program.

7.9   Consider the linear program (7.19) appearing in Section 7.4.

(a) Write down its dual.

(b) Using the concepts of the Dantzig-Wolfe Decomposition Algorithm, write down the full master problem.

(c) Write down the steps of the Dantzig-Wolfe Decomposition Algorithm to solve this problem.

(d) How does this compare to the Benders Decomposition Algorithm applied to the primal problem (7.19)?

7.10 Formulate the analog of the LP given in Example 7.7 for the case of $n = 4$. Solve this problem on a computer. Compare the number of iterations with what the Klee-Minty result predicts. If they are not the same, how do you account for the difference?

7.11 Read up on the Tower of Hanoi puzzle and model the process of solving it in such a way that each state of the system (arrangement of n rings on n pegs) is associated with a "code vector" representing an integer in the set $\{0, 1, \ldots, 2^n - 1\}$.

7.12 *A little art project.* Make a 16×16 grid and label the rows and columns $0,1,2, \ldots, 15$. Think of the 256 cells of the grid as entries of a 16×16 distance matrix. Let i and j correspond to vertices of the 4-dimensional cube labeled according to the Gray code. The entry in the (i, j) cell is to represent the shortest distance from point i to point j by moving along edges of the cube.

(a) Find each entry of the distance matrix described above. (Each entry will be one of the following five numbers: 0,1,2,3,4.)

(b) Select a different color (or shading pattern) to represent each distinct distance in the matrix. "Paint" (or shade) each cell of the grid according to the color (shading pattern) associated with the distance recorded in that cell.

(c) Comment on the structure of the matrix.

Suggestion: Do this exercise for the 3-dimensional case using Figure 7.4 as a guide. Then generalize your solution to 4 dimensions.

8. NLP MODELS AND APPLICATIONS

Overview

We now turn to the study of *nonlinear programming*, also known as *nonlinear optimization*. In this chapter we discuss the differences between linear and nonlinear programming and state the form of the general nonlinear optimization model. We will see that in contrast to linear programming, it *does* make sense to talk about unconstrained nonlinear optimization problems.

The principal aim of this chapter is to identify several classes of nonlinear programming problems and illustrate the practical importance of the subject through a few simple examples. In practice there are certain classes of nonlinear optimization problems having a rather methodological source, namely as optimization subproblems arising within algorithms for other nonlinear programming problems.

Many textbooks on optimization (including this one) provide rather little information on where problems of maximization and minimization actually come up[1]. This is especially true of *nonlinear* optimization problems. The reason for this is that many applications tend to require extensive background knowledge on the reader's part; moreover, there is much to say about theory and algorithms. The situation is very different, however, in the periodical literature. There are literally hundreds of scholarly journals containing articles on applications of nonlinear optimization. These publications are usually directed to an audience that needs little or no introduction to the area of application. This book cannot really remedy this shortcoming, but it does give some useful examples of—and pointers to—nonlinear optimization problems encountered in the real world. In addition, the Bibliography and the Internet can be consulted for further reading on applications of nonlinear optimization.

8.1 Nonlinear programming

A linear program is a constrained optimization problem in which the objective function and constraints are all linear and there is at least one linear

---

[1]Of course there are exceptions, some of them being [164], [26], [10], [18], [19], [189], [8], and [122], all of which provide a broad range of science and engineering applications.

© Springer Science+Business Media LLC 2017
R.W. Cottle and M.N. Thapa, *Linear and Nonlinear Optimization*,
International Series in Operations Research & Management Science 253,
DOI 10.1007/978-1-4939-7055-1_8

inequality constraint. An optimization problem in continuous variables in which one or more of these functions is *not* linear is called a *nonlinear program*. If only its objective function is nonlinear in a constrained optimization problem, it is said to be *linearly constrained*.

General differences between linear and nonlinear programming

As you know, the study of *optimization* is mainly about maximization and minimization. Part I of this book covers linear programming (optimization) in which a given linear function is to be maximized or minimized subject to linear constraints. A linear program has the following characteristics:

L1. Linear objective function.

L2. Linear constraints involving at least one inequality.

L3. Typically many variables.

L4. Solvability by finite (and other types of) algorithms.

By way of contrast, nonlinear programs may—but need not—have any of these properties. Some salient features of nonlinear programs are:

N1. The presence of at least one nonlinear function.

N2. One or more variables all of which are continuous.

N3. Inequality constraints, equality constraints, or no constraints.

N4. Properties of the functions, which may include continuity, differentiability, or convexity.

N5. Occasionally intricate optimality criteria.

N6. Convergent (but not usually finite) solution algorithms with associated rates of convergence, such as linear, superlinear, and quadratic.

Convexity of functions

In N4 above, we mentioned the special property of convexity in connection with functions. For the sake of readers not already familiar with this property, we give a brief definition here.

Let X be a nonempty convex set in R^n and let $f : X \longrightarrow R$. Then f is said to be *convex* on X if it satisfies the inequality

$$f(\lambda u + (1 - \lambda)v) \le \lambda f(u) + (1 - \lambda)f(v) \tag{8.1}$$

for all u and v belonging to X and all $\lambda \in [0, 1]$. If this inequality is satisfied strictly when $u \ne v$ and $0 < \lambda < 1$, then the function f is said to be *strictly convex* on X. Other criteria for convexity of functions are discussed in Section A.7.

The negative of a (strictly) convex function is said to be *(strictly) concave*. Every affine function is both convex and concave, but not strictly so. No other kind of function has this property.

The study of convex sets and functions has a long history and an enormous literature. We shall have more to say on this subject in Section 11.4, but our treatment is necessarily limited. See the Appendix as well as [18] and the references therein for background on convex functions and related matters.

General form of optimization problems

Optimization problems of the type considered in this book can be posed in the form

$$
\begin{aligned}
\text{minimize} \quad & f(x) \\
\text{subject to} \quad & c_i(x) \le 0, \;\; i \in \mathcal{I} \\
& c_i(x) = 0, \;\; i \in \mathcal{E} \\
& x \in \Omega.
\end{aligned}
\tag{8.2}
$$

Although the problem is expressed here as one of minimizing the objective function f, the conversion to maximization is simple because

$$\min_{x \in S} f(x) = -\max_{x \in S}(-f(x)). \tag{8.3}$$

The symbols $\mathcal{I}$ and $\mathcal{E}$ represent *index sets*. The notations are chosen to correspond to the words "inequalities" and "equations," respectively. The statement $x \in \Omega$ indicates the space in which the variables take values. In our case Ω is usually R^n, a constraint that can be—and normally is—omitted from the problem statement.

We now refer back to points N1–N6 listed above. The first one means that in a nonlinear optimization problem, at least one of the functions f or c_i must be nonlinear.[2] That said, it is possible for either or both of the index sets $\mathcal{I}$ and $\mathcal{E}$ to be empty. The terminology goes as follows:

- Unconstrained nonlinear optimization : $\mathcal{E} = \emptyset$, $\mathcal{I} = \emptyset$.

- Inequality-constrained nonlinear optimization : $\mathcal{E} = \emptyset$, $\mathcal{I} \neq \emptyset$.

- Equality-constrained nonlinear optimization : $\mathcal{E} \neq \emptyset$, $\mathcal{I} = \emptyset$.

- Mixed-constrained nonlinear optimization : $\mathcal{E} \neq \emptyset$, $\mathcal{I} \neq \emptyset$.

Note, by the way, that it is possible for a nonlinear optimization problem to have a linear (or, more correctly, affine) objective function as long as it has at least one nonlinear constraint.

Point N4 above refers to the fact that it is necessary to take account of function properties, such as continuity, (orders of) differentiability, convexity, etc. Point N5 has to do with the question of how we identify optimal solutions. This is best left for a later discussion, but for now recall that the case of unconstrained optimization of differentiable functions is treated in multivariate differential calculus. Points N5 and N6 are connected with the iterative nature of nonlinear programming algorithms, which generate sequences of trial solutions and objective function values. Ideally, these sequences converge to feasible solutions that yield (locally) optimal objective function values. If so, it is of interest to know the rate of convergence, that is, how rapidly this occurs—and, for that matter, how much computation is involved in each iteration.

---

[2]Generally speaking, one would prefer *not* to have any nonlinear functions amongst the equality constraints c_i, $i \in \mathcal{E}$, but sometimes the inclusion of such nonlinear constraints cannot be avoided.

Models

As stated earlier, book such as this can never adequately cover the applications of optimization. The subject is extremely broad. This being so, we give only a sample of nonlinear optimization models that arise in nonlinear programming.

It is possible to classify optimization problems according to their fields of application, such as agriculture, the many engineering disciplines, economics, finance, etc. But for our purposes, it is more useful to organize the examples by their structure. For a start, we consider models according to whether their constraints are linear or nonlinear. (Of course, to be classified as nonlinear optimization problems, linearly constrained models must have nonlinear objective functions.) We shall also classify nonlinear optimization models according to whether their constraints are all equations or whether the constraints include inequalities (and possibly no equations). It has to be said that constrained optimization problems with no inequalities of any kind at all are much less prevalent than those with at least some inequalities (possibly combined with equations as well). Other classification schemes involve more special features of the functions; the main representatives of this kind are quadratic functions. Thus, an optimization problem with a quadratic objective function and linear constraints (equations, inequalities, or both) is called a *quadratic programming problem* or *quadratic program* for short.[3]

8.2   Unconstrained nonlinear programs

Unconstrained minimization problems typically have the form

$$
\begin{aligned}
\text{minimize} \quad & f(x) \\
\text{subject to} \quad & x \in R^n
\end{aligned}
$$

and are often written as

$$
\min_{x \in R^n} f(x) \quad \text{or} \quad \min\{f(x) : x \in R^n\}.
$$

The optimization problems encountered in elementary differential calculus are of this sort. Recall that sometimes the objective function may have

---

[3]The further abbreviation QP is also widely used for this class of problems.

many local minima. Cases where a global minimum is to be found and a plethora of local minima exist can be quite challenging.

The linear least-squares problem

Linear least-squares or linear regression is a widely used statistical models. The method is used to fit data to a function that is linear in the model parameters to be estimated. It comes under nonlinear optimzation because the technique minimizes a quadratic function of these parameters, which is the sum of the squares of the errors of the fit.

The linear function used to relate the parameters (or unknown coefficients) with the output can be written as:

$$b = a_1 x_1 + a_2 x_2 + \cdots + a_n x_n, \tag{8.4}$$

where the b (output) and $a_1, a_2, \ldots, a_n$ (inputs) are determined through experiments or surveys, and the x_j, for $j = 1, 2, \ldots, n$ are the unknown parameters to be determined. Often the coefficient of the first parameter x_1 is assumed to be 1 or, alternatively, an additional term x_0 is added to the above relation.

Suppose that, for a given set of inputs, a process is run several times, say m, and, for each run i, the output is recorded:

Run	Input Vector	Output
i	$(a_{i1}, a_{i2}, \ldots, a_{in})$	b_i

Using the above-mentioned data in (8.4) results in the following set of m equations.

$$b_i = a_{i1} x_1 + a_{i2} x_2 + \cdots + a_{in} x_n \quad \text{for } i = 1, 2, \ldots, m.$$

In practice $m \gg n$, and the problem is to determine the parameters x_j that best fit the data specified by an *overdetermined* system of equations. The resulting function can then be used to predict the output based on an arbitrary set of inputs.

The least-squares approach to finding values for the variables in such a system is to minimize the sum of the squares of the differences of each output from its predicted output. For $i = 1, \ldots, m$ define

$$\varepsilon_i = b_i - (a_{i1} x_1 + a_{i2} x_2 + \cdots + a_{in} x_n).$$

The linear least-squares problem is then

$$\text{minimize} \quad \sum_{i=1}^{m} \varepsilon_i^2.$$

The problem can be stated concisely in matrix form; such a statement also makes the solution process easier to follow. To do this, let

$$A_{i\bullet} = [a_{i1}\ a_{i2}\ \cdots\ a_{in}],$$
$$x = (x_1, x_2, \ldots, x_n),$$
$$b = (b_1, b_2, \ldots, b_m),$$
$$\varepsilon = (\varepsilon_1, \varepsilon_2, \ldots, \varepsilon_m).$$

Thus, we minimize

$$\sum_{i=1}^{m} \varepsilon_i^2 = \varepsilon^T \varepsilon = (b - Ax)^T (b - Ax) = b^T b - 2b^T Ax + x^T A^T Ax.$$

The expression on the right-hand side of this equation is a quadratic function of the n variables $x_1, \ldots, x_n$.

To belong to the class of *linear* least-squares problems, the function has to be linear in the model parameters x_j as is the case in (8.4), but there are no restrictions on the a_j. So, for example, this approach can be used to fit polynomials

$$b = x_0 + a_1 x_1 + a_2^2 x_2 + \cdots + a_n^n x_n.$$

The relations can contain any functions of the α_j. For example,

$$b = a_1 x_1 + a_1 a_2 x_2 + a_1 a_2 a_3 x_3 \quad \text{and} \quad b = \log(a_1) x_1 + \log(a_2) x_2$$

are linear functions of the x_j.

The nonlinear least-squares problem

As we have just seen, the linear least-squares model fits data to a function that is *linear* in the parameters to be estimated. In nonlinear least-squares problems, the task is to fit data to a function that is *nonlinear* in the parameters to be estimated. An example of the latter is

$$b = a_1 \sin(x_1) + a_2 \sin(x_2) + \cdots + a_n \sin(x_n).$$

In general, the problem is to fit a functional form

$$b = f(x_1, x_2, \ldots, x_n; a_1, \ldots, a_k), \tag{8.5}$$

where (some of) the parameters x_j enter nonlinearly in the function f.

As in the linear least-squares case, suppose that for a given set of inputs, a process is run several times, say m, and, for each run, the outputs are recorded. For each set of inputs $a_i = (a_{i1}, a_{i2}, \ldots, a_{ik})$, for $i = 1, \ldots, m$, we record an output b_i. This results in a set of m equations:

$$b_i = f(x_1, x_2, \ldots, x_n; a_{i1}, \ldots, a_{ik}) \quad \text{for} \quad i = 1, \ldots, m.$$

For each such i, define

$$\varepsilon_i = b_i - f(x_1, x_2, \ldots, x_n; a_{i1}, \ldots, a_{ik}).$$

The nonlinear least-squares problem is then

$$\text{minimize} \quad \sum_{i=1}^{m} \varepsilon_i^2.$$

Sometimes simple transformations can be used to reduce the nonlinear functional form to a linear form, in which case linear least-squares methodology can be used. For example, the function

$$\bar{\beta} = e^{x_1 t}$$

can be transformed to

$$\ln(\bar{\beta}) = x_1 t.$$

Letting $\beta = \ln(\bar{\beta})$ we obtain

$$\beta = x_1 t,$$

a linear form in x_1. However, the error of the fit (i.e., the objective function value) is not necessarily the same for the nonlinear least-squares model and its transformation to a linear least-squares model.

Example 8.1: A NONLINEAR LEAST-SQUARES PROBLEM. As we all know, automobiles lose a large part of the value during the first year; then in the second year they lose less value than in the first year but more than the value lost in the third year, etc. Suppose we assume that the best way to model this is to assume that the value of an automobile decreases over time according to an exponential functional form. A nonlinear model to represent this is

$$v = x_1 + (p - x_1)e^{-x_2 t},$$

where p is the purchase price, t is the age of the vehicle in years, x_1 and x_2 are parameters to be determined. The parameter x_1 is the residual value at the end of the life of the vehicle. This nonlinear least-squares problem happens to be one that cannot be transformed to a linear least-squares problem.

Location problems in the plane

A type of problem that comes up in many settings and many versions is called a *location problem*. The following is just one simple instance.

Suppose we have a set of n "consumers" at known locations and want to choose the location of a "service facility" for these consumers in such a way that the sum of the distances from the service facility to the consumers is minimum. Examples of this arise in situations where the location of a manufacturing plant, a fire station, or a switching center is to be determined.

Formulation of the location problem

Assume the n consumers have known coordinates (x_i, y_i) for $i = 1, \ldots, n$. Then we seek a point $(x, y) \in R^2$ such that the sum of the distances from (x, y) to each of the points (x_i, y_i) is minimum.

Using the usual Euclidean distance function[4] and placing no restrictions on the location (x, y) of the service facility, we wish to

$$\text{minimize} \sum_{i=1}^{n} \sqrt{(x - x_i)^2 + (y - y_i)^2}.$$

The case of $n = 3$

As a mathematical problem, the case of $n = 3$ was stated by Pierre de Fermat (1601-1665):

---

[4]See Appendix A for background on norms, metrics, and notions of distance.

Given three points in the plane, find a fourth point such that the sum of its distances to the three points is a minimum.

This problem has a beautiful solution that can be found using methods of plane geometry:

The sides of the triangle (formed by the three given points) subtend angles of 120° from the solution point.

As noted by Kuhn [117], this problem (with $n = 3$) has a long history and is often mistakenly identified as "Steiner's problem," an attribution that seems to date back to the book *What is Mathematics?* by Courant and Robbins [39]. Although this name is incorrect, it persists, particularly in connection with its generalization in which a set of one or more points are sought so that the spanning tree on the union of the two sets (those given and those computed) has minimum total length. In the economics literature, the location problem stated above is called the "(Generalized) Weber Problem" after A. Weber, the author of the book [188] on the theory of industrial location.

Variations on the location problem

The location problem can be posed in many ways.

- With a positive "weight" w_i on the i-th term

$$\text{minimize} \ \sum_{i=1}^{n} w_i \sqrt{(x - x_i)^2 + (y - y_i)^2}$$

  in the case of the Euclidean distance function. A large positive weight tends to "penalize" the corresponding term for being large. For example, if $w_1 \gg 1$ and $w_2 = \cdots w_n = 1$, the model emphasizes the importance of having the solution (x, y) be as close as possible to point (x_1, y_1).

- With different distance measures (not Euclidean). In certain cases, it is appropriate to calculate distance a different way. For instance, in an urban setting (e.g., Manhattan) where the streets are laid out in a more or less grid-like way, the distance between two points may have to be measured by what is called the ℓ_1 (or Manhattan) distance measure. Thus, instead of measuring the distance from (x, y) to (x_i, y_i) as $\sqrt{(x - x_i)^2 + (y - y_i)^2}$ the distance in the "Manhattan distance function" would be $|x - x_i| + |y - y_i|$, hence the analogous (weighted) objective would be

$$\text{minimize} \quad \sum_{i=1}^{n} w_i \left\{ |x - x_i| + |y - y_i| \right\}.$$

Figure 8.1 indicates how the Euclidean and Manhattan distance functions differ.

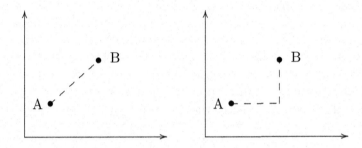

Figure 8.1: Euclidean and Manhattan distances.

- With restrictions on the admissible locations. It may be that only a certain—possibly finite—set of physical locations is feasible. If the set is finite and small, it may be possible just to use the model and do an exhaustive search for the best location. In continuous optimization, another type of restriction might be one in which certain regions of the plane are prohibited. For example the location of a manufacturing plant in the middle of a city park would presumably not be allowed. If the set of candidate locations for the facility is finite, but large, this might not be the most appropriate approach. But notice that either way, when the set of eligible sites from which to choose is finite, the problem is no longer one of *continuous* optimization and hence falls outside the class of problems under consideration here. This discrete problem is studied in combinatorial optimization. It is known to be a hard problem. One of the approaches used in "solving" it is by so-called approximation methods.

- With points in higher-dimensional spaces. One can easily envision applications (for example, in electronics) in which the given points lie in (restricted regions of) 3-dimensional space. The appropriate measure of distance might also be of the ℓ_1 type and possibly weighted.

Projectile problems

Here you will find a couple of related applications of unconstrained optimization; each of them is concerned with finding the optimal value of a single variable. These are given even though everyone who has ever studied elementary differential calculus will have seen many applications of such optimization problems.

The first problem is based on a discussion in a textbook by K.G. Murty [145, p. 530]. This unconstrained univariate optimization model seeks to determine the range of a cannon, defined as the maximum horizontal distance to which a projectile can be propelled. The problem is to choose the firing angle, θ, that is, the angle of the cannon's barrel above horizontal. Clearly

$$0^\circ \leq \theta \leq 90^\circ.$$

The horizontal distance traversed by the projectile (shell) is denoted $d(\theta)$. This is the function to be maximized. But what is the form of the objective function $d(\theta)$?

Let's say the mass of the projectile is 50kg and its initial velocity is $v_0 = 800m/s$. At velocity v the drag at that time is assumed to be $0.015v^2$. After the application of a bit of Newtonian physics and mathematics, one arrives at the expression

$$d(\theta) = 3333.33 \ln(1 + 0.24t \cos\theta)$$

for the range as a function of θ where

$$t = \frac{800}{g} \sin\theta + \sqrt{\frac{2}{g} \left[\frac{(800\sin\theta)^2}{2g} + 1 + 5\sin\theta \right]}.$$

Is this really an unconstrained optimization problem? It is perfectly clear that the optimal θ is *less than* 90°. But what about having $\theta = 0°$? To settle this, one can compare $d(0)$ with $d(\theta)$ for arbitrarily small values of θ. From such an analysis, one sees that the optimal value of θ lies in the open $0° < \theta < 90°$. Accordingly, instead of $\theta \in R$, we actually have $\theta \in \Omega = (0, 90)$, that is, $\{\theta : 0 < \theta < 90\}$ (where these values are measured in degrees). Any purported "solution" that fails to lie in this set is called "extraneous" and is rejected, so in this sense, the optimization problem is unconstrained.

Another application of this sort comes from the track and field event known as the shot put in which a heavy ball (the shot) is "put" (essentially pushed from shoulder height). In ancient times, the object put was a stone, but for the last century or so it has been (in effect) a cannon ball weighing 16 pounds (7.26 kg) for men and 8 pounds, 13 ounces (4 kg) for women. The object of this sport is to make the shot go as far as possible from the edge of the ring within which the shot-putter must remain until the judge calls the mark. Factors that influence the length of the throw are the release height, the speed, and the angle (above horizontal) at which the shot is released. In an interesting book called *Mathematics in Sport* [185, p. 34] the author, M.S. Townend, discusses this application with a model that is much the same as in the artillery example above except that it takes account of the shot-putter's height, or more precisely, the height from which the shot it released and the velocity of the throw. Holding release height and velocity constant, the author analyses the optimal angle as an unconstrained optimization problem.[5]

In a section titled "Projectile Motion in a Gravitational Field" Paul J. Nahin [146, Section 5.4] gives further sports applications of this sort as well some history of the gunnery application dating back to work of Edmond Halley (1656–1742).

Other applications

In the overview to this chapter, we commented on the difficulty of presenting nonlinear optimization applications due to the need for specialized technical background. There are, however, textbooks written for audiences with such backgrounds. One of these is [5] which is largely addressed to readers in electrical engineering and computer science. It contains a chapter on applications of unconstrained optimization, notably point-pattern matching (for recognition of cursive handwriting), inverse kinematics for robotic

---

[5]Experimental evidence indicates that the variables affecting the range of the throw are not independent, see [99].

manipulators, and design of digital filters.

8.3　Linearly constrained nonlinear programs

Constrained nonlinear optimization problems come up in a wide range of economic, engineering, and applied science work. In this section and the next, we exhibit just a few of these to get across the idea. Others will be found in many of the references listed in this book.

Here we discuss problems of two types. The first is of the form

$$
\begin{aligned}
& \text{minimize} && f(x) \\
& \text{subject to} && Ax = b.
\end{aligned}
\tag{8.6}
$$

The second is of the form

$$
\begin{aligned}
& \text{minimize} && f(x) \\
& \text{subject to} && Ax = b \\
& && x \geq 0.
\end{aligned}
\tag{8.7}
$$

The constraints of the latter problem are just like those of a linear program in *standard form*. In fact, just as in linear programming, any linearly constrained problem can be brought to this form by the inclusion of slack or surplus variables and the conversion of free variables to nonnegative variables. Thus, insofar as linearly constrained problems are concerned, it is not particularly restrictive to focus attention on this class of problems.

Within this category, quadratic programming problems stand out as especially important. In these problems, the quadratic objective function may or may not be convex, but when it is, the problem is often called a *convex quadratic program*. The problems illustrated in the next example are of that sort. In this case, the presence of inequalities introduces a combinatorial aspect. Generally speaking, these problems cannot be solved in closed form as (8.8) can. This makes it necessary to use iterative algorithms to solve them. A remarkable feature of convex quadratic programs is that they can be solved with algorithms that *terminate* in a (possibly large) finite number of iterations, either with a globally optimal solution or with the information that the objective function is unbounded below over the feasible region. (A similar assertion can be made for nonconvex quadratic programs, but in most cases algorithms for solving these can at best only be relied upon to produce local minimizers.)

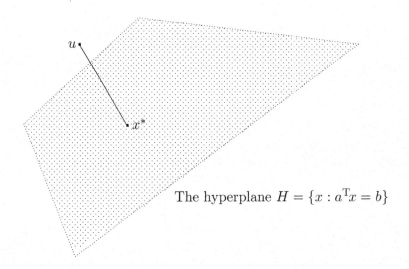

Figure 8.2: Nearest-point problem.

The nearest-point (or projection) problem

Many optimization problems are concerned with some notion of "closeness," and the following simple example is just such a case. Suppose H is a hyperplane in R^n, i.e., the solution set of a linear equation $a^T x = b$ where a is a nonzero n-vector and b denotes a scalar. There are circumstances in which one seeks the point x^* on H that is nearest (in the Euclidean sense) to a given point u that does *not* lie on H. When $n = 2$, and H is just a line, this problem is readily solved by the methods of plane geometry we learn in high school. In higher dimensions, particularly those so high that our geometric intuition fails us, the way to obtain the desired point is not as obvious. Thus, we solve the problem

$$\begin{aligned} \text{minimize} \quad & \tfrac{1}{2}\|x - u\|^2 \\ \text{subject to} \quad & a^T x = b. \end{aligned} \tag{8.8}$$

A unique solution x^* to the above problem exits. It is called the *projection* of u onto H; see Figure 8.2.

As we will see in Chapter 11, the "Method of Lagrange Multipliers" can be used to show that the problem (8.8) has the *closed-form solution*

$$x^* = u + a \left(\frac{b - a^{\mathrm{T}} u}{a^{\mathrm{T}} a} \right). \tag{8.9}$$

Nearest-point (or projection) problems come up in various settings, one of which is motion-planning algorithms (see, for example, [156, pp. 294–346] and [177]) which call for the projection of a gradient vector g onto the nullspace[6] of a matrix A. In that case, we would have the problem

$$\begin{aligned} \text{minimize} \quad & \tfrac{1}{2}\|x - g\|^2 \\ \text{subject to} \quad & Ax = 0. \end{aligned} \tag{8.10}$$

In some equality-constrained problems the variables are naturally sign restricted, but this is implicitly handled through the definition of the objective function. Our next example is of this type.[7]

An entropy maximization problem

Analytical studies of "traffic" in transportation planning, World Wide Web usage, and other fields are often accomplished through the application of optimization models like

$$\begin{aligned} \text{maximize} \quad & -\sum_j x_j \ln x_j \\ \text{subject to} \quad & \sum_j a_{ij} x_j = b_i, \quad x_j > 0 \text{ for all } j. \end{aligned} \tag{8.11}$$

There are several things to be said about problems of this type.

1. In "traffic" applications, the constraints $Ax = b$ typically represent *flow conservation conditions* in very large networks.

2. The objective function as stated represents *entropy*.

---

[6]Interest in the nullspace of a matrix A stems from the fact that the difference of any two vectors that satisfy $Ax = b$ is an element of the nullspace of A. If an algorithm requires all the generated iterates to satisfy $Ax = b$, then two successive iterates x_k and x_{k+1} will differ by an element of the nullspace of A. Indeed, we will have $x_{k+1} = x_k + \alpha_k p_k$ where α_k is a scalar and $Ap_k = 0$.

[7]For another example, see [84, pp. 278–280].

3. The definition of the objective function makes it "unattractive" to consider variables whose values are not positive. We can, therefore, disregard these restrictions in an algorithm provided the starting point is a strictly positive vector and the iterates all satisfy the other constraints $(Ax = b)$.

4. The *negative* of the objective function is strictly convex over the interior of the nonnegative orthant (i.e., the set commonly denoted R^n_{++}), so it is a reasonable function to minimize.

5. The objective function is *separable*. This is a helpful property for a nonlinear function.

8.4 Quadratic programming

Among nonlinear programming problems, perhaps the most common type falls under the heading of quadratic programming (QP). All such problems have a quadratic objective function and affine constraints. Let us interpret this statement in terms of the formulation given in equation (8.2). A *quadratic function* is one of the form[8]

$$f(x) = \tfrac{1}{2}x^{\mathrm{T}}Hx + u^{\mathrm{T}}x + d \tag{8.12}$$

where the (symmetric) matrix H, the vector u, and the scalar d are all constant. Because an additive constant such as d does not affect the location of an optimal solution, it is usually omitted during the optimization process, but it should not be forgotten when the actual optimal value of the objective is reported.

The affine constraints corresponding to the index sets $\mathcal{I}$ and $\mathcal{E}$ may be represented by the linear inequality systems

$$c_i(x) \le A_i.x - b_i \le 0, \qquad i \in \mathcal{I},$$
$$c_i(x) = A_i.x - b_i = 0, \qquad i \in \mathcal{E}.$$

Among the inequality constraints associated with index set $\mathcal{E}$ there could be some that are just *simple bounds* $\ell_j \le x_j \le u_j$ on the variables. As in linear programming, bound constraints are often treated separately from

---

[8]The presence of the factor $\tfrac{1}{2}$ is useful when the gradient and Hessian are computed.

the rest of the inequalities. Note that the quadratic programming problem, so defined, is an extension of the linear programming problem; in an LP, the matrix H is zero.

Estimating a data table

Data are often presented in the form of tables, which is to say, rectangular arrays, much like spreadsheets. But collecting the data might be expensive or problematic. In some cases, the row and column sums (which are called *marginals*) are known even though the entries in the individual cells are not. The type of optimization problem we have in mind here is of that sort. The general setup goes like this. We are given a known $m \times n$ table (or matrix) A whose entries a_{ij} are nonnegative. We seek an $m \times n$ table X with nonnegative entries x_{ij} and a given set of positive marginals, say $r_1, \ldots, r_m$ for the rows and $c_1, \ldots, c_n$ for the columns. The conditions on the marginals of the table X translate into the following system of linear constraints.

$$\sum_{j=1}^{n} x_{ij} = r_i, \quad i = 1, \ldots, m$$

$$\sum_{i=1}^{m} x_{ij} = c_j, \quad j = 1, \ldots, n$$

$$x_{ij} \geq 0 \quad \text{for all } i, j.$$

Notice that the constraints of Example 1.1 are of exactly this form and are consistent provided that $r_1 + r_2 + \cdots + r_m = c_1 + c_2 + \cdots + c_n$.

In the nonlinear programming model under consideration, the criterion for choosing the entries of the table X calls for minimizing the sum of squared differences of known and unknown table entries, that is

$$\sum_{i=1}^{m} \sum_{j=1}^{n} (x_{ij} - a_{ij})^2.$$

Sometimes *weights* $w_{ij} > 0$ are attached to the individual terms $(x_{ij} - a_{ij})^2$ in this objective function. The larger the weight, the more essential it is to minimize the corresponding squared difference.

Having $m \times n$ positively weighted squared variables in it, the objective function displayed above is quadratic and convex rather than just linear, so it qualifies as a *nonlinear* function, and thus we have a convex nonlinear programming problem described in Section 11.4. See [35] for further discussion of this model.

Finding the convex hull of a point set in the plane

The convex hull of a given nonempty finite set $\mathcal{P}$ of points is a polytope, that is, a compact polyhedral set. If the given points all lie in a plane, say R^2, and the set consists of at least 2 points, its boundary will consist of a finite set of line segments called *facets*. A point that is incident to two adjacent facets of different slope is called a *breakpoint*.

This problem[9] often arises in computational geometry. Suppose you need to find the facets and the extreme points of the convex hull of a given set $\mathcal{P}$ of $n + 2$ distinct points in R^2, say

$$\mathcal{P} = \{p_i = (x_i, y_i) \in R^2 : i = 0, 1, \ldots n, n+1\}$$

where $n = 10^4$. For this purpose, you probably would not want to plot the points of $\mathcal{P}$ by hand.

The convex hull will have a *lower envelope* and an *upper envelope*. Each of these is a subset of the boundary of the convex hull you are looking for. The lower envelope will be a piecewise-linear convex function L. The upper envelope will be a piecewise-linear concave function U. The problem can be broken into two parts: one for the lower envelope and one for the upper envelope. The approaches to solving these two parts are analogous. Each piece of either envelope will be a line segment between two of the given points.

Finding the lower envelope L. The points of $\mathcal{P}$ can be ordered lexicographically. Among all points having the same x-coordinate, only the one with the smallest y-coordinate is of interest in finding the lower envelope of the convex hull. (For the upper envelope, only the one with the largest y-coordinate is of interest.) In the following discussion, we assume that this bit of preprocessing has already been carried out and that (for some n) there are $n+2$ points (x_i, y_i) given for which $x_i < x_{i+1}$ for all $i = 0, 1, \ldots, n$.

A general approach. Given a general lower-envelope problem for a set of

---

[9]Based on [37, pp. 11–13].

$n + 2$ points satisfying $x_i < x_{i+1}$ for $i = 0, \ldots, n$, define $z_i = y_i - L(x_i)$. At this stage, the function L is unknown, but we do know that $y_i \geq L(x_i)$, hence $z_i \geq 0$ for all i. Moreover, we must have $z_0 = z_{n+1} = 0$. Next define $\alpha_i = 1/(x_{i+1} - x_i)$ and $\beta_i = \alpha_i(y_{i+1} - y_i)$ for $i = 0, 1, \ldots, n$. The scalar β_i represents the slope of the line segment $[p_i, p_{i+1}]$.

Using these constants, define q with components $q_i = \beta_i - \beta_{i-1}$ for $i = 1, \ldots n$ and the $n \times n$ matrix M with entries given by

$$m_{ij} = \begin{cases} \alpha_{i-1} + \alpha_i & \text{if } i = j \\ -\alpha_i & \text{if } j = i+1 \\ -\alpha_j & \text{if } j = i-1 \\ 0 & \text{otherwise} \end{cases}$$

As you will see later in this book, the system (q, M) given by

$$q + Mz \geq 0, \quad z \geq 0, \quad z^{\mathrm{T}}(q + Mz) = 0 \tag{8.13}$$

represents the optimality conditions of the quadratic programming problem

$$\text{minimize} \quad F(z) = q^{\mathrm{T}}z + \tfrac{1}{2}z^{\mathrm{T}}Mz \quad \text{subject to} \quad z \geq 0. \tag{8.14}$$

In the present context, problems (8.13) and (8.14) are equivalent. (See Exercise 11.14.) Relying on this fact and an algorithm to solve either problem, one can compute the lower envelope L and the upper envelope U of the convex hull. See the exercises at the end of this chapter for further development of this model and its application.

The isotonic regression problem

A commonly encountered problem in statistics is to find values for the variables $x_1, \ldots, x_n$ that satisfy the inequalities

$$x_1 \leq x_2 \leq \cdots \leq x_n \tag{8.15}$$

and minimize the function

$$f(x_1, \ldots, x_n) = \tfrac{1}{2}\sum_{j=1}^{n} w_j(x_j - a_j)^2 \tag{8.16}$$

where the "weights" $w_1, \ldots, w_n$ are given positive constants and the values $a_1, \ldots, a_n$ are given as well. In this instance $f(x_1, \ldots, x_n) \geq 0$ for all $(x_1, \ldots, x_n) \in R^n$.

In the isotonic regression problem, there are no equality constraints, hence $\mathcal{E} = \emptyset$. The form of the constraints given in (8.15) stresses the fact that the values of the variables are nondecreasing as the subscript increases. For two consecutive variables, say x_i and x_{i+1}, there is an inequality

$$c_i(x_1, \ldots, x_n) = x_i - x_{i+1} \leq 0, \quad i = 1, \ldots, n-1.$$

As we see, the constraints consist of $n - 1$ homogeneous linear inequalities in n variables. In matrix form, the constraints look like

$$\begin{bmatrix} 1 & -1 & & & \\ & 1 & -1 & & \\ & & \ddots & \ddots & \\ & & & \ddots & \ddots \\ & & & & 1 & -1 \end{bmatrix} \begin{bmatrix} x_1 \\ x_2 \\ \vdots \\ \vdots \\ x_n \end{bmatrix} \leq \begin{bmatrix} 0 \\ 0 \\ \vdots \\ \vdots \\ 0 \end{bmatrix}.$$

The matrix A of coefficients on the left-hand side of the inequality system has $n-1$ rows and n columns. Be sure to notice that A has special structure.[10] All the nonzero entries of this matrix are either 1 or -1, and these lie along a band (of width 2).

To relate the objective function (8.16) to the general definition of a quadratic function as stated in (8.12), we put

$$H = \begin{bmatrix} w_1 & & & \\ & w_2 & & \\ & & \ddots & \\ & & & w_n \end{bmatrix}, \quad u = \begin{bmatrix} -w_1 a_1 \\ -w_2 a_2 \\ \vdots \\ -w_n a_n \end{bmatrix}, \quad \text{and} \quad d = \frac{1}{2} \sum_{j=1}^{n} w_j a_j^2.$$

What is this problem really about? Its objective function seeks to find values $x_1, \ldots, x_n$ that lie as close as possible to some given values $a_1, \ldots, a_n$ whereas its constraints impose the isotonicity condition $x_i - x_{i+1} \leq 0$ for $i = 1, \ldots, n-1$. It is easy to see that if the given scalars $a_1, \ldots, a_n$ constitute a feasible solution to this problem, then $(x_1, \ldots, x_n) = (a_1, \ldots, a_n)$ is an optimal solution, for in that case, $f(x) = f(a) = 0$ which is the *unconstrained* minimum value of this objective function.

Isotonic regression problems have special properties that facilitate their solution by several methods. One thing is evident: the feasible region described by (8.15) is a polyhedral cone and as such is nonempty.

---

[10]This banded matrix has what is called *upper bidiagonal structure*.

Among the most important of the many quadratic programming appli-cations is the one we discuss next.

The portfolio selection problem

As in the pioneering publication of H.M. Markowitz [131] in 1952,[11] we imagine an investor (usually a large investor such as a mutual fund or an insurance company) that has an amount of money to invest in n assets (stocks, bonds, etc.). Assuming perfect divisibility, the investor wishes to decide on the percentage, x_j, of the total amount to be invested in the j-th security ($j = 1, \ldots, n$). For the purposes of this discussion, the decision variables must satisfy the conditions

$$e^{\mathrm{T}}x = 1, \ x \geq 0. \tag{8.17}$$

The vector x is called a *portfolio*. In addition to the requirement (8.17), there could be others. For instance, there could be upper bounds and lower bounds on the individual variables or on sums of variables. The latter might be included to achieve or to limit the percentage of holdings belonging to a particular industry (electronics, automotive, utilities, etc.) or geographical regions (Asia, Europe, North America, etc.). Linear inequality restrictions of this type are quite readily incorporated in the model.

Two aspects of the investment program are of major importance. One is the *expected return*; the other is the *risk* as measured here by the variance of the return.[12] The most desirable portfolio, of course, would be one which simultaneously maximizes the return and minimizes the risk. Unfortunately, such a portfolio need not exist. Therefore another criterion for choosing the portfolio must be used.

The expected return for the portfolio x is given by

$$E(x) = \sum_{j=1}^{n} r_j x_j = r^{\mathrm{T}}x \tag{8.18}$$

---

[11]Markowitz's work is further elaborated in [132] and [133]. The portfolio selection problem is discussed in many other books and scholarly papers; see for example [171]. Markowitz was one of three recipients of the 1990 Nobel Prize in Economic Science.

[12]In the literature of this subject, risk is usually defined as the standard deviation of the return. The standard deviation is the square root of the variance. From the optimization standpoint, the variance is more convenient to work with than the standard deviation.

where r_j is the average return on the j-th security. The r_j are calculated on the basis of past history, or may be extrapolated in some other way. The variance of the return is given by the *quadratic form*

$$V(x) = x^{\mathrm{T}} D x \qquad (8.19)$$

in which D is the symmetric "covariance matrix." The calculation of the coefficients in V (i.e., the entries of the matrix D) is likewise a separate issue. Nonetheless, it is well known from probability theory that the matrix D is positive semidefinite. This means $V(x) = x^{\mathrm{T}} D x \geq 0$ for all $x \in R^n$.

The feasible portfolios (those satisfying all the imposed linear constraints) form a set P. The set of pairs $(E(x), V(x))$ where x is a feasible portfolio also form a set $\Omega \subset R^2$. The investor would like to find $\bar{x} \in P$ such that

$$(-E(\bar{x}), V(\bar{x})) \leq (-E(x), V(x)) \quad \text{for all } x \in P. \qquad (8.20)$$

One rather natural way to overcome the obstacle presented by the nonexistence of a solution to (8.20) is to combine the two objective functions into one. This is done by letting λ denote a nonnegative parameter and writing

$$f_\lambda(x) = -\lambda E(x) + V(x) = -\lambda r^{\mathrm{T}} x + x^{\mathrm{T}} D x. \qquad (8.21)$$

For a fixed value of λ, minimizing $f_\lambda(x)$ subject to $x \in P$ is a (convex) quadratic program. Letting λ vary over the nonnegative reals, one obtains what is called a *parametric (convex) quadratic programming problem* also known as the *portfolio selection problem*. Each value of λ can be viewed as a weight attached to the expected return. In particular, if $\lambda = 0$, the minimand consists of the variance only. As λ is increased, ever more importance is attached to the expected return; as $\lambda \to \infty$, the expected return is maximized and the variance is ignored.

There is an intimate relationship between the solutions $\bar{x}(\lambda)$ of the portfolio selection problem and certain boundary points of Ω. A pair $(\bar{E}, \bar{V}) \in \Omega$ is said to be an *efficient point* of Ω if there is no pair (E, V) in Ω such that either $E > \bar{E}$ and $V \leq \bar{V}$ or else $E \geq \bar{E}$ and $V < \bar{V}$. An *efficient portfolio* is one for which the associated pair $(\bar{E}, \bar{V})$ is an efficient point of Ω. The efficient points of Ω constitute what is called the *efficient frontier*. They can be generated by solving the portfolio selection problem.

Even though it is theoretically possible to solve the portfolio selection problem as a parametric (convex) quadratic programming problem, and thereby to generate $x(\lambda)$ for each $\lambda \geq 0$, this approach has not met with much favor. One customary approach is to solve the quadratic program repeatedly

$$
\begin{aligned}
&\text{minimize} \quad -\lambda r^{\mathrm{T}}x + x^{\mathrm{T}}Dx \\
&\text{subject to} \quad e^{\mathrm{T}}x = 1, \ x \geq 0
\end{aligned}
\tag{8.22}
$$

for all λ in a *finite set* of selected values. We will consider the formulations of other approaches as exercises in this chapter.

Least distance to a polyhedron

Recall that a polyhedral set or polyhedron is the set of all solutions of a finite system of linear inequalities. Any such system can be converted to what we call standard form (see (8.7)). Some of the reasons for being interested in this problem of finding the least distance to a polyhedron are analogous to those associated with the equality-constrained problem. But here is another: deciding whether the polyhedral set

$$
X = \{x : Ax = b, \ x \geq 0\}
$$

is nonempty and, if so, to obtain a vector in it. Problems like this are normally solved with the Phase I Procedure of the Simplex Method for linear programming which, in the successful outcome, also produces a *basic solution* or equivalently, an extreme point of X. Our aim here is simply to exhibit another formulation of the feasibility problem.

Now to determine whether $X \neq \emptyset$, we consider the set

$$
Y = \{y : y = Ax, \ x \geq 0\}.
$$

The question, in a nutshell, is whether b belongs to Y, because $X \neq \emptyset$ if and only if $b \in Y$. The set Y is the cone spanned by the columns of A. It, too, is a polyhedral set. The vector b belongs to Y if and only if its distance from Y is zero. Thus, we solve the QP

$$
\begin{aligned}
&\text{minimize} \quad \tfrac{1}{2}\|y - b\|^2 \\
&\text{subject to} \quad Ax = y, \ x \geq 0.
\end{aligned}
\tag{8.23}
$$

Taking account of the constraints, we can write this equivalently as

$$
\begin{aligned}
&\text{minimize} \quad \tfrac{1}{2}\|Ax - b\|^2 \\
&\text{subject to} \quad \quad x \geq 0.
\end{aligned}
\tag{8.24}
$$

In a formal sense, this is a special case of (8.7) in which the constraints are absent. (They have been accounted for in the objective function.) A unique optimal solution x^* to this problem must exist. It will yield the point $y^* = Ax^* \in Y$ that is nearest to b. If $\|y^* - b\| = 0$, then the vector x^* that solves (8.24) is an element of X. See [33].

Note that in (8.24) we may write

$$\tfrac{1}{2}\|Ax - b\|^2 = \tfrac{1}{2}x^TA^TAx - b^TAx + \tfrac{1}{2}b^Tb.$$

We turn now to another quadratic programming problem.

Constrained least-squares problems

Regression problems involve the estimation of parameters in a setting where measurements are made on functions whose values depend partially on some known parameter values. In general, the model has the form $y = \varphi(u, x)$ where $u \in R^k$ and $x \in R^n$. The components of u are *input parameters*, whereas those of x are unknown *model parameters* that need to be determined. The purpose, of course, is to devise a predictive instrument that would enable one to forecast the output y arising from an input vector u. But the model is not very useful while the vector x is still unknown. The optimization problem involved here is to determine a suitable vector x^* by measuring the outputs y_i that arise when inputs u_i are used. This measurement takes the form

$$y_i = \varphi(u_i, x^*) + \varepsilon_i, \quad i = 1, \ldots, m$$

where the $\varepsilon_i \in R$ are "random errors."

By far the most familiar scheme for estimating x in such problems is the *least-squares method*, especially in the case where the function $\varphi(u, x)$ is of the form u^Tx; this is called the *linear least-squares model*.[13]

In the general least-squares approach, one minimizes the sum of the squared errors, i.e., the function

$$\sum_{i=1}^{m} (y_i - \varphi(u_i, x))^2 = \sum_{i=1}^{m} \varepsilon_i^2.$$

---

[13]The case where the model is affine, i.e., of the form $u^Tx^* + x_{n+1}^*$ can be handled by taking $u_{i,n+1} = 1$ for $i = 1, \ldots, m$.

As stated, this is an unconstrained minimimization problem in the variables $x_1, \ldots, x_n$. In the linear least-squares case, we minimize $(y - Ux)^{\mathrm{T}}(y - Ux)$, where y is the m-vector of outputs y_i and U is the $m \times n$ matrix whose i-th row is the input vector $(u_i)^{\mathrm{T}}$. Computing the gradient of this quadratic function and setting it equal to zero gives rise to the *normal equations*

$$U^{\mathrm{T}}Ux - U^{\mathrm{T}}y = 0.$$

In some applications the variables x_j are required to satisfy constraints, such as nonnegativity in which case, one gets what is called the *nonnegative least-squares problem*. For a discussion of these and other least-squares problems see [123]. See also [65] for a specific application of nonnegative least-squares in acoustics.

The chemical equilibrium problem

The problem[14] at hand is to determine the molecular composition of the equilibrium state of a gaseous mixture containing m different types of atoms. Although these can combine into many chemically possible molecular species, only the standard types that occur in measurable amounts are considered in the model. We use the following notation:

b_i = the number of atomic weights of atom type i present in the mixture,

x_j = the number of moles of molecular species j present in the mixture,

$\bar{x}$ = $\sum_j x_j$, the total number of moles of gas in the mixture

a_{ij} = the number of atoms of type i in a molecule of species j.

The unknowns in the problem are the x_j which are required to be nonnegative. By definition, their sum is $\bar{x}$. The *mass balance* equations are

$$\sum_{j=1}^{n} a_{ij}x_j = b_i, \quad i = 1, \ldots, m.$$

It is known that determining the equilibrium composition of the gaseous mixture is equivalent to finding values of the mole numbers x_j that satisfy

---

[14]See Dantzig, Johnson, and White [42]. The subject is also treated in Dantzig [40, pp. 481–482] and Bracken and McCormick [19, pp. 46–49].

the mass balance equations and minimize the total free energy of the mixture which is given by

$$G(x_1, \ldots, x_n) = \sum_{j=1}^{n} c_j x_j + \sum_{j=1}^{n} x_j \ln(x_j/\bar{x})$$

$$= \sum_{j=1}^{n} c_j x_j + \bar{x} \sum_{j=1}^{n} (x_j/\bar{x}) \ln(x_j/\bar{x})$$

$$= \bar{x} \left[\sum_{j=1}^{n} c_j (x_j/\bar{x}) + \sum_{j=1}^{n} (x_j/\bar{x}) \ln(x_j/\bar{x}) \right].$$

The coefficients c_j are the values of the Gibbs free energy function of the atomic species at a given temperature plus the natural logarithm of the pressure in atmospheres.

Through a change of variables $u_j = x_j/\bar{x}$ we obtain

$$G(u_1, \ldots, u_n) = \bar{x} \left[\sum_{j=1}^{n} c_j u_j + \sum_{j=1}^{n} u_j \ln(u_j) \right].$$

This is a convex minimization problem because the objective function is convex and the constraints are linear. In one (very early) solution method, a structural property of the problem is used to show that a piecewise-linear approximation to the objective function can be used to obtain an approximate solution via *linear programming* (see Dantzig [40, pp. 481–482]).

A discrete-time production-inventory problem

Consider a manufacturing company that makes an item for which the demand is known in each of K successive periods. The company wishes to develop a production schedule in order to meet the demand at least cost.[15] During any period, the demand can be met from inventory on hand at the beginning of the period or from production that occurs during the period. Assume that production capacity is limited in each period by the same amount, b. (No more that b units can be produced in a single period.) An

---

[15]This discussion is based on [9, p. 5].

individual worker can produce p units of the product per period. Assume that extra labor can be hired as needed and fired if it is not needed. But changes in labor force size are undesirable. To discourage changes in the labor force from one period to the next, a cost proportional to the square of the difference in labor force in successive periods is imposed. There is also a cost associated with carrying inventory forward from one period to the next. The problem is to meet the demand d_k in each of the K periods at least cost. To do this an optimal schedule of inventory and labor force size is to be determined.

It is assumed that the initial inventory I_0 and the initial labor force L_0 are known. The corresponding *unknown* numbers for period k will be denoted I_k and L_k.

The optimization problem can be expressed as

$$\text{minimize} \quad \sum_{k=1}^{K} (c_1 u_k^2 + c_2 I_k)$$

subject to

$$\left. \begin{aligned} L_k &= L_{k-1} + u_k \\ I_k &= I_{k-1} + pL_{k-1} - d_k \\ 0 &\leq L_k \leq b/p \\ I_k &\geq 0 \end{aligned} \right\} \quad \text{for } k = 1, \ldots, K.$$

In this formulation, the variables u_k are auxiliary, and they are *free*, i.e., they are not sign restricted. Indeed, if $u_k < 0$ then there will be a reduction in labor force from period $k-1$ to period k. By using these auxiliary variables as above, one obtains an objective function having a simple form. Indeed, the problem as stated is a quadratic program.

Discrete-time control problems in general

The general set-up for discrete-time control problems goes as follows. One thinks of an entity called a *system* in each of K time periods. In each of these periods, the system is in a particular *state* quantitatively described by a vector y_k called a *state vector*. By choosing a *control vector*, u_k, the decisionmaker changes the state of the system from y_{k-1} to y_k, in accordance with a recursive functional relationship of the form

$$y_k = y_{k-1} + \varphi(y_{k-1}, u_k), \quad k = 1, \ldots, K. \tag{8.25}$$

For this to make sense, an initial state vector y_0 must be given, and the vectors y_k must be of the same dimension. The sequence of vectors generated in this fashion is called a *trajectory*. In problems of this class, different trajectories result in different consequences, some better than others. These consequences may be reckoned in terms of cost or time. Because the control vectors determine the trajectory, one wants to choose them optimally.

There may well be restrictions on the choices of control vectors as well as on the state vectors. Abstractly, these are expressed by relations like

$$\left. \begin{array}{c} u_k \in U_k \\ y_k \in Y_k \end{array} \right\}.$$

There may also be a relationship that links all the states and controls together. This can be expressed by a constraint of the form

$$\psi(y_0, y_1, \ldots, y_K, u_1, \ldots, u_K) \in \mathcal{D},$$

where $\mathcal{D}$ is some prescribed set. The task at hand is to minimize an objective function

$$f(y_0, y_1, \ldots, y_K, u_1, \ldots, u_K)$$

subject to the recursion (8.25) and the constraints which could very well be nonlinear.

8.5   Nonlinearly constrained nonlinear programs

In this section we exhibit three examples of nonlinear optimization problems having nonlinear constraints. All of them are based on examples appearing in other books. Our account of these examples is necessarily sketchy; it is intended only to give a little taste of how nonlinearly constrained optimization can also arise in engineering design problems.

The first example of this section is based on the discussion in [9, pp. 8–9]. For a somewhat more detailed account of this problem, see [22, pp. 375–377].

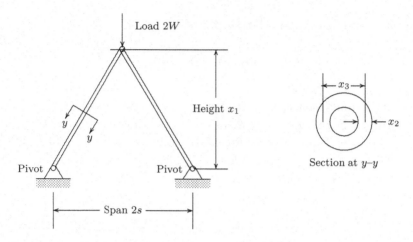

Figure 8.3: A two-bar truss and a cross-section of one bar.

A simple design problem: A two-bar truss

Consider a structure involving two steel tubes pinned at one end (the top) and fixed at two pivot points at the other end (the bottom), see Figure 8.3. The span between the pivot points is fixed at $2s$. A vertical load $2W$ is to be applied at the top of the structure.

The problem is to determine optimal values for the following variables

x_1 = truss height

x_2 = tube thickness

x_3 = average tube diameter

so that the truss will support the load $2W$ and have minimum total weight given by $2\pi\rho x_2 x_3 (s^2 + x_1^2)^{1/2}$ where the scalar constant ρ denotes the density of the material.

For suitably defined parameters b_1, b_2, b_3, b_4, the following constraints must be satisfied:

Space limitation

$$x_1 \leq b_1.$$

Diameter to thickness ratio

$$x_3/x_2 \leq b_2 \qquad \text{(or equivalently } x_3 - b_2 x_2 \leq 0\text{)}.$$

Compression stress must not exceed steel stress

$$W(s^2 + x_1^2)^{1/2} \leq b_3 x_1 x_2 x_3.$$

No buckling

$$W(s^2 + x_1^2)^{3/2} \leq b_4 x_1 x_3 (x_2^2 + x_3^2).$$

Nonnegativity of variables

$$x_j \geq 0, \quad j = 1, 2, 3.$$

Notice that this problem has nonlinearities in both the objective function and the constraints. Given that the load and the span are positive, it follows from the constraints that none of the variables can be zero.

Optimization over an ellipsoid

Let Q be a symmetric positive definite matrix of order n, and let μ be a positive constant. Then the set

$$\{x : x^T Q x \leq \mu\} \tag{8.26}$$

is called an *ellipsoid*. (The closed *unit ball* in n-space is a special case in which $Q = I$ and $\mu = 1$.) In some situations, it is required to minimize or maximize an affine function over an ellipsoid. For example, a variant of the portfolio selection problem is to maximize expected return (8.18) subject to an upper bound on risk (8.19). Thus, the problem of interest might be

$$\text{maximize} \quad c^T x + \delta$$
$$\text{subject to} \quad x^T Q x \leq \mu.$$

A slightly more elaborate problem might also include linear constraints and an ellipsoid

$$\{x : (x - x_0)^T Q (x - x_0) \leq \mu\}$$

centered at a point x_0.

An optical design problem

We turn now to the last example in this chapter. The material is based on the discussion in Section 3.3 of the book [151] which in turn is based on an article [61] in a technical journal. The problem at hand is to design a "Cooke triplet," a configuration of three lenses as illustrated in Figure 8.4.[16]

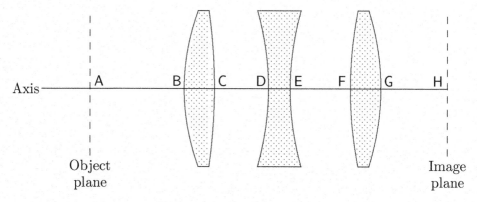

Figure 8.4: Cooke triplet.

Though diagrammed as planar, the lenses are supposed to be spherical and symmetric with respect to the axis. There are 8 points identified as A, B, ..., H; the plane in which the object is assumed to lie is fixed. This implies that position of the image plane cannot be chosen independently of certain other variables: the radii of the 6 spheres and the 6 consecutive separations A-B, B-C, ..., F-G along the axis. So far we have accounted for 12 variables. But there are 6 more: the refractive indexes of the 3 lenses and their measures of dispersion. The design problem will require the selection of values for these 18 independent variables.

Using a technique called "ray tracing," one assembles data on different sorts of optical *errors*. These errors can be described by functions f_i of the lens characteristics. Suppose there are n of these kinds of errors under consideration (and hence n corresponding functions f_i). Then the assignment of values to $x_1, \ldots, x_m$, the independent variables,[17] leads to values for the

---

[16]Also known as "air-spaced triplets," Cooke triplets were developed in 1894 by H.D. Taylor who was employed by T. Cooke and Sons of York, England, makers of astronomical telescopes [119, p. 55].

[17]In the previous paragraph we had $m = 18$, but we are now being a little more general.

errors. This relationship is expressed by

$$\eta_i = f_i(x_1, \ldots, x_m), \quad i = 1, \ldots, n.$$

The basic problem is to choose values of the design variables so as to reduce these errors to a minimum (according to some measure). But, as with so many engineering design problems, the variables must also satisfy some constraints. For example, there are thickness constraints on the components of the lens, limitations on the allowable separations, variations in the refractive indexes, etc. These give rise to constraints symbolically expressed as

$$c_j(x_1, \ldots, x_m) \leq 0, \quad j = 1, \ldots, k.$$

As noted in [151, p. 43] some of these constraints can be "highly nonlinear."

The formulation of this design problem is completed by selecting an objective function aiming to minimize the errors. That function would typically be the positively weighted sum of the squares of the error measures η_i. This could make it a nonlinearly constrained *nonlinear* least-squares problem.

8.6 Exercises

8.1   When H is a real $n \times n$ matrix with entries h_{ij} $(i, j = 1, 2, \ldots n)$, the function

$$q(x) = x^{\mathrm{T}} H x = \sum_{i=1}^{n} \sum_{j=1}^{n} h_{ij} x_{ij}$$

is a quadratic form. Our definition of the quadratic programming problem involves an objective function of the form

$$f(x) = \tfrac{1}{2} x^{\mathrm{T}} H x + c^{\mathrm{T}} x + d,$$

see (8.12). (The constant $1/2$ in front of the quadratic form $x^{\mathrm{T}} H x$ is there for computational convenience as mentioned earlier.) Suppose you encounter a quadratic function $q(x)$ which does not make use of the scalar coefficient $1/2$ as in our formulation. An example of such a case appears in the variance function that comes up in the portfolio selection problem, specifically as

$$V(x) = x^{\mathrm{T}} D x,$$

where $D = D^T$ is the covariance matrix associated with the returns r_i, $i = 1, 2, \ldots, n$. Explain how a quadratic function $q(x)$ that lacks the coefficient $1/2$ can be transformed into a quadratic function $\bar{q}(x)$ that does have the coefficient $1/2$ and takes the same values as $q(x)$.

8.2   (a) Formulate a version of the portfolio selection problem in which the risk is to be minimized subject to the portfolio constraint and the requirement that the expected return shall be at least a given amount r^*.

      (b) If only the constraints mentioned in part (a) are present, what is the largest value that r^* can reasonably attain?

      (c) Discuss the formulation you gave in part (a) in terms of the general form of an optimization problem in equation (8.2).

8.3   Another way to formulate a portfolio selection problem is to consider maximizing the return subject to an upper limit ρ^* on the risk (variance).

      (a) What would you expect the largest attainable risk to be?

      (b) Discuss your formulation in terms of the general form of an optimization problem in equation (8.2).

8.4   Consider a nonempty set $\mathcal{M}$ given as the set of all solutions of a system of linear equations, say $Ax = b$, where A is $m \times n$ with $m < n$. Formulate the problem of finding the point $\bar{x} \in \mathcal{M}$ that is the nearest in the Euclidean distance sense to a given point y that does not belong to $\mathcal{M}$.

8.5   Consider the nonlinear equation $\beta = x_1 e^{x_2 t}$.

      (a) Show how to transform this to a form that is linear in the parameters x_1 and x_2.

      (b) Can this be solved as an unconstrained linear least-squares problem?

8.6   Suppose you want to design a right circular cone whose volume is one cubic meter. Formulate, in terms of one-dimensional optimization, the problem of finding the dimensions that should be used so as to make the cone have minimum surface area. [Note: You may assume that the material of which the cone is made has negligible thickness. Here are some formulas for the surface area and volume of a right circular cone with radius r, altitude h, and slant height l: $S = \pi r l$, $V = \pi r^2 h$.]

8.7   Consider the matrices

$$A = \begin{bmatrix} 3 & 1 & -3 & 3 \\ -2 & 2 & 5 & 0 \\ 0 & -1 & 2 & 2 \\ 4 & 0 & 1 & -5 \end{bmatrix} \quad \text{and} \quad B = \begin{bmatrix} -2 & -4 & 2 & 0 & 4 \\ 5 & 3 & 3 & 5 & -2 \\ 0 & 5 & -1 & 3 & 3 \\ 4 & 2 & 2 & -1 & 3 \end{bmatrix}.$$

Define $\mathcal{A}$ and $\mathcal{B}$ to be the convex hulls (see page 44) of the columns of A and B, respectively. Formulate a nonlinear programming problem by which you can decide whether the sets $\mathcal{A}$ and $\mathcal{B}$ are strongly separable, i.e., there exists a hyperplane $\mathcal{H}$ in 4-space such that points of $\mathcal{A}$ and $\mathcal{B}$ lie on opposite sides of $\mathcal{H}$ and are at a positive distance from it.

8.8   A variation on the location problem discussed in this chapter is the following. Let $p_1, p_2, \ldots, p_n$ be a set of $n > 2$ points in the plane R^2. Suppose the coordinates (x_i, y_i) of each point p_i are given.

(a) Formulate the problem of finding the smallest closed disk that contains the points $p_1, p_2, \ldots, p_n$ as a nonlinear program.

(b) Describe a real-world problem for which you think such a model would be appropriate. Explain why this is so.

8.9   Consider an optimization problem of the form

$$\text{minimize} \quad F(x, y) = \sum_{j=1}^{n} F_j(x_j, y_j) = \sum_{j=1}^{n} (c_j x_j + f_j y_j)$$

$$\text{subject to} \qquad\qquad Ax = b$$
$$x \geq 0$$
$$x \leq y * \hat{x}$$
$$y \geq 0$$
$$y \leq e$$
$$y * (e - y) = 0$$

where $f > 0$ is a given n-vector, $\hat{x} > 0$ is a known vector of upper bounds on the x-variables, e is a vector of all ones, and the symbol $*$ between two vectors denotes the vector of corresponding componentwise products.[18]

This is a model for the linearly constrained *fixed charge* problem which is often encountered in practice.

(a) Interpret the constraints and the objective function in this model.

(b) Is this a linear programming problem? Is the objective function linear? Is it continuous?

8.10   A symmetric $n \times n$ binary matrix $A = [a_{ij}]$ represents an undirected graph $G = (V, E)$ having n vertices. In particular, $a_{ij} = 1$ if and only if i is adjacent to j, which is to say, the edge (i, j) belongs to E. If i and j are not adjacent vertices, we write $a_{ij} = 0$. We assume here that no vertex is adjacent to itself. A nonempty subset σ of V is called a *clique* if for all distinct i and j in σ, the vertices i and j are adjacent. This can be expressed by saying that all off-diagonal elements of $A_{\sigma\sigma}$ equal 1, in which case the corresponding subgraph $G(\sigma)$ is said to be *complete*. A clique σ is said to be *maximal* in G if there is no vertex $k \in V \setminus \sigma$ that is adjacent

---

[18] More generally, if P and Q are two matrices of the same order, the matrix given by $[p_{ij} q_{ij}]$ is called their *Hadamard product*, named after the French mathematician Jacques Hadamard (1865–1963).

to all the vertices in σ. As shown by the graph G corresponding to the adjacency matrix

$$A = \begin{bmatrix} 0 & 1 & 1 & 1 \\ 1 & 0 & 1 & 0 \\ 1 & 1 & 0 & 0 \\ 1 & 0 & 0 & 0 \end{bmatrix},$$

a graph may contain maximal cliques of different sizes (i.e., cardinality of its vertex set σ). A clique whose cardinality is the largest of all maximal cliques in the graph G is called a *maximum clique*. The cardinality of a maximum clique in G is called the *clique number* of G, denoted $\omega(G)$.

When $G = (V, E)$ is a graph with adjacency matrix A, there is a complementary graph $\bar{G} = (V, \bar{E})$ with adjacency matrix $\bar{A}$ whose off-diagonal entries satisfy the equation

$$\bar{a}_{ij} = 1 - a_{ij}, \quad \text{for all } i \neq j.$$

A related concept is that of an *independent set* in the graph $G = (V, E)$. A subset τ of V is *independent* if for all distinct pairs of vertices i and j in τ there is no edge $(i, j) \in E$. Thus, σ is a clique in G if and only if it is an independent set in $\bar{G}$. An independent set is *maximal* if it not not possible to enlarge it and still obtain an independent set. The maximal independent set in a graph having the largest cardinality is called a *maximum independent set*, and its cardinality is denoted $\alpha(G)$.

A theorem of Motzkin and Straus [140] states that the clique number $\omega(G)$ of an undirected graph G with adjacency matrix A can be determined by solving the quadratic program

$$(P) \qquad \text{maximize} \quad x^{\mathrm{T}}Ax \quad \text{subject to} \quad e^{\mathrm{T}}x = 1, \ x \geq 0$$

and that, in fact, subject to these constraints the optimal value of the function $x^{\mathrm{T}}Ax$ is $(\omega(G) - 1)/\omega(G)$. If σ is the vertex set of a maximum clique in G, and $|\sigma| = \omega(G)$ is its cardinality, then $\bar{x}_\sigma = (1/\omega(G))e_\sigma$ is an optimal solution of the quadratic program.

(a) Interpret the problem of finding a maximum clique in G in terms of principal submatrices of its adjacency matrix A.

(b) Interpret the problem of finding a maximum independent set in G in terms of principal submatrices of its adjacency matrix A.

(c) For a given $n > 2$, let $\bar{x}$ be an optimal solution of the quadratic program (P). What would you expect its coordinates to be? Justify your answer.

8.11 Consider the following tiny example of a finite set $\mathcal{P} \subset R^2$ whose convex hull (see page 251) is to be found. Let the points of $\mathcal{P}$ be given by the rows of the following 8×2 matrix:

$$P = \begin{bmatrix} 0 & -0.49 \\ 1 & -1.03 \\ 2 & -0.73 \\ 2 & 0.25 \\ 3 & 0.40 \\ 4 & -0.29 \\ 5 & 0.79 \\ 6 & -0.89 \end{bmatrix}.$$

Note that in this simple case, there are two points having the same x-coordinate. The one with the larger y-coordinate can be disregarded in the search for the lower envelope. This will result in the smaller set represented by the matrix

$$P_L = \begin{bmatrix} 0 & -0.49 \\ 1 & -1.03 \\ 2 & -0.73 \\ 3 & 0.40 \\ 4 & -0.29 \\ 5 & 0.79 \\ 6 & -0.89 \end{bmatrix}.$$

It is worth noting that in this instance, the x-coordinates of successive points now differ by 1.

(a) Form the appropriate matrix P_U corresponding to the upper envelope of the point set $\mathcal{P}$.

(b) What would the condition $q \geq 0$ mean in terms of the unknown function L?

(c) Form the vector q and the matrix M corresponding to the 7 points listed in the example P_L.

(d) Plot the 7 points listed in P_L and then plot the lower and upper envelopes of the convex hull.

(e) In terms of the methodology indicated so far, what would it mean if the lower and upper envelopes do not have the same left and right endpoints?

(f) What are some of the properties of the corresponding matrix M in the example? In general?

(g) Write out and interpret the problems (8.13) and (8.14) in the context of this application.

(h) Can the system (q, M) in (8.13) have more than one solution? Why?

(i) Assume you have solved the equivalent problems stated in (g). How would you go about identifying the facets of the lower (upper) envelope and the extreme points of the convex hull of the given point set?

(j) If (x_i, y_i) lies on the line segment $[(x_{i-1}, y_{i-1}), (x_{i+1}, y_{i+1})]$ can it be a breakpoint of the lower envelope?

8.12 (Based on Dorfman [55].) Consider a firm in the joint production of m commodities, and that the price of each commodity is a known decreasing function p_i of the output y_i. Let these functions be of the form

$$p_i(y_i) = f_i - g_i y_i, \quad i = 1, \ldots, m$$

where the f_i and the g_i are positive constants. This can be expressed in the form

$$p(y) = f = Gy$$

where G is the diagonal matrix with diagonal elements $g_1, \ldots, g_m$. The output vector y is obtained as the result of a linear production corresponding to a matrix H. This means that

$$y = Hx.$$

Assume that the entries h_{ij} of H represent the output of commodity i per dollar (or other monetary (unit) spent on production process j. The net revenue is the gross revenue minus the direct cost of production. In particular,

$$\begin{aligned} r(x) &= p^T y - e^T x \\ &= f^T y - y^T G y - e^T x \\ &= f^T H x - x^T H^T G H x - e^T x \\ &= c^T x - x^T D x \end{aligned}$$

where $c^T = f^T H - e^T$, and $D = H^T G H$.

(a) What sort of function is r?

(b) Suppose the (monetary) inputs x_j are constrained by a system of linear inequalities

$$Ax \le b, \text{ and } x \ge 0.$$

Interpret the individual elements of the matrix A and the vector b.

(c) The manufacturer's aim is to maximize the function r subject to the constraints stated in part (b). What sort of problem is this?

9. UNCONSTRAINED OPTIMIZATION

Overview

The previous chapter presented a small sample of nonlinear optimization problems that might occur in practice; some of these had constraints on their variables and others did not. From this chapter onwards we will discuss ways to solve various types of optimization problems stated in terms of minimization. This is not really restrictive for we have already shown in (8.3) how to convert maximization problems to minimization problems. Our discussion will emphasize algorithms to solve nonlinear minimization problems and develop ways to determine if, in practice, the algorithms described are behaving as expected. Moreover, our discussion will strive to provide an understanding of optimality criteria, how they motivate the development of algorithms, and how they are used in determining if an optimal solution has been found.

The nonlinear programming algorithms we discuss are not always guaranteed to find a global minimum; instead the algorithms terminate at a stationary point which usually is at least a local minimum. We will examine conditions which, when satisfied, imply that a global (unique) minimum has been found.

A nonlinear programming algorithm generates a sequence of trial solutions (iterates x_k) utilizing information available at the current iteration to obtain the next trial solution (iterate x_{k+1}). As in linear programming, the iterates are usually chosen so as to move closer to a solution in a monotonic way.

Optimization algorithms ordinarily require a starting point, an initial iterate. When an iterate is at hand, there are three questions that need to be answered. Is this iterate what we call an optimal solution? If so, the algorithm stops. If not, then in which direction should we search for an improvement? How far should we go in that direction? Once the new iterate is computed, it is tested for optimality, and if it fails the test, the latter two questions must be answered anew.

In this and later chapters the reader may occasionally encounter some unfamiliar terms and concepts. We recommend consulting the Appendix for background information.

© Springer Science+Business Media LLC 2017

R.W. Cottle and M.N. Thapa, *Linear and Nonlinear Optimization*,
International Series in Operations Research & Management Science 253,
DOI 10.1007/978-1-4939-7055-1_9

9.1   Generic Optimization Algorithm

In this book we only consider algorithms for "smooth functions," by which we mean ones that are twice-continuously differentiable. We denote the class of such functions by C^2. We shall see that for nonlinear programs with smooth functions all the algorithms considered in this book have steps of the following kind.

Algorithm 9.1 (Generic Optimization Algorithm)   Let $f(x)$, $x \in R^n$ be the function to be minimized. Choose a starting point x_0 and set the counter $k = 0$.

G1. *Test for convergence.* Determine whether the current iterate x_k satisfies whatever criteria have been established for a "solution" (local minimum). If the criteria for convergence are satisfied, declare a solution and terminate the algorithm.

Comment: The test is performed first, because even a user-provided starting point may already be a solution.

G2. *Determine a search direction.* If the current iterate x_k does not satisfy the convergence criteria, then seek a direction[1] p_k such that $g_k^T p_k < 0$, where $g_k = g(x_k)$ is the *gradient vector* at x_k. This implies that in the neighborhood of x_k, along the halfline with direction p_k that starts at x_k, the function f has values less than the current function value $f_k = f(x_k)$.

Comment: The search direction will usually be chosen to be a direction of decrease of the objective function value; such a direction is called a *descent direction*. The primary difference between most algorithms lies in their determination of search directions. Some techniques will result in fewer iterations at the expense of a higher computational effort per iteration and possibly larger computer memory requirements.

G3. *Determine a steplength.* The steplength is a positive scalar α_k that governs how far to go in this direction from x_k. Choose the scalar α_k such that $x_k + \alpha_k p_k$ yields a decrease in the function being minimized.

---

[1]Sometimes the word *direction* is reserved for a nonzero vector having 2-norm equal to 1. This often comes up in discussions of *directional derivatives*. In nonlinear optimization, it is customary to use the term *descent direction* for a vector p_k such that $p_k^T g_k < 0$ without insisting that p_k be of unit length. But if desired, the normalization can be carried out by the scaling: $(1/\|p_k\|)p_k$.

Comment: Simply choosing $x_{k+1} = x_k + p_k$, will not always produce a decrease in the objective function value, hence an α_k is determined such that $x_k + \alpha_k p_k$ results in a decrease in the objective function. Since x_k and p_k are known, *the determination of α_k is a one-dimensional problem.*

G4. *Update the iterate.* Having computed p_k and α_k, determine the next iterate $x_{k+1} = x_k + \alpha_k p_k$. Update other quantities as necessary for the specific algorithm in question.

G5. *Next iteration.* Let $k \leftarrow k + 1$ and return to Step G1.

9.2   Optimality conditions for univariate minimization

Local and global minima

In univariate optimization one is working with the real number line R and subsets of it called *intervals*. By an *open interval* (a, b) we mean the set of all values between a and b but excluding a and b. It is traditional to denote open intervals by parentheses around the numbers[2]. On the other hand, square brackets are used for *closed intervals*, as in $[a, b]$ which denotes an interval that includes (a, b) and its end points a and b. Intervals can be half-open (or half-closed). For example, $\theta \in [0, \pi/2)$ says that θ can take on values from 0 up to but not including $\pi/2$. However the intervals are specified, the numbers a and b are called the *endpoints* of the interval, and it is always the case that $-\infty < a \leq b < +\infty$. Sometimes it is appropriate to consider unbounded intervals such as $[0, +\infty)$, which simply means all nonnegative numbers.

In optimization, when we speak of a *minimizer* of f, what we have in mind is a feasible solution $x = x^*$ that yields a local or global *minimum value* of the objective function f. We next define local and global minimizers of functions; similar definitions hold for maximizers.

A point x^* is said to be a *global minimizer* of a function f if $f(x^*) \leq f(x)$ for all x. A point x^* is said to be a *strict global minimizer* of a function f if $f(x^*) < f(x)$ for all x.

---

[2]This should not be confused with the similar notation for vectors in R^2. The meaning should be clear from the context.

A point x^* is said to be a *local minimizer* of a function f if for some $\delta > 0$, we have $f(x^*) \leq f(x)$ for all $x \in [x^* - \delta, x^* + \delta]$. A point x^* is said to be a *strict local minimizer* of a function f if for some $\delta > 0$, we have $f(x^*) < f(x)$ for all $x \in [x^* - \delta, x^* + \delta]$.

These concepts pertain to functions of one variable (univariate functions) and, with suitable modifications, to those of many variables (multivariate functions).

First-order conditions for a minimum

Suppose we have a differentiable univariate function f. The sign of the first derivative $f'(x)$ at $x = \bar{x}$, a point in the domain of f, indicates how the function values $f(x)$ change near $\bar{x}$. Thus if $f'(\bar{x}) > 0$, then the function is increasing at $\bar{x}$; if $f'(\bar{x}) \geq 0$, then the function is nondecreasing at $\bar{x}$. Analogously if $f'(\bar{x}) < 0$, then the function is decreasing at $\bar{x}$; if $f'(\bar{x}) \leq 0$, then the function is nonincreasing at $\bar{x}$. If, $f'(\bar{x}) = 0$, then the function is neither increasing nor decreasing at $\bar{x}$; such an $\bar{x}$ is called a *stationary point*[3] of f.

The minimization of differentiable univariate functions is a standard problem in first-year calculus. There, we learn that if x^* is a local minimizer of a function f and if f is differentiable at x^*, then the first derivative of f must *vanish* at x^*. That is, we must have

$$f'(x^*) = 0.$$

But, as we also know, the vanishing of the first derivative is insufficient for deciding whether a point x^* is a local minimizer, a local maximizer or neither. This is illustrated in the following set of examples.

Example 9.1: INSUFFICIENCY OF FIRST-ORDER INFORMATION. Define the following four functions:

(a) $f_1(x) = x^2$;   (b) $f_2(x) = -x^2$;

(c) $f_3(x) = x^3$;   (d) $f_4(x) = 3x^2 - 2x^3$.

The graphs of these functions are given in Figure 9.1 below.

---

[3]Some authors use the term *critical point* for this.

The first derivative of each of these functions vanishes at $x = 0$. The first derivative of f_4 also vanishes when $x = 1$. In the case of f_1, it is clear that $x = 0$ is a local (indeed, global) minimizer, whereas it is a local (global) maximizer for f_2. The function f_3 has an inflection point at $x = 0$, so that point is neither a minimizer nor a maximizer. Finally, f_4 has a local minimum at $x = 0$ and a local maximum at $x = 1$. This function f_4 has no global minimum or maximum.

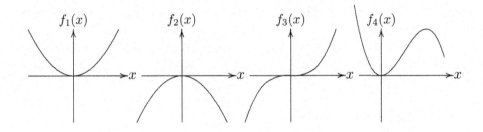

Figure 9.1: Insufficiency of first-order conditions.

Determining a local minimum of a differentiable function usually necessitates the finding of a stationary point. But (b), (c), and (d) in Example 9.1 tell us that simply finding a stationary point of the function is not guaranteed to produce a local minimum (or a local maximum). So is there a guarantee? Is there a way to distinguish between local minima and maxima? The answer is yes, and it involves second-order conditions as discussed below.

Second-order conditions for a minimum

We assume that the reader is familiar with the notion of continuity and derivatives of a function. A function is *continuously differentiable* if its first derivative exists and is continuous. It is *twice-continuously differentiable* if its second derivative exists and is continuous. Recall that if a function is twice-continuously differentiable, then the function and its first derivative must be continuous.

In the following, we denote the statement that f is twice-continuously differentiable on the open interval (a, b) by writing $f \in C^2(a, b)$.

Second-order necessary conditions

If $f \in C^2(a, b)$ and $x^* \in (a, b)$ is a local minimizer of f, then

$$f'(x^*) = 0 \quad \text{and} \quad f''(x^*) \geq 0.$$

Note that the condition on the second derivative says that the first derivative is *nondecreasing* at x^*. Unfortunately, this necessary condition for a local minimum is too weak to be a sufficient condition as well. This can be seen from the Example 9.1(c). The function f_3 given there has a stationary point $x^* = 0 \in (-1, 1)$ or any open interval containing this point. Moreover, its second derivative $f_3''(0) = 0$. Thus, the necessary conditions of local optimality are satisfied at $x^* = 0$, even though this point is not actually a local minimizer (or maximizer). A sufficient condition for local optimality does exist, however.

Second-order sufficient conditions

If $f \in C^2(a, b)$, then $x^* \in (a, b)$ is a *strict* local minimizer of f if

$$f'(x^*) = 0 \quad \text{and} \quad f''(x^*) > 0.$$

There are analogous conditions for relative maxima. In Exercise 9.1 we ask you to supply them.

Note that a point x^* satisfying $f'(x^*) = 0$ and $f''(x^*) = 0$ might actually be a local minimizer (or even a strict local minimizer), but still fail to satisfy the *sufficient* condition for this. A good example of this is $f(x) = x^4$ over $(-1, 1)$. (See Exercise 9.2.)

9.3   Finite termination versus convergence of algorithms

When applied to a linear program, the Simplex Algorithm *terminates* in a finite number of iterations.[4] As we have seen, this number could be large, but it is finite. Most nonlinear programming algorithms are designed to generate a sequence of iterates that *converge* to a "solution." One hopes that as the iterates get "close" to a solution, the convergence will occur rapidly. The following discussion is intended to make this notion more precise and help determine how rapidly a sequence is converging to a solution.

---

[4]There are other algorithms for linear programming that generate infinite sequences of iterates. See Chapter 14.

Let $\{x_k\}$ be a sequence with limit point x^*, written as $\lim_{k\to\infty} x_k = x^*$. We assume that the elements of the sequence are distinct and that all are different from x^*. The latter assumption is needed to make sense of the quotient in the following definition.

The *order* of convergence of the sequence is the largest r for which there exists a number $\gamma \in [0, \infty)$ such that

$$\lim_{k\to\infty} \frac{|x_{k+1} - x^*|}{|x_k - x^*|^r} = \gamma.$$

The number r is called the *(asymptotic) order of convergence* and γ is called the *asymptotic error constant*. See [84, pp. 56–58] and [155, §9.1].

The standard terminology regarding rates of convergence goes this way:

- When $r = 1$ and $\gamma \neq 0$, the rate of convergence is said to be *linear*.

- When $r = 2$, it is said to be *quadratic*.

- The number r is not required to be an integer. Thus, when $1 < r < 2$, the rate of convergence is said to be *superlinear*. When $\gamma = 0$ and $r = 1$, the rate convergence is also *superlinear*.

- In the case where $\gamma = 0$ and $r < 1$, the rate of convergence is called *sublinear*.

It should be clear that quadratic convergence is the most desirable of those mentioned above. This is because when $|x_k - x^*|$ is less than 1 (as it must eventually be if the sequence converges), its square is even smaller. This means that the x_k converge to x^* faster (fewer iterations) in the quadratic case than in the linear or superlinear case.

Example 9.2: RATES OF CONVERGENCE. The following examples of sequences of real numbers are contrived to illustrate different rates of convergence.

1. Let $0 < c < 1$ be given. The sequence $\{x_k\} = \{c^{2^k}\}$ converges to $x^* = 0$. The asymptotic order of convergence in this case is $r = 2$

(quadratic rate of convergence), and the asymptotic error constant is 1. This can be shown as follows:

$$\lim_{k\to\infty}\frac{c^{2^{k+1}}-0}{(c^{2^k}-0)^2}=\lim_{k\to\infty}\frac{c^{2^{k+1}}}{c^{2^{k+1}}}=1.$$

2. Let $c>0$ be given. The sequence $\{x_k\}=\{c^{2^{-k}}\}$ converges to $x^*=1$. The asymptotic order of convergence is $r=1$ (linear rate of convergence), and the asymptotic error constant is $1/2$.

3. The sequence $\{x_k\}=\{1/k^k\}$ converges superlinearly. The limit x^* is 0, as is the asymptotic error constant γ. Indeed

$$\lim_{k\to\infty}\frac{(\frac{1}{k+1})^{k+1}-0}{(\frac{1}{k})^k-0}=\lim_{k\to\infty}\left(\frac{k}{k+1}\right)^k\cdot\frac{1}{k+1}=0.$$

9.4 Zero-finding methods

Since local minima and maxima of differentiable functions occur at stationary points, it is appropriate to devote some attention to the subject of *zero-finding*, that is, finding an $x=x^*$ for which $f(x^*)=0$; such an x^* is called a *zero* of f. For example, zero-finding comes up in the process of minimizing or maximizing a continuously differentiable function. In this case the function for which a zero is sought is the first derivative.

Not all real numbers can be represented exactly on a computer, and subsequent calculations may result in small errors. Thus, when searching for a zero of a continuous function f on a computer, we usually have to settle for a value x that is very close to a zero x^* of that function. Thus, we will be satisfied if our algorithm provides an $\bar{x}$ such that

$$\bar{x}\in[x^*-\delta,x^*+\delta]$$

for some "small" tolerance δ where $f(x^*)=0$. Equivalently, this can be stated as

$$x^*\in[\bar{x}-\delta,\bar{x}+\delta].$$

A zero of a continuous function f is said to be *bracketed* in an interval $[a,b]$ if f changes sign there[5]. For example Figure 9.2 indicates that

---

[5]There exist functions, such as $f(x)=x^2$, whose zeros are not bracketed in this sense in any interval.

$f(a)f(b) < 0$ so that the interval $[a, b]$ brackets a zero of f. In general, an interval in which a zero of a function f is *known* to lie is called an *interval of uncertainty.*

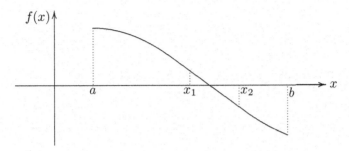

Figure 9.2: Bisection Method; zero of f bracketed.

The Bisection Method

The Bisection Method finds a zero of a function by systematically reducing the interval of uncertainty by a factor of 2 using function-value comparisons. It requires starting with an interval that is known to bracket a zero of the function.

The procedure works as follows. Suppose that a zero of f is bracketed in the interval $[a, b]$. Then $f(a)f(b) < 0$. The function is evaluated at $c = (a+b)/2$, the midpoint of the interval. If the function value is zero at c, the algorithm terminates with c as a zero of f; otherwise, a new interval of uncertainty is obtained by replacing a or b by c depending on whether $f(a)$ or $f(b)$ has the same sign as $f(c)$; that is, if $f(a)f(c) > 0$, replace a by c, otherwise replace b by c. The procedure is repeated with the new interval.

With the above procedure, at each iteration the interval is halved. The procedure terminates either when a zero is found or the length of the interval is "acceptably small."

Algorithm 9.2 (Bisection Method) *Find a zero x^* of a continuous univariate function g.*

By knowledge of the function or trial and error determine an interval $[a, b]$, such that $f(a)f(b) < 0$, i.e., the zero is bracketed in $[a, b]$. Determine the midpoint c and evaluate $f(c)$. Let $\delta > 0$ be a predefined tolerance within which a solution will be acceptable.

1. If $|b - a| \leq \delta$ or $f(c) = 0$, stop and report the zero as $x^* = c$.

2. If $f(a)f(c) > 0$, let $a \leftarrow c$ else let $b \leftarrow c$ to create a new interval.

3. Find the midpoint c of the new interval and evaluate $f(c)$.

4. Go to Step 1.

Example 9.3: BISECTION METHOD USED TO FIND A SQUARE ROOT.

Suppose that we wish to find the square root of a number, say $s = 12$. Define a function whose zero will be the square root of s, that is, $f(x) = x^2 - s$. In this case, it is obvious that we can set $a = 3$ and $b = 4$; the function values at these points are $f(a) = -3$ and $f(b) = 4$. Thus, the zero is bracketed in the interval $[a, b] = [3, 4]$. We now perform the iterations of the Bisection Method, choosing $\delta = 10^{-6}$.

At the midpoint $c = (a + b)/2 = (3 + 4)/2 = 3.5$, we obtain $f(c) = 3.5 \times 3.5 - 12 = 0.25$. Hence we set $b = 3.5$ and leave a unchanged. At the next midpoint $c = (3 + 3.5)/2 = 3.25$, we obtain $f(c) = -1.4375$. We now set $a = 3.25$ and leave b unchanged. Continuing in this manner results in the iterations as shown in the tabular form below:

a	b	c	$f(c)$	$|a - b|$
3	4	3.5	0.250000	1
3	3.5	3.25	−1.437500	0.5
3.25	3.5	3.375	−0.609375	0.25
3.375	3.5	3.4375	−0.183594	0.125
3.4375	3.5	3.46875	0.032227	0.0625
3.4375	3.46875	3.453125	−0.075928	0.03125
$\vdots$	$\vdots$	$\vdots$	$\vdots$	$\vdots$

As long as the computation of the function value at any x has the correct sign and a zero of the function is bracketed in the starting interval, the bisection method is guaranteed to find its zero x^* within any specified tolerance δ. It is straightforward to see that it takes about $\log_2((b-a)/\delta)$

evaluations of c to find such an interval. The initial interval $b-a$ gets halved at each iteration. After n iterations the interval is $(b-a)/2^n$, therefore we would like an n such that

$$\frac{b-a}{2^n} \leq \delta$$

or, equivalently,

$$2^n \geq \frac{b-a}{\delta}$$

or, taking base-2 logarithms of both sides, we obtain

$$n \geq \log_2 \left(\frac{b-a}{\delta} \right).$$

The algorithm is optimal for the class of functions that change sign in $[a, b]$ in the sense that it yields the smallest interval of uncertainty for a specified number of function evaluations. Exercise 9.10 asks you to show that the convergence rate of the Bisection Method is linear with the asymptotic constant equal to $1/2$.

As with all numerical algorithms, care must be exercised during the computations. For example, it is possible that on a computer, the computed midpoint c may not lie in the interval $[a, b]$. Such a case is illustrated by the following example.

Example 9.4: COMPUTATION OF A MIDPOINT. Suppose that we are performing calculations on a machine that can represent at most three digits and that anything beyond 3 digits is truncated. We will use the symbol $fl(x)$ to mean floating-point calculation of x. Suppose that at some iteration of the Bisection Method we have the interval $[a, b] = [0.802, 0.804]$. Then calculating c, we obtain

$$c = fl((0.802 + 0.804)/2) = fl(1.606/2) = 1.60/2 = 0.800,$$

which is not in the original interval! In this case, the calculations can be reorganized to yield a midpoint within the interval; that is, we compute $c = a + (b-a)/2$. This results in

$$c = fl(0.802) + fl((0.804 - 0.802)/2) = 0.802 + 0.002/2 = 0.803\,.$$

See Exercise 9.12 on how compute the midpoint when the endpoints of the interval have opposite signs.

Newton's Method

The underlying idea of this method (and several others) is to approximate f by a "simple" function whose zero can be easily calculated, and then use the calculated zero of the simple function as an *estimate* of the zero of the function f itself. If the estimate is not a zero of f, the process is repeated. A simple function that comes to mind is a linear function; its graph in this univariate case will be a straight line.

We evaluate the zero of the linear function that represents the tangent line at the current estimate of the zero of f to get a new estimate thereof. If the current estimate of the zero is x_k, we calculate the equation of the line tangent to the curve at $(x_k, f(x_k))$. For this, we use the point-slope formula of analytic geometry to obtain the equation $y = f(x_k) + (x - x_k)f'(x_k)$, where $f'(x_k) \neq 0$. The next estimate is the value x_{k+1} that makes $y = 0$. To put this another way: $(x_{k+1}, 0)$ is the point in R^2 where the tangent line meets the horizontal axis, namely

$$x_{k+1} = x_k - \frac{f(x_k)}{f'(x_k)}. \tag{9.1}$$

Algorithm 9.3 (Newton's Method for univariate zero finding)   *Find a zero x^* of a differentiable univariate function f.*

Let $\delta > 0$ be a predefined tolerance such that the algorithm will terminate when a point x_k has been found that satisfies $|f(x_k)| \leq \delta$. Pick a point x_0, evaluate $f(x_0)$, and set $k = 0$.

1. If $|f(x_k)| \leq \delta$, stop and report a zero of f as $x^* = x_k$.

2. Determine $p_k = -f(x_k)/f'(x_k)$.

3. Set the steplength $\alpha_k = 1$.

4. Let $x_{k+1} = x_k + \alpha_k p_k = x_k - f(x_k)/f'(x_k)$, and compute $f(x_{k+1})$.

5. Let $k \leftarrow k + 1$ and return to Step 1.

An example of this procedure is sketched below in Figure 9.3.

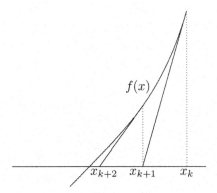

Figure 9.3: Newton's Method.

Example 9.5: NEWTON'S METHOD FOR FINDING A SQUARE ROOT. As we have seen in Example 9.3, we can find the square root of a number $s = 12$ by finding the zero of a function $f(x) = x^2 - s$. Substituting the latter for $f(x)$ and $f'(x) = 2x$ in (9.1) we obtain

$$x_{k+1} = x_k - \frac{f(x_k)}{f'(x_k)} = x_k - \frac{x_k^2 - s}{2x_k} = \frac{1}{2}\left(x_k + \frac{s}{x_k}\right),$$

which is the classical iterative procedure for computing square roots.

Starting with $x_0 = 3$ as the first iterate, it is easy to verify that (to 6 decimal places) the iterates converging to the square root of 12 are:

$$3,\ 3.5,\ 3.464286,\ 3.464102,\ 3.464102\,.$$

The convergence rate can be seen by examining the sequence $|x_k - x^*|$, where, to 12 decimal places, x^* is 3.464101615138.

Iteration	$|x_k - x^*|$	$\dfrac{|x_{k+1} - x^*|}{|x_k - x^*|^2}$
1	4.64E−01	
2	3.59E−02	1.67E−01
3	1.84E−04	1.43E−01
4	4.89E−09	1.44E−01
5	0.00E+00	0.00E+00

From column 2 of the above table[6] we can see that an iterate's distance from the solution is roughly changing by square of the previous iterate's distance from the solution, implying a quadratic rate of convergence. This is confirmed by column 3, which shows the ratio $|x_{k+1} - x^*|/|x_k - x^*|^2$ converging to around 0.144.

Convergence rate

If $f'(x^*) \neq 0$, and the starting point is "sufficiently" close to x^*, then Newton's Method converges *quadratically* to x^*. More formally, the following theorem holds.

Theorem 9.1 (Local convergence of Newton's Method)   *Let f be a univariate function that is continuously differentiable on R. Assume that $f(x^*) = 0$ for some x^* and that $f'(x^*) \neq 0$. Then there exists an open interval S containing x^* such that, for any x_0 in S, the Newton iterates*

$$x_{k+1} = x_k - f(x_k)/f'(x_k)$$

are well defined, remain in S, and converge superlinearly to x^*. Moreover, if

$$|f'(x) - f'(x^*)| \leq \kappa|x - x^*| \quad for\ all\ x \in S,$$

where $\kappa > 0$ is a constant, then the sequence $\{x_k\}$ converges quadratically to x^*.

The problem with Newton's Method is that its quadratic convergence rate is only local. Thus, as discussed in [84] and many other books, if x_0 is not close enough to x^*, the Newton iterates may actually *diverge*. Such an instance is depicted in Figure 9.4, for the function $f(x) = 1 - 1/x$. Hence Newton's Method must be used with care; it is not an infallible technique for zero-finding. Often Newton's Method is combined with the Bisection Method to obtain guaranteed convergence; in this case the method no longer converges quadratically.

In some practical problems, f' may be difficult or impossible to compute; in such cases Newton's Method cannot be used. A related method that does not require $f'(x)$ is described next.

---

[6]Here we are using exponent notation where, for example, 4.64E–01 means 4.64×10^{-1}.

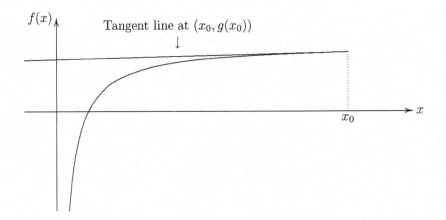

Figure 9.4: Divergence of Newton's Method.

The Secant Method

This method, like Newton's Method, also approximates the function by a straight line. Instead of using a tangent line, it uses a line determined by two points on the graph of the function. The use of such a line can be thought of as replacing the derivative $f'(x_k)$ by the finite-difference formula $(f(x_k) - f(x_{k-1}))/(x_k - x_{k-1})$. Using this approximation, the iterates are defined by the formula

$$x_{k+1} = x_k - f(x_k)\frac{x_k - x_{k-1}}{f(x_k) - f(x_{k-1})}. \tag{9.2}$$

The advantage of this method over Newton's Method is that it is no longer necessary to find the first derivative of f, which might be difficult to determine analytically.

Convergence rate

Unfortunately, with this modification of Newton's Method the property of quadratic convergence is lost. However, it can be shown that the iterates converge superlinearly; see, for example, Stewart [180].

Algorithm 9.4 (Secant Method for univariate zero finding)   *Find a zero* x^* *of a univariate differentiable function* f.

Let $\delta > 0$ be a predefined tolerance such that the algorithm will terminate when a point x_k has been found that satisfies $|f(x_k)| \leq \delta$. Pick distinct points x_0 and x_1 and evaluate $f(x_1)$. Set $k = 1$.

1. If $|f(x_k)| \leq \delta$, stop and report a zero of f as $x^* = x_k$.

2. Determine $p_k = -f(x_k) \dfrac{x_k - x_{k-1}}{f(x_k) - f(x_{k-1})}$.

3. Set the steplength $\alpha_k = 1$.

4. Let $x_{k+1} = x_k + \alpha_k p_k = x_k - f(x_k) \dfrac{x_k - x_{k-1}}{f(x_k) - f(x_{k-1})}$, and compute $f(x_{k+1})$.

5. Let $k \leftarrow k + 1$ and return to Step 1.

The Secant Method is illustrated in Figure 9.5.

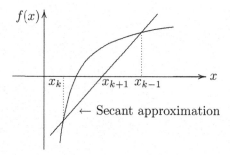

Figure 9.5: Secant Method.

Example 9.6: SECANT METHOD FOR FINDING A SQUARE ROOT. As we have seen in Example 9.3, we can find the square root of a number $s = 12$ by finding the zero of a function $f(x) = x^2 - s$. Substituting for $f(x)$ in (9.2) we obtain

$$x_{k+1} = x_k - f(x_k) \frac{x_k - x_{k-1}}{f(x_k) - f(x_{k-1})}$$

$$= x_k - (x_k^2 - s) \frac{x_k - x_{k-1}}{x_k^2 - s - (x_{k-1}^2 - s)}$$

$$= x_k - \frac{(x_k^2 - s)}{x_k + x_{k-1}}$$

$$= \frac{x_k x_{k-1} + s}{x_k + x_{k-1}}.$$

Starting with $x_0 = 3$ and $x_1 = 4$ as the first two iterates, it is easy to verify that, to 6 decimal places, the iterates converging to the square root are

$$3, \ 4, \ 3.4285714, \ 3.461538, \ 3.4641148, \ 3.464102, \ 3.464102 \,.$$

As in Newton's Method, we can verify the convergence rate by examining the sequence $|x_k - x^*|$, recall that x^* to 12 decimal places is 3.464101615138.

Iteration	$|x_k - x^*|$	$\dfrac{|x_{k+1} - x^*|}{|x_k - x^*|}$
1	3.55E$-$02	
2	2.56E$-$03	7.21E$-$02
3	1.32E$-$05	5.16E$-$03
4	4.89E$-$09	3.70E$-$04
5	9.33E$+$15	1.91E$+$06

From column 2 of the above table we can see that distance of an iterate from the solution is improving at a rate better than linear but not as rapidly as by the square of the previous iterate's distance. Column 3 shows the ratio $|x_{k+1} - x^*|/|x_k - x^*|$ converging to 0.0, which implies that a superlinear convergence rate is being achieved.

Like Newton's Method, the Secant Method can diverge. Figure 9.6 illustrates such behavior. In the next section we describe a method which prevents such divergence.

The Method of False Position (Regula Falsi)

One way to prevent divergence of the Secant Method is to maintain an interval of uncertainty that always brackets a zero of the function, similar to what is done in the Bisection Method. The secant line is constructed as the line that passes through the end points of the interval of uncertainty. Thus, at each iteration, the current iterate replaces one of the end points of the interval from the previous iteration. That is, if $f(x_{k+1})$ has the same sign as $f(x_k)$, then x_{k+1} and x_{k-1} are retained; otherwise x_{k+1} and x_k are retained. Of course, in order for this to work, we must start the algorithm with two points x_0 and x_1 such that $f(x_0)f(x_1) < 0$.

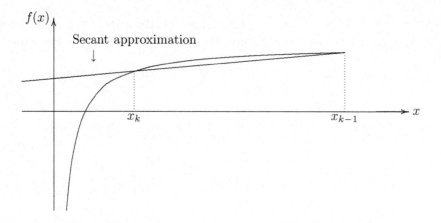

Figure 9.6: Divergence of the Secant Method.

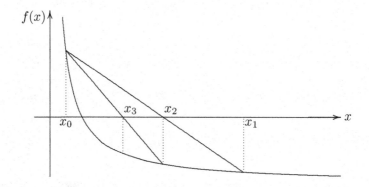

Figure 9.7: Regula Falsi.

Convergence rate

Unfortunately, although this modification guarantees convergence, the rate of convergence can be extremely slow. It can be shown that the rate of convergence is linear with the asymptotic constant arbitrarily close to unity. Thus the Method of False Position can perform worse than the Bisection Method. Figure 9.7 shows an example of poor convergence, where the initial point x_0 is never discarded.

Algorithm 9.5 (Regula Falsi Method for univariate zero finding) *Find a zero x^* of a univariate differentiable function f.*

Let $\delta > 0$ be a predefined tolerance such that the algorithm will terminate when a point x_k has been found that satisfies $|f(x_k)| \leq \delta$. Pick points x_0 and x_1 such that $f(x_0)f(x_1) < 0$. Set $k = 1$.

1. If $|f(x_k)| \leq \delta$, stop and report a zero of f as $x^* = x_k$.

2. Determine $p_k = -f(x_k)\dfrac{x_k - x_{k-1}}{f(x_k) - f(x_{k-1})}$.

3. Set the steplength $\alpha_k = 1$.

4. Let $x_{k+1} = x_k + \alpha_k p_k = x_k - f(x_k)\dfrac{x_k - x_{k-1}}{f(x_k) - f(x_{k-1})}$, and compute $f(x_{k+1})$.

5. If $f(x_{k+1})f(x_{k-1}) < 0$, set $x_k = x_{k-1}$.

6. Let $k \leftarrow k + 1$ and return to Step 1.

Example 9.7: REGULA FALSI METHOD FOR FINDING A SQUARE ROOT.

We apply the Regula Falsi Method to the function in Example 9.3. The computation of the next iterate is by the formula used in Example 9.6.

We start with $x_0 = 3$ and $x_1 = 4$ as the first two iterates that bracket a zero of the function $f(x) = x^2 - 12$. The iterations are shown shown in the tabular form below:

x_{k-1}	x_k	$f(x_{k-1})$	$f(x_k)$	x_{k+1}	$f(x_{k+1})$
3.000000	4.000000	−3.000000	4.000000	3.428571	−0.244898
3.428571	4.000000	−0.244898	4.000000	3.461538	−0.017751
3.461538	4.000000	−0.001275	4.000000	3.463918	−0.001275
3.463918	4.000000	−0.017751	4.000000	3.464088	−0.000092
3.464088	4.000000	−0.000092	4.000000	3.464101	−0.000007
3.464101	4.000000	−0.000007	4.000000	3.464102	0.000000

As before, we can verify the convergence rate by examining the sequence $|x_k - x^*|$, where x^* to 12 decimal places is 3.464101615138.

Iteration	$\lvert x_k - x^* \rvert$	$\dfrac{\lvert x_{k+1} - x^* \rvert}{\lvert x_k - x^* \rvert}$
1	3.55E−02	
2	2.56E−03	7.21E−02
3	1.84E−04	7.18E−02
4	1.32E−05	7.18E−02
5	9.49E−07	7.18E−02
6	6.81E−08	7.18E−02
7	9.77E−15	1.43E−07

In this case, the superlinear convergence is not obvious. If we compute the quantities for one more iteration, we see that $\lvert x_{k+1}-x^* \rvert / \lvert x_k-x^* \rvert$ in column 3 appears to be going to zero, thus indicating superlinear convergence.

Safeguarded Fast Methods for univariate zero finding

As we have seen, Newton's Method and the Secant Method converge quickly to the solution but, in some cases, can actually diverge. We can prevent divergence of the Secant Method by making sure that a zero is always bracketed in an interval whose end points are used to construct the secant line. However, this bracketing can degrade the convergence rate.

An approach for Newton's Method (and the Secant Method) is to combine it with one that is guaranteed to converge, such as the Bisection Method. That is, we apply Newton's Method, unless at some iteration, the iterates start to diverge. At that point we would switch to the Bisection Method.

9.5   Univariate minimization

The first-order necessary condition for x^* to be a minimizer of a differentiable function f is that $f'(x^*) = 0$. Thus, there is a similarity between the problem of minimization and zero finding. The techniques used to solve the minimization problem are analogous to those of the zero finding problem, although they are more complicated.

To develop an analogy to the Bisection Method, we need some conditions that ensure that there is a proper minimum in a given interval. For this purpose, we introduce the concept of *unimodality*. A function is said to be *unimodal* in $[a, b]$ if there exists a unique $x \in [a, b]$ such that either:

1. $f(x)$ is strictly decreasing in $[a, x)$ and strictly increasing in $[x, b]$; or

2. $f(x)$ is strictly decreasing in $[a, x]$ and strictly increasing in $(x, b]$.

There is another way to define unimodality. The function $f(x)$ is *unimodal* in $[a, b]$ if there exists a unique $x^* \in [a, b]$ such that, for any x_1 and x_2 belonging to $[a, b]$ satisyfing $x_1 < x_2$, the following hold:

1. If $x_2 < x^*$, then $f(x_1) > f(x_2)$.

2. If $x_1 > x^*$, then $f(x_1) < f(x_2)$.

The practicality of the latter definition is that if f is unimodal, then it is possible to reduce the interval of uncertainty by comparing the values of f at two interior points. This idea is illustrated in Figure 9.8 where the height of the symbol $\bullet$ is used to indicate a function value.

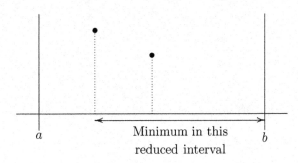

Figure 9.8: Univariate minimization.

In order to devise algorithms for finding the minimizer x^* of f based on this property, we must specify how to choose the two interior points at each iteration. Clearly, it would be most efficient (in terms of evaluations of f) to choose the points such that one of the two points can be re-used in the next iteration of the algorithm.

The "optimum" set of points, in the sense that there is a maximum reduction in the interval of uncertainty for a given number of function evaluations, is based on the sequence of Fibonacci[7] numbers which we describe next.

Fibonacci Search

The numbers of the *Fibonacci sequence* $\{F_k\}$ satisfy the recursion

$$F_k = F_{k-1} + F_{k-2}, \qquad F_0 = F_1 = 1.$$

The first nine values of the Fibonacci sequence are $1, 1, 2, 3, 5, 8, 13, 21, 34$. These numbers are used to specify the placement of points within the interval, depending on how many function evaluations are allowed.

Suppose that only two evaluations of f are to be used. The original interval may be considered to be of length F_2 $(= 2)$, and the points are placed at positions corresponding to F_0 and F_1. Because $F_0 = F_1 = 1$ we use instead 1 and $1+\varepsilon$, where $\varepsilon > 0$ is "small." Then, regardless of the shape of the graph of f (see Figure 9.9), the interval is reduced to approximately half its original length.

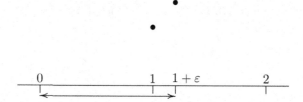

Figure 9.9: Fibonacci Search—2 function evaluations.

If three function evaluations are allowed, then the original interval is considered to be of length F_3 $(= 3)$, and the first two points are placed at F_2 and F_1 (as shown in Figure 9.10). Depending on the values of $f(x_1)$ and $f(x_2)$, the reduced interval will either be $[a, x_2]$ or $[x_1, b]$, and the problem

---

[7]Leonardo Fibonacci, also known as Leonardo of Pisa, was born in 1175 and died around 1250. He is said to have been the greatest Western mathematician of the Middle Ages. Author of the books *Liber Abaci* (1202) and *Practica Geometriae* (1220), Fibonacci (meaning "son of Bonacci") discovered this interesting sequence of numbers that bears his name. Fibonacci is also credited with helping to promote the use of the Hindu-Arabic system of numeration. See the book by Posamentier and Lehrman [157].

then is equivalent to that with two function evaluations, since one function value is already available. With three evaluations, the interval of uncertainty is reduced to approximately $1/3$ of its initial size.

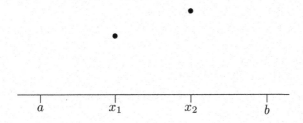

Figure 9.10: Fibonacci Search—3 function evaluations.

The procedure can be extended to the case where n function evaluations are allowed, in which case the interval of uncertainty is reduced to approximately $1/F_n$ times its initial size. The general procedure follows.

Algorithm 9.6 (Fibonacci Search) *Find a "small" interval of uncertainty that contains a minimizer x^* of a univariate unimodal function f.*

Let $[a, b]$ be an interval in which a minimum of $f(x)$ is estimated to lie in. Either pick $n + 1$, the desired number of function evaluations, or pick the desired reduction to the interval of uncertainty by some F_{n+1}; that is, pick n such that at the end of the algorithm a minimizer will lie in the interval $(b - a)/F_{n+1}$. Set $k = n$, $l = a$, $u = b$, $\Delta = u - l$, and $\varepsilon > 0$ (a small number).

1. *Determine points at which to evaluate the function.* These are given by
$$x_{k-1} = l + \frac{F_{k-1}}{F_{k+1}}\Delta \quad \text{and} \quad x_k = l + \frac{F_k}{F_{k+1}}\Delta.$$
If $x_{k-1} > x_k$, interchange x_k and x_{k-1}. If $k = 1$, set $x_k = x_{k-1} + \varepsilon$.

2. *Reduce the interval of uncertainty.*

   (a) If $f(x_k) > f(x_{k-1})$ set $u = x_k$.
   (b) If $f(x_k) \leq f(x_{k-1})$ set $l = x_{k-1}$.

   Set $\Delta = u - l$.

3. *Update.* Set $k \leftarrow k - 1$. If $k = 0$, stop; otherwise go to Step 1.

As will be shown in the examples to follow, one of the evaluation points in each iteration will be in the same position as in the previous one. Hence at each iteration, after the first, only one more function evaluation is needed.

In using this procedure, the strategy changes depending on the number of function evaluations to be used. A simpler procedure, the Fibonacci Golden Section Search, is described in the next subsection.

Example 9.8: FIBONACCI SEARCH—2 FUNCTION EVALUATIONS. Let us consider the function $f(x) = x^3/3 - 12x$. Note that the first derivative of this function is $g(x) = x^2 - 12$, which is the same function that we found the zero of in Section 9.4. We guess that a minimum of the function lies in the interval $[2.5, 5.5]$. If only two function evaluations are to be considered, then the Fibonacci interval length is $F_n = 2$ and we evaluate the function at $F_0 = 1$ and $F_1 = 1 + \varepsilon$. For our interval, this means we evaluate $f(x)$ at $x_0 = 2.5 + (3 \times 1/2) = 4$ and, say, $x_1 = 4.100$ as shown below. (We have chosen a large value for ε so that it is easy to distinguish between x_0 and x_1 in Figure 9.11.) From the figure we can see that the interval of uncertainty is reduced to half the original interval, i.e., to $[2.5, 4.0]$.

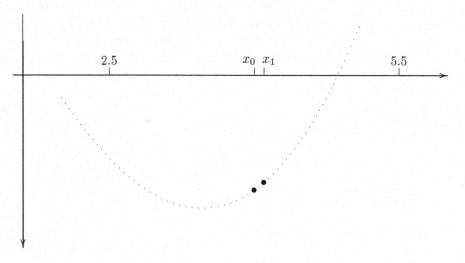

Figure 9.11: Fibonacci Search—2 function evaluations.

Example 9.9: FIBONACCI SEARCH—4 FUNCTION EVALUATIONS. Once again consider the function $f(x) = x^3/3 - 12x$. We guess that the interval in which a minimum of the function is known to lie be $[2.5, 5.5]$. Suppose that we are now willing to evaluate the function at four points. Then, the Fibonacci interval length is $F_4 = 5$. We start by evaluating the function at $F_3 = 3$ and $F_2 = 2$. For our interval, this means we evaluate $f(x)$ at $x_3 = 2.5 + (3 \times 3/5) = 4.30$ and $x_2 = 2.5 + (2 \times 3/5) = 3.70$ as shown n Figure 9.12 below. The new interval of uncertainty is then $[2.5, 3.7]$. Next we evaluate the function at $F_2 = 2$ and $F_1 = 1$; however, we already have the value at F_2. Hence we evaluate the function at one more point, $x_1 = 2.5 + (3/5) = 3.1$; this results in Figure 9.13 below. The new interval of uncertainty is then $[3.1, 3.7]$. Next we evaluate the function at $F_1 = 1$ and $F_0 = 1 + \varepsilon$; however, we already have the value at F_1. Hence we evaluate the function at one more point, $x_0 = 3.16$; this results in Figure 9.14. The interval of uncertainty is reduced to one-fifth of the original interval, i.e., to $[2.5, 4.0]$.

The Golden Section Search

In the Golden Section Search, the procedure for picking the points is based on the limiting behavior of the Fibonacci sequence. It can be shown that

$$\lim_{n \to \infty} \frac{F_{n-1}}{F_n} = \frac{2}{1 + \sqrt{5}} = \frac{1}{\phi} = \tau \approx .6180$$

where ϕ is known as the golden mean[8], golden section, golden ratio, or the divine proportion. The golden mean satisfies the quadratic equation $\phi^2 - \phi - 1 = 0$. By substitution we can verify that τ satisfies the quadratic equation $\tau^2 + \tau - 1 = 0$.

Using this, if the initial interval is $[0, 1]$, the two points would be placed at τ and $(1 - \tau)$. Note that, no matter which point is discarded, one of the old points will be in the correct position with respect to the new interval (see Figure 9.15), since τ satisfies

$$\frac{1 - \tau}{\tau} = \frac{\tau}{1}.$$

---

[8]This number $\phi = (1 + \sqrt{5})/2$, or its rational approximation 8/5, has long been used to its aesthetic appeal: by the Egyptians in the design of pyramids, the Greeks in architecture, Renaissance artists in paintings, in the design of the Notre Dame Cathedral. In modern times, it has been used to predict the movement of stock prices. For a monograph on the golden mean, see [100].

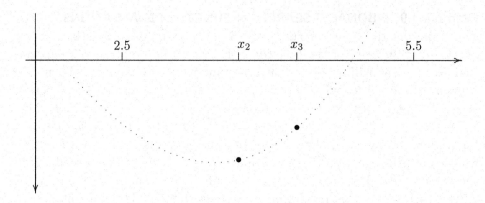

Figure 9.12: Fibonacci Search—4 function evaluations, iteration 1.

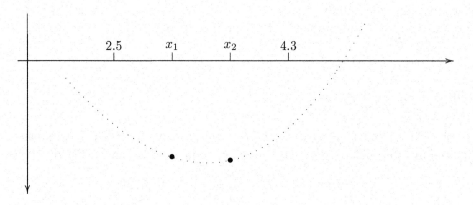

Figure 9.13: Fibonacci Search—4 function evaluations, iteration 2.

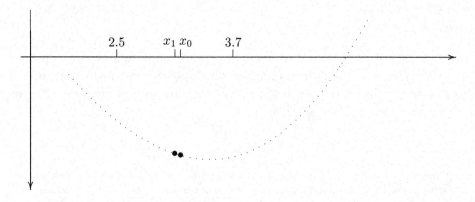

Figure 9.14: Fibonacci Search—4 function evaluations, iteration 3.

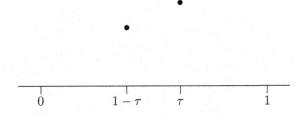

Figure 9.15: Golden Section Search.

The above equation is a statement about proportions.

With the Golden Section Search there is a guaranteed reduction of the interval of uncertainty by a factor of τ at every step, thus the length of the interval converges linearly to zero. Given a $\delta > 0$, the number of iterations, n, required to reduce an initial interval of uncertainty Δ to approximately δ, is determined as follows:

$$\tau^n \Delta \leq \delta$$
$$\tau^n \approx \frac{\delta}{\Delta}$$
$$n \approx \ln\left(\frac{\delta}{\Delta}\right) \times \frac{1}{\ln(\tau)}. \tag{9.3}$$

Algorithm 9.7 (Fibonacci Golden Section Search) *Find a "small" interval of uncertainty that contains a minimizer x^* of a univariate unimodal function f.*

Let $[a, b]$ be an interval in which a minimum of $f(x)$ is estimated to lie in. Pick $n+1$, the desired number of function evaluations, such that at the end of the algorithm a minimizer will lie in an interval of size $\delta > 0$; see (9.3). Set $k = n$, $l = a$, $u = b$, $\Delta = u - l$, and $\varepsilon > 0$ (a small number).

1. *Determine points at which to evaluate the function.* These are given by
$$x_{k-1} = l + (1 - \tau)\Delta \quad \text{and} \quad x_k = l + \tau\Delta.$$
   If $x_{k-1} > x_k$, interchange x_k and x_{k-1}. If $k = 1$, set $x_k = x_{k-1} + \varepsilon$.

2. *Reduce the interval of uncertainty.*

(a) If $f(x_k) > f(x_{k-1})$, set $u = x_k$.

(b) If $f(x_k) \leq f(x_{k-1})$, set $l = x_{k-1}$.

Set $\Delta = u - l$.

3. *Update.* Set $k \leftarrow k - 1$. If $k = 0$, stop; otherwise go to Step 1.

Similar to the Fibonacci Search, as will be shown in the example to follow, one of the evaluation points in each iteration will be in the same position as in the previous one. Hence at each iteration, after the first, only one more function evaluation is needed.

Example 9.10: FIBONACCI SEARCH—GOLDEN SECTION SEARCH. Once again consider the function $f(x) = x^3/3 - 12x$; an initial interval in which a minimum of the function is known to lie as $[2.5, 5.5]$. We will consider two iterations of the Golden Section Search. The function $f(x)$ is to be evaluated at $F_\tau = 0.6180$ and $F_{1-\tau} = 0.3820$. For our interval, this means we evaluate $f(x)$ at $x_0 = 4.3540$ and say $x_1 = 3.6460$ as shown below.

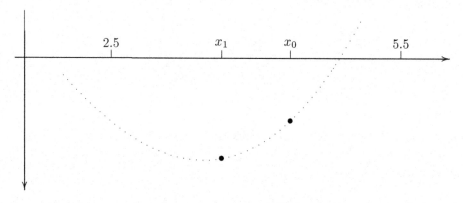

Figure 9.16: Fibonacci Search—Golden Section Search, iteration 1.

From the figure we can see that the interval of uncertainty is reduced to $[2.5, x_0] = [2.5000, 4.3540]$, which is $\tau \approx .6180$ times the original interval. For the next iteration we repeat the process. The function has already been evaluated at the point x_1, which now corresponds to F_τ. Thus, we evaluate the function at x_1 corresponding to the point $F_{1-\tau}$.

Now the interval of uncertainty is reduced to $[x_2, x_0] = [3.2082, 4.3540]$, and the point x_1 is at $F_{1-\tau}$ scaled with respect to this interval.

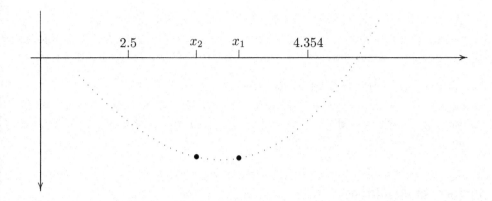

Figure 9.17: Fibonacci Search—Golden Section Search, iteration 2.

9.6 Optimality conditions for multivariate minimization

In a sense, the problem of minimizing a function of many variables is analogous to that of minimizing a function of single variable. For a start, there are first- and second-order necessary conditions of optimality similar to those of the univariate case. Naturally, the presence of many variables complicates matters somewhat.

Let us suppose we wish to minimize a smooth function f of $n \geq 1$ real unconstrained variables. By saying f is "smooth" we mean $f \in C^2$ on R^n.

Because we use them repeatedly, we write[9] $g(x)$ for the *gradient* vector and $G(x)$ for the *Hessian* matrix of the function f at x. The gradient is defined by the column n-vector of first partial derivatives of the function f:

$$g(x) = \left(\frac{\partial f(x)}{\partial x_1}, \ \frac{\partial f(x)}{\partial x_2}, \ \ldots, \ \frac{\partial f(x)}{\partial x_n} \right).$$

The Hessian is the $n \times n$ matrix of second partial derivatives of the function f. If this function has continuous second derivatives (i.e., it is twice-continuously differentiable), the Hessian matrix must be symmetric. The ij-th element of the Hessian is

$$G_{ij}(x) = \frac{\partial^2 f(x)}{\partial x_i \, \partial x_j}.$$

---

[9]Notice that this way of writing the gradient and Hessian makes no reference to the function f. This is fine when only one function's gradient or Hessian is under discussion. A more elaborate notational scheme is required for cases involving many functions.

As for the univariate case, a point $x^* \in R^n$ is said to be a *global minimizer* of a function f if $f(x^*) \leq f(x)$ for all $x \in R^n$. A point $x^* \in R^n$ is said to be a *strict global minimizer* of a function f if $f(x^*) < f(x)$ for all $x \in R^n$.

A point $x^* \in R^n$ is said to be a *local minimizer* of a function f if for some $\delta > 0$, we have $f(x^*) \leq f(x)$ for all $x \in R^n$ such that $\|x - x^*\| \leq \delta$. A point $x^* \in R^n$ is said to be a *strict local minimizer* of a function f if for some $\delta > 0$, we have $f(x^*) < f(x)$ for all $x \in R^n$ such that $\|x - x^*\| \leq \delta$.

First-order conditions

The multivariate analog of the stationarity condition for a local minimum x^* of a differentiable function is just the vanishing of the gradient vector:

$$g(x^*) = 0.$$

And as we know, this first-order stationarity condition is only necessary. It is not strong enough to distinguish minima from maxima or from other sorts of points. Again, this is where the second-order conditions come in.

Second-order conditions

Second-order necessary conditions

If x^* is a local minimum of the function $f \in C^2$, then

$$g(x^*) = 0 \quad \text{and} \quad s^\mathrm{T} G(x^*)s \geq 0 \quad \text{for all } s \in R^n.$$

The positive semidefiniteness condition on $G(x^*)$ is similar to the nonnegativity of the second derivative in the univariate case. In fact, if $f(x) = f_1(x_1) + \cdots + f_n(x_n)$, that is, f is the sum of n univariate functions[10], then $G(x^*)$ will be a diagonal matrix whose diagonal entries are just the (ordinary) second derivatives $f_1''(x_1^*), \ldots, f_n''(x_n^*)$, and the positive semidefiniteness of $G(x^*)$ boils down to the nonnegativity of the diagonal entries $f_1''(x_1^*), \ldots, f_n''(x_n^*)$.

Just as in the univariate case where we have witnessed the *insufficiency* of these necessary conditions, we obtain sufficient conditions for local optimality by imposing a stronger condition on the Hessian matrix.

---

[10]Such functions are said to be *separable*.

Second-order sufficient conditions

For $f \in C^2$, x^* is a *strict* local minimizer of f if

$$g(x^*) = 0, \quad \text{and} \quad s^{\mathrm{T}}G(x^*)s > 0 \quad \text{for all } s \in R^n, \ s \neq 0.$$

Here too, the positive definiteness condition on $G(x^*)$ is analogous to the second derivative being positive in the univariate case.

9.7 Methods for minimizing smooth unconstrained functions

In this section we discuss a couple of methods for minimizing smooth multivariate functions in the absence of constraints[11]. In particular, we look at what are called *descent methods*; they are so called because they generate sequences of iterates (trial solutions) for which the corresponding objective functions values constitute a monotonically decreasing sequence. Thus the key feature of these vectors is that

$$f(x_{k+1}) \leq f(x_k) \quad \text{for all } k = 0, 1, \ldots.$$

An apparently better property of the sequence would be *strict* decrease of the function values, i.e., a sequence satisfying $f(x_{k+1}) < f(x_k)$ for all $k = 0, 1, \ldots$. However, as we shall observe below, this is not enough to guarantee convergence to a minimum.

Warning. Notice that in the inequality just displayed, we have *vectors* (x_k and x_{k+1}) being distinguished by subscripts. In the literature it is fairly customary to use such notation even though these symbols would ordinarily denote individual coordinates of a single vector x. This practice occasionally requires a bit of mental agility—or else further notational apparatus.

Discussion on the Generic Optimization Algorithm

A Generic Optimization Algorithm was described on page 272. Included in this process are the computation of the objective function, its gradient, and Hessian matrix. (The latter two may need to be obtained numerically

---

[11]There are, of course, unconstrained minimization problems in which the objective function is not smooth. For example, such problems are treated in Part II of [17].

rather than analytically.) We frequently have to provide a *starting point* x_0, preferably a "good" one. Another extremely important part of the algorithmic process is to decide what we mean by "convergence" and what we mean by a "solution." Ideally, we would like to generate a sequence of vectors $x_0, x_1, \ldots, x_K$ such that x_K satisfies the sufficient conditions for a strict local minimum, and K is small. (We might also need to find a global minimum, rather than merely a strict local minimum, but it might not be possible to verify that a given local minimum is a global minimum.) Some algorithms merely strive to satisfy the first-order stationarity condition $g(x) = 0$. If this is done without recourse to second-order information, it may be necessary to adopt a generous definition of the term "solution." Even then, there is a question of just how "close" to a stationary point an iterate would have to be in order to declare convergence. This and possibly other numerical "tolerances" need to be decided before the computation proceeds.

In this discussion, we let $\{x_k\}$ denote the sequence of vectors generated by the iterative process of an optimization algorithm. The corresponding function values, gradient vectors, and Hessian matrices will be denoted f_k, g_k, and G_k, respectively. That is, for all $k = 0, 1, \ldots$

$$f_k = f(x_k), \quad g_k = g(x_k), \quad \text{and} \quad G_k = G(x_k).$$

For the sake of "brevity," we these abbreviations throughout the book.

Example 9.11: CONVERGENCE TO A NON-MINIMUM. It would seem that generating a sequence of iterates x_k for which the corresponding f_k decrease strictly would produce a minimum when the function has one. The following (univariate) examples show that this is not the case.

1. Consider the function $f(x) = x^2$ and the sequence $\{x_k\}$ where $x_k = (-1)^k \left(\frac{1}{2} + 2^{-k}\right)$. In this case, the function values decrease strictly from one iteration to the next, but the iterates themselves do not converge to a limit. Rather, the sequence of iterates has two *accumulation points* (limits of subsequences) namely $-\frac{1}{2}$ and $+\frac{1}{2}$, neither of which is a local minimum. The function clearly has one minimum: $x^* = 0$.

2. Now take $f(x) = e^x$ and evaluate it at each point of the sequence $\{x_k\}$ where $x_k = 2^{-k}$. In this case, $x_k \to x^* = 0$. For this sequence of iterates, $f_k \to 1$. But this is clearly not a minimum of the function e^x. In this example, the function $f(x)$ is bounded below (by 0) and $f(x_k)$

is strictly decreasing at each iteration and yet x^* is not a minimum of $f(x)$ for $x \in R$.

Although the functions used in these two examples are univariate, we could combine them to yield a bivariate function and a convergent sequence of points $(x_k, y_k) \in R^2$ for which the associated objective function values fail to converge to a minimum.

Conditions for convergence to a stationary point

We have assumed that we are minimizing a smooth multivariate function f. Under these circumstances, if the Generic Optimization Algorithm is to converge[12] to a local minimum, x^*, then x^* must be a stationary point of f. Thus, we seek a solution of $g(x) = 0$. Notice that this is actually a system of n equations in n unknowns, the components of x. In general, such a system will be nonlinear in the components of x and would require an iterative algorithm to solve it.

If $f_{k+1} \leq f_k$ and $f(x)$ is bounded below, then, by a result of real analysis, the sequence $\{f_k\}$ converges to something; however, the sequence $\{f_k\}$ may not converge to the minimum of f as we have just seen in part 1 of Example 9.11. In order to guarantee that $\{x_k\}$ converges to a minimizer of f, i.e., x^*, it will be necessary to show that

$$\lim_{k \to \infty} \frac{g_k^T p_k}{\|p_k\|} = 0,$$

and that

$$\lim_{k \to \infty} g_k = 0.$$

We next define three key features that are used in a convergence result for the Generic Optimization Algorithm. We assume that $f \in C^2$ and is bounded below.

1. *Forcing function.* A mapping $\sigma : [0, \infty) \to [0, \infty)$ is called a *forcing*

---

[12]Strictly speaking, it is not the algorithm that converges but rather the sequence of iterates it generates. Regrettably, there is little point in trying to change this pervasive misuse of language.

function if for any sequence $\{t_k\} \subset [0, \infty)$,

$$\lim_{k \to \infty} \sigma(t_k) = 0 \quad \text{implies} \quad \lim_{k \to \infty} t_k = 0.$$

2. *Sufficient decrease.* If, for the iteration

$$x_{k+1} = x_k + \alpha_k p_k, \quad ||p_k|| \neq 0, \quad k \geq 0$$

we have

$$f(x_k) - f(x_{k+1}) \geq \sigma \left(\frac{|g_k^{\mathrm{T}} p_k|}{||p_k||} \right), \quad k \geq 0,$$

for some forcing function σ, then

$$\lim_{k \to \infty} \frac{g_k^{\mathrm{T}} p_k}{||p_k||} = 0.$$

Given p_k, the next iterate depends on the steplength α_k. Thus the steplength algorithms must be designed to ensure sufficient decrease.

3. *Gradient-related search directions.* Given the sequence $\{x_k\}$, the sequence $\{p_k\}$ of nonzero vectors is *gradient related* to the sequence $\{x_k\}$ if there is a forcing function σ such that

$$\frac{|g_k^{\mathrm{T}} p_k|}{||p_k||} \geq \sigma \left(||g_k|| \right).$$

For descent algorithms, the descent direction p_k must satisfy $p_k^{\mathrm{T}} g_k < 0$; thus it is important for the p_k to be bounded away from orthogonality to the gradient and for the term $|g_k^{\mathrm{T}} p_k| / ||p_k||$ to go to zero only when the gradient goes to zero.

Here we outline how the above can be used to ensure that the Generic Optimization Algorithm converges to a stationary point. If the steplength algorithm provides sufficient decrease at each iteration, $\lim_{k \to \infty} g_k^{\mathrm{T}} p_k / ||p_k|| = 0$. If the search directions p_k are gradient related to x_k, then this implies that $0 \geq \lim_{k \to \infty} \sigma \left(||g_k|| \right)$ for some forcing function σ. From the definition of a forcing function, it follows that $\lim_{k \to \infty} ||g_k|| = 0$.

Lemma 9.2 Convergence of the Generic Optimization Algorithm   *Let $f(x)$, $x \in R^n$ be the function to be minimized. The iterates from the Generic Optimization Algorithm 9.1 result in $\lim_{k \to \infty} x_k = x^*$ such that*

$$g(x^*) = \lim_{k \to \infty} ||g(x_k)|| = 0$$

provided the following conditions hold:

1. *f is twice-continuously differentiable;*

2. *the level set*
$$S_{f(x_0)} = \{x \in S : f(x) \le f(x_0)\}$$
   is closed and bounded (i.e., compact);

3. *f maintains a sufficient decrease at each iteration;*

4. *the search direction sequence p_k is gradient related to x_k.*

Global convergence

In the lemma above, the starting point x_0 can be chosen arbitrarily as long as the level set $S_{f(x_0)}$ is closed and bounded. Accordingly, the *convergence is said to be global*, though it has to be recognized that *the limit point is not necessarily a global minimizer* of the objective function. In fact, it is only guaranteed to be a stationary point.

9.8 Steplength algorithms

We have already seen different methods for univariate minimization. In multivariate minimization, each iteration of the Generic Optimization Algorithm requires solving a univariate minimization problem. However, in such cases it rarely makes sense to find the exact minimum of a function along the direction p_k. All we need (provided that we have a gradient-related algorithm) is to guarantee that we have a "sufficient decrease." Algorithms used to find a better point along the descent direction are called *steplength algorithms*. Incidentally, steplength algorithms are often called *linesearches*; accordingly, we may feel free to use the two terms interchangeably.

In this section, we describe a number of steplength algorithms that satisfy the conditions stated on page 304. For a proof of Lemma 9.2 in conjunction with the steplength algorithms described below, refer to Ortega and Rheinboldt [155, p. 479].

Before proceeding to describe the various steplength algorithms, let's pause to refer to the three-part Figure 9.18 below. The top part depicts a portion of the surface given by the function f. It also indicates the set of points on that surface corresponding to the restriction of f to points of the form $x_0 + \alpha p$ where p is regarded as a search direction. The middle part is a graph of the function $\phi(\alpha) := f(x_0 + \alpha p)$. The bottom part indicates some *isovalue contours*[13] of the function f and the line in R^n passing through x_0 and having direction p. We can now derive some optimality conditions pertaining to the univariate function

$$\phi(\alpha) = f(x_0 + \alpha p),$$

which is the restriction of the function f to the line given by the equation $\xi = x_0 + \alpha p$. This is illustrated in Figure 9.18. From the chain rule of differentiation, we have

$$\phi'(\alpha) = g(x_0 + \alpha p)^\mathrm{T} p,$$

$$\phi''(\alpha) = p^\mathrm{T} G(x_0 + \alpha p) p.$$

(In general, the quantity $g(x_k + \alpha p_k)^\mathrm{T} p_k$ is called the *projected gradient* along p_k at iteration k.) If the steplength α^* is a minimum of $\phi(\alpha)$, then

$$\phi'(\alpha^*) = 0 \text{ and } \phi''(\alpha^*) \geq 0$$

must hold. This implies that

$$g(x_0 + \alpha^* p)^\mathrm{T} p = 0. \tag{9.4}$$

Note that this equation expresses the fact that the gradient of f must be orthogonal to the direction p at a minimum α^* of the restricted function $\phi(\alpha)$.

Next we provide a brief summary of five techniques for determining the steplength parameter α_k at iteration k. We then move on to a description of a *simple* technique, which, though not computationally efficient, is easy to understand and implement.

1. *Minimization condition.* In this method we choose α_k so as to minimize $\phi(\alpha)$.

2. *The Curry condition.* In this case we choose α_k as the smallest positive root of

$$\phi'(\alpha) = g(x_k + \alpha p_k)^\mathrm{T} p_k = 0.$$

---

[13]These are the sets in R^n over which f has constant value.

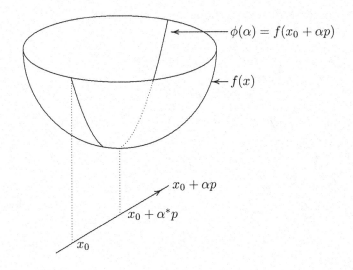

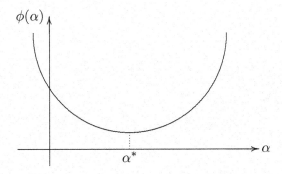

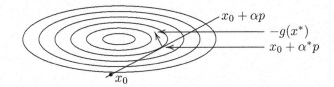

Figure 9.18: Function restricted to the direction p.

That is, α_k is the smallest positive stationary point of the projected gradient. In general, this criterion is not equivalent to the minimization condition, because there may be several stationary points.

3. *The Altman condition.* For a fixed constant η satisfying $0 \le \eta < 1$, choose α_k as the smallest positive root of

$$\phi'(\alpha) - \eta\phi'(0) = 0, \quad 0 \le \eta < 1.$$

That is, the projected gradient at x_{k+1} is equal to some fraction of the projected gradient at x_k. Note that if $\eta = 0$, the Altman condition is equivalent to the Curry condition.

4. *The Goldstein conditions.* The previous three criteria discussed ways to choose exactly one value of α_k, and one would therefore (rightly) suspect that they are not very useful as practical criteria for this purpose. The Goldstein conditions allows a *range* of acceptable values for choosing α_k.

Let μ_1 and μ_2 be given such that $0 < \mu_1 < \mu_2 < 1$. Consider the lines at x_k, with slopes equal $\mu_1\phi'(0)$ and $\mu_2\phi'(0)$, where $\phi'(0) = g_k^{\mathrm{T}}p_k$. The *tangent line* at x_k is defined as

$$t(\alpha) = f(x_k) + \alpha\phi'(0) = \phi(0) + \alpha\phi'(0).$$

Using μ_1 and μ_2, we define the two lines

$$t_1(\alpha) = f(x_k) + \alpha\mu_1\phi'(0) = \phi(0) + \alpha\mu_1\phi'(0),$$
$$t_2(\alpha) = f(x_k) + \alpha\mu_2\phi'(0) = \phi(0) + \alpha\mu_2\phi'(0).$$

Clearly, $t(\alpha) < t_2(\alpha) < t_1(\alpha)$. Any scalar $\alpha = \alpha_k$ is said to satisfy the Goldstein conditions provided α_k satisfies

$$t_2(\alpha_k) \le \phi(\alpha_k) \le t_1(\alpha_k).$$

5. *The Wolfe conditions.* Wolfe proposed conditions for determining a suitable α_k based on two fixed parameters μ and η that satisfy

$$0 < \mu \le \eta < 1.$$

We say that a steplength $\alpha > 0$ belongs to $\Gamma(\eta, \mu)$ if

$$\left|g(x_k + \alpha p_k)^{\mathrm{T}}p_k\right| \le \eta\left|g(x_k)^{\mathrm{T}}p_k\right|,$$
$$f(x_k + \alpha p_k) \le f(x_k) + \alpha\mu g(x_k)^{\mathrm{T}}p_k.$$

These two conditions restated in terms of our function ϕ are

$$|\phi'(\alpha)| \le \eta|\phi'(0)|, \qquad\qquad (9.5)$$

$$\phi(\alpha) \le \phi(0) + \alpha\mu\phi'(0). \qquad\qquad (9.6)$$

The parameter μ is restricted to be between 0 and $1/2$. A safeguarded procedure based on finding an $\alpha \in \Gamma(\eta, \mu)$ is summarized as follows. Given a descent direction p_k at x_k, apply a safeguarded univariate minimization procedure to the function $\phi(\alpha)$. Terminate the search at the first point α_k such $\alpha_k \in \Gamma(\eta, \mu)$. In practice, the parameter μ is chosen as a small value (say $\mu = 10^{-4}$). Then, an α that satisfies (9.5) almost always satisfies (9.6).

Using the discussion on page 303, one can show that the Generic Optimization Algorithm will converge if the sequence $\{x_k\}$ is defined by $x_{k+1} = x_k + \alpha_k p_k$ where p_k is a descent direction and $\alpha_k \in \Gamma(\eta, \mu)$. Steplength algorithms based on Wolfe's conditions are currently considered to be the best available because they terminate with a suitable α_k when some very relaxed conditions are met.

Note that the value of η determines the accuracy with which the one-dimensional minimization is performed. When $\eta = 0$ the steplength procedure is called an *exact linesearch*; when η is "small," the steplength procedure is called an *accurate linesearch*. The second condition above ensures that a suitable reduction in f has taken place.

Example 9.12: USE OF MINIMIZATION CONDITION ON A QUADRATIC.
Suppose we are minimizing a quadratic function $f(x) = \frac{1}{2}x^T H x + c^T x$, where H is an $n \times n$ symmetric and positive definite matrix. In this case, the gradient is $g(x) = Hx + c$ and the Hessian is $G(x) = H$. Let x_k be the current iterate with corresponding search direction p_k. Then we need to determine the step to the next iterate. In using the minimization condition (see page 306), we need to determine $\alpha = \alpha_k$, that is, a minimizer for $\phi(\alpha) = f(x_k + \alpha p_k)$. The first derivative of $\phi(\alpha)$ is

$$\begin{aligned}
\phi'(\alpha) &= g(x_k + \alpha p_k)^T p_k \\
&= (H(x_k + \alpha p_k) + c)^T p_k \\
&= x_k^T H p_k + \alpha p_k^T H p_k + c^T p_k \\
&= g_k^T p_k + \alpha p_k^T H p_k.
\end{aligned}$$

Solving $\phi'(\alpha) = 0$ for $\alpha = \alpha_k$, we obtain

$$\alpha_k = \frac{-g_k^T p_k}{p_k^T H p_k}.$$

The above α_k is the minimizer for $f(x_k + \alpha p_k)$ because the Hessian is positive definite. Thus, for a quadratic, a linesearch is not necessary because a closed-form solution is available when using the minimization condition.

A simple steplength algorithm

We next describe an approach that is easy to follow and implement. The basic idea is that we start by assuming $\alpha = 1$ (the reasons for this choice will become apparent later when we discuss convergence properties of various algorithms based on the choice of p_k). If the initial choice of α does not result in a decrease, we halve it and continue in this manner. That is, we compute the steplength by the following algorithm.

Algorithm 9.8 (Simple Steplength Algorithm)   At iteration k, x_k, $f(x_k)$, $g(x_k)$, and p_k are given. Choose a small value for μ, for example let $\mu = 10^{-4}$ and do the following:

1. Set $\alpha = 1$.

2. If $f(x_k + \alpha p_k) \leq f(x_k) + \mu \alpha g_k^T p_k$, set $\alpha_k = \alpha$ and stop.

3. Set $\alpha = \alpha/2$ and go to Step 2.

Example 9.13: COMPUTATION OF α.   We illustrate the process of finding a steplength on Rosenbrock's function[14], which is

$$f(x) = 100(x_2 - x_1^2)^2 + (1 - x_1)^2,$$

with

$$g(x) = \begin{bmatrix} -400x_1(x_2 - x_1^2) - 2(1 - x_1) \\ 200(x_2 - x_1^2) \end{bmatrix}.$$

---

[14]Rosenbrock's original function has been generalized to more than 2 variables. One version of it is $f(x) = \sum_{i=1}^{n-1} 100(x_{i+1} - x_i^2)^2 + (1 - x_i)^2$

Suppose that the current iterate is $x_k = (0.2, 1.0)$. The function value and gradient at this point are $f_k = f(x_k) = 92.8$ and $g_k = g(x_k) = (-78.4, 192)$. We choose the search direction to be the negative of the gradient; i.e., $p_k = -g_k = (78.4, -192)$. The steplength algorithm then proceeds as follows, starting with $\alpha = 1$. Let $\mu = 0.1$. The function value at $x_k + \alpha p_k = (78.6, -191)$ is $f_k = 4.06 \times 10^9$ which is greater than $f_k + \mu \alpha g^T p = -4208.26$. Next we set $\alpha \leftarrow \alpha/2 = 1/2$, and compute $f(x_k + (1/2)p)$ and test once again for convergence. The steps are summarized in the table below

α	$x_k + \alpha p_k$	$f(x_k + \alpha p_k)$	$f_k + 0.1\alpha g_k^T p_k$
1	$(78.6, -191)$	4.06×10^9	-4208.26
$1/2$	$(39.4, -95)$	2.71×10^8	-2057.73
$1/4$	$(19.8, -47)$	1.93×10^7	-982.46
$1/8$	$(10.0, -23)$	1.51×10^6	-444.83
$1/16$	$(5.1, -11)$	1.37×10^5	-176.02
$1/32$	$(2.65, -5)$	1.45×10^4	-41.61
$1/64$	$(1.425, -2)$	1624.77	25.60
$1/128$	$(0.8125, -0.5)$	134.63	59.20
$1/256$	$(0.50625, 0.25)$	0.2477	76.00

The steplength algorithm terminates after 9 iterations with $\alpha_k = 1/256$.

9.9   Exercises

9.1   State the second-order conditions for a local maximum of a function $f \in C^2(a, b)$ at a point $x^* \in (a, b)$.

9.2   Verify that $x^* = 0$ is a strict global minimizer of the function $f(x) = x^4$ over $(-1, 1)$.

9.3   Contrast the idea about bracketing a zero x^* of $g(x)$ by finding an interval $[\bar{x} - \delta, \bar{x} + \delta]$ (for some $\bar{x}$ and some small positive δ) with that of finding a point $\bar{x}$ such that $g(\bar{x})$ is "nearly zero," i.e., $|g(\bar{x})|$ is "small."

9.4   Apply, in turn, the Bisection Method, Newton's Method, the Secant Method, and the Regula Falsi Method to find a zero of the polynomial $g(x) = x^3 - 7x^2 + 10x - 6$ over the interval $[2, 6]$.

9.5   Apply, in turn, the Bisection Method, Newton's Method, the Secant Method, and the Regula Falsi Method to find zeros of the following two functions

$$f(x) = \phi^2 - \phi - 1$$
$$h(x) = \tau^2 + \tau - 1.$$

(Note: The above functions are those used in the Fibonaccci Golden Section Search.)

9.6   Implement the following four algorithms for use on a computer:

(a) Bisection Method 9.2.
(b) Newton's Method 9.3.
(c) Secant Method 9.4.
(d) Regula Falsi Method 9.5.

9.7   Implement and test the Simple Steplength Algorithm 9.8 to find α for $f(x + \alpha p)$, where $f(x) = 10x_1^2 + 20x_2^2 + 20x_1 + 20x_2$ and $p = (-2, -1)$. Further test it on a univariate optimization problem of your choice. Examine its behavior by executing the algorithm using different initial values for μ, for example, $\mu = 0.1, 0.01, 0.001$, and 0.0001.

9.8   *Quadratic interpolation* can be used in linesearches. In this exercise we explore how to derive such methods. A quadratic in one variable has the form

$$q(x) = ax^2 + bx + c.$$

Given $a > 0$ its minimizer is at $x^* = -b/2a$. What remains is to construct the quadratic and then specify its minimizer. We have three unknowns a, b, and c. So we can solve for them with three appropriate equations. In the following parts we ask you to do this under different sets of conditions.

(a) Function values at three points are known. Given function values $f_i = f(x_i)$ at three points x_i for $i = 1, 2, 3$, find closed-form solutions for a, b, c, x^*.
(b) Function values at two points and the derivative at one point are known. Given $f_1 = f(x_1)$, $f_2 = f(x_2)$, and $f_1' = f'(x_1)$, find closed-form solutions for a, b, c, x^*.
(c) Function value at one point and the derivative at two point are known. Given $f_1 = f(x_1)$, $f_1' = f'(x_1)$, and $f_2' = f'(x_2)$, find closed-form solutions for a, b, c, x^*.

9.9   *Cubic Interpolation.* In commercial solver implementations, instead of the Simple Steplength Algorithm 9.8, cubic interpolation methods are used. In this exercise we explore how to derive such methods. A cubic in one variable is of the form:

$$h(x) = ax^3 + bx^2 + cx + d.$$

We have four unknowns a, b, c, and d. So we can solve for them with four appropriate equations. In the following parts we ask you to do this under different sets of conditions.

(a) Determine x^*, a minimizer for $h(x)$. Use the second derivative to specify the conditions under which x^* is a minimizer of $h(x)$.

(b) Function value at two points and the derivative at two point are known. Given $f_1 = f(x_1)$, $f_2 = f(x_2)$, $f_1' = f'(x_1)$, $f_2' = f'(x_2)$, find closed form solutions for a, b, c, d, x^*.

(c) Function values at three points and the derivative at one point are known. Given $f_1 = f(x_1)$, $f_2 = f(x_3)$, $f_3 = f(x_3)$, and $f_1' = f'(x_1)$, find closed form solutions for a, b, c, d, x^*.

9.10   Show that, with $c \geq 0$, the sequence $y_k = c^{2^{-k}}$ has a linear convergence rate with $\gamma = 1/2$.

9.11   Consider the function $f(x) = x \ln x$ where x denotes a *positive* real variable.

(a) Compute the first and second derivatives of f over its domain.

(b) Plot (graph) the values of $f(x)$ for $x \in (0, 3]$.

(c) What is $\lim_{x \downarrow 0} f(x)$?

(d) Assume that the function f is extended to the entire real line R by setting

$$f(x) = \begin{cases} +\infty & \text{if } x < 0 \\ \lim_{x \downarrow 0} f(x) & \text{if } x = 0 \end{cases}.$$

Determine the global minimizer of f on R.

9.12   Suppose that you are performing calculations on a machine that can represent at most three digits and that anything beyond 3 digits is truncated. How would you compute the midpoint of $[-0.802, 0.804]$?

9.13   *Fibonacci Search Algorithm.*

(a) Implement the Fibonacci Search Algorithm 9.6.

(b) Test your algorithm on $f(x) = x^2 + 4x + 3$.

(c) Show that at the end of Step 2, one of the points is in the same position in the interval and hence the function need not be evaluated there. That is,

- If u is set to x_k, then x_{k-1} has the same value for the next iteration.

- If l is set to x_{k-1}, then x_k has the same value for the next iteration.

9.14   *Fibonacci Golden Section Search Algorithm.*

(a) Implement the Fibonacci Golden Section Search Algorithm 9.7.

(b) Test your algorithm on $f(x) = x^2 + 4x + 3$.

(c) Show that at the end of Step 2, one of the points is in the same position in the interval and hence the function need not be evaluated there. That is,

- If u is set to x_k, then x_{k-1} has the same value for the next iteration.

- If l is set to x_{k-1}, then x_k has the same value for the next iteration.

9.15   *Comparing Fibonacci Search with the Golden Section Search.* A well-known relation between Fibonacci numbers and the golden mean $\phi = (1 + \sqrt{5})/2$, for $n > 10$ is

$$F_n \approx \frac{\phi^{n+1}}{\sqrt{5}}.$$

(a) Use the above formula to compute F_n for $n = 12$. Compare it to the actual value of F_{12} from the Fibonacci sequence.

(b) Given an initial interval Δ what is the interval of uncertainty after n iterations of the Fibonacci Search Algorithm 9.6?

(c) Given an initial interval Δ what is the interval of uncertainty after n iterations of the Golden Search Algorithm 9.7?

(d) How does the reduction in the interval of uncertainty for the Golden Section Algorithm 9.7 compare to that achieved by the the Fibonacci Search Algorithm 9.6? What would be an advantage of using the Golden Section Search Algorithm 9.7?

9.16   Being tempted by various car advertisements, you decide to lease a new expensive car for three years. Since everything is always negotiable, you decide to negotiate the best possible lease structure; that is, the lowest possible payments over the three years that you have use of the car. Knowing full well that a smooth-talking salesman is going to do his best to charge you as much as possible, you decide to do some research. From the Internet you find that the dealer invoice is $74,000 (including destination charges). The Manufacturer's Suggested Retail Price (MSRP) is $78,500. You first start by negotiating the price down to $76,000 and then tell the dealer that you would like to lease the car at the advertised 4% annual interest rate. The dealer pulls out his calculator and tells you that you need to make the advertised $5,000 down payment and then pay $995 per month for 36 months plus the sales tax on the monthly payment as well as on the down payment (but not on the total car price). Being fully aware that the parameters are initial price ($P = \$76,000$), down payment ($D = \$5,000$), annual interest rate ($r = .04$), lease term ($n = 36$), lease payment x, and value (V) of the car at the end of lease, you ask for the final value of the car; the salesman informs you that it is \$39,000 and will be put into your contract as a buy option at the end of the lease. Not willing to trust the dealer, you decide to believe the initial price of the car and determine the interest rate r that the dealer is charging you. Having done your research, you know that the lease payments are determined by the following equation:

$$P = D + \sum_{k=1}^{n} \frac{x}{(1+r/12)^k} + \frac{V}{(1+r/12)^n}.$$

Solve for the interest rate and determine the actual interest rate that the dealer is charging you. Is it the same as the specified interest rate of $r = 0.04$? Comment on your results.

9.17   (Based on Exercise 4 in Hancock [92, p. 69].) Suppose x, y, and z satisfy the equation $z^2 = 2a\sqrt{x^2 + y^2} - x^2 + y^2$. Find the maximum and minimum values of z as a function of x and y.

9.18   (Exercise 5 in Hancock, [92, p. 69].) Find the minimum value of u, where $u = (x^2 + y^2)^{1/3}$.

9.19   Let $A \in R^{m \times n}$ and $b \in R^m$ be given. A frequently encountered (unconstrained) problem is to find $x \in R^n$ to

$$\text{minimize } ||Ax - b||^2,$$

where $||\cdot||$ represents the Euclidean norm.

(a) Explain what this problem means in geometric terms.
(b) Give a necessary (first-order) condition of optimality for this problem. Is this condition also sufficient for optimality in this case? Explain.
(c) Does the optimization problem at hand have a unique solution? Explain why this is so.
(d) Can you give a closed-form solution to the problem? Specify any assumptions you may need.

9.20   Consider the function

$$f(x_1, x_2) = ax_1^2 + bx_2^2 + cx_1 + dx_2,$$

with $a > 0$ and $b < 0$.

(a) Show that $f(x_1, x_2)$ has only one stationary point.
(b) Show that this stationary point is neither a minimizer nor maximizer for f.
(c) With $a = 1$, $b = -3$, $c = 3$, $d = 6$, draw a few contour lines of f.

9.21   (Based on Exercise 4 in Fletcher [68, p. 30].) The function $f(x_1, x_2) = x_1^5 + (x_1^2 - x_2)^2$ has only one stationary point (i.e., solution of $\nabla f(x_1, x_2) = 0$). Find it and show that it is neither a local maximizer nor a local minimizer.

9.22   Find a global minimum of the function

$$f(x_1, x_2, x_3) = 6x_1^2 + 4x_2^2 + 2x_3^2 + 2x_1x_2 + 2x_1x_3 + 2x_2x_3 - 11x_1 - 12x_2 - 9x_3.$$

9.23   (Based on Exercise 2.8 in Nocedal and Wright [152].) Let $f(x_1, x_2) = (x_1 + x_2^2)^2$. Consider the point $\bar{x} = (1, 0)$; show that the search direction $\bar{p} = (-1, 1)$ is a descent direction and find all solutions of the steplength problem

$$\min_{\alpha > 0} f(\bar{x} + \alpha\bar{p}).$$

10. DESCENT METHODS

Overview

In Chapter 9 we discussed optimality conditions, described a generic optimization algorithm and summarized several steplength algorithms, these being for univariate optimization. In this chapter we present several methods for unconstrained minimization of smooth *multivariate* functions, i.e., for problems of the form

$$\min_{x \in R^n} \ f(x),$$

where $n > 1$. In addition to the fact that such problems can arise naturally, another reason for looking at unconstrained optimization problems is that they occur as a part of some constrained optimization algorithms, i.e., as subproblems. As an illustration of this statement see the Penalty Function Algorithm 13.1.

The different methods described here are named according to the choice of the search direction p_k for Step G2 of the Generic Optimization Algorithm 9.1. We will first discuss the Steepest-descent Method and Newton's Method, both of which are centuries old. After that we will move on to Quasi-Newton Methods and the Conjugate-gradient Method which date from more recent times.

For this class of problems, the Steepest-descent Method is the only one that can be shown to have guaranteed global convergence. Unfortunately, it converges very slowly on most problems and is not practical. Newton's Method, on the other hand, when started close to a solution, has a very fast convergence rate but could require extensive computations during each iteration. Furthermore, Newton's Method requires second-derivative information which may not be readily available. Quasi-Newton Methods attempt to strike a balance between these two methods; they have a good convergence rate and do not require explicit second derivatives. Commercial solvers tend to favor Quasi-Newton Methods. The Conjugate-gradient Method requires less computation than a Quasi-Newton Method but its convergence rate is much slower. However, for certain very large problems it may be the best choice. In general, the choice of method (or combination of methods) depends on the problem at hand.

© Springer Science+Business Media LLC 2017
R.W. Cottle and M.N. Thapa, *Linear and Nonlinear Optimization*,
International Series in Operations Research & Management Science 253,
DOI 10.1007/978-1-4939-7055-1_10

10.1   The Steepest-descent Method

The *steepest-descent direction*, $p = -g$, is obviously a gradient-related search direction. Incorporation of such a search direction in the Generic Optimization Algorithm results in the Steepest-descent Method. This method is the simplest and probably the most well known for this class of problems. As stated above, it is, in fact guaranteed to converge, but has a serious flaw, namely, that its convergence can be extremely slow.

The steepest-descent direction

We derive the steepest-descent direction by examining the behavior of the function at the current iterate be x_k. Consider the first-order Taylor series expansion of f about x_k, along an arbitrary search direction p:

$$f(x_k + p) = f(x_k) + g(x_k)^{\mathrm{T}} p + \cdots$$

The steepest-descent direction p is the one along which the function f sustains the largest possible decrease to first order. To find such a p, we would select one that minimizes the linear term $g_k^{\mathrm{T}} p$ in the Taylor series. Clearly any p such that $g_k^{\mathrm{T}} p < 0$ can be chosen, and then we can take an arbitrarily large multiple of it. Hence we need to impose some normalization on the direction p. Thus, the steepest-descent direction vector p is chosen to be one that solves the minimization problem

$$\min_{p} \frac{g_k^{\mathrm{T}} p}{||p||} \tag{10.1}$$

for some norm $|| \cdot ||$. The solution depends on the norm used. Suppose, we define the norm of x in terms of a symmetric positive definite matrix C; i.e.,

$$||x||_C^2 = x^{\mathrm{T}} C x; \tag{10.2}$$

then (see Exercise 10.1) the solution to the minimization problem 10.1 is

$$p_k = -C^{-1} g_k. \tag{10.3}$$

The *steepest-descent direction* is defined to be the direction obtained when the Euclidean norm (2-norm) is used in the above minimization problem; this is the same as choosing $C = I$. That is, the *steepest-descent direction* is:

$$p_k = -g_k. \tag{10.4}$$

An algorithm based on this is as follows.

Algorithm 10.1 (Steepest-descent Method) *Find a minimizer of a differentiable function $f : R^n \to R$ using the steepest-descent direction.*

Set $k = 0$ and let x_0 be given. Compute the function value $f_0 = f(x_0)$ and the gradient $g_0 = g(x_0)$.

1. *Test for convergence.* If the gradient satisfies

$$||g_k|| \leq \sqrt{\varepsilon_M},$$

   where ε_M is the machine precision (around 10^{-15} for double-precision computations), then report $x^* = x_k$ as a solution and stop.

2. *Compute a search direction.* Set the steepest-descent direction as

$$p_k = -g_k.$$

3. *Compute a steplength.* Compute the steplength using one of the algorithms described in Section 9.8, for example, the Simple Steplength Algorithm 9.8.

4. *Update the iterate and return to Step 1.* Using p_k and α_k, determine the next iterate

$$x_{k+1} = x_k + \alpha_k p_k$$
$$f_{k+1} = f(x_{k+1})$$
$$g_{k+1} = g(x_{k+1}).$$

Return to Step 1 with the iteration counter k incremented by 1.

In commercial solver implementations the test for convergence is more than one check, and the tolerances for the tests are chosen with care based on the properties of the function being minimized. In all the algorithms discussed in this book, we use a simplified convergence test; for further details on tests for convergence, see Gill et al. [84].

The following should be noted about the steepest-descent direction:

1. It is obviously a descent direction. Indeed,

$$p_k^T g_k = -g_k^T g_k < 0.$$

2. It is a gradient-related direction with forcing function $\sigma(t) = t$.

3. Any of the steplength criteria presented earlier in Section 9.8 may be used to get guaranteed convergence of the Steepest-descent Method.

The above-mentioned three properties may lead us to believe that the steepest-descent direction could be used to develop an effective algorithm. Unfortunately, this is not so because the *Steepest-descent Method has a very poor rate of convergence.*[1] In order to analyze the convergence of the Steepest-descent Method, we will consider its behavior on a quadratic function, because, at a local minimum, a smooth function can be well approximated by a quadratic. Before doing so, we consider some properties of quadratic functions.

Approximation by a quadratic function

From the Taylor series expansion of a smooth function $f(x)$ at $x = \bar{x}$, we know that f can be well-approximated by a quadratic function in a small neighborhood of $\bar{x}$. Accordingly, it is of interest to first analyze properties of quadratic functions. Consider

$$q(x) = \tfrac{1}{2}x^T H x + c^T x,$$

where H is an $n \times n$ symmetric Hessian matrix, and c is a given n-vector. The function q has a stationary point at $x = \bar{x}$ if and only if $\nabla q(\bar{x}) = 0$, which is to say $x = \bar{x}$ is a solution of the system of equations

$$H x = -c. \tag{10.5}$$

The characteristics of the matrix H determine what we can say about the set of solutions of the system of equations (10.5). As illustrated in Figure 10.1, the following cases can arise.

---

[1] In discussing the Steepest-descent Method, Bonnans et al. [17, p. 31] go so far as to say that it "should be **forbidden**" (the boldface is their's).

(a) *H is positive definite.* The system of equations has a unique solution that minimizes the quadratic function.

(b) *H is positive semidefinite and* rank $(H) < n$. If the equations (10.5) are consistent (i.e., c belongs to the column space of H), then infinitely many minimizers exist.

(c) *H is negative definite, negative semidefinite, or indefinite.* There exists a direction along which the quadratic function $q(x)$ is unbounded below.

To learn about the function's behavior about its stationary point $\bar{x}$, we now examine its behavior in a small neighborhood of $\bar{x}$. For any vector p and scalar α, the function q evaluated at $x + \alpha p$ can be written as

$$q(x + \alpha p) = q(x) + \alpha p^{\mathrm{T}}(Hx + c) + \tfrac{1}{2}\alpha^2 p^{\mathrm{T}} H p. \qquad (10.6)$$

At a stationary point $\bar{x}$, it follows from (10.5) and (10.6) that

$$q(\bar{x} + \alpha p) = q(\bar{x}) + \tfrac{1}{2}\alpha^2 p^{\mathrm{T}} H p. \qquad (10.7)$$

Let λ_j and u_j denote the j-th eigenvalue and normalized eigenvector, respectively, of the matrix H; then from (10.7) we get

$$q(\bar{x} + \alpha u_j) = q(\bar{x}) + \tfrac{1}{2}\alpha^2 \lambda_j, \quad \text{since } u_j^{\mathrm{T}} u_j = 1. \qquad (10.8)$$

Equation (10.8) implies that we can get information about the change in the function q when moving from $\bar{x}$, in the direction of an eigenvector. Hence we can predict the behavior of a nonlinear function f in a small neighborhood of a stationary point $\bar{x}$ by considering the eigensystem[2] of the Hessian matrix $G(x^*)$.

When all the eigenvalues of H are positive (that is, when H is positive definite), $x = \bar{x} = x^*$ is the unique global minimizer of q. In this case, the contours of q are ellipsoids whose principal axes are in the directions of the eigenvectors of H, with lengths proportional to the reciprocals of the square roots of the corresponding eigenvalues.

Because of the simple form of the quadratic function, it is possible to derive many useful relationships that hold among the function and gradient

---

[2] That is, the eigenvalues and eigenvectors (see page 544).

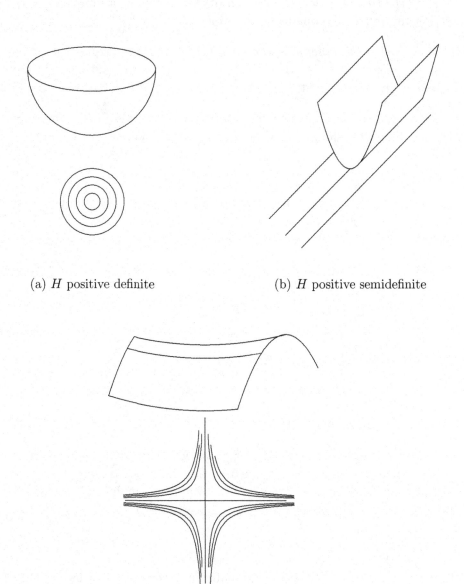

(a) H positive definite   (b) H positive semidefinite

(c) H indefinite

Figure 10.1: Quadratic functions and their contours.

values. For example, in the sequel, we use the following relationships. If $\bar{x}$ is a stationary point of q, we have

$$q(x) - q(\bar{x}) = \tfrac{1}{2}g(x)^\mathrm{T}H^{-1}g(x). \tag{10.9}$$

$$q(x) - q(\bar{x}) = \tfrac{1}{2}(x - H^{-1}c)^\mathrm{T}H(x - H^{-1}c). \tag{10.10}$$

Moreover, for *any* two vectors x and y in R^n

$$\begin{aligned} q(y) - q(x) &= \tfrac{1}{2}(y - x)^\mathrm{T}H(y - x) + (y - x)^\mathrm{T}(Hx + c) \\ &= \tfrac{1}{2}(y - x)^\mathrm{T}H(y - x) + (y - x)^\mathrm{T}g(x). \end{aligned} \tag{10.11}$$

The reader is asked to establish these in Exercise 10.3.

Convergence rate of the Steepest-descent Method

Consider a quadratic function $q(x) = \tfrac{1}{2}x^\mathrm{T}Hx + c^\mathrm{T}x$, with H positive definite and symmetric. Substituting the steepest-descent direction $p_k = -g_k$ in the Generic Optimization Algorithm, we find that the iterates take the form

$$x_{k+1} = x_k - \alpha_k g_k,$$

where $g_k = Hx_k + c$. Since the function under consideration is a quadratic, specifically $f(x) = q(x)$, the α_k that minimizes $q(x_k - \alpha g_k)$ is explicitly given by

$$\alpha_k = \frac{g_k^\mathrm{T}g_k}{g_k^\mathrm{T}Hg_k}. \tag{10.12}$$

With the steepest-descent iteration, the α_k as given above, and the relations (10.9), (10.10), and (10.11) derived for a quadratic, we obtain

$$f(x_{k+1}) - f(x_k) = \frac{-\tfrac{1}{2}(g_k^\mathrm{T}g_k)^2}{g_k^\mathrm{T}Hg_k}.$$

Then, using the expression $f(x) - f(x^*) = \tfrac{1}{2}g(x)^\mathrm{T}H^{-1}g(x)$ from (10.9), we get

$$\frac{f(x_{k+1}) - f(x^*) - [f(x_k) - f(x^*)]}{f(x_k) - f(x^*)} = \frac{-(g_k^\mathrm{T}g_k)^2}{(g_k^\mathrm{T}Hg_k)(g_k^\mathrm{T}H^{-1}g_k)}.$$

Thus,

$$f(x_{k+1}) - f(x^*) = \left[1 - \frac{(g_k^\mathrm{T}g_k)^2}{(g_k^\mathrm{T}Hg_k)(g_k^\mathrm{T}H^{-1}g_k)}\right][f(x_k) - f(x^*)].$$

Next we apply the *Kantorovich inequality* which states that for any nonzero vector y,

$$\frac{(y^\mathrm{T}y)^2}{(y^\mathrm{T}Hy)(y^\mathrm{T}H^{-1}y)} \geq \frac{4\lambda_1\lambda_n}{(\lambda_1 + \lambda_n)^2}, \tag{10.13}$$

where λ_1 is the smallest eigenvalue of H, and λ_n is the largest eigenvalue of H. (See Exercise 10.2.) Using this bound, we have

$$f(x_{k+1}) - f(x^*) \leq \left(1 - \frac{4\lambda_1\lambda_n}{(\lambda_1 + \lambda_n)^2}\right)[f(x_k) - f(x^*)]$$

$$= \frac{(\lambda_1 - \lambda_n)^2}{(\lambda_1 + \lambda_n)^2}[f(x_k) - f(x^*)].$$

Thus,

$$f(x_{k+1}) - f(x^*) \leq \frac{(\kappa - 1)^2}{(\kappa + 1)^2}[f(x_k) - f(x^*)],$$

where $\kappa = \lambda_n/\lambda_1$. The quantity κ is the *condition number* of H in the 2-norm. See page 546 for a discussion on condition numbers.

In this case, the asymptotic error constant is

$$\gamma = \frac{(\kappa - 1)^2}{(\kappa + 1)^2}, \tag{10.14}$$

and r, the asymptotic order of convergence, equals 1. This implies that the rate of convergence of the Steepest-descent Method on a quadratic function is *linear*. For linear convergence, the asymptotic error constant is the factor of reduction in the distance from the solution (or approximation error) at each iteration. From (10.14) we can see that this constant γ can be very close to 1 even when the Hessian matrix is not ill-conditioned. For example, if the condition number of the Hessian is $\kappa = 100$, then $\gamma = (99/101)^2 \approx 0.96$. This implies that there is negligible progress towards the solution at each subsequent iteration because of the bound $f(x_{k+1}) - f(x^*) \leq \gamma[f(x_k) - f(x^*)]$. This theoretical result implies that the Steepest-descent Method would require hundreds of iterations to make very little progress toward the solution (see Example 10.2 for an instance of this). A pictorial representation of the behavior of the Steepest-descent Method on a quadratic function appears in Figure 10.2.

The rate of convergence of the method when applied to more general nonlinear functions (in particular, nonquadratic functions) is likewise poor.

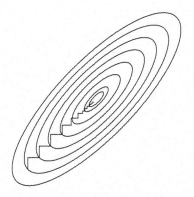

Figure 10.2: Path of the Steepest-descent Method on a quadratic.

We illustrate the behavior of the Steepest-descent Method on two simple quadratics in R^2. In the first example, the quadratic has a Hessian with a condition number of 1, and in the second the Hessian has a condition number of 100, so the Hessian is not a badly conditioned; nevertheless the Steepest-descent Method takes a very large number of iterations in this case.

Example 10.1: STEEPEST-DESCENT METHOD WITH $\kappa = 1$. Consider the quadratic function $f(x) = \frac{1}{2}x^\mathrm{T} H x + c^\mathrm{T} x$ with

$$H = \begin{bmatrix} 1 & 0 \\ 0 & 1 \end{bmatrix}, \quad c = \begin{bmatrix} 1 \\ 1 \end{bmatrix}.$$

In this case the eigenvalues are obviously $\lambda_1 = \lambda_2 = 1$ and so $\kappa = \lambda_2/\lambda_1 = 1$. The unique minimizer for this is $x^* = (-1, -1)$. In order to apply the Steepest-descent Method, we pick a starting point, say $x_0 = (0, 0)$. We next compute the gradient at x_0 as

$$g(x_0) = Hx_0 + c = \begin{bmatrix} 1 & 0 \\ 0 & 1 \end{bmatrix} \begin{bmatrix} 0 \\ 0 \end{bmatrix} + \begin{bmatrix} 1 \\ 1 \end{bmatrix} = \begin{bmatrix} 1 \\ 1 \end{bmatrix}.$$

The convergence test is not satisfied, and so we set the steepest-descent direction $p_0 = -g(x_0)$. The steplength is computed using the minimization condition (see Example 9.12) as

$$\alpha_0 = \frac{-g_0^\mathrm{T} p_0}{p_0^\mathrm{T} H p_0} = \frac{g_0^\mathrm{T} g_0}{g_0^\mathrm{T} g_0} = 1,$$

because $H = I$. The next iterate is computed as

$$x_1 = x_0 + \alpha_0 p_0 = \begin{bmatrix} 0 \\ 0 \end{bmatrix} + \begin{bmatrix} -1 \\ -1 \end{bmatrix} = \begin{bmatrix} -1 \\ -1 \end{bmatrix}.$$

It is straightforward to verify that the gradient at x_1 is zero, and hence we stop and report $x^* = x_1$. In this case the Steepest-descent Method converges to the solution in one iteration. Clearly, in this particular case, the same result would be obtained if we were to use the Simple Steplength Algorithm 9.8.

Example 10.2: STEEPEST-DESCENT METHOD WITH LARGER κ. Here we consider the quadratic function $f(x) = \frac{1}{2}x^{\mathrm{T}}Hx + c^{\mathrm{T}}x$, with

$$H = \begin{bmatrix} \kappa & 0 \\ 0 & 1 \end{bmatrix}, \quad c = \begin{bmatrix} 1 \\ 1 \end{bmatrix}.$$

The unique minimizer for this function is $x^* = (-1/\kappa, -1)$. In order to apply the Steepest-descent Method, we set the termination criterion to be that the norm of the gradient must be less than $\tau_g = 10^{-7}$ (which is less than the square root of machine precision) and set the Simple Steplength Algorithm 9.8 parameter to be $\mu = 10^{-1}$. Set $\kappa = 100$ (in Example 10.1, κ was 1.0), and pick a starting point, say $x_0 = (0, 0)$. In this case, the algorithm takes 736 iterations to find the optimal solution. In fact, if we set $\kappa = 1000$, the algorithm would take over 7500 iterations to find the optimal solution. Even with $\kappa = 10$, the algorithm takes 65 iterations to find the optimal solution.

When $\kappa = 100$ the last four values of the ratio $\bar{\gamma} = |(f_{k+1} - f^*)/(f_k - f^*)|$ are: 1.0, 1.0. 0.88, 1.0. The ratio $\bar{\gamma}$ becomes 1.0 due to numerical errors as the values get close to the solution.

If we were to change τ_g to be 10^{-6}, the last four values for $\bar{\gamma}$ would be 0.940, 0.991, 0.925, 0.954. Recall that the computed upper bound (see page 324) on the asymptotic constant was $\gamma = 0.96$. If we choose $\kappa = 1000$, the computed upper bound on the asymptotic constant γ is 0.996; the algorithm quickly gets to this bound and then $\bar{\gamma}$ becomes 1.0.

Example 10.3: ZIG-ZAG PATTERN OF STEEPEST-DESCENT METHOD. On examining the iterates for the Steepest-descent Method applied to the quadratic in Example 10.2 we can easily spot the zig-zag pattern of the iterates. The table below has the iterate values and search direction values for six intermediate iterations.

k	x_1	x_2	p_1	p_2
254	-0.0099	-0.9966	-0.0052	-0.0034
255	-0.0100	-0.9967	0.0029	-0.0033
256	-0.0099	-0.9968	-0.0062	-0.0032
257	-0.0100	-0.9968	0.0035	-0.0032
258	-0.0099	-0.9969	-0.0075	-0.0031
259	-0.0100	-0.9970	0.0042	-0.0030

The zig-zag pattern of the iterates is evident; the x_1 components alternates between two values, while the x_2 component stays relatively constant and is approaching the optimal value slowly. The first component of the search direction alternates in sign, and this results in the iterates having the observed pattern.

10.2   Newton's Method

The Newton direction

In the last section, we saw that the Steepest-descent Method typically takes a very large number of iterations to converge. Thus, clearly, alternative strategies are needed to define a search direction p_k. An obvious strategy is to use a quadratic approximation of the objective function at the current point x_k and devise methods that work efficiently for quadratic functions. To do this, we use the Taylor series expansion of the function to three terms to obtain a "quadratic model" of the function around x_k, i.e.,

$$f(x_k + p) \approx f(x_k) + p^{\mathrm{T}}g(x_k) + \tfrac{1}{2}p^{\mathrm{T}}G(x_k)p \qquad (10.15)$$

which has an error term of the order of $||p||^3$. Within the context of the Generic Optimization Algorithm, it is helpful to formulate the quadratic function in equation (10.15) in terms of the search direction p (the step to the minimum) rather than the predicted minimum itself. The minimum of the right-hand side of (10.15) will be achieved if p_k is the solution of the problem

$$\text{minimize} \quad Q(p) = p^{\mathrm{T}}g_k + \tfrac{1}{2}p^{\mathrm{T}}G_k p, \tag{10.16}$$

where $g_k = g(x_k)$ is a fixed vector and $G_k = G(x_k)$ is a fixed symmetric matrix at iteration k. Recall that a stationary point $p = p_k$ of $Q(p)$ is given by the solution of the linear system of equations

$$G_k p_k = -g_k. \tag{10.17}$$

It is customary to refer to any minimization algorithm that uses (10.17) to define a search direction p_k as *Newton's Method*.

Convergence properties of Newton's Method

From our discussion on quadratic functions, recall that if G_k is positive definite, the solution to (10.17) gives the step to the exact minimum of the quadratic model function specified in (10.15). Furthermore, the direction is scaled so that the steplength is unity. Therefore, we expect good convergence from Newton's Method when the quadratic function is a good approximation to the nonlinear function being minimized.

In particular, for a general nonlinear function f, Newton's Method converges quadratically to x^* if x_0 is sufficiently close to x^*, the Hessian matrix is positive definite at x^*, and the steplengths $\{\alpha_k\}$ converge to unity. The proofs for results such as these can be found in Ortega and Rheinboldt [155]. Under slightly more restrictive conditions, in the multivariate case, we can show convergence of Newton's Method in a manner analogous to that of the univariate case. We state corresponding theorems as follows.

Theorem 10.1 (Convergence of Newton's Method)   *Let $g : R^n \to R^n$ be continuously differentiable. Let Ω be an open set containing a point x^* such that $g(x^*) = 0$ and $\nabla g(x^*) = G(x^*)$ is nonsingular.*

(a) *For any x_0 in Ω, the Newton iterates*

$$\begin{aligned} x_{k+1} &= x_k + p_k, \\ G(x_k)p_k &= -g(x_k) \end{aligned} \tag{10.18}$$

for $k = 0, 1, 2, \ldots$, are well defined, remain in Ω, and converge to x^*.

(b) *The above sequence $\{x_k\}$ converges superlinearly to x^*. Moreover, if*

$$||G(x) - G(x^*)|| \leq \kappa ||x - x^*||, \quad x \in \Omega,$$

for some constant $\kappa > 0$, the sequence converges quadratically to x^*.

The outstanding local convergence properties of Newton's Method make it a potentially attractive algorithm for unconstrained minimization. A further benefit of the availability of the second derivatives is that sufficient conditions for a minimum can be verified. In fact, Newton's Method is often regarded as a standard against which other algorithms are measured. However, as in the one-dimensional case, difficulties and even failure can occur if the quadratic approximation does not have a positive definite Hessian matrix.

If G_k is positive definite, the search direction p_k obtained using equations (10.17) is a *descent direction*, since

$$g_k^T p_k = -g_k^T G_k^{-1} g_k < 0.$$

Furthermore, such a direction p_k is *gradient related* to x_k if G_k is positive definite for all x_k, since there exist constants $0 < \beta \leq \gamma$ such that

$$||G_k|| \leq \gamma,$$
$$y^T G_k y \geq \beta y^T y, \quad y \neq 0;$$

that is, γ is an upper bound on the largest eigenvalue of $G(x)$, and β is a lower bound on the smallest eigenvalue of $G(x)$. Using norm inequalities and some algebraic manipulations we obtain

$$|g_k^T p_k| \geq (\beta/\gamma)||g_k|| \, ||p_k||. \tag{10.19}$$

This also implies that if the condition number of G_k is bounded by a constant, then the directions p_k are bounded away from orthogonality to the negative gradient. However, even if G_k is positive definite, the quadratic model may be a poor approximation to the objective function—in particular, $f(x_k + p_k)$ may exceed $f(x_k)$, which violates the descent condition in Step 3 of the Generic Optimization Algorithm. Thus, to construct a convergent algorithm, a steplength procedure must be included according to one of the acceptable criteria discussed in Chapter 9. When Newton's Method is used with a steplength algorithm, it is sometimes termed a *damped Newton's*

Method because the "natural" steplength of unity is not taken. Finally, note that the Newton direction defined by equation (10.17) is a "steepest-descent" direction with respect to the norm defined by

$$||y||_{G_k}^2 = y^\mathrm{T} G_k y;$$

see the discussion on page 318.

Solving for a search direction when the Hessian is positive definite

When the Hessian is positive definite, the search direction p_k, as computed in (10.17), will be a descent direction. The solution of the system of equations for such a p_k can be found by several methods. We will look at one such method, namely one that uses the *Cholesky*[3] *factorization* of the Hessian matrix. Such a factorization always exists for a symmetric positive definite matrix, as stated in the following classical theorem.

Theorem 10.2 (Cholesky factorization)   *Let G be any $n \times n$ symmetric positive definite matrix. There exists a factorization of G of the form*

$$G = \bar{L}\bar{L}^\mathrm{T},$$

where $\bar{L}$ is a nonsingular lower-triangular matrix. Equivalently,

$$G = LDL^\mathrm{T}. \tag{10.20}$$

where L is a unit lower-triangular matrix (a lower-triangular matrix with all diagonal elements being equal to 1) and D is a diagonal matrix with all diagonal elements strictly positive.

Any matrix G satisfying (10.20), where L is unit lower-triangular and D is a diagonal matrix with positive diagonal elements d_j, must be symmetric and positive definite. From a computational standpoint this means that if we were to factorize a matrix G and obtain such an L and D as its factors, then G would be positive definite. If G is not symmetric and positive definite, then, in general, there does not exist a Cholesky factorization.

---

[3]For the life and work of Major André Cholesky, see [23] and [127, pp. 346–348].

Example 10.4: NON-EXISTENCE OF A CHOLESKY FACTORIZATION.
Consider the following matrix which is symmetric but not positive definite.

$$G = \begin{bmatrix} 0 & 1 \\ 1 & 2 \end{bmatrix}.$$

Clearly, G has no Cholesky factorization.

Algorithm 10.2 (Cholesky Factorization) *Let G be a symmetric $n \times n$ positive definite matrix.*

1. *Initialize.* Set the column index $j = 1$.

2. *Compute the j-th diagonal element of D.*

$$d_j = g_{jj} - \sum_{s=1}^{j-1} d_s l_{js}^2. \tag{10.21}$$

   If $j = n$ terminate with the Cholesky factors having been found.

3. *Compute the j-th column of L.*

$$l_{ij} = \frac{1}{d_j} \left(g_{ij} - \sum_{s=1}^{j-1} d_s l_{is} l_{js} \right) \quad \text{for } i = j+1, \ldots, n. \tag{10.22}$$

4. *Update.* Set $j \leftarrow j + 1$ and go to Step 2.

Using the above algorithm, the computation of the Cholesky factors requires $n^3/6 + O(n^2)$ arithmetic operations. Because of their symmetry, the factors require only $(n/2)(n+1)$ storage locations. The factors can overwrite the relevant portions of G as they are computed, and hence no additional storage beyond that used by G is required.

Example 10.5: CHOLESKY FACTORIZATION. Let

$$G = \begin{bmatrix} 2 & 4 & 6 \\ 4 & 11 & 24 \\ 6 & 24 & 70 \end{bmatrix}.$$

1. Set $j = 1$. Compute the diagonal element d_1 by the formula (10.21); that is,

$$d_1 = g_{11} = 2.$$

Next compute the first column of L according to the formula (10.22); that is,

$$l_{21} = g_{21}/d_1 = 4/2 = 2$$
$$l_{31} = g_{31}/d_1 = 6/2 = 3.$$

2. Set $j \leftarrow j + 1 = 2$. We determine the next diagonal element by

$$d_2 = g_{22} - d_1 l_{21}^2 = 11 - 2 \times 2^2 = 3.$$

Next the second column of L is computed

$$l_{32} = (g_{32} - d_l l_{31} l_{21})/d_2 = (24 - 2 \times 3 \times 2)/3 = 4.$$

3. Set $j \leftarrow j + 1 = 3$. Determine the diagonal element

$$d_3 = g_{33} - d_1 l_{31}^2 - d_2 l_{32}^2 = 70 - 2 \times 3^2 - 3 \times 4^2 = 4.$$

The factoization is now complete and given by

$$\begin{bmatrix} 2 & 4 & 6 \\ 4 & 11 & 24 \\ 6 & 24 & 70 \end{bmatrix} = \begin{bmatrix} 1 & 0 & 0 \\ 2 & 1 & 0 \\ 3 & 4 & 1 \end{bmatrix} \begin{bmatrix} 2 & 0 & 0 \\ 0 & 3 & 0 \\ 0 & 0 & 4 \end{bmatrix} \begin{bmatrix} 1 & 2 & 3 \\ 0 & 1 & 4 \\ 0 & 0 & 1 \end{bmatrix}.$$

The elements of the Cholesky factors are bounded, thus the factorization is inherently numerically stable. This follows from the property of positive definiteness which implies that each diagonal element of G is positive. Equations (10.21) and (10.22) imply that for every $j \in \{1, 2, \ldots, n\}$,

$$g_{jj} = \sum_{s=1}^{j} l_{js}^2 d_s \quad (\text{where } l_{ss} = 1).$$

Since $g_{jj} > 0$ and $d_s > 0$, it follows that the elements l_{js} and d_s are bounded by g_{jj}, the original diagonal elements of G_k.

Once the Cholesky factorization is obtained, the search direction p_k can be computed in $n^2 + O(n)$ arithmetic operations as follows:

1. Solve $Ly = -g$ ($n^2/2 + O(n)$ operations).
2. Solve $Dz = y$ (n operations).
3. Solve $L^T p_k = z$ ($n^2/2 + O(n)$ operations).

Solving for a search direction when the Hessian is not positive definite

So far, we have described how to obtain a descent search direction when G_k is positive definite at each iteration k. While we know that G_k will be positive definite at a strict local minimum, there is no guarantee for a general nonlinear function that G_k will be positive definite at every iteration k. If G_k is not positive definite, then there is no longer a guarantee that the p_k obtained as solution to $G_k p_k = -g_k$ will be a descent direction. Furthermore, there is no guarantee that the Cholesky factors of G_k exist.

When G_k is *not* positive definite, there are several possible approaches for obtaining a descent direction. These include:

1. Switch to the Steepest-descent Method. This is usually not a good idea due to the poor convergence rate of the Steepest-descent Method.

2. If G_k is nonsingular, let $p_k = -G_k^{-1} g_k$ (the usual definition, but the system is solved by a method other than the Cholesky factorization) if $p_k^T g_k < 0$, and let $p_k = G_k^{-1} g_k$ otherwise. That is, if it is not a descent direction, simply go the opposite way.

3. Construct a direction of *negative curvature*. That is, a direction p_k such that $p_k^T G_k p_k < 0$.

4. If G_k is not "sufficiently" positive definite, construct a "related" positive definite matrix $\bar{G}_k$, and solve for the search direction using $\bar{G}_k p_k = -g_k$. We discuss this next.

Several methods have been proposed for constructing a related positive definite matrix $\bar{G}_k$. Here we describe a numerically stable approach due to Gill and Murray (see [84]). The idea behind this is to construct $\bar{G}_k$ implicitly by creating a modified-Cholesky factorization of the matrix G_k. With this approach the elements of L and D are computed subject to two requirements that would be satisfied if the matrix were positive definite to start with:

1. All elements of D are strictly positive.

2. The elements of the factors have a uniform bound.

Algorithm 10.3 (Modified-Cholesky Factorization) *Given a symmetric matrix G_k, find a Cholesky factorization; if necessary, adjust the diagonal elements of G_k so that the factorization is that of a "nearby" positive definite symmetric matrix.*

The algorithm utilizes the auxiliary quantities $c_{ij} = l_{ij}d_j$. These numbers may be stored in the same array as G_k until they are overwritten by the appropriate elements of L.

Set $\delta = \max\{\varepsilon_M, \|G_k\|\varepsilon_M\}$, where ε_M is the machine precision.

1. *Compute bound on elements of the factors.* Set $\beta^2 = \max\{\gamma, \xi/\nu, \varepsilon_M\}$, where $\nu = \max\{1, \sqrt{n^2 - 1}\}$, and γ and ξ are, respectively, the maximum magnitudes of the diagonal and off-diagonal elements of G_k.

2. *Initialize.* Set the column index $j = 1$. Define $c_{ii} = g_{ii}$, $i = 1, \ldots, n$.

3. *Perform row and column interchanges.* Find an index q such that

$$|c_{qq}| = \max\{\,|c_{ii}| : j \le i \le n\,\}.$$

If $q \ne j$, interchange all the information corresponding to rows and columns q and j of G_k.

4. *Compute c_{ij} and find one with maximum absolute value.*

$$c_{ij} = g_{ij} - \sum_{s=1}^{j-1} l_{js}c_{is} \quad \text{for } i = j+1, \ldots n$$

$$\theta_j = \max\{|c_{ij}| : j+1 \le i \le n\}.$$

If $j = n$, set $\theta_j = 0$.

5. *Compute the j-th diagonal element of D.* Set $d_j = \max\{\delta, |c_{jj}|, \theta_j^2/\beta^2\}$, and define the diagonal modification $e_j = d_j - c_{jj}$. If $j = n$ terminate with the modified-Cholesky factors having been found.

6. *Compute the $(j+1)$-st row of L.* First update the prospective diagonal elements

$$c_{ii} = c_{ii} - c_{ij}(c_{ij}/d_j) \quad \text{for } i = j+1, \ldots, n.$$

Next determine

$$l_{j+1,s} = c_{j+1,s}/d_s \quad \text{for } s = 1, \ldots, j.$$

Set $j \leftarrow j + 1$ and go to Step 3.

It can be shown that modified-Cholesky factorization is a numerically stable method that produces a positive definite matrix differing from the original symmetric matrix only in its diagonal elements. The diagonal modification done during the factorization process is optimal in the sense that an a priori bound on the norm of E, the diagonal modification, is minimized subject to the requirement that "sufficiently" positive definite matrices are left unaltered. If no diagonal modifications are made, then the matrix being factored is positive definite.

Example 10.6: MODIFIED-CHOLESKY FACTORIZATION. Suppose that the matrix to be factorized is

$$G_k = \begin{bmatrix} 25.0000 & 5.0000 & 4.0000 & 0.0000 \\ 5.0000 & 12.0000 & 0.0000 & 0.0000 \\ 4.0000 & 0.0000 & 0.4000 & 1.0000 \\ 0.0000 & 0.0000 & 1.0000 & 0.2000 \end{bmatrix}.$$

Let $\varepsilon_M = 10^{-14}$. Using the Modified-Cholesky Factorization Algorithm we start by computing the 1-norm of the matrix G_k, which is $\|G_k\| = 34$, and setting

$$\delta = \max\{\varepsilon_M, \|G_k\|\varepsilon_M\} = \max\{10^{-14}, 34 \times 10^{-14}\} = 34 \times 10^{-14}.$$

Next we compute the bound β^2 on the elements of the factors. We first compute

$$\nu = \max\{1, \sqrt{n^2 - 1}\} = 15, \quad \gamma = 25, \quad \xi = 5,$$

where γ and ξ are, respectively, the maximum magnitudes of the diagonal and off-diagonal elements of G_k. Then

$$\beta^2 = \max\{\gamma, \xi/\nu, \varepsilon_M\} = 25.$$

During the course of the algorithm, the computed quantities will be stored back in the original matrix G_k; for simplicity of exposition we will illustrate the algorithm by *not* overwriting the elements of G_k, and instead storing all the computed quantities in a separate matrix C. Assign $c_{ii} = g_{ii}$ for $i = 1, \ldots, 4$.

1. Set $j = 1$. No column and row interchanges are necessary since $|c_{11}| = \max\{|c_{ii}| \text{ for } i = 1, \ldots, 4\}$. Next we compute

$$c_{i1} = g_{i1} \quad \text{for } i = 2, 3, 4.$$

At this stage

$$C = \begin{bmatrix} 25.0000 & & & \\ 5.0000 & 12.0000 & & \\ 4.0000 & & 0.4000 & \\ 0.0000 & & & 0.2000 \end{bmatrix}.$$

Next we compute θ_1 which is used to decide whether or not to modify the diagonal element d_1. We find that

$$\theta_1 = \max\{|c_{21}|, |c_{31}|, |c_{41}|\} = 5.$$

We determine the diagonal element by

$$d_1 = \max\{\delta, |c_{11}|, \theta_1^2/\beta^2\} = \max\{34 \times 10^{-14}, 25, 25/25\} = 25.$$

In this case there is no diagonal modification and so $e_1 = 0$. Next we update the remaining diagonal elements according to the formula $c_{ii} = c_{ii} - c_{i1}(c_{i1}/d_1)$ for $i = 2, 3, 4$. That is,

$$c_{22} = 12 - 5 \times 5/25 = 11$$
$$c_{33} = 0.4 - 4 \times 4/25 = -0.24$$
$$c_{44} = 0.2 - 0 = 0.2.$$

Now we compute $l_{21} = c_{21}/d_1 = 5/25 = 0.2$. We do not compute the other lower-triangular factors in column 1 just yet because we need the c_{i1} for $i > 2$ for other calculations. The updated matrix C with l_{21} inserted in the position occupied by c_{21} is

$$C = \begin{bmatrix} 25.0000 & & & \\ 0.2000 & 11.0000 & & \\ 4.0000 & & -0.24000 & \\ 0.0000 & & & 0.2000 \end{bmatrix}.$$

2. We set $j = 2$. Once again, no column and row interchanges are necessary since $|c_{22}| = \max\{|c_{ii}|, \text{ for } i = 2, 3, 4\}$. Next we compute

$$c_{32} = g_{32} - \sum_{s=1}^{1} l_{2s}c_{3s} = -0.8$$

$$c_{42} = g_{42} - \sum_{s=1}^{1} l_{2s}c_{4s} = 0.0.$$

The updated C matrix is

$$C = \begin{bmatrix} 25.0000 \\ 0.2000 & 11.0000 \\ 4.0000 & -0.8000 & -0.24000 \\ 0.0000 & 0.0000 & & 0.2000 \end{bmatrix}.$$

We compute

$$\theta_2 = \max\{|c_{32}|, |c_{42}|\} = 0.8.$$

We next determine the diagonal element by

$$d_2 = \max\{\delta, |c_{22}|, \theta_2^2/\beta^2\} = \max\{34 \times 10^{-14}, 11, (.8)^2/25\} = 11.$$

In this case also there is no diagonal modification and so $e_2 = 0$. Next we update the remaining diagonal elements according to the formula $c_{ii} = c_{ii} - c_{i2}(c_{i2}/d_2)$ for $i = 3, 4$. That is,

$$c_{33} = c_{33} - c_{32} \times c_{32}/d_2 = 0.4 - 0.8 \times 0.8/11 = -0.2982$$
$$c_{44} = c_{44} - c_{42} \times c_{42}/d_2 = 0.2 - 0.0 = 0.2.$$

Next we compute l_{31} and l_{32} by

$$l_{31} = c_{31}/d_1 = \quad 0.4/25 = 0.16$$
$$l_{32} = c_{32}/d_2 = -0.8/11 = -0.0727.$$

The updated matrix C with l_{31} and l_{32} inserted in the positions occupied by c_{31} and c_{32} is

$$C = \begin{bmatrix} 25.0000 \\ 0.2000 & 11.0000 \\ 0.1600 & -0.0727 & -0.2982 \\ 0.0000 & & & 0.2000 \end{bmatrix}.$$

3. Set $j = 3$. Once again, no column and row interchanges are necessary since $|c_{33}|$ is the larger of $|c_{33}|$ and $|c_{44}|$. Next we compute

$$c_{43} = g_{43} - \sum_{s=1}^{2} l_{3s}c_{4s} = 1.0 - 0.0 = 1.0.$$

The updated C matrix now is

$$C = \begin{bmatrix} 25.0000 \\ 0.2000 & 11.0000 \\ 0.1600 & -0.0727 & -0.2982 \\ 0.0000 & 0.0000 & 1.0000 & 0.2000 \end{bmatrix}.$$

We compute $\theta_3 = \max\{|c_{43}|\} = 1.0$. We determine the diagonal element by

$$d_3 = \max\{\delta, |c_{33}|, \theta_3^2/\beta^2\} = \max\{34 \times 10^{-14}, 0.2982, (1.0)^2/25\}$$
$$= 0.2982.$$

In this case the diagonal is modified from -0.2982 to 0.2982, that is, the diagonal modification is $e_3 = 0.5964$. Next we update the remaining diagonal elements as

$$c_{44} = c_{44} - c_{43} \times c_{43}/d_3 = 0.2 - 1.0/0.2982 = -3.1537.$$

Next we compute l_{4i} for $i = 1, \ldots, 3$ by

$$l_{41} = c_{41}/d_1 = 0.0$$
$$l_{42} = c_{42}/d_2 = 0.0$$
$$l_{43} = c_{43}/d_3 = 1.0/0.2982 = 3.3537.$$

The updated matrix C with l_{4i} inserted is

$$C = \begin{bmatrix} 25.0000 & & & \\ 0.2000 & 11.0000 & & \\ 0.1600 & -0.0727 & 0.2982 & \\ 0.0000 & 0.0000 & 3.3537 & -3.1537 \end{bmatrix}.$$

4. Set $j = 4$. Because $j = n$ we set $\theta_4 = 0.0$ and then determine the diagonal element

$$d_4 = \max\{\delta, |c_{44}|, \theta_4^2/\beta^2\} = \max\{34 \times 10^{-14}, 3.1537, 0\} = 3.1537.$$

In this case the diagonal is modified from -3.1537 to 3.1537, that is, the diagonal modification is $e_4 = 6.3073$.

The final factors are:

$$L = \begin{bmatrix} 1.0000 & & & \\ 0.2000 & 1.0000 & & \\ 0.1600 & -0.0727 & 1.0000 & \\ 0.0000 & 0.0000 & 3.3537 & 1.0000 \end{bmatrix}$$

and

$$D = \begin{bmatrix} 25.0000 & & & \\ & 11.0000 & & \\ & & 0.2982 & \\ & & & 3.1537 \end{bmatrix}.$$

These are the factors of the related symmetric positive definite matrix obtained by modifying the diagonal elements of the original matrix, that is

$$
\bar{G}_k =
\begin{bmatrix}
25.0000 & 5.0000 & 4.0000 & 0.0000 \\
5.0000 & 12.0000 & 0.0000 & 0.0000 \\
4.0000 & 0.0000 & 0.4000 & 1.0000 \\
0.0000 & 0.0000 & 1.0000 & 0.2000
\end{bmatrix}
+
\begin{bmatrix}
0.0000 & & & \\
& 0.0000 & & \\
& & 0.5964 & \\
& & & 6.3073
\end{bmatrix}
$$

$$
=
\begin{bmatrix}
25.0000 & 5.0000 & 4.0000 & 0.0000 \\
5.0000 & 12.0000 & 0.0000 & 0.0000 \\
4.0000 & 0.0000 & 0.9964 & 1.0000 \\
0.0000 & 0.0000 & 1.0000 & 6.5073
\end{bmatrix}.
$$

Modified-Newton Method

It can be shown that the use of a modified-Cholesky factorization with Newton's Method results in global convergence of the method, see [84]. Besides the fast convergence properties of Newton's Method, the presence of the Hessian allows us to check the second-order conditions of optimality. The algorithm below incorporates the modified-Cholesky factorization.

Algorithm 10.4 (Modified-Newton Method) *Find a minimizer of a twice-continuously differentiable function $f : R^n \rightarrow R$.*

Set $k = 0$ and let x_0 be given. Compute the gradient vector $g_0 = g(x_0)$ and Hessian matrix $G_0 = G(x_0)$.

1. *Test for convergence.*

   (a) If the gradient does not satisfy $\|g_k\| \leq \sqrt{\varepsilon_M}$, go to Step 2.
   (b) If the above is satisfied, then compute the modified-Cholesky factorization of the Hessian as $G_k = L_k \bar{D}_k L_k^T$.

      i. If the above factorization has the diagonal modification as zero, then the Hessian is positive definite. In this case report $x^* = x_k$ as the minimizer.
      ii. If the above factorization has nonzero diagonal modifications, then report $x^* = x_k$ as a stationary point.[4]

---

[4]When the Hessian is indefinite, it is possible for the algorithm to proceed along a direction of negative curvature; discussion of this technique is outside the scope of this book. See, for example, [84].

2. *Compute a search direction.* Compute the modified-Newton direction p_k by solving
$$L_k \bar{D}_k L_k^T p_k = -g_k.$$

3. *Compute a steplength.* Compute the steplength using one of the algorithms described in Section 9.8, for example, the Simple Steplength Algorithm 9.8.

4. *Update the iterate.* Using p_k and α_k, determine the next iterate
$$x_{k+1} = x_k + \alpha_k p_k$$
$$g_{k+1} = g(x_{k+1})$$
$$G_{k+1} = G(x_{k+1}).$$

Return to Step 1 with the iteration counter k incremented by 1.

Example 10.7: MODIFIED-NEWTON METHOD USED ON A QUADRATIC FUNCTION. When a Modified-Newton Method is applied to the quadratic functions in Examples 10.1 and 10.2, it converges, as expected, in exactly one iteration with a steplength $\alpha = 1.0$. This is in stark contrast to the Steepest-descent Algorithm which takes over 7500 iterations when applied to the quadratic of Example 10.2 having $\kappa = 100$.

Example 10.8: MODIFIED-NEWTON METHOD USED ON ROSENBROCK FUNCTION. We introduced the Rosenbrock function in Example 9.13 on page 310.
$$f(x) = 100(x_2 - x_1^2)^2 + (1 - x_1)^2.$$

This nonconvex function is of particular interest in optimization because its minimum is in a long curved valley making it difficult for many algorithms to converge to it. The gradient and Hessian for Rosenbrock's function are:
$$g(x) = \begin{bmatrix} -400x_1(x_2 - x_1^2) - 2(1 - x_1) \\ 200(x_2 - x_1^2) \end{bmatrix}$$

and
$$G(x) = \begin{bmatrix} -400(x_2 - x_1^2) + 800x_1^2 + 2.0 & -400x_1 \\ -400x_1 & 200 \end{bmatrix}.$$

The global minimizer for this problem is $x^* = (1,1)$. In order to apply the Modified-Newton Method, we set the termination criterion to be that the norm of the gradient must be less than $\tau_g = 10^{-7}$ and set the Simple Steplength Algorithm 9.8 parameter to be $\mu = 10^{-1}$. The Modified-Newton Method converges in $k = 14$ iterations. The Steepest-descent Algorithm applied to this function, did not converge in $k = 10,000$ iterations.

The last few iterations when using the Modified-Newton Method are

k	α	$\|x_k - x^*\|$	$|(f_k - f^*)|$
11	1.0000	1.1956e-002	3.1775e-005
12	1.0000	4.7203e-004	1.0681e-007
13	1.0000	2.3390e-006	1.2201e-012
14	1.0000	1.9327e-011	1.8009e-022

We can see that for the final two iterations the iterates' distance from the solution is roughly changing by the square of the previous iterate's distance from the solution, implying a quadratic rate of convergence.

From Theorem 10.1 we know that there is an open set S such that the steplengths converge to 1.0, once an iterate $x_k \in S$; in this case the steplengths are 1.0 for several iterations before convergence. In fact if the steplengths do not converge to 1.0 for the Modified-Newton Method when the assumptions of Theorem 10.1 hold, then there is a high probability that the Hessian computation is incorrect.

Furthermore, from Theorem 10.1, for Newton's Method to exhibit a quadratic rate of convergence, the steplengths must be precisely 1.0 as the iterates approach a solution. In the literature, the search direction p_k in such a case is called the *Newton direction*.

In practice an optimal solution will not be known in advance, so the convergence rates cannot be verified. However, once an optimal solution is obtained, then it be used to verify that the convergence rate is being achieved. If the convergence rate is not achieved, or is not close to being achieved, then there could be an error in the Hessian matrix due either to incorrect derivation or a some programming error. In practice a finite-difference approximation to the Hessian is computed at several points to verify the accuracy of the analytic representation of the Hessian.

Example 10.9: IMPACT OF ERRORS IN MODIFIED-NEWTON METHOD.
Typically an optimization algorithm will not converge when even small errors
are present in the computation of the gradient or the function value; this is
because its steplength algorithm fails to produce a decrease in the function
value. One reason for this failure is that the search direction determined by
the gradient is not really a direction of decrease for the function, and this
causes the steplength to terminate with the error condition that an iterate
having a lower function value has not been found.

For instance, when the gradient calculation is deliberately modified so
as to introduce an error as small as -10^{-6} in the second component of the
gradient for Rosenbrock's function, Newton's Method fails to find an optimal
solution because its steplength algorithm fails to produce a decrease in the
function value.

Errors in computing the Hessian are typically less harmful than those in
computing the gradient or function values, and in a large number of practical cases the Modified-Newton Algorithm will succeed in finding an optimal
solution but after many more iterations. In this case, the search directions
are in fact descent directions, just not the analytically correct descent directions. When there are such errors, the convergence rates typically do not
show quadratic convergence and often the steplengths are also not converging to 1.0 as the theory predicts.

If we deliberately change the second diagonal component of the Hessian
for Rosenbrock's function to be 100.0, instead of 200.0, the Modified-Newton
Method takes almost 3000 iterations to converge. In this case the steplengths
do converge to 1.0, but quadratic convergence is not observed as seen in the
table below for a few iterations near the solution.

k	α	$\|x_k - x^*\|$	$|(f_k - f^*)|$
2890	1.0000	2.3280e-007	1.0823e-014
2891	1.0000	2.3163e-007	1.0715e-014
2892	1.0000	2.3047e-007	1.0608e-014
2893	1.0000	2.2932e-007	1.0502e-014
2894	1.0000	2.2817e-007	1.0397e-014
2895	1.0000	2.2703e-007	1.0293e-014
2896	1.0000	2.2589e-007	1.0190e-014
2897	1.0000	2.2476e-007	1.0088e-014

Finite-difference approximations of the Hessian

In some situations it may be difficult to obtain an analytical representation of the Hessian matrix. In other cases, the Hessian may be available but its computation may be prohibitively expensive. One possible approach is to use finite-differencing of the gradients to approximate the Hessian. To see how to do this, consider the Taylor's series expansion of the gradient g about the point x, namely

$$g(x + hv) = g(x) + hG(x)v + O(h^2),$$

where h is a scalar. As h becomes "small," the second- and higher-order terms become much smaller than the rest. Thus, assuming that $O(h^2)$ is small, we can write

$$G(x)v \approx \frac{g(x + hv) - g(x)}{h}. \qquad (10.23)$$

Next we note that if we take $v = e_j$, where e_j is the j-th unit vector, then the left-hand side of (10.23) is the the j-th column of $G(x)$. Thus, by sequentially choosing v to be the columns $e_1, e_2, \ldots, e_n$ of the identity matrix in sequence, we can generate the columns of the Hessian matrix. It can be shown that as $h \to 0$, such an approximation would be exact. However, as discussed below, if h is too small, numerical errors may cause the procedure to fail.

Even though the calculations are performed with a small h, chosen small enough to balance the numerical errors against accuracy of determining the Hessian, this will, in general, result in a Hessian approximation that is not symmetric. If we let $\bar{G}$ denote the matrix generated by the finite-difference operation, then we can obtain a symmetric approximation by setting

$$G = \frac{1}{2}(\bar{G} + \bar{G}^T).$$

The quadratic rate of convergence of Newton's Method will only be maintained with the finite-difference approximation described above if the Hessian approximation becomes accurate as the solution is approached. This implies that *finite-difference interval* h must be chosen with care. Numerically, of course, a very small value of h would make the approximation G useless because of rounding errors. It should be small enough to give the required convergence rate, yet large enough to ensure that the finite difference approximation yields a good approximation of the Hessian. While, in theory, there are many such considerations, in practice, the choice of h

less critical than it might seem for the success of these methods. Numerical experiments given by Gill, Murray and Picken (1972), see [81], indicate that provided h is chosen in the interval $[\sqrt{\varepsilon_M}, \sqrt[3]{\varepsilon_M^2}]$ the convergence results of Newton's Method are almost invariant with respect to h. Newton's Method implemented with a finite-difference approximation to the Hessian is called a *Discrete Newton Method*.

When implemented well, Discrete Newton Methods behave much like Newton's Method. Since we already have g_k, the computation of a finite-diffference approximation to the Hessian, requires n additional gradient evaluations at each iteration.

In practice, unless it is impossible to represent the Hessian analytically, it turns out that for large n, Discrete Newton Methods tend to be less useful than Quasi-Newton Methods (to be discussed in the next section). However, when the Hessian matrices are *sparse*, advantage can be taken of this sparsity to significantly reduce the number of gradient evaluations required to obtain the Hessian approximation.

In applications, it is often the case that the special structure of the objective function $f(x)$ leads to a sparse Hessian whose sparsity pattern is known in advance or is one that can be determined. In such cases it is possible to obtain a finite-difference approximation to the Hessian $G(x)$ in fewer than the n gradient evaluations mentioned above. The following two examples illustrate the process on a diagonal and a tridiagonal Hessian matrix.

Example 10.10: DIAGONAL HESSIAN. Consider the simple case when the $n \times n$ Hessian $G(x)$ is a diagonal matrix for all x. Define the vector v in equation (10.23) to be the vector of all ones. Then, because all the off-diagonal elements are zero, it follows that $Gv = (G_{11}, G_{22}, \ldots, G_{nn})$. In this special case we can obtain a finite-difference approximation to the Hessian in one gradient evaluation using equation (10.23).

Example 10.11: TRIDIAGONAL HESSIAN. Suppose that $G(x)$ is tridiagonal; in such a case the nonzero/zero structure of the matrix is as shown in Figure 10.3, where the entries indicated by the • symbol are nonzeros. Consider a subset of a 10×10 tridiagonal matrix comprised of columns $1, 4, 7, 10$. Notice that there is exactly one Hessian element in each row of this submatrix. Based on this observation, we choose the vector $v = v_1$ of

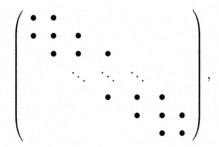

Figure 10.3: Tridiagonal Hessian matrix

equation (10.23) as one with ones in positions $1, 4, 7, 10$ and zeros elsewhere, i.e.,

$$v_1 = (1, 0, 0, 1, 0, 0, 1, 0, 0, 1).$$

It is easy to see that this results in

$$Gv_1 = (G_{11}, G_{21}, G_{34}, G_{44}, G_{54}, G_{67}, G_{77}, G_{87}, G_{10,10}).$$

In a similar manner we can choose

$$v_2 = (0, 1, 0, 0, 1, 0, 0, 1, 0, 0), \quad \text{and} \quad v_3 = (0, 0, 1, 0, 0, 1, 0, 0, 1, 0),$$

and thus obtain the remaining Hessian elements. In the case of any $n \times n$ tridiagonal Hessian matrix, an analogous choice of three vectors provides the entire Hessian in three gradient evaluations.[5]

In general when the sparsity pattern is known (but not necessarily as simple as in the above examples), it is also possible to obtain an approximation of the Hessian in fewer than n gradient evaluations.

Example 10.12: FINITE-DIFFERENCE MODIFIED-NEWTON METHOD. In Example 10.8 we computed the Hessian analytically. Here we use a finite-difference approximation to compute the Hessian and apply the Modified-Newton Method to the Rosenbrock function described in Example 9.13 on page 310. To calculate the Hessian we choose the finite-difference interval $h = \varepsilon_M^{0.6}$; this is in the interval $[\sqrt{\varepsilon_M}, \sqrt[3]{\varepsilon_M^2}]$ described on page 344.

---

[5]It turns out that the finite-difference approximation to this Hessian can actually be obtained in two gradient evaluations; see Powell and Toint [162].

As before, we set the termination criterion to be that the norm of the gradient must be less than $\tau_g = 10^{-7}$ and set the Simple Steplength Algorithm 9.8 parameter to be $\mu = 10^{-1}$. The Modified-Newton Method using a finite difference approximation of the Hessian also converges in 14 iterations; the iterates are not very different from those found by using the analytically computed Hessian. The last few iterations in this case are:

k	α	$\|x_k - x^*\|$	$\|(f_k - f^*)\|$
11	1.0000	1.1956e-002	3.1775e-005
12	1.0000	4.7203e-004	1.0680e-007
13	1.0000	2.3388e-006	1.2198e-012
14	1.0000	1.9201e-011	1.7919e-022

10.3 Quasi-Newton Methods

Newton's Method (modified to take into account Hessians that are not positive definite) has a fast rate of convergence to the solution. However, it requires the computation of the Hessian matrix or a finite-difference approximation to it at every iteration of the algorithm. The Steepest-descent Method, on the other hand, does not require any second-derivative information and is guaranteed to converge. Unfortunately, its convergence can be extremely slow to the point where it appears not to converge at all. The idea behind Quasi-Newton Methods is to mimic the properties of Newton's Method without requiring second derivative information or finite-differencing.

In 1959, Davidon published a technical report at Argonne National Laboratory (see [49]) introducing a method, which he called a *variable-metric method*, for unconstrained minimization problems. His approach was based on building up the Hessian approximation based on the computed values of $f(x)$ and $g(x)$ as the iterations proceed. Following his original paper, there was an explosion of interest in such variable-metric methods, which are called Quasi-Newton Methods.

If an approximation B_k to the Hessian G_k is known, the search direction at each iteration of a Quasi-Newton Algorithm can be computed as in Newton's Method with the approximate Hessian B_k used in place of G_k. That is, we solve

$$B_k p_k = -g_k.$$

Quasi-Newton Methods determine such a B_k in a systematic manner. The theory of Quasi-Newton Methods is based on the fact that the Hessian along the step taken is approximately equal to the difference in gradients. To see this, let x_k be the current iterate and let $x_{k+1} = x_k + s_k$ be the next iterate, where $s_k = \alpha_k p_k$. Using the Taylor series expansion,

$$g(x_k + s_k) = g(x_k) + G_k s_k + \cdots$$

or

$$G_k s_k \approx g(x_k + s_k) - g(x_k). \tag{10.24}$$

It is easy to show that if the function is a quadratic $f(x) = \frac{1}{2} x^\mathrm{T} H x + c^\mathrm{T} x$, then the above relationship is satisfied *exactly* at each iteration. The idea behind these methods, then, is to determine an approximation B_{k+1} to the Hessian G_k at each iteration such that the above condition is satisfied; i.e.,

$$B_{k+1} s_k = y_k \tag{10.25}$$

where $y_k = g_{k+1} - g_k$. The equation (10.25) is known as the *Quasi-Newton condition*.

These methods typically start with an initial Hessian approximation B_0 equal to the identity matrix. In this case, the initial direction is a steepest-descent direction $p_0 = -g_0$. Once x_1 has been computed, a new Hessian approximation B_1 is computed as an update of B_0:

$$B_1 = B_0 + U_0$$

such that B_1 satisfies the Quasi-Newton condition (10.25). In general, after x_{k+1} has been computed, a new Hessian approximation B_{k+1} is obtained as

$$B_{k+1} = B_k + U_k$$

where U_k, the updating matrix at the end of iteration k, is chosen so as to make B_{k+1} satisfy (10.25). The particular choice of U_k has given rise to several methods.

The Quasi-Newton condition (10.25) specifies a constraint on the new approximation B_{k+1} along a direction. As a result, one may expect (though not be guaranteed) that the updating matrix U_k will be of low rank. With this in mind, researchers have developed Quasi-Newton updates of rank one or two. For minimizing C^2 functions, the true Hessian matrix will be symmetric, and it is natural to seek a symmetric Hessian approximation. Further, it is highly desirable that the Hessian approximations be positive definite because then the solution of $B_k p_k = -g_k$ will result in a descent direction. We next describe three symmetric updates that satisfy the property of positive definiteness. For details and other updates, see [84].

Symmetric rank-one update

In 1965, Broyden proposed a rank-one update which did not satisfy the property of symmetry. Let u and v be any two vectors, then define an update of the form

$$B_{k+1} = B_k + uv^{\mathrm{T}}.$$

For the matrix B_{k+1} to satisfy the Quasi-Newton condition (10.25), we mulitply the above on the right by $s_k = x_{k+1} - x_k$, thereby obtaining

$$B_{k+1}s_k = (B_k + uv^{\mathrm{T}})s_k = y_k$$

or

$$u(v^{\mathrm{T}}s_k) = y_k - B_k s_k.$$

For any v, such that $v^{\mathrm{T}}s_k \neq 0$, the vector u is given by $(y_k - B_k s_k)/v^{\mathrm{T}}s_k$, and the rank-one modification from B_k to B_{k+1} is given by

$$B_{k+1} = B_k + \frac{1}{v^{\mathrm{T}}s_k}(y_k - B_k s_k)v^{\mathrm{T}}. \tag{10.26}$$

Let w be a vector satisfying $w^{\mathrm{T}}s_k = 0$. Note that in this case we can add to B_k any number of rank-one matrices of the form zw^{T}, where z is any vector, and the Quasi-Newton condition will be satisfied. Thus, if $n > 1$, there are infinitely many ways to modify B_k so that B_{k+1} satisfies (10.25) as long as u is in the direction $y_k - B_k s_k$. In the Broyden update, v is chosen to be s_k. However, in this case the updated Hessian approximation B_{k+1} will not be symmetric. To preserve symmetry, v must be a multiple of u; hence choosing $v = y_k - B_k s_k$, we get the symmetric rank-one update

$$B_{k+1} = B_k + \frac{1}{(y_k - B_k s_k)^{\mathrm{T}}s_k}(y_k - B_k s_k)(y_k - B_k s_k)^{\mathrm{T}} \tag{10.27}$$

where $y_k - B_k s_k \neq 0$ and $(y_k - B_k s_k)^{\mathrm{T}}s_k \neq 0$.

DFP update

This popular update is named after its discoverers Davidon, Fletcher, and Powell, see [69]. In this case the updated Hessian approximation is obtained as

$$B_{k+1} = B_k - \frac{1}{s_k^\mathrm{T} B_k s_k}(B_k s_k)(B_k s_k)^\mathrm{T} + \frac{1}{y_k^\mathrm{T} s_k} y_k y_k^\mathrm{T} + (s_k^\mathrm{T} B_k s_k) w_k w_k^\mathrm{T} \quad (10.28)$$

where

$$w_k = \frac{1}{y_k^\mathrm{T} s_k} y_k - \frac{1}{s_k^\mathrm{T} B_k s_k} B_k s_k. \quad (10.29)$$

Among the early discoveries after Davidon's paper [49], the DFP update was the first successful one for solving unconstrained optimization problems.

BFGS update

On practical problems, the BFGS update has been the most successful. It is named after Broyden, Fletcher, Goldfarb, and Shanno who each independently discovered it around 1970; see [25], [67], [86], and [178]. It is given by

$$B_{k+1} = B_k - \frac{1}{s_k^\mathrm{T} B_k s_k}(B_k s_k)(B_k s_k)^\mathrm{T} + \frac{1}{y_k^\mathrm{T} s_k} y_k y_k^\mathrm{T}. \quad (10.30)$$

The BFGS update clearly satisfies the property of symmetry. Furthermore, it can be shown that as long as $y_k^\mathrm{T} s_k > 0$, it is also guaranteed to result in positive definite B_{k+1}. (Note that from equation (10.25), it is easy to see that if B_{k+1} is positive definite, then $y_k^\mathrm{T} s_k > 0$ because $s_k^\mathrm{T} B_k s_k = y_k^\mathrm{T} s_k$.) It can be shown that $y_k^\mathrm{T} s_k > 0$ if a steplength is chosen to satisfy Wolfe's conditions (see page 308).

One-parameter family of updates

For the DFP update described above, it is easy to see that $w_k^\mathrm{T} s_k = 0$; hence any multiple of $w_k w_k^\mathrm{T}$ may be added to equation (10.28) without affecting satisfaction of the Quasi-Newton condition (10.25). Doing this results in the one-parameter family of updates

$$B_{k+1} = B_k - \frac{1}{s_k^\mathrm{T} B_k s_k}(B_k s_k)(B_k s_k)^\mathrm{T} + \frac{1}{y_k^\mathrm{T} s_k} y_k y_k^\mathrm{T} + \phi_k(s_k^\mathrm{T} B_k s_k) w_k w_k^\mathrm{T}. \quad (10.31)$$

Notice that when the parameter ϕ_k is chosen to be zero, (10.31) reduces to the BFGS update, and when ϕ_k is chosen to be 1, (10.31) is the DFP update. Thus, the one-parameter family of updates can be written in terms of the DFP and the BFGS update as a convex combination

$$B_{k+1} = (1 - \phi_k)B_k^{BFGS} + \phi_k B_k^{DFP}. \tag{10.32}$$

In 1972, Dixon showed (see [52] and [53]) that when exact line searches are used in the algorithms, all the members of the one-parameter family of updates (10.32) produce the same iterates. Hence all the differences in numerical experiments on the updates can be attributed to differences in the determination of the steplength. Dixon's result also demonstrated that the *elements of a Quasi-Newton approximation to the Hessian will not necessarily be the same as those of the true Hessian.*

Convergence properties

When applied to a strictly convex quadratic function, all the Quasi-Newton Methods based on the one-parameter family of updates will converge to the minimum in a finite number of iterations (typically $n + 1$). This is called *quadratic termination*, but it should not be confused with quadratic rate of convergence. In fact, it can be shown that, under certain conditions, Quasi-Newton Methods converge superlinearly; see Powell [159] and [161].

A Quasi-Newton Method

In a Quasi-Newton Method we solve a system of equations to obtain a search direction. In this system, the Hessian matrix of Newton's Method is replaced by the Quasi-Newton approximation B_k. The algorithm is summarized below. It makes use of the modified-Cholesky Factorization for numerical stability; the use of the Cholesky factorization for Quasi-Newton Methods was first proposed by Gill and Murray [79].

Algorithm 10.5 (Quasi-Newton Method)   *Find a minimizer of a twice-continuously differentiable function $f : R^n \to R$.*

Set $k = 0$ and let x_0 be given. Compute the gradient $g_0 = g(x_0)$ and set the starting Quasi-Newton Hessian approximation to be $B_0 = I$, where I is the identity matrix.

1. *Test for convergence.* If the gradient satisfies $\|g_k\| \leq \sqrt{\varepsilon_M}$, report $x^* = x_k$ as the solution.

2. *Compute a search direction.* Solve $B_k p_k = -g_k$ for p_k as follows:

   (a) Obtain a modified-Cholesky factorization of B_k as $L_k \bar{D}_k L_k^T$.

   (b) Find the Quasi-Newton direction p_k by solving

   $$L_k \bar{D}_k L_k^T p_k = -g_k.$$

3. *Compute a steplength.* Compute the steplength using one of the algorithms described in Section 9.8, for example, the Simple Steplength Algorithm 9.8.

4. *Update the iterate.* Using p_k and α_k, compute

   $$x_{k+1} = x_k + \alpha_k p_k,$$
   $$g_{k+1} = g(x_{k+1}),$$
   $$B_{k+1} = B_k + U_k,$$

   where U_k is one of the Quasi-Newton updates. Return to Step 1 with the iteration counter $k \leftarrow k + 1$.

Example 10.13: SYMMETRIC RANK-ONE UPDATE. As before, we set the termination criterion to be that the norm of the gradient must be less than $\tau_g = 10^{-7}$ and set the Simple Steplength Algorithm 9.8 parameter to be $\mu = 10^{-1}$. The initial Hessian approximation B_0 is set to the identity matrix, but any symmetric positive definite matrix will suffice.

On the quadratic of Example 10.1, the Quasi-Newton Algorithm 10.5 using a symmetric rank-one update converges in one iteration. In this case the search direction is the steepest-descent direction, which is also the step (Newton direction) to the minimum of the quadratic function. When applied to the quadratic of Example 10.2, the method finds the optimal solution in two steps as expected by the property of quadratic termination.

The Quasi-Newton Algorithm 10.5 starts with the identity as the initial Hessian approximation. Thus the initial search direction is the steepest-descent direction. For the Rosenbrock function, at the point $x_0 = (0,0)$, the initial search direction is

$$p_0 = -g_0 = (2.0, \ 0.0).$$

On applying the Simple Steplength Algorithm, we obtain $\alpha = 1/16 = 0.0625$, and the next iterate is

$$x_1 = x_0 + \alpha_0 p_0 = (0.125, \ 0.0).$$

The function value and gradient at this point are

$$f_1 = 0.7904 \quad \text{and} \quad g_1 = (-0.9687, \ -3.1250).$$

The termination condition is not satisfied so we compute the next Hessian approximation. To do this we first compute the quantities

$$s_0 = x_1 - x_0 = \alpha_0 p_0 = 0.0625 \times (2.0, \ 0.0) = (0.125, \ 0.0)$$
$$y_0 = g_1 - g_0 = (-0.9687, \ -3.1250) - (-2.0, \ 0.0)$$
$$= (1.0313, \ -3.1250)$$
$$B_0 s_0 = \begin{bmatrix} 1 & 0 \\ 0 & 1 \end{bmatrix} \begin{bmatrix} 0.125 \\ 0.000 \end{bmatrix} = \begin{bmatrix} 0.125 \\ 0.000 \end{bmatrix}$$
$$y_0 - B_0 s_0 = (0.9063, \ -3.1250).$$

Next, using (10.27), we compute

$$B_1 = B_0 + \frac{1}{(y_0 - B_0 s_0)^{\mathrm{T}} s_0}(y_0 - B_0 s_0)(y_0 - B_0 s_0)^{\mathrm{T}}$$
$$= \begin{bmatrix} 8.2500 & -25.0000 \\ -25.0000 & 87.2069 \end{bmatrix}.$$

Using this updated Hessian approximation, we solve $B_1 p_1 = -g_1$ to obtain

$$p_1 = (0.5293 \ 0.1250).$$

Continuing in this manner, we see that the last few iterations are as follows:

k	α	$\dfrac{\|x_{k+1} - x^*\|}{\|x_k - x^*\|}$	$\dfrac{\|f_{k+1} - f^*\|}{\|f_k - f^*\|}$
22	1.0000	7.0122e-004	2.0580e-007
23	1.0000	6.0151e-005	1.7720e-009
24	1.0000	1.7192e-005	9.5724e-011
25	1.0000	1.1494e-008	1.0812e-016
26	1.0000	1.5579e-010	1.2548e-020

The last two columns show that the ratios are converging to zero. Thus, the Quasi-Newton Algorithm using a symmetric rank-one update exhibits superlinear convergence.

At the optimal solution found, the Quasi-Newton Hessian obtained using the DFP update and the true Hessian are respectively

$$B_{26} = \begin{bmatrix} 801.9767 & -399.9894 \\ -399.9894 & 199.9952 \end{bmatrix} \quad \text{and} \quad H = \begin{bmatrix} 802.0000 & -400.0000 \\ -400.0000 & 200.0000 \end{bmatrix}.$$

Example 10.14: DFP UPDATE. We use the initial conditions and tolerances described in Example 10.13. On the quadratic of Example 10.1, the Quasi-Newton Algorithm 10.5 using a DFP update converges in one iteration. In this case the search direction is the steepest-descent direction, which is also the step (Newton direction) to the minimum of the quadratic function. When applied to the quadratic of Example 10.2, the method, using a steplength algorithm with the minimization condition (see page 306), finds the optimal solution in two steps as expected by the property of quadratic termination.

The Quasi-Newton Algorithm 10.5 starts with the identity as the initial Hessian approximation. Thus the initial search direction is the steepest-descent direction. For the Rosenbrock function, at the point $x_0 = (0, 0)$, the initial search direction is

$$p_0 = -g_0 = (2.0, \ 0.0).$$

On applying the Simple Steplength Algorithm, we obtain $\alpha = 1/16 = 0.0625$, and the next iterate is

$$x_1 = x_0 + \alpha_0 p_0 = (0.125, \ 0.0).$$

The function value and gradient at this point are

$$f_1 = 0.7904 \quad \text{and} \quad g_1 = (-0.9687, \, -3.1250).$$

The termination condition is not satisfied so we compute the next Hessian approximation. To do this we first compute the quantities

$$s_0 = x_1 - x_0 = \alpha_0 p_0 = 0.0625 \begin{bmatrix} 2.0 \\ 0.0 \end{bmatrix} = \begin{bmatrix} 0.125 \\ 0.000 \end{bmatrix}$$

$$y_0 = g_1 - g_0 = \begin{bmatrix} -0.9687 \\ -3.1250 \end{bmatrix} - \begin{bmatrix} -2.0 \\ 0.0 \end{bmatrix} = \begin{bmatrix} 1.0313 \\ -3.1250 \end{bmatrix}$$

$$B_0 s_0 = \begin{bmatrix} 1 & 0 \\ 0 & 1 \end{bmatrix}^{\mathrm{T}} \begin{bmatrix} 0.125 \\ 0.000 \end{bmatrix} = \begin{bmatrix} 0.125 \\ 0.000 \end{bmatrix}$$

$$y_0^{\mathrm{T}} s_0 = \begin{bmatrix} 1.0313 \\ -3.1250 \end{bmatrix}^{\mathrm{T}} \begin{bmatrix} 0.125 \\ 0.000 \end{bmatrix} = 0.1289$$

$$s_0^{\mathrm{T}} B_0 s_0 = \begin{bmatrix} 0.125 \\ 0.000 \end{bmatrix} \begin{bmatrix} 0.125 \\ 0.000 \end{bmatrix} = 0.0156$$

$$w_0 = \frac{1}{y_0^{\mathrm{T}} s_0} y_0 - \frac{1}{s_0^{\mathrm{T}} B_0 s_0} B_0 s_0 = \frac{1}{0.1289} \begin{bmatrix} 1.0313 \\ -3.1250 \end{bmatrix} - \frac{1}{0.0156} \begin{bmatrix} 0.125 \\ 0.000 \end{bmatrix}$$

$$= \begin{bmatrix} 0.0000 \\ -24.2424 \end{bmatrix}.$$

Next we compute B_1 using (10.28):

$$B_1 = B_0 - \frac{1}{s_0^{\mathrm{T}} B_0 s_0} (B_0 s_0)(B_0 s_0)^{\mathrm{T}} + \frac{1}{y_0^{\mathrm{T}} s_0} y_0 y_0^{\mathrm{T}} + (s_0^{\mathrm{T}} B_0 s_0) w_0 w_0^{\mathrm{T}}$$

$$= \begin{bmatrix} 8.2500 & -25.0000 \\ -25.0000 & 85.9403 \end{bmatrix}.$$

Using this updated Hessian approximation we solve $B_1 p_1 = -g_1$ to obtain

$$p_1 = \begin{bmatrix} 0.5952 \\ 0.0625 \end{bmatrix}.$$

Continuing in this manner, we see that the last few iterations are

k	α	$\dfrac{\lVert x_{k+1} - x^* \rVert}{\lVert x_k - x^* \rVert}$	$\dfrac{\lvert f_{k+1} - f^* \rvert}{\lvert f_k - f^* \rvert}$
21	1.0000	1.0614e-004	2.3731e-009
22	1.0000	5.3406e-006	2.8564e-011
23	1.0000	1.5085e-007	2.4824e-013
24	1.0000	2.6616e-008	1.8765e-016
25	1.0000	7.3823e-010	1.1020e-019

The last two columns show that the ratios are converging to zero. Thus, the Quasi-Newton Algorithm using a DFP update exhibits superlinear convergence.

At the optimal solution found, the Quasi-Newton Hessian obtained using the DFP update and the true Hessian are respectively

$$
B_{25} = \begin{bmatrix} 808.5848 & -403.3835 \\ -403.3835 & 201.7385 \end{bmatrix} \quad \text{and} \quad H = \begin{bmatrix} 802.0000 & -400.0000 \\ -400.0000 & 200.0000 \end{bmatrix}.
$$

Example 10.15: BFGS UPDATE. We use the initial conditions and tolerances described in Example 10.13. On the quadratic of Example 10.1, the Quasi-Newton Algorithm 10.5 using a BFGS update converges in one iteration. In this case the search direction is the steepest-descent direction, which is also the step (Newton direction) to the minimum of the quadratic function. When applied to the quadratic of Example 10.2, the method, using a steplength algorithm with the minimization condition (see page 306), finds the optimal solution in two steps as expected by the property of quadratic termination.

The Quasi-Newton Algorithm 10.5 starts with the identity as the initial Hessian approximation. Thus the initial search direction is the steepest-descent direction. For the Rosenbrock function, at the point $x_0 = (0,0)$, the initial search direction is

$$
p_0 = -g_0 = (2.0, \ 0.0).
$$

On applying the Simple Steplength Algorithm, we obtain $\alpha = 1/16 = 0.0625$, and the next iterate is

$$
x_1 = x_0 + \alpha_0 p_0 = (0.125, \ 0.0).
$$

The function value and gradient at this point are

$$f_1 = 0.7904 \quad \text{and} \quad g_1 = (-0.9687, \ -3.1250).$$

The termination condition is not satisfied so we compute the next Hessian approximation. To do this we first compute the quantities

$$s_0 = x_1 - x_0 = \alpha_0 p_0 = 0.0625 \times (2.0, \ 0.0) = (0.125, \ 0.0)$$

$$y_0 = g_1 - g_0 = (-0.9687, \ -3.1250) - (-2.0, \ 0.0) = (1.0313, \ -3.1250)$$

$$B_0 s_0 = \begin{bmatrix} 1 & 0 \\ 0 & 1 \end{bmatrix} \begin{bmatrix} 0.125 \\ 0.000 \end{bmatrix} = \begin{bmatrix} 0.125 \\ 0.000 \end{bmatrix}$$

$$y_0^T s_0 = (1.0313, \ -3.1250) \begin{bmatrix} 0.125 \\ 0.0 \end{bmatrix} = 0.1289$$

$$s_0^T B_0 s_0 = (0.125, \ 0.0) \begin{bmatrix} 0.125 \\ 0.0 \end{bmatrix} = 0.0156.$$

Next, using (10.30), we compute

$$B_1 = B_0 - \frac{1}{s_0^T B_0 s_0}(B_0 s_0)(B_0 s_0)^T + \frac{1}{y_0^T s_0} y_0 y_0^T$$

$$= \begin{bmatrix} 8.2500 & -25.0000 \\ -25.0000 & 76.7576 \end{bmatrix}.$$

Using this updated Hessian approximation we solve $B_1 p_1 = -g_1$ to obtain

$$p_1 = (18.4829, \ 6.0606).$$

Continuing in this fashion, we see that the last few iterations are as follows:

k	α	$\dfrac{\|x_{k+1} - x^*\|}{\|x_k - x^*\|}$	$\dfrac{|f_{k+1} - f^*|}{|f_k - f^*|}$
19	1.0000	2.5855e-003	2.6360e-005
20	1.0000	2.2045e-003	1.2417e-006
21	1.0000	1.2525e-004	3.1674e-009
22	1.0000	1.0673e-006	3.8116e-013
23	1.0000	3.9881e-010	1.5958e-018

The last two columns show that the ratios are converging to zero. Thus, the Quasi-Newton Algorithm with DFP update has superlinear convergence.

At the optimal solution found, the Quasi-Newton Hessian obtained using the DFP update and the true Hessian are respectively

$$B_{25} = \begin{bmatrix} 802.7427 & -400.3556 \\ -400.3556 & 200.1703 \end{bmatrix} \quad \text{and} \quad H = \begin{bmatrix} 802.0000 & -400.0000 \\ -400.0000 & 200.0000 \end{bmatrix}.$$

Impact of errors in a Quasi-Newton Method

Recall our discussion in Example 10.9 where we noted that typically an optimization algorithm will not converge when errors are present in the computation of the gradient or the function value; this is because its steplength algorithm fails to produce a decrease in the function value. When the gradient calculation is deliberately modified so as to introduce an error as small as -10^{-6} in the second component of the gradient for Rosenbrock's function, the Quasi-Newton Algorithm utilizing the BFGS update fails to find an optimal solution because its steplength algorithm fails to produce a decrease in the function value at iteration 23.

Solving for a Quasi-Newton search direction

The early implementations of Quasi-Newton Methods updated the inverse Hessian approximations. Working with the inverse has its drawbacks in terms of numerical stability. Furthermore, it is not easy to recover if numerical errors cause a loss of positive definiteness. It turns out that it is possible to update the modified-Cholesky factorization with efficiency similar to that obtained when working with inverse Hessian updates.

Inverse Hessian updates

The first effective Quasi-Newton Method due to Davidon, Fletcher and Powell computed the search direction as $p_k = -D_k g_k$, where D_k is the inverse Hessian approximation that satisfies the Quasi-Newton condition

$$D_{k+1} y_k = s_k. \tag{10.33}$$

The DFP update in this case is

$$D_{k+1} = D_k - \frac{1}{y_k^T D_k y_k} (D_k y_k)(D_k y_k)^T + \frac{1}{y_k^T s_k} s_k s_k^T. \tag{10.34}$$

Updating the factors of the Hessian matrix

Gill and Murray (see [84, p. 42]) showed how to update the Cholesky factors of B_k to obtain those of B_{k+1} following a rank-one change, so that the corresponding number of operations is similar to that associated with retaining an approximation to the Hessian inverse. If the Cholesky factors of the Hessian approximation are available at each iteration, then $O(n^2)$ operations are required to obtain the search direction similar to that required with the matrix product $D_k g_k$. An advantage of having the Cholesky factors available is that we can easily determine whether or not the Hessian approximation is positive definite. Updating procedures can be designed to guarantee that the diagonal elements of the Cholesky factors will stay positive. Such algorithms are described in Gill, Golub, Murray, and Saunders [80].

10.4   The Conjugate-gradient Method

In 1952, Hestenes and Stiefel (see [94]) developed the Conjugate-gradient Method, an iterative method for solving a large linear system $Ax = b$, where A is an $n \times n$ positive definite and symmetric matrix. It turns out that if there are no rounding errors, then the Conjugate-gradient Method requires no more than n iterations to obtain the solution. However, on a computer, the calculations are in finite precision thereby resulting in rounding errors that typically cause the algorithm to take many more than n iterations to converge to a solution. Yet there are special cases, where, in spite of rounding errors, the algorithm converges in fewer than n iterations. This tends to be true of problems where the matrix has eigenvalues that are clustered into sets containing eigenvalues of similar magnitude (see Gill et al. [84]).

In 1964, Fletcher and Reeves [70] suggested that the Conjugate-gradient Method can be applied to minimization. Since its discovery, there have been considerable improvements in techniques for applying the Conjugate-gradient Method, but the performance of these methods has been inferior to rival techniques. For example, there has been quite a bit of development of Modified-Newton Methods for nonlinear programs with sparse Hessian matrices. However, there is a class of problems for which Conjugate-gradient Methods are currently the only techniques that can be applied. One such class is that for which the Hessian matrix is large but not sparse. Another such class is where the system being solved is very large; applications of this arise in interior-point methods (see Chapter 14, for example). Another area for possible application of Conjugate-gradient Methods is the situation where an exact solution of a system of equations is not required and a few iterations of a Conjugate-gradient Method will give the required approximate solution; examples of this also arise in interior-point methods for optimization problems. To describe such methods, we start by defining conjugate directions.

A set of nonzero vectors $\{p_k\}$ are said to be *conjugate*[6] with respect to a positive definite symmetric matrix G if

$$p_i^{\mathrm{T}} G p_j = 0, \quad \text{for } i \neq j.$$

In this case, conjugacy can be interpreted as a generalized orthogonality and we can easily show that a set of conjugate directions is necessarily a linearly independent set of vectors.

Conjugate-gradient Method applied to a quadratic function

Conjugate-gradient Methods also possess the property of quadratic termination. That is, if the search directions are conjugate and an exact line-search is used, then the method will terminate with an optimal solution in at most n iterations for a quadratic function with a positive definite and symmetric Hessian G. More precisely, if the iterates $x_{k+1} = x_k + \alpha_k p_k$ have search direction p_k conjugate with respect to all previous search directions $p_1, p_2, \ldots, p_{k-1}$, and α_k is the step to the minimum along the direction p_k, we get termination in a finite number of steps. This result is proved by establishing that

$$p_j^{\mathrm{T}} g_{i+1} = 0 \quad \text{for } j = 1, \ldots, i. \tag{10.35}$$

Then at iteration $i = n$, the gradient is orthogonal to n linearly independent search directions implying that it is zero.

We next describe the method for a quadratic function with a positive definite symmetric Hessian matrix G. The iterations are typically started by setting the initial search direction to be the steepest descent direction,

$$p_1 = -g_1;$$

---

[6]Conjugacy could be defined with respect to any matrix G that is not necessarily positive definite and symmetric. In this case the vectors $\{p_k\}$ would need to satisfy the additional condition $p_i^{\mathrm{T}} G p_i \neq 0$ for all i.

The search direction in subsequent iterations is computed by

$$p_i = -g_i + \gamma_i p_{i-1},$$

where

$$\gamma_i = \frac{y_{i-1}^{\mathrm{T}} g_i}{y_{i-1}^{\mathrm{T}} p_{i-1}}, \qquad (10.36)$$

and $y_{i-1} = g_i - g_{i-1}$. For a quadratic function, the following alternative definitions can be used:

$$\text{either} \quad \gamma_i = \frac{y_{i-1}^{\mathrm{T}} g_i}{\|g_{i-1}\|_2^2} \quad \text{or} \quad \gamma_i = \frac{\|g_i\|_2^2}{\|g_{i-1}\|_2^2}.$$

We have seen earlier that a quadratic approximation is typically used to develop algorithms for smooth nonlinear functions because close to a minimum, the function "looks" like a quadratic. A Conjugate-gradient Method, when applied to a general nonlinear function need not terminate in n, or fewer, iterations. Hence, it may make sense to restart the method with a steepest-descent direction every n iterations. In practice, restarting is critical to the success of Conjugate-gradient Methods and various schemes have been developed for doing this.

Linear Conjugate-gradient Method

We next describe the Linear Conjugate-gradient Method which is used for solving a system of equations $Ax = b$, where the matrix A is positive definite and symmetric. From the standpoint of solving systems of linear equations, the vector $r = Ax - b$ is traditionally called the *residual* vector. Note that doing this is equivalent to finding the minimum of a strictly convex quadratic function $f(x) = (1/2)x^{\mathrm{T}}Ax - b^{\mathrm{T}}x + \kappa$, because the solution occurs where the gradient $g(x) = Ax - b$ (i.e., the residual vector r) is zero.

Algorithm 10.6 (Linear Conjugate-gradient Method) *Solve a square system of equations $Ax = b$ where the coefficient matrix A is positive definite and symmetric; equivalently, minimize the strictly convex quadratic function $f(x) = \frac{1}{2}x^{\mathrm{T}}Ax - b^{\mathrm{T}}x + \kappa$.*

Set $k = 1$. Let $\gamma_0 = 0$, $p_0 = 0$. Choose x_1 arbitrarily (but wisely) and compute the residual (gradient) $r_1 = Ax_1 - b$.

1. *Test for convergence.* If $||r_k||_2 \leq \sqrt{\varepsilon_M}$, then report $x^* = x_k$ as the solution.

2. *Compute the Conjugate-gradient direction.* Set $p_k = -r_k + \gamma_{k-1}p_{k-1}$.

3. *Compute the steplength.*

$$\alpha_k = \frac{||r_k||_2^2}{p_k^T A p_k}.$$

4. *Update.* Using p_k and α_k, compute

$$x_{k+1} = x_k + \alpha_k p_k$$
$$r_{k+1} = r_k + \alpha_k A p_k$$
$$\gamma_k = \frac{||r_{k+1}||_2^2}{||r_k||_2^2}.$$

Return to Step 1 with the iteration counter $k \leftarrow k + 1$.

Preconditioned Conjugate-gradient Method

Over the years there has been considerable research and development on Conjugate-gradient Methods; among these are *preconditioned* Conjugate-gradient Methods (see [173]) which have done well in practice. Furthermore, the advent of interior-point methods for constrained optimization has generated renewed interest in Conjugate-gradient Methods for approximately solving the systems of equations that arise in such algorithms. See, for example, [111].

10.5 Exercises

10.1   *Steepest-descent direction.* Find the vector p that solves

$$\min_{p} \frac{g^T p}{||p||_C},$$

where C is a symmetric positive definite matrix. Do this as follows:

(a) Show that the above problem is equivalent to the maximization problem

$$\max_{p} \frac{|g^T p|}{||p||_C}.$$

(b) Use the Cauchy-Schwartz inequality (A.17) to derive the inequality

$$(g^Tp)^2 \leq (g^TC^{-1}g)(p^TCp).$$

(c) Use the above to show that when $g^Tp < 0$, the solution of the problem

$$\min_{p \in R^n} \frac{g^Tp}{||p||_C}$$

is given by $p = -C^{-1}g$.

10.2   Let $H \in R^{n \times n}$ be a positive definite symmetric matrix with eigenvalues $\lambda_1 \leq \cdots \leq \lambda_n$. As stated earlier in the chapter, the *Kantorovich inequality* states that for any nonzero vector $x \in R^n$

$$\frac{(x^Tx)^2}{(x^THx)(x^TH^{-1}x)} \geq \frac{4\lambda_1\lambda_n}{(\lambda_1 + \lambda_n)^2}.$$

(a) Let H and its corresponding inverse be

$$H = \begin{bmatrix} 4 & 0 \\ 0 & 7 \end{bmatrix} \quad \text{and} \quad H^{-1} = \begin{bmatrix} 1/4 & 0 \\ 0 & 1/7 \end{bmatrix}.$$

The eigenvalues of H are $\lambda_1 = 4$ and $\lambda_2 = 7$. For $x = (1,1)$ show that the Kantorovich inequality holds.

(b) Show that for all vectors x of norm equal to 1, the Kantorovich inequality can be written in the form

$$(x^THx)(x^TH^{-1}x) \leq \frac{1}{4}\left[\left(\frac{\lambda_1}{\lambda_n}\right)^{1/2} + \left(\frac{\lambda_n}{\lambda_1}\right)^{1/2}\right]^2.$$

(c) Is the assumption made above in (b) restrictive? Justify your answer.
(d) Prove that the Kantorovich inequality holds in general.

10.3   Verify that relations (10.9), (10.10), and (10.11) are true.

10.4   *Steepest descent.* Consider the quadratic function $q(x) = \frac{1}{2}x^THx + c^Tx$ where

$$H = \begin{bmatrix} 1 & 0 & 0 \\ 0 & 9 & 0 \\ 0 & 0 & 100 \end{bmatrix} \quad \text{and} \quad c = (1,1,1).$$

(a) Algebraically find the vector x^* that minimizes $q(x)$.

(b) Starting at $x_0 = (0, 0, 0)$, apply the Steepest-descent Method to $q(x)$ and compute the ratio

$$\frac{q(x_{k+1}) - q(x^*)}{q(x_k) - q(x^*)}.$$

Use the minimization condition to compute the value of α at each iteration. Do this for four iterations if doing the calculations manually (or more iterations if you have written a program to do this).

10.5   Define the function

$$f_c(x) = cx_1^2 + x_2^2.$$

(a) Starting with the point $x = (10, 10)^T$, carry out three iterations of the Steepest-descent Method to minimize $f_c(x)$ when $c = 2$, $c = 10$, and $c = 100$.

(b) Do the same with Newton's Method.

10.6   *Newton's Method.* Consider the quadratic of Exercise 10.4 above.

(a) Starting with $x_0 = (0, 0, 0)$, apply Newton's Method.

(b) How many iterations did it take? Explain your answer.

(c) Write down the formula for α that minimizes $\phi(\alpha) = f(x + \alpha p)$ and show that, for Newton's Method applied to any strictly convex quadratic, this value is $\alpha^* = 1$.

(d) A symmetric positive definite matrix H has a square root; that is, $H = H^{1/2}H^{1/2}$. Define $x = H^{-1/2}y$ and transform the $q(x)$ problem to a $q(y)$ problem.

(e) Use the Steepest-descent Method on $q(y)$. Show that the resulting steepest-descent search direction for $q(y)$ is the same as the Newton direction for $q(x)$ in the sense that y_1 obtained results in the same x_1 when Newton's Method is applied with α as the minimizer of $\phi(\alpha)$.

10.7   *Gradient-related direction.* Let G_k be symmetric and positive definite and let the search direction p_k be obtained using equations $G_k p_k = -g_k$. As discussed on page 329, there exist constants $0 < \beta \leq \gamma$ such that

$$\|G_k\|_2 \leq \gamma$$
$$y^T G_k y \geq \beta y^T y, \quad y \neq 0,$$

that is, γ is an upper bound on the largest eigenvalue of $G(x)$ and β is a lower bound on the smallest eigenvalue of $G(x)$. Show that this implies (using norm inequalities and some algebraic manipulations):

$$|g_k^T p_k| \geq (\beta/\gamma)\|g_k\|\,\|p_k\|.$$

Hint: Review the definition of the 2-norm and show that

$$||G_k^{-1}||_2 \geq \gamma^{-1}.$$

Also note that

$$||Ay|| \leq ||A|| \, ||y||$$

from the definition of a vector-induced norm.

10.8   Let $\mathcal{A}$ and $\mathcal{B}$ denote the convex hulls referred to in Exercise 8.7. Find the points $\bar{x} \in \mathcal{A}$ and $\bar{y} \in \mathcal{B}$ such that

$$||\bar{x} - \bar{y}|| \leq ||x - y||, \quad \text{for all} \quad x \in \mathcal{A}, \ y \in \mathcal{B}.$$

Note that by squaring the objective function $||x - y||$, you can formulate this problem as a convex quadratic program.

(a) Is the solution $\bar{x}$, $\bar{y}$ unique? If so, why?
(b) Does $\bar{x} = \bar{y}$? What do you conclude from this?
(c) Is it possible to put a pair of parallel hyperplanes between $\mathcal{A}$ and $\mathcal{B}$? If so, how far apart can they be?

10.9   *Steepest-descent Method—implementation.* Implement and test your implementation of the Steepest-descent Method as follows:

(a) Implement the Steepest-descent Algorithm 10.1. Modify the convergence check so that the algorithm terminates without a solution if the number of iterations reaches a pre-specified maximum value denoted by K_{max}.
(b) Examine the solution process; at the very least, print out the following quantities at each iteration: $k, N_f, N_g, f_k, ||g_k||, ||x_k - x^*||$, $||f_k - f^*||, \alpha_k$; where, N_f and N_g are the cumulative number of function and gradient evaluations at the end of iteration k.
(c) Apply the algorithm to the following functions
    (i)   The quadratic function $f(x) = \frac{1}{2}x^{\mathrm{T}}Hx + c^{\mathrm{T}}x$, with

$$H = \begin{bmatrix} 1000 & 0 & 0 \\ 0 & 1000 & 0 \\ 0 & 0 & 1000 \end{bmatrix}, \quad c = \begin{bmatrix} 100 \\ 100 \\ 100 \end{bmatrix}.$$

    (ii) The quadratic function $f(x) = \frac{1}{2}x^{\mathrm{T}}Hx + c^{\mathrm{T}}x$, with

$$H = \begin{bmatrix} 10 & 0 & 0 \\ 0 & 100 & 0 \\ 0 & 0 & 1000 \end{bmatrix}, \quad c = \begin{bmatrix} 100 \\ 100 \\ 100 \end{bmatrix}.$$

(iii) Rosenbrock's function

$$f(x) = 100(x_2 - x_1^2)^2 + (1 - x_1)^2$$

with $K_{max} = 500$ and $x_0 = (.2, 1)$. The Rosenbrock function has a unique minimum at $x^* = (1, 1)$.

(iv) Powell's Badly Scaled Function (see [158])

$$f(x) = (10^4 x_1 x_2 - 1)^2 + (e^{-x_1} + e^{-x_2} - 1.0001)^2$$

with $K_{max} = 500$ and $x_0 = (0, 1)$. The function has a unique minimum at approximately $x^* = (1.098 \times 10^{-5}, 9.106)$.

(d) *Analysis of your results.*

(i)   Explain the difference in behavior on the two quadratic functions.
(ii) Do the convergence rates on the functions match expectations? Explain your answer based on the theory.
(iii) Is there a pattern to the iterates?
(iv) Comment on the steplengths obtained as the iterations proceed.

(e) Suppose that you modify the gradient computation on the first quadratic to return an incorrect gradient; that is, introduce a small error. How does this impact the behavior of the algorithm?

10.10   *Modified-Cholesky Factorization—implementation.*

(a) Implement the Modified-Cholesky Algorithm *without* row and column interchanges.

(b) Implement the Modified-Cholesky Algorithm *with* row and coluumn interchanges.

(c) Test your implementations on the matrix used in Example 10.6.

(i)   Print out the factors and the diagonal modification.
(ii) Multiply the factors and verify that the resulting matrix is the sum of the original matix and the diagonal modification.
(iii) Implement a function to do the above verification automatically.

(d) Test your implementation on the following two matrices, the first of which is positive definite.

$$H = \begin{bmatrix} 2 & -2 & 0 \\ -2 & 5 & -2 \\ 0 & -2 & 2 \end{bmatrix} \quad \text{and} \quad \widehat{H} = \begin{bmatrix} 2 & -2 & 0 \\ -2 & 5 & -2 \\ 0 & -2 & -6 \end{bmatrix}.$$

Verify your answer by the using the function you implemented in part (c).

10.11   *Modified-Newton's Method—implementation.* Implement and test your implementation of the Modified-Newton Method as follows:

(a) Implement the Modified-Newton Algorithm 10.4. Modify the convergence check so that the algorithm terminates without a solution if the number of iterations reaches a pre-specified maximum value denoted by K_{max}.

(b) Examine the solution process; at the very least, print out the following quantities at each iteration: $k, N_f, N_g, f_k, ||g_k||, ||x_k - x^*||,$ $||f_k - f^*||, \alpha_k$; where, N_f and N_g are the cumulative number of function and gradient evaluations at the end of iteration k.

(c) Apply the algorithm to the functions in Exercise 10.9.

(d) *Analysis of your results.*

   (i)   Examine the behavior on the two quadratic functions and explain why you expect the algorithm to converge in one iteration.
   (ii) Do the convergence rates on the functions match expectations? Explain your answer based on the theory.
   (iii) Comment on the steplengths obtained on the Rosenbrock function as the iterations proceed. Does the sequence of steplengths behave as expected based on the theory.
   (iv) What would happen if you ignored (set to 0) the diagonal adjustment in the Modified-Cholesky Algorithm?

(e) Suppose that you modify the Hessian computations to return an incorrect Hessian. How does this impact the behavior of the algorithm?

(f) Suppose that you modify the gradient computation to return an incorrect gradient; that is, introduce a small error. How does this impact the behavior of the algorithm?

10.12   *Modified-Newton's Method with Finite-Differencing—implementation.*

(a) Repeat Exercise 10.11, using a finite-differencing approach to computing the Hessian.

(b) Try running the same algorithm, but instead of using the recommended finite-difference interval, use both smaller and larger intervals and examine the behavior of the algorithm.

10.13   *Quasi-Newton Method—implementation.* Implement and test your implementation as follows:

(a) Implement the different versions of the Quasi-Newton Algorithm described in this chapter. Modify the convergence check so that the algorithm terminates without a solution if the number of iterations reaches a pre-specified maximum value denoted by K_{max}.

(b) Examine the solution process; at the very least, print out the following quantities at each iteration: $k, N_f, N_g, f_k, ||g_k||, ||x_k - x^*||,$ $||f_k - f^*||, \alpha_k$; where, N_f and N_g are the cumulative number of function and gradient evaluations at the end of iteration k.

(c) Apply the algorithm to the functions in Exercise 10.9.

(d) *Analysis of your results.*

   (i) Verify that the Quasi-Newton condition is satisfied at each iteration.

   (ii) Examine the behavior of these methods on the two quadratic functions versus Rosenbrock's function. Do the methods satisfy the property of quadratic termination? Explain why or why not.

   (iii) Do the convergence rates on the functions match expectations? Explain your answer based on the theory.

   (iv) Comment on the steplengths obtained on the Rosenbrock function as the iterations proceed. Does the behavior match expectations?

   (v) On successful termination of each of your Quasi-Newton Algorithms is the approximate Hessian obtained equal to the true Hessian? Does this match expectations? Explain why or why not.

   (vi) Did the Modified-Cholesky Algorithm result in any modifications to the factors? Does this match expectations? What would happen if you ignored (set to 0) the diagonal adjustment in the Modified-Cholesky Algorithm?

(e) Suppose that you modify the Quasi-Newton update to an incorrect Hessian. How does this impact the behavior of the algorithm?

(f) Suppose that you modify the gradient computation to return an incorrect gradient; that is, introduce a small error. How does this impact the behavior of the algorithm?

(g) For the quadratic functions, replace the steplength algorithm by an exact linesearch (see Example 9.12).

   (i) Compare the behavior of the Quasi-Newton Methods using an exact linesearch to the ones implemented above.

   (ii) Does the implementation with an exact linesearch satisfy the property of quadratic termination?

10.14 *Linear Conjugate-gradient Method—implementation.* Implement and test your implementation as follows:

(a) Implement the Linear Conjugate-gradient Method described in this chapter. Modify the convergence check so that the algorithm terminates without a solution if the number of iterations reaches a pre-specified maximum value denoted by K_{max}.

(b) Examine the solution process; at the very least, print out the following quantities at each iteration: $k, N_f, N_g, f_k, ||g_k||, ||x_k - x^*||$, $||f_k - f^*||$, α_k; where, N_f and N_g are the cumulative number of function and gradient evaluations at the end of iteration k.

(c) Apply the algorithm to the two quadratic functions in Exercise 10.9.

(d) *Analysis of your results.*

    (i) Is each search direction conjugate with respect to the previous search directions?

    (ii) Examine the behavior on the two quadratic functions. Does the algorithm satisfy the property of quadratic termination? Explain why or why not.

    (iii) Do the convergence rates on the functions match expectations? Explain you answer based on the theory.

(e) Suppose that you modify the gradient computation to return an incorrect gradient; that is, introduce a small error. How does this impact the behavior of the algorithm?

(f) Apply the method to Rosenbrock's function and Powell's function after replacing the steplength calculation by the Simple Steplength Algorithm 9.8 implemented in Exercise 9.7.

10.15   Test your implementations on additional functions such as those specified in [138].

11. OPTIMALITY CONDITIONS

Overview

This chapter is concerned with material that is central to the theory of non-linear optimization: the conditions (equations and inequalities) that must hold at a local minimizer of a (smooth) nonlinear optimization problem. These are called *necessary conditions of optimality*. In some (but by no means all) cases, these same conditions are also *sufficient for optimality*, meaning that a point satisfying them is guaranteed to be a local minimizer of the problem. In this chapter we discuss these and other conditions aimed at identifying local minimizers.

We have already seen two examples of optimality conditions: the feasibility and complementary slackness conditions of linear programming (see Section 5.1) and the optimality conditions of unconstrained optimization (see Section 9.6). The content of this chapter extends the optimality conditions covered earlier in this book to constrained nonlinear optimization. It should be emphasized that these topics have a long history and are well documented in scientific journals, scholarly monographs, and textbooks. We have attempted to keep our discussion as accessible as possible.

Optimality conditions for optimization problems are useful for two reasons. First, they inform our search for a local minimizer by telling us what conditions such a point would have to satisfy. Second, they help us to decide whether a given point is or is not a locally optimal solution. In favorable instances, they can also assure us that a local minimizer is a global minimizer, which is usually what we want. Such favorable instances normally hold when the minimand and feasible region are convex[1]; see Exercise 11.26. Many more properties of convex functions are discussed in the Appendix.

The chapter closes with a discussion of an elementary duality theory for nonlinear programming. Due to the nature of the functions involved, this duality theory of nonlinear programming is more complicated than its linear programming counterpart. Nevertheless, when restricted to the linear case, this nonlinear programming duality theory reduces to that of linear programming.

---

[1] A feasible region can be convex even though the constraints defining it are not all convex, as in the case of $\{x \in R : x \geq 0, \ (x-1)^3 \leq 0\}$.

© Springer Science+Business Media LLC 2017

R.W. Cottle and M.N. Thapa, *Linear and Nonlinear Optimization*,
International Series in Operations Research & Management Science 253,
DOI 10.1007/978-1-4939-7055-1_11

The initial approach to duality presented here (and in most introductory treatments of this subject) should in effect be regarded mainly as an *application* of the optimality theory discussed earlier in this chapter. As in LP, the duality theory will involve a *pair* of optimization problems, one called the *primal problem*, the other called the *dual problem*. In our presentation, we state the primal problem in a rather general way. Later we restrict it, thereby permitting some conclusions that are stronger and more in line with what are considered essential to a genuine duality theory.

11.1   Statement of the problem

The general form of the problems under consideration here is

$$
\begin{aligned}
\text{minimize} \quad & f(x) \\
\text{subject to} \quad & c_i(x) \geq 0, \quad i \in \mathcal{I} \\
& c_i(x) = 0, \quad i \in \mathcal{E} \\
& x \in R^n.
\end{aligned}
\tag{11.1}
$$

When we speak of *nonlinear optimization*, we mean that at least one of the functions in the problem (11.1) is nonlinear.

The index sets $\mathcal{I}$ and $\mathcal{E}$ are assumed to be finite and disjoint (i.e., to have no elements in common). But it should be noted that either or both could be vacuous (empty). When $\mathcal{I} = \emptyset$ and $\mathcal{E} = \emptyset$, we have an *unconstrained optimization problem*. Other variants are the *equality-constrained optimization problem* in which $\mathcal{I} = \emptyset$ and the *inequality-constrained optimization problem* in which $\mathcal{E} = \emptyset$. If both index sets are nonempty, we take $\mathcal{I} = \{1, \ldots, m\}$ and $\mathcal{E} = \{m+1, \ldots, m+\ell\}$. When $\mathcal{I}$ is empty and $\mathcal{E}$ is not, we take m to be zero, thereby making $\mathcal{E} = \{1, \ldots, \ell\}$.

The formulation (11.1) is rich enough to encompass the linear programming problem. Many problem classes take their names from properties of the functions used to specify them. For instance, the *quadratic programming problem* has a quadratic objective function f and affine constraint functions c_i, $i \in \mathcal{I} \cup \mathcal{E}$. In a *convex programming problem*, the objective function f and the constraint functions c_i, $i \in \mathcal{I}$ are required to be concave (or the $-c_i(x)$ are required to be convex) while the c_i, $i \in \mathcal{E}$ are required to be affine.[2]

---

[2]Such functions are convex, of course, but the solution set of a system of nonlinear *equations* is not usually convex even when the functions involved are convex.

11.2   First-order optimality conditions

In this section, we develop necessary conditions of local optimality. We assume that all the functions defining the problem (11.1) are continuously differentiable. Here the functions need not be smooth in the restrictive sense of having continuous *second* partial derivatives. Hence we are assuming they are of class C^1 rather than of class C^2. Because these results involve only first derivatives, they are called *first-order optimality conditions*. We begin with a "classical" case.

Equality-constrained nonlinear optimization

The classical equality-constrained case is of the form

$$\begin{aligned}
\text{minimize} \quad & f(x) \\
\text{subject to} \quad & c_i(x) = 0, \quad i \in \mathcal{E} \\
& x \in R^n.
\end{aligned} \tag{11.2}$$

Here we have $\mathcal{E} = \{1, \ldots, \ell\}$. As already indicated, we also assume the functions f and $c_1, \ldots, c_\ell$ are continuously differentiable. We denote by c the column ℓ-vector $(c_1, c_2, \ldots, c_\ell)$.

If the constraints $c_i(x) = 0$ were not present, we would have the unconstrained optimization problem for which we know that the vanishing of the gradient vector $\nabla f(x^*)$ is a necessary condition for x^* to be a local minimizer of f. How can we take account of the constraints?

There is a special case in which the behavior of $f(x)$ for x satisfying the constraints $c_i(x) = 0$, $i \in \mathcal{E}$ is quite straightforward. This is where the constraints of (11.2) are affine and linearly independent. This means the constraints are of the form

$$c(x) = Ax - b = 0, \quad A \in R^{\ell \times n}, \quad \text{and} \quad \text{rank}(A) = \ell. \tag{11.3}$$

Under these circumstances, after possibly reordering the columns of A, we can write

$$A = [B \, N] \quad \text{and} \quad x = (x_B, x_N)$$

where B is a basis in A. We can then write

$$x_B = B^{-1}b - B^{-1}Nx_N, \tag{11.4}$$

and use this to substitute for x_B in the objective function. Doing this, we obtain

$$f(x) = f(x_B, x_N) = f(B^{-1}b - B^{-1}Nx_N, x_N) = F(x_N). \qquad (11.5)$$

The function F is now an *unrestricted* differentiable function of the variables in the subvector x_N. Thus,

$$\nabla F(x_N) = (-B^{-1}N)^T \nabla_B f(x) + \nabla_N f(x) = 0$$

is a necessary condition of optimality for the unconstrained problem. If x_N can be found, (11.4) can be used to recover the subvector x_B.

Example 11.1: EQUALITY-CONSTRAINED PROBLEM. In this example we illustrate how to explicitly reduce an equality-constrained optimization problem to an unconstrained optimization problem. Consider the following simple optimization problem

$$\begin{aligned}
\text{minimize} \quad & x_1^2 + x_2^2 \\
\text{subject to} \quad & x_1 - x_2 = 1.
\end{aligned}$$

From the single equality constraint we obtain $x_2 = x_1 - 1$ (equivalently, we could have made x_1 the basic variable). Substituting for x_2 in the above optimization problem we obtain

$$\text{minimize} \quad 2x_1^2 - 2x_1 + 1,$$

an unconstrained optimization problem in x_1.

The observations above pertaining to linearly independent affine constraints are primarily of theoretical interest. From the algorithmic standpoint the elimination procedure is unnecesary as we shall see in Chapter 12.

In the expression (11.4), x_B is dependent whereas x_N is independent. Moreover, in this affine case, the relationship is globally valid. In other words, if $\bar{x}$ is any point satisfying the system $Ax = b$, then $\bar{x}_B$ and $\bar{x}_N$ satisfy (11.4). The same equation represents the dependent variables in terms of the independent variables. This statement cannot be made when the problem has nonlinear equality constraints.

To arrive at what can be said, we mention an important mathematical concept: the $\ell \times n$ *Jacobian matrix* defined as

$$J_c(x) = \left[\frac{\partial c_i(x)}{\partial x_j}\right]. \tag{11.6}$$

The rows of $J_c(x)$ are the transposes of the gradients $\nabla c_i(x)$. When $m = n$, the determinant of the Jacobian matrix $J_c(x)$ is called the *Jacobian* of c at x. Notice that if c is the mapping defined in (11.3), then $J_c(x) = A$.

The regularity condition LICQ

Let $\bar{x}$ be a vector satisfying the constraints of (11.2). If $J_c(\bar{x})$ has linearly independent rows, we say that $\bar{x}$ is a *regular point* of c and that the *linear independence constraint qualification* (LICQ) is satisfied at $\bar{x}$. As will be seen in Example 11.2, the LICQ is not always satisfied everywhere. The LICQ is an example of what is called a *regularity condition*. More will be said about this and other first-order regularity conditions (constraint qualifications) later.

The LICQ plays a key role in an important result: the *Implicit Function Theorem*. We state it here in a manner more closely aligned with the present application than what you might find in the mathematics literature. In essence, this theorem gives sufficient conditions for existence of a function that *locally* expresses a set of basic variables in terms of nonbasic variables, i.e., a nonlinear analog of (11.4). For a proof of this classical theorem, see, for instance, Rudin [172, pp. 195–197] or Krantz and Parks [115].

Theorem 11.1 (Implicit Function Theorem)   *Suppose $c : R^n \to R^\ell$ is a mapping of class C^1 and suppose $\bar{x}$ is a point in R^n that satisfies the equation $c(\bar{x}) = 0$. If c satisfies the LICQ at $\bar{x}$, and (after a possible reordering of the variables) $\bar{x} = (\bar{x}_B, \bar{x}_N)$ where $\bar{x}_B \in R^\ell$ and the $\ell \times \ell$ matrix $(J_c(\bar{x}))._B$ is nonsingular, then there is an open neighborhood U of $\bar{x}_N$ and a unique continuously differentiable mapping φ from U to R^ℓ such that $\varphi(\bar{x}_N) = \bar{x}_B$ and $c(\varphi(x_N), x_N) = 0$ for all x_N in U.*

Theorem 11.1 is used in proving Theorem 11.2 the main result of this section which is sometimes called the *Lagrange Multiplier Rule*. The idea for Theorem 11.2 was first presented in a book by Lagrange published in

1788 (See Pulte [163]). For a detailed modern proof of the theorem, see, for example, Bertsekas [12, p. 259]. See also Fraser [73].

Theorem 11.2 (Lagrange Multiplier Rule) *If x^* is a local minimizer for the equality-constrained problem (11.2) and c satisfies the LICQ at x^*, then there exist numbers $\lambda_1^*, \ldots, \lambda_\ell^*$ such that*

$$\nabla f(x^*) - \sum_{i=1}^{\ell} \lambda_i^* \nabla c_i(x^*) = 0 \qquad (11.7)$$

and

$$c_i(x^*) = 0, \quad i = 1, \ldots, \ell. \qquad (11.8)$$

Taken together, (11.7) and (11.8) say that (x^*, λ^*) is a stationary point of the *Lagrangian function*

$$L(x, \lambda) = f(x) - \lambda^\mathsf{T} c(x) = f(x) - \sum_{i=1}^{\ell} \lambda_i c_i(x). \qquad (11.9)$$

associated with the optimization problem (11.2).

The numbers λ_i in (11.9) are called *Lagrange Multipliers*, and the application of the above necessary conditions of local optimality in the equality-constrained problem is called the *Method of Lagrange Multipliers*[3]. Note how the stationary point condition $\nabla L = 0$ gives $\ell + n$ equations in $\ell + n$ unknowns $x_1, \ldots, x_n$ and $\lambda_1 \ldots, \lambda_\ell$. These $\ell + n$ equations facilitate the determination of the decision variables x_j and the multipliers λ_i.

Equation (11.7) implies that both $\nabla f(x^*)$ and $-\nabla f(x^*)$ lie in the column space of the transpose of the Jacobian matrix $J_c(x^*)$. The following example shows that the conclusion of Theorem 11.2 (the Lagrange Multiplier Rule) need not hold when the LICQ is not satisfied.

Example 11.2: THE NEED FOR THE LICQ IN THEOREM 11.2. In the problem

$$\begin{array}{ll} \text{minimize} & x_1 \\ \text{subject to} & x_1^2 + (x_2 - 1)^2 - 1 = 0 \\ & x_1^2 + (x_2 + 1)^2 - 1 = 0, \end{array}$$

---

[3]This approach was developed by J.L. Lagrange (1736-1813). A great mathematician, Joseph-Louis Lagrange was baptized as Giuseppe Lodovico Lagrangia and is claimed by the Italians as well as the French.

$x^* = (0,0)$ is optimal (for *any* objective) since it is the only feasible solution. But $\nabla f(x^*) = (1,0)$ is not in the column space of the transpose of the Jacobian matrix $J_c(x^*)$. Hence the conclusion of Theorem 11.2 does not hold in this case, meaning that the theorem needs to include a regularity condition such as the LICQ.[4]

In the next example, we use of the Method of Lagrange Multipliers and a simple transformation to obtain a closed-form solution of an elementary geometric optimization problem. As it happens, this problem has an affine objective function and a single nonlinear constraint. In such circumstances, the LICQ is not an issue.

Example 11.3: USING THE METHOD OF LAGRANGE MULTIPLIERS. Consider the problem of minimizing a nonconstant affine function over a sphere. The problem is of the form

$$\begin{aligned} \text{minimize} \quad & d^T x + \kappa \\ \text{subject to} \quad & \|x - a\|^2 = \rho^2. \end{aligned}$$

It is clear that the solution exists by virtue of the continuity of the objective function and the compactness of the feasible region (sphere). In fact, it is intuitively clear that the solution is the point where the halfline in the direction $-d$ meets the sphere. Nonetheless, we proceed with the Method of Lagrange Multipliers.

To tackle this problem, it is helpful to transform it to something a bit simpler. In the first place, the analysis is eased by assuming that $a = 0$ or equivalently by changing variables so that $y = x - a$. The objective function becomes $d^T y + \kappa + d^T a$, and the constraint becomes $y^T y - \rho^2 = 0$. Next, we (temporarily) disregard the additive constant $\kappa + d^T a$ since it plays no role in finding the minimizer. It affects only the optimal function value, not the location of the optimal solution. (Another simplification that could be made is to assume that the 2-norm of d is 1, but we really need not do so.)

The Lagrangian function for the transformed problem is

$$L(y, \lambda) = d^T y - \lambda \left(y^T y - \rho^2 \right).$$

[4]As pointed out by Hestenes [93, p. 43], an exceptional case should be noted here. When the functions $c_i(x)$ in (11.8) are all affine, there is no need to assume that the LICQ is satisfied in (Lagrange's) Theorem 11.2. See Exercise 11.6.

The stationarity conditions are then

$$d - 2\lambda y = 0$$

$$y^{\mathrm{T}}y - \rho^2 = 0.$$

The first of these two equations gives

$$y = (1/2\lambda)d. \tag{11.10}$$

We can now write

$$\rho^2 = y^{\mathrm{T}}y = (1/4\lambda^2)d^{\mathrm{T}}d.$$

It follows from this equation that

$$\lambda = \lambda^* = \pm\frac{\|d\|}{2\rho}.$$

As a consequence of this and (11.10) we find that

$$y = \pm\frac{\rho d}{\|d\|}.$$

Since $\rho > 0$ (it is the radius of the sphere) and we are minimizing, we take the minus sign in the formula above. We can now easily transform this answer back in terms of the original variables x. We obtain

$$x = x^* = a + y = a - \frac{\rho d}{\|d\|}.$$

Example 11.4: THE METHOD OF LAGRANGE MULTIPLIERS AGAIN. Let us return to the nearest-point (projection) problem appearing on page 247:

$$\text{minimize} \quad \tfrac{1}{2}\|x - u\|^2$$
$$\text{subject to} \quad a^{\mathrm{T}}x = b.$$

This problem has a unique optimal solution, x^*. It is a quadratic program with the (strictly convex) objective function $\tfrac{1}{2}x^{\mathrm{T}}x - u^{\mathrm{T}}x + \tfrac{1}{2}u^{\mathrm{T}}u$. So the Lagrangian is

$$L(x, \lambda) = \tfrac{1}{2}x^{\mathrm{T}}x - u^{\mathrm{T}}x + \tfrac{1}{2}u^{\mathrm{T}}u - \lambda(a^{\mathrm{T}}x - b),$$

and at an optimal solution we must have

$$x - u - \lambda a = 0. \tag{11.11}$$

Since $a^Tx = b$, we have, after multiplying Equation (11.11) by a^T,

$$a^Tx - a^Tu - \lambda a^Ta = b - a^Tu - \lambda a^Ta = 0.$$

From the latter equation, we see that the solution is

$$\lambda = \lambda^* = \frac{1}{a^Ta}(b - a^Tu).$$

After substituting of λ^* in (11.11), the solution of the projection problem is given by

$$x = x^* = u + a\left(\frac{b - a^Tu}{a^Ta}\right). \tag{11.12}$$

This establishes the claim (made on page 247) about the closed-form solution of this problem.

Inequality-constrained nonlinear optimization

We now turn to problems of the form

$$\begin{aligned} \text{minimize} \quad & f(x) \\ \text{subject to} \quad & c_i(x) \geq 0, \quad i \in \mathcal{I} \\ & x \in R^n. \end{aligned} \tag{11.13}$$

In the present circumstances, we assume $\mathcal{I} = \{1, \ldots, m\}$ and continue to assume that all the functions are differentiable on R^n. It would actually be enough for the functions to be differentiable on an open subset of R^n containing the feasible region, but, in practice, that set may be difficult to determine.

Active constraints

A constraint $c_i(x) \geq 0$ is said to be *active* at $\bar{x}$ if $c_i(\bar{x}) = 0$. For a feasible vector $\bar{x}$, let $\mathcal{A}(\bar{x})$ be the set of subscripts of all constraints that are active at $\bar{x}$. (We mention in passing that $\mathcal{A}(\bar{x})$ and $\text{supp}(c(\bar{x}))$ are complementary subsets of $\mathcal{I}$.)

The regularity condition LICQ again

For the inequality-constrained problem above, we say that the LICQ is satisfied at a feasible point $\bar{x}$ if the gradients of the constraints that are active at $\bar{x}$ are linearly independent. Another way to phrase this idea is to say that the LICQ holds at $\bar{x}$ if the vectors $\nabla c_i(\bar{x})$ for $i \in \mathcal{A}(\bar{x})$ are linearly independent. The transposes of these vectors form a submatrix of the Jacobian matrix $J_c(\bar{x})$; that submatrix has linearly independent rows if and only if the LICQ is satisfied.

The need for some sort of regularity condition will be illustrated after we present a key theorem on first-order optimality conditions. For now, we simply remark that there are other regularity conditions—all of them less restrictive than the LICQ—which were actively studied for many years. Although much could be said about this subject, we limit our discussion of it to a small portion of what is known. Two valuable references for further study of this subject are Avriel [7] and Bazarra, Sherali, and Shetty [9].

First-order necessary conditions of optimality

Theorem 11.3 *If x^* is a local minimizer for the inequality-constrained problem (11.13) and the LICQ is satisfied by the set of all active constraints at x^*, then there exists $\lambda^* \in R_+^m$ such that $\lambda_i^* c_i(x^*) = 0$ for all $i = 1, \ldots, m$ and x^* is a stationary point of*

$$L(x, \lambda^*) = f(x) - \sum_{i=1}^{m} \lambda_i^* c_i(x). \tag{11.14}$$

It is helpful to put the conclusion of this result into the form of the following mixed system:

$$\nabla f(x) - \sum_{i=1}^{m} \lambda_i \nabla c_i(x) = 0$$

$$\lambda_i \geq 0, \quad i = 1, \ldots, m \tag{11.15}$$

$$\lambda_i c_i(x) = 0, \quad i = 1, \ldots, m.$$

The theorem states that if x^* is a local minimizer of (11.13) and the LICQ is satisfied by the constraints c_i at that point, then there exists a nonnegative vector $\lambda^* = (\lambda_1^*, \ldots, \lambda_m^*)$ such that together x^* and λ^* satisfy the system (11.15). These are *first-order necessary conditions of optimality*. They hold under suitable regularity conditions (of which there are many), one of which is the LICQ used in the theorem above.

The first-order necessary conditions of optimality given in (11.15) are known as the *Karush-Kuhn-Tucker (KKT) conditions*. But it would not be quite correct to call the theorem above the KKT Theorem because Karush (1939) and Kuhn and Tucker (1951) did not use the LICQ as a regularity condition.[5] The constraint qualification they used is more difficult to comprehend (and verify) than the LICQ; this is why we have deferred discussion of it up to this point.

The *Karush-Kuhn-Tucker constraint qualification* (KKTCQ) can be put as follows: Let $x = \bar{x}$ be a feasible point for the inequality-constrained problem (11.13) in which all the functions are differentiable. Assume $\mathcal{A}(\bar{x}) \neq \emptyset$. We say the KKTCQ is satisfied at $\bar{x}$ if for every nonzero solution v of the inequality system

$$v^{\mathrm{T}} \nabla c_i(\bar{x}) \geq 0 \quad \text{for all } i \in \mathcal{A}(\bar{x}),$$

there exists a differentiable curve $\gamma : [0, 1] \to R^n$ whose image is contained in the feasible region such that

$$\gamma(0) = \bar{x}, \quad \gamma'(0) = \kappa v \quad \text{for some } \kappa > 0.$$

This is another instance of a first-order constraint qualification.

If $x = \bar{x}$ is a feasible solution of (11.13) and there exists a vector $\lambda = \bar{\lambda}$ such that $(\bar{x}, \bar{\lambda})$ satisfies the KKT conditions, we could refer to $(\bar{x}, \bar{\lambda})$ as a *KKT pair* and call $\bar{x}$ a *KKT stationary point*.

The proof of the actual KKT Theorem (i.e., Theorem 11.3 with the LICQ replaced by the KKTCQ) rests on the fact that, when x^* is a local minimizer of problem (11.13), the homogeneous linear inequality system

$$v^{\mathrm{T}} \nabla f(x^*) < 0 \tag{11.16}$$

$$v^{\mathrm{T}} \nabla c_i(x^*) \geq 0 \qquad \text{for all } i \in \mathcal{A}(x^*) \tag{11.17}$$

cannot have a solution. Farkas's Lemma then guarantees the existence of scalars $\lambda_i^* \geq 0$, $i \in \mathcal{A}(x^*)$, such that

$$\nabla f(x^*) = \sum_{i \in \mathcal{A}(x^*)} \lambda_i^* \nabla c_i(x^*).$$

---

[5]For some historical background on these matters, see the publications by Kuhn [117], Cottle [34], and Giorgi and Kjeldsen [85].

Setting $\lambda_i^* = 0$ for all $i \notin \mathcal{A}(x^*)$, we find that the KKT conditions stated in (11.15) hold.

There is another well-known theorem on first-order necessary conditions of optimality called Fritz John's Theorem [102][6]. It, too, is based on an alternative theorem, in this case the one due to P. Gordan [89] that was stated in Exercise 5.11. John's Theorem requires no constraint qualification at all. As a consequence, there are circumstances when the conclusion of the theorem is too weak to be useful. Yet there are others (for example, the LICQ) in which John's Theorem produces essentially same conditions as Theorem 11.3. We develop these statements through Exercise 11.18.

Inequalities and the form of the Lagrangian

Sometimes an inequality constraint of a minimization problem is most naturally expressed in the form $c_i(x) \leq 0$. In such cases, we could define $\bar{c}_i(x) = -c_i(x)$ and use $\bar{c}_i(x) \geq 0$ thereby putting the constraint in the form (11.13). An equivalent way to handle this situation is to leave the inequality in its original form and instead modify the Lagrangian to account for the different type of inequality constraints. It is straightforward to see that this only requires changing the sign in front of λ_i in the formula for the Lagrangian. Examples 11.5 and 11.6 utilize this idea.

Geometrical interpretation of the KKT conditions

The stationary point condition

$$\nabla f(\bar{x}) - \sum_{i=1}^{m} \lambda_i \nabla c_i(\bar{x}) = 0$$

and the sign restriction $\lambda \geq 0$ imply that the vector $\nabla f(\bar{x})$ is a nonnegative linear combination of the gradients $\nabla c_i(\bar{x})$. The complementary slackness conditions $\lambda_i c_i(\bar{x}) = 0$ for all i imply that only the gradients of active constraints need to be used in this nonnegative linear combination. Figure 11.1 will help to visualize the idea. Note that in this figure, $\mathcal{A}(\bar{x}) = \{1, 3\}$.

---

[6]As Kuhn [117, p. 15] rightly points out, essentially the same theorem appears in the M.S. thesis of William Karush [108] nearly a decade before that of Fritz John. One difference between the two results is that Karush's entire thesis deals with finite-dimensional spaces only whereas in John's theorem, the indices (subscripts) of the constraint functions are drawn from a compact metric space; see Cottle [30].

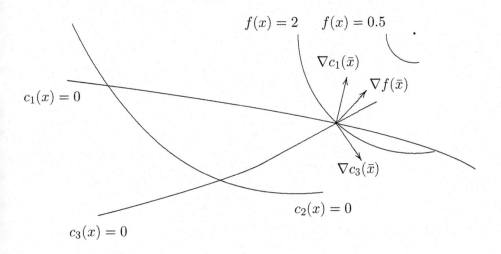

$$f(x) = 2 \qquad f(x) = 0.5$$

$$\nabla c_1(\bar{x})$$

$$\nabla f(\bar{x})$$

$$c_1(x) = 0$$

$$\nabla c_3(\bar{x})$$

$$c_2(x) = 0$$

$$c_3(x) = 0$$

Figure 11.1: Geometrical interpretation of the KKT conditions.

Example 11.5: USING THE KKT CONDITIONS. This example is a simple variant of the one in Example 11.3. This time we consider the minimization of a nonconstant affine function over a closed ball. Let us take the problem to be

$$\begin{aligned} \text{minimize} \quad & d^{\mathrm{T}}x + \kappa \\ \text{subject to} \quad & \|x - a\|^2 \le \rho^2. \end{aligned}$$

We begin with a few preliminary observations. First, there is only one constraint, and this constraint has a nonempty compact solution set. Since the objective function is continuous, it attains its minimum and maximum values over the feasible region. This means that a minimizer x^* exists. If a minimizer x^* exists and $\mathcal{A}(x^*) = \emptyset$, then there is no need to discuss the LICQ. In that case x^* minimizes $d^{\mathrm{T}}x + \kappa$ over an open set, which implies that the gradient of the objective function must vanish at x^*. This, however, would imply, first, that $d = 0$, and, second, that the objective function is constant, a contradiction. So in fact $\mathcal{A}(x^*) \ne \emptyset$. Ordinarily, this would bring up the question of whether the LICQ is satisfied by the constraint at x^*. But because there is only one constraint, and the gradient of this constraint is nonzero[7], we are assured that the LICQ holds; linear

---

[7]If the gradient of the constraint were zero, it would follow that $x^* = a$ which is impossible unless $\rho = 0$ (i.e., the ball reduces to a point), in which case the objective function is constant on the feasible region.

dependence is impossible in this case. For this problem the Lagrangian is

$$L(x, \lambda) = d^{\mathrm{T}}x + \kappa + \lambda \left(\|x - a\|^2 - \rho^2 \right),$$

and the KKT conditions are

$$d + 2\lambda(x - a) = 0$$
$$\lambda \geq 0$$
$$\lambda(x^{\mathrm{T}}x - 2a^{\mathrm{T}}x - a^{\mathrm{T}}a - \rho^2) = 0.$$

It is clear that $\lambda \neq 0$; otherwise $d = 0$ which is ruled out by our hypothesis on the nonconstancy of the objective function. Having $\lambda > 0$ enables us to write

$$x - a = -(1/2\lambda)d.$$

Taking the transposes of both sides and carrying out the obvious multiplications, we get

$$\rho^2 = (x - a)^{\mathrm{T}}(x - a) = (1/4\lambda^2)d^{\mathrm{T}}d,$$

from which we find that

$$\lambda = \lambda^* = \pm\frac{\|d\|}{2\rho}.$$

The alternative $-\|d\|/(2\rho)$ is clearly impossible, so $\lambda = \|d\|/(2\rho)$. From this and the stationarity condition, we can write

$$x = x^* = -\frac{1}{2\lambda}d + a = -\frac{\rho}{\|d\|}d + a.$$

It now follows that the point x^* above is the solution to this optimization problem.

This rather belabored discussion of a very straightforward optimization problem illustrates the point that the KKT conditions (along with the feasibility condition) can sometimes be used to determine both the Lagrange multiplier and the optimal solution.

Example 11.6: USING THE KKT CONDITIONS ON A HARDER PROBLEM.
Let us now consider a slightly more challenging problem: the minimization of a nonconstant affine function on the intersection of two closed balls. Let u and v be given n-vectors.

$$
\begin{aligned}
\text{minimize} \quad & a^{\mathrm{T}}x + \kappa \\
\text{subject to} \quad & \|x - u\|^2 \leq \rho_1^2 \\
& \|x - v\|^2 \leq \rho_2^2,
\end{aligned}
$$

where ρ_1 and ρ_2 are the respective radii of the two balls. In this case, the feasible region will certainly be compact, but there is a question of feasibility. Do these closed balls intersect? They do if and only if the distance between their centers does not exceed the sum of their radii, i.e., satisfies

$$\|u - v\| \leq \rho_1 + \rho_2.$$

We assume that this inequality holds. In this model we have $\leq$ constraints and

$$c_1(x) = (x - u)^{\mathrm{T}}(x - u) - \rho_1^2$$
$$c_2(x) = (x - v)^{\mathrm{T}}(x - v) - \rho_2^2,$$

for which the gradients are given by

$$\nabla c_1(x) = 2(x - u)$$
$$\nabla c_2(x) = 2(x - v).$$

If at any point x, these two gradients are linearly dependent, it means that the three points x, u, v are collinear (i.e., lie on a straight line).

Let x^* be a locally optimal solution to the stated problem. Just as in Example 11.5, $\mathcal{A}(x^*)$ cannot be empty. We therefore have three cases:

$$\mathcal{A}(x^*) = \{1\}, \quad \mathcal{A}(x^*) = \{2\}, \quad \mathcal{A}(x^*) = \{1, 2\}.$$

In the first case, the first ball lies in the interior of the second ball. In the second case, the second ball lies in the interior of the first ball. Checking for these two conditions is a matter of analytic geometry. For example, if $\|u - v\| + \rho_1 < \rho_2$, then the first ball lies in the interior of the second. The other case is analogous. In these two cases, the approach to finding x^* is much the same as in Example 11.5 where we were able to solve the problem in closed form. We move on to the third case, wherein a closed-form solution seems out of reach.[8]

We are now assuming that both constraints are active at x^*. If $\nabla c_1(x^*)$ and $\nabla c_2(x^*)$ are linearly independent, Theorem 11.3 says that there will exist scalars $\lambda_1 = \lambda_1^*$ and $\lambda_2 = \lambda_2^*$ such that

$$a + 2\lambda_1^*(x^* - u) + 2\lambda_2^*(x^* - v) = 0$$
$$\lambda_i^* \geq 0, \quad i = 1, 2$$
$$\lambda_i^* c_i(x^*) = 0, \quad i = 1, 2.$$

---

[8]Note that in dimensions above 2, the intersection of two spheres is another sphere rather than a pair of points as it is in the case of 2 dimensions.

If $\nabla c_1(x^*)$ and $\nabla c_2(x^*)$ are linearly dependent (as they would be if the two balls were tangent to each other as in Example 11.2), then it can happen that $-a$ is not a nonnegative linear combination of the gradients $\nabla c_1(x^*)$ and $\nabla c_2(x^*)$. In this case, there would be no solution of the KKT conditions (even though x^* is a local minimizer for the problem).

An exceptional case

We mentioned earlier that, in addition to the LICQ, there are other regularity conditions. The one used by Karush, Kuhn and Tucker in conjunction with (11.13) allows one to show that when the functions $c_i(x)$, $i = 1, \ldots, m$ are *all affine*, the KKT constraint qualification is always satisfied. Hence *in the all-affine case, no regularity condition is required in order to assert the existence of the Lagrange multipliers in* (11.15).

Example 11.7: KKT CONDITIONS FOR QUADRATIC PROGRAMMING.
One of the most frequently used applications of the KKT conditions occurs in the case of the quadratic programming (QP) problem which can be stated in numerous ways, the variety stemming from the nature of the constraints. In this example, we have only linear inequality constraints, but in applications there can be equations or a mixture of linear equations and linear inequalities. No matter what the form of the QP problem, there is no need to impose a regularity condition, as we have noted in the preceding paragraph.

Consider a QP of the form

$$\begin{aligned}\text{minimize} \quad & \tfrac{1}{2}x^{\mathrm{T}}Hx + a^{\mathrm{T}}x \\ \text{subject to} \quad & Ax \geq b,\end{aligned} \tag{11.18}$$

where, as usual, the matrix H is symmetric. To put this in the form of (11.13), we write $c(x) = Ax - b$. The Lagrangian for this problem is then

$$L(x, \lambda) = \tfrac{1}{2}x^{\mathrm{T}}Hx + a^{\mathrm{T}}x - \lambda^{\mathrm{T}}(Ax - b),$$

and the KKT conditions are

$$Hx + a - A^{\mathrm{T}}\lambda = 0$$
$$\lambda \geq 0$$
$$\lambda^{\mathrm{T}}(Ax - b) = 0.$$

As we know, these first-order conditions—along with the feasibility of x—are necessary conditions of local optimality. There are second-order conditions for QP that are both necessary and sufficient for optimality. These will be discussed in Section 11.3.

The stationarity condition $Hx + a - A^{\mathrm{T}}\lambda = 0$ can sometimes be used to "reduce" the optimality conditions by eliminating x and obtaining a system of conditions in λ alone. That type of system is an instance of a *Linear Complementarity Problem*. (See [37].)

11.3   Second-order optimality conditions

Just as in the case of *unconstrained* optimization, there are conditions involving second derivatives that help in identifying local minima (or maxima) of *constrained* optimization problems. There are necessary conditions, and there are sufficient conditions (which are not always the same).[9] We discuss these matters in this section, beginning with a little terminology.

Feasible directions

Let X be a proper subset of R^n. At this stage, we are not specifying the manner in which X is defined. In particular, there are no explicitly stated constraints like those of (11.1) in the picture. Let $\bar{x}$ be a given element of X. A nonzero vector $p \in R^n$ is called a *feasible direction* at $\bar{x}$ if there exists a positive scalar $\bar{\alpha}$ such that $\bar{x} + \alpha p \in X$ for all $\alpha \in [0, \bar{\alpha}]$. This condition requires that *all* points on some closed line segment of positive length beginning at $\bar{x}$ belong to X. If $\bar{x}$ belongs to the *interior* of X, then all directions are feasible at $\bar{x}$, whereas this is not the case when $\bar{x}$ belongs to the boundary of X.

When augmented by the zero vector (which, properly speaking, is not a direction), the set of all feasible directions at a feasible point is a *cone*, denoted $\mathcal{F}(\bar{x})$. In this regard, see Exercise 11.19 for an interesting example.

Theorem 11.4 (Second-order necessary conditions)   *If the function f is twice-continuously differentiable on a set X in R^n, and $x^* \in X$ is a local minimizer of f, then for any feasible direction p at x^**

---

[9]Much of the material presented in this section is developed in the important early paper by McCormick [134]. See also Fiacco and McCormick [64].

(i) $\nabla f(x^*) \cdot p \geq 0$

(ii) if $\nabla f(x^*) \cdot p = 0$, then $p^T \nabla^2 f(x^*) p \geq 0$.

It is worthwhile commenting on the meaning of this theorem. Condition (i) means that there are no *feasible* directions of descent at x^*. Condition (ii) says that all directions p perpendicular to the gradient of f at x^* make the quadratic form associated with the Hessian matrix take on a nonnegative value. The set of all such directions constitutes an $(n-1)$-dimensional linear space. So condition (ii) can be interpreted as saying that the quadratic form associated with the Hessian matrix $\nabla^2 f(x^*)$ is positive semidefinite on the aforementioned linear space. This is *not* the same thing as saying the Hessian matrix $\nabla^2 f(x^*)$ is truly positive semidefinite however.

It follows from the above theorem that if x^* is a local minimizer of f belonging to the *interior* of X, then for *all $p \in R^n$*

(i) $\nabla f(x^*) \cdot p = 0$

(ii) $p^T \nabla^2 f(x^*) p \geq 0$.

In this case, condition (ii) means that $\nabla^2 f(x^*)$ must be positive semidefinite.

Next we present some analogous sufficient conditions for optimality.

Theorem 11.5 (Second-order sufficient conditions)   *Let f be a twice-continuously differentiable function on the set $X \subset R^n$. Let $\bar{x}$ be an interior point of X. If*

(i) $\nabla f(\bar{x}) = 0$

(ii) $\nabla^2 f(\bar{x})$ *is positive definite,*

then $\bar{x}$ is a strict local minimizer of f on X.

Looking back, we find that the theorems stated on necessary conditions and sufficient conditions for local minima are very much like those found in the study of unconstrained optimization. The difference has to do with what search directions are allowed. In the unconstrained case, $X = R^n$, and all search directions are feasible.

When inequalities and/or equations are used to define the feasible region X as in (11.1), one has additional conditions with which to work. To illustrate how this happens, we first consider an equality-constrained problem.

$$\begin{aligned} \text{minimize} \quad & f(x) \\ \text{subject to} \quad & c_i(x) = 0, \quad i \in \mathcal{E} \\ & x \in R^n. \end{aligned} \tag{11.19}$$

For concreteness (and ease of notation), assume that $\mathcal{E} = \{1, \ldots, \ell\}$. Let $c(x) = (c_1(x), \ldots, c_\ell(x))$ be differentiable on R^n. So c is a mapping from R^n to R^ℓ. Let $\bar{x}$ satisfy the system of equations $c(x) = 0$. Recall that $J_c(\bar{x})$ denotes the Jacobian matrix, see (11.6). The solutions of the system

$$J_c(\bar{x})p = 0$$

form a linear subspace of R^n called the *tangent space* $T(\bar{x})$ at $\bar{x}$. The name is appropriate inasmuch as every solution p of the above system is perpendicular to each of the gradient vectors associated with the constraint functions. The tangent space $T(\bar{x})$ is just the null space of the Jacobian matrix, $J_c(\bar{x})$.

The optimality conditions for (11.19) in the theorems that follow involve the gradients of the objective function f and the gradients of the constraint functions c_i, $i = 1, \ldots, \ell$ as in Theorems 11.2 and 11.3. It is natural, therefore, that they include an appropriate regularity condition. The simplest, most practical, regularity condition is the LICQ.

Theorem 11.6 (Second-order necessary conditions: equality case)   *If x^* is a local minimizer of the minimization problem (11.19) in which f and the c_i are twice-continuously differentiable and the LICQ holds at x^*, then there exists a vector λ^* such that*

(i) $\nabla f(x^*) - \sum_{i=1}^{m} \lambda_i^* \nabla c_i(x^*) = 0,$

(ii) $p^T \left[\nabla^2 f(x^*) - \sum_{i=1}^{m} \lambda_i^* \nabla^2 c_i(x^*) \right] p \geq 0 \quad \text{for all } p \in T(x^*).$

The corresponding theorem on sufficient conditions is the following.

Theorem 11.7 (Second-order sufficient conditions: equality case)   *Let*
$\bar{x}$ *be a feasible solution of the minimization problem (11.19) in which f and
the c_i are twice-continuously differentiable. Suppose there exists a vector $\bar{\lambda}$
such that*

(i) $\nabla f(\bar{x}) - \sum_{i=1}^{m} \bar{\lambda}_i \nabla c_i(\bar{x}) = 0,$

(ii) $p^{\mathrm{T}} \left[\nabla^2 f(\bar{x}) - \sum_{i=1}^{m} \bar{\lambda}_i \nabla^2 c_i(\bar{x}) \right] p > 0$ *for all nonzero* $p \in T(\bar{x}),$

then $\bar{x}$ is a strict local minimizer of f.

There are analogous optimality criteria for problems of the more general
form (11.1). Again, in the case of necessary conditions of optimality, we
take the LICQ as the regularity condition.

Theorem 11.8 (Second-order necessary conditions: general mixed case)
*Let x^* be a feasible solution of the minimization problem (11.1) in which f
and all the c_i are twice-continuously differentiable. If x^* is a local minimizer at which the LICQ holds, then there exist scalars λ_i^* ($i \in \mathcal{E}$) and
$\mu_i^* \geq 0$ ($i \in \mathcal{I}$) such that $\mu_i^* c_i(\bar{x}) = 0$ for all $i \in \mathcal{I}$ and*

(i) $\nabla f(x^*) - \sum_{i \in \mathcal{E}} \lambda_i^* \nabla c_i(x^*) - \sum_{i \in \mathcal{I}} \mu_i^* \nabla c_i(x^*) = 0,$

(ii) $p^{\mathrm{T}} \left[\nabla^2 f(x^*) - \sum_{i \in \mathcal{E}} \lambda_i^* \nabla^2 c_i(x^*) - \sum_{i \in \mathcal{I}} \mu_i^* \nabla^2 c_i(x^*) \right] p \geq 0$ *for all* $p \in T(x^*).$

Finally, sufficient conditions for local optimality in the problem (11.1)
are given by

Theorem 11.9 (Second-order sufficient conditions: general mixed case)
*Let $\bar{x}$ be a feasible solution of the minimization problem (11.1) in which f
and all the c_i are twice-continuously differentiable. If there exist scalars λ_i
($i \in \mathcal{E}$) and $\mu_i \geq 0$ ($i \in \mathcal{I}$) such that $\mu_i c_i(\bar{x}) = 0$ for all $i \in \mathcal{I}$ and*

(i) $\nabla f(\bar{x}) - \sum_{i \in \mathcal{E}} \lambda_i \nabla c_i(\bar{x}) - \sum_{i \in \mathcal{I}} \mu_i \nabla c_i(\bar{x}) = 0,$

(ii) $p^{\mathrm{T}} \left[\nabla^2 f(\bar{x}) - \sum_{i \in \mathcal{E}} \lambda_i \nabla^2 c_i(\bar{x}) - \sum_{i \in \mathcal{I}} \mu_i \nabla^2 c_i(\bar{x}) \right] p > 0$ *for all* $0 \neq p \in T(\bar{x}),$

then $\bar{x}$ is a strict local minimizer of f.

The second-order sufficient conditions presented above may raise a question in the reader's mind: why are they not just like the necessary conditions? The same sort of question came up in the study of unconstrained optimization following Example 9.1. In the present case, we can consider the univariate function $f(x) = x^3$ on the interval $[-1, 1]$. The point $\bar{x} = 0$ satisfies the first- and second-order *necessary* conditions for optimality, but it is clearly not a local minimum; this is why a stronger condition is imposed; but notice that with it comes the added feature that the local minimum is isolated.

Nevertheless, there are nonlinear optimization problems such as quadratic programming in which the first- and second-order conditions for optimality are the same.

Theorem 11.10 (Second-order necessary and sufficient conditions in QP)
Let $\bar{x}$ be a feasible solution of the quadratic programming problem (11.18), and let $\mathcal{F}(\bar{x})$ be the cone of feasible directions at $\bar{x}$. Then $\bar{x}$ is a local minimum for (11.18) if, and only if,

(i) $(Q\bar{x} + a)^{\mathrm{T}}p \geq 0$ *for all $p \in \mathcal{F}(\bar{x})$,*

(ii) $p^{\mathrm{T}}Qp \geq 0$ *for all $p \in \mathcal{F}(\bar{x}) \cap (Q\bar{x} + a)^{\perp}$.*

A few words about this theorem[10] are in order. First, if the point $\bar{x}$ is interior to the feasible region, then all directions are feasible, and condition (i) holds with equality, i.e., the gradient vanishes at $\bar{x}$ as usual. In this case, condition (ii) says that Q is positive semidefinite, again as usual. Second, if $\bar{x}$ belongs to the boundary of the feasible region, then the cone of feasible directions will be of the form

$$\mathcal{F}(\bar{x}) = \{p : A_i.p \geq 0 \text{ for all } i \in \mathcal{A}(\bar{x})\}.$$

Under these circumstances, condition (i) says that the gradient vector $Q\bar{x} + a$ makes an angle of at most $\pi/2$ radians with every feasible direction. The inequality $p^{\mathrm{T}}Qp \geq 0$ in condition (ii) does not express the positive semidefiniteness of Q because the place where it holds is a cone properly contained in the ambient space R^n. In general, a quadratic form that is nonnegative over a cone K is said to be K-*copositive*. (The prefix K is normally dropped when $K = R^n_+$.) Checking K-copositivity is normally not an easy task.

---

[10] The theorem is attributed to Contesse [29]; it builds on earlier work of Ritter [165] and Majthay [128], among others.

Verifying second-order sufficient conditions

Theorems 11.7 and 11.9 give sufficient conditions for a solution of the first-order conditions to be a strict local minimizer. To make use of them, one needs a method for checking the restricted positive definiteness condition stated in these theorems. This involves the partial Hessian of the appropriate Lagrangian function L at a particular point. In the discussion to follow, depending on the case at hand, we use the notation H to represent one of the much more detailed expressions

$$\nabla^2 f(\bar{x}) - \sum_{i=1}^{m} \lambda_i \nabla^2 c_i(\bar{x}) \quad \text{or} \quad \nabla^2 f(\bar{x}) - \sum_{i\in\mathcal{E}} \lambda_i \nabla^2 c_i(\bar{x}) - \sum_{i\in\mathcal{I}} \mu_i \nabla^2 c_i(\bar{x}).$$

In addition, we need a convenient way to describe the vectors belonging to the tangent space. The latter is the null space of the Jacobian matrix. Let us call this matrix $\hat{A}$. What we want to know is whether $\hat{A}p = 0$ and $p \neq 0$ implies $p^{\mathrm{T}}Hp > 0$.

It is not restrictive to assume that the rows of the matrix $\hat{A}$ are linearly independent. By permuting the columns of $\hat{A}$ if necessary, we may assume that $\hat{A} = [B \ N]$ where B is a basis in $\hat{A}$. This leads to corresponding partitioning of p and H. Thus, we write

$$p = (p_B, p_N) \quad \text{and} \quad H = \begin{bmatrix} H_{BB} & H_{BN} \\ H_{NB} & H_{NN} \end{bmatrix}.$$

Since $Bp_B + Np_N = 0$, we have $p_B = -B^{-1}Np_N$. Writing

$$p^{\mathrm{T}}Hp = \begin{bmatrix} p_B \\ p_N \end{bmatrix}^{\mathrm{T}} \begin{bmatrix} H_{BB} & H_{BN} \\ H_{NB} & H_{NN} \end{bmatrix} \begin{bmatrix} p_B \\ p_N \end{bmatrix},$$

we can eliminate p_B from the expression. This results in an *unconstrained* quadratic form in p_N. To write this, it will help to let $\bar{N} = -B^{-1}N$. We then obtain

$$p_N^{\mathrm{T}}\bar{N}^{\mathrm{T}}H_{BB}\bar{N}p_N + p_N^{\mathrm{T}}\bar{N}^{\mathrm{T}}H_{BN}p_N + p_N^{\mathrm{T}}H_{NB}\bar{N}p_N + p_N^{\mathrm{T}}H_{NN}p_N =$$
$$p_N^{\mathrm{T}}\left(\bar{N}^{\mathrm{T}}H_{BB}\bar{N} + 2H_{NB}\bar{N} + H_{NN}\right)p_N$$

as the formula for $p^{\mathrm{T}}Hp$ restricted to the tangent space $T(x^*)$. The question then boils down to whether the matrix $\bar{N}^{\mathrm{T}}H_{BB}\bar{N} + 2H_{NB}\bar{N} + H_{NN}$ is positive definite. It can be shown that under the assumptions stated above, the crucial matrix for determining whether the point $\bar{x}$ is a strict local minimizer is given by

$$
H_{NN} - \begin{bmatrix} N^{\mathrm{T}} & H_{NB} \end{bmatrix} \begin{bmatrix} 0 & B \\ B^{\mathrm{T}} & H_{BB} \end{bmatrix}^{-1} \begin{bmatrix} N \\ H_{BN} \end{bmatrix}.
$$

This matrix is known as the *Schur complement* of

$$
\begin{bmatrix} 0 & B \\ B^{\mathrm{T}} & H_{BB} \end{bmatrix}
$$

in the matrix

$$
\begin{bmatrix} 0 & B & N \\ B^{\mathrm{T}} & H_{BB} & H_{BN} \\ N^{\mathrm{T}} & H_{NB} & H_{NN} \end{bmatrix}.
$$

See [32] for background on properties of the Schur complement.

11.4 Convex programs

Optimization problems that call for the minimization of a convex function over a convex set constitute an important class of mathematical programs. They go by various names: *convex programs, convex programming problems,* and *convex optimization problems.* See Boyd and Vandenberghe [18] and Bertsekas [13]. Section A.7 covers some basic properties of convex sets and convex functions. Here we discuss optimality criteria for convex programs. The theme of convex programs continues in Section 11.5 where we touch on duality theory for nonlinear programming problems.

In line with the notational convention set forth in Section 11.1, we express the convex programming problem as

$$
\begin{aligned}
\text{minimize} \quad & f(x) \\
\text{subject to} \quad & c_i(x) \geq 0, \quad i \in \mathcal{I} \\
& x \in X,
\end{aligned}
\tag{11.20}
$$

where $\mathcal{I} = \{1, \ldots, m\}$, f is a convex function, $c = (c_1, \ldots, c_m)$ is a vector of concave functions, and X is a convex subset of R^n on which these functions are well defined. The set X could be the solution set of a system of linear equations and/or inequalities, or it could be R^n itself; these are just two of the most common possibilities. The concavity of the functions c_i assures that the feasible region is convex.

Ordinarily, we take particular interest in convex programs with further properties such as first- or second-order differentiability of their defining functions. However, as we shall see in the following discussion and in the subsection below, there are circumstances where differentiability need not be assumed. But, if no differentiability is assumed, one may rightly wonder what sort of conditions can used to characterize an optimal solution of (11.20). Sometimes (for example, in theoretical arguments) it is enough to postulate this optimality.

The Lagrangian function associated with the convex programming problem (11.20) is

$$L(x, y) = f(x) - y^{\mathrm{T}} c(x), \quad x \in X, \ y \geq 0. \qquad (11.21)$$

Note that, as formulated, $L(\cdot, y)$ is a convex function of $x \in X$. The corresponding *saddle point problem* is to find $(\bar{x}, \bar{y})$ such that $\bar{x} \in X$ and $\bar{y} \geq 0$ such that

$$L(\bar{x}, y) \leq L(\bar{x}, \bar{y}) \leq L(x, \bar{y}) \text{ for all } x \in X, \text{ and all } y \geq 0. \qquad (11.22)$$

It is plain to see that a saddle point $(\bar{x}, \bar{y})$ of the Lagrangian minimizes $L(\cdot, \bar{y})$ over X and maximizes $L(\bar{x}, \cdot)$ over R^m_+.

If there exists a nonnegative vector $\bar{y}$ such that $(\bar{x}, \bar{y})$ is a saddle point of L over $X \times R^m_+$, then $\bar{x}$ must be an optimal solution of (11.20). Indeed this much is true even if (11.20) is not a convex program. The argument goes as follows. The vector $\bar{x}$ is feasible for (11.20), for if $c(\bar{x})$ has a negative component, then the inequality $L(\bar{x}, y) \leq L(\bar{x}, \bar{y})$ for all $y \geq 0$ cannot hold. Moreover, $\bar{y}^{\mathrm{T}} c(\bar{x}) = 0$ since

$$-y^{\mathrm{T}} c(\bar{x}) \leq -\bar{y}^{\mathrm{T}} c(\bar{x}) \leq 0 \ \text{ for all } y \geq 0,$$

and the value 0 is attainable. The vector $\bar{x}$ is a global minimizer, for otherwise there exists a vector $\tilde{x}$ such that $c(\tilde{x}) \geq 0$ and

$$f(\tilde{x}) - \bar{y}^{\mathrm{T}} c(\tilde{x}) \leq f(\tilde{x}) < f(\bar{x}) = f(\bar{x}) - \bar{y}^{\mathrm{T}} c(\bar{x}) \leq f(\tilde{x}) - \bar{y}^{\mathrm{T}} c(\tilde{x})$$

which is obviously impossible.

The property of being a saddle point of the Lagrangian function L is clearly very strong. It yields a *sufficient* condition for a vector to be a global minimizer using no differentiability assumption, no regularity condition, and no convexity assumption. To obtain *necessary* conditions, we normally need to make some sort of regularity and convexity assumption as illustrated below.

Example 11.8: NONEXISTENCE OF A SADDLE POINT. This example exhibits a nonlinear program having a globally optimal solution but no saddle point for the Lagrangian function. Consider the (convex programming) problem

$$\begin{aligned} \text{minimize} \quad & -x_1 \\ \text{subject to} \quad & -x_1^2 + x_2 \geq 0 \\ & \quad - x_2 \geq 0. \end{aligned}$$

This problem has a unique global minimizer, namely $\bar{x} = (0,0)$. Since $0 \geq x_2 \geq x_1^2 \geq 0$, $\bar{x}$ is the only feasible solution. The associated Lagrangian function is

$$L(x_1, x_2, y_1, y_2) = -x_1 + y_1(x_1^2 - x_2) + y_2 x_2.$$

Note that $L(\bar{x}, y) = 0$ for all y. If the Lagrangian had a saddle point $(\bar{x}, \bar{y})$, (11.22) would yield

$$0 \leq -x_1 + \bar{y}_1(x_1^2 - x_2) + \bar{y}_2 x_2 = L(x, \bar{y}) \quad \text{for all } x.$$

Let $x_2 = 0$. If x_1 is positive and sufficiently small, we get $L(x, \bar{y}) < 0$, whereas for $(\bar{x}, \bar{y})$ to be a saddle point, we must have $0 = L(\bar{x}, \bar{y}) \leq L(x, \bar{y}) < 0$ which is a contradiction.

The Slater condition

In the realm of convex programs, there is a special constraint qualification which dates back to the very early days of nonlinear optimization.[11] The

---

[11]Slater's constraint qualification first appeared in a Cowles Commission Discussion Paper in 1950. It appeared again as a RAND Corporation research memorandum [179] in 1951; it was reissued by the Cowles Commission in 1959 and 1980; and appears in Giorgi and Kjeldsen [85, pp. 293–306]. As far as we know, it was never published in the periodical literature. Apart from having written one of the most famous unpublished papers in the field, Morton Slater can probably be credited with having popularized the use of the phrase "(fill in subject) revisited" as the title of a research work.

constraints of the optimization problem (11.20) satisfy the *Slater condition* if there exists a feasible point $\bar{x}$ such that $c_i(\bar{x}) > 0$ for all $i \in \mathcal{I}$.

The differentiable convex program given in Example 11.8 satisfies neither the Slater condition nor the LICQ.

The following theorem expresses a key fact about convex programs (11.20) for which the Slater condition is satisfied.

Theorem 11.11   *If (11.20) is a convex program for which the constraints satisfy the Slater condition, then a feasible point $\bar{x}$ is an optimal solution of the problem if and only if there exists a vector $\bar{y}$ such that $(\bar{x}, \bar{y})$ is a saddle point of the Lagrangian function (11.21) on $X \times R_+^m$.*

The Slater condition is of interest because when it holds, the vector $\bar{y}$ mentioned in Theorem 11.11 is guaranteed to exist. As will be seen from the discussion in Section A.7, the existence of the vector $\bar{y}$ is obtained by invoking a *separation theorem* relating to a pair of appropriately defined convex sets.

Differentiability again

Assuming first-order differentiability, Kuhn and Tucker [118] developed their necessary and sufficient conditions conditions for a problem like (11.20) having $X = R_+^n$. They did so by way of a somewhat more abstract saddle point problem (which they called the *saddle value problem*): Given $\varphi : R_+^n \times R_+^m \to R$, find $(\bar{x}, \bar{y}) \in R_+^n \times R_+^m$ such that

$$\varphi(\bar{x}, y) \leq \varphi(\bar{x}, \bar{y}) \leq \varphi(x, \bar{y}) \quad \text{for all} \quad x \in R_+^n \text{ and all } y \in R_+^m. \quad (11.23)$$

Kuhn and Tucker[12] first showed the following necessary conditions:

If $(\bar{x}, \bar{y})$ is a saddle point of φ over $R_+^n \times R_+^m$, then

$$\nabla_x \varphi(\bar{x}, \bar{y}) \geq 0, \quad \bar{x} \cdot \nabla_x \varphi(\bar{x}, \bar{y}) = 0, \quad \bar{x} \geq 0 \quad (11.24)$$

$$\nabla_y \varphi(\bar{x}, \bar{y}) \leq 0, \quad \bar{y} \cdot \nabla_y \varphi(\bar{x}, \bar{y}) = 0, \quad \bar{y} \geq 0. \quad (11.25)$$

---

[12]Karush [108] did not discuss the saddle value problem and did not mention convexity.

To establish sufficient conditions for $(\bar{x}, \bar{y})$ to be a saddle point, Kuhn and Tucker used these necessary conditions augmented by local versions of the gradient inequality for convex and concave functions, respectively (see Section A.7, especially Equation (A.18)). Their result goes as follows.

If $(\bar{x}, \bar{y}) \in R_+^n \times R_+^m$ satisfies (11.24), (11.25), and

$$\varphi(\bar{x}, \bar{y}) + (x - \bar{x}) \cdot \nabla_x \varphi(\bar{x}, \bar{y}) \leq \varphi(x, \bar{y}) \text{ for all } x \geq 0 \text{ in a neighborhood of } \bar{x}$$

$$\varphi(\bar{x}, \bar{y}) + (y - \bar{y}) \cdot \nabla_y \varphi(\bar{x}, \bar{y}) \geq \varphi(\bar{x}, y) \text{ for all } y \geq 0 \text{ in a neighborhood of } \bar{y},$$

then $(\bar{x}, \bar{y})$ is a saddle point of φ over $R_+^n \times R_+^m$.

When the function φ is taken to be the Lagrangian L of the convex program (11.20), as would be the traditional approach, the set X might not be R_+^n in which case there would be a corresponding change in the conditions (11.24). It might simply be

$$\nabla_x L(\bar{x}, \bar{y}) = 0$$

which corresponds to the first-order optimality condition (11.14). The first of the conditions given in (11.25) expresses the feasibility of $\bar{x}$ in the convex program (11.20), whereas the last two conditions correspond to the last two of (11.15) known as *complementary slackness conditions*.

11.5   Elementary duality theory for nonlinear programming

In Chapter 5 we discussed the duality theory for linear programming. There are also duality theorems for nonlinear programming; some of them are presented in this section. As may be expected, duality theory in nonlinear programming is more challenging than it is in linear programming. One explanation for this is that linear functions possess two properties (namely, convexity and differentiability) that might need to be assumed in the nonlinear case. The need for satisfaction of a constraint qualification is another issue that comes up in nonlinear programming, but not in linear programming. The complications of the functions involved may add further difficulties.

Whether it is the linear or nonlinear case, however, duality is of strong interest for several reasons. First, it provides a way of lower-bounding the value of the primal objective function (the minimand) over the feasible region. In algorithms, this can be useful in deciding whether a current iterate yields an objective function value that is nearly optimal. A second reason for interest in duality theory is the possibility that the dual problem might be easier to solve than the primal problem. If this is the case, and if a solution to the primal problem can be recovered from a solution to the dual problem, then it would be advantageous to solve the dual rather than the primal.

Lagrangian duality theory

We begin with a duality theory based upon the Lagrangian function associated with the underlying primal problem. From the standpoint of assumptions, this is a very general theory which may be compared with the formulation in terms of saddle value problems.

We take (11.1), that is,

$$
\text{(P)} \quad
\begin{aligned}
\text{minimize} \quad & f(x) \\
\text{subject to} \quad & c_i(x) \geq 0, \quad i \in \mathcal{I} \\
& c_i(x) = 0, \quad i \in \mathcal{E} \\
& x \in R^n.
\end{aligned}
$$

to be the *primal* nonlinear programming problem. One sometimes sees the statement $x \in R^n$ expressed more generally as $x \in X$ where X is a *subset* of R^n. This would not rule out the possibility that $X = R^n$ of course. Writing $x \in X$ makes it possible to impose certain types of constraints without explicitly writing them down. Nevertheless, we stick with the problem statement as given above. The steps to create a Lagrangian dual of the above problem are as follows:

Step 1: Form the Lagrangian associated with problem (P).

$$
L(x, u, v) = f(x) - \sum_{i \in \mathcal{I}} u_i c_i(x) - \sum_{i \in \mathcal{E}} v_i c_i(x),
$$

where $u \geq 0$.

Step 2: Define the function

$$
\theta(u, v) = \inf_{x \in R^n} L(x, u, v).
$$

In principle, the evaluation of the function θ entails a minimization problem for every pair (u, v). (This may make it seem impractical to deal with, but there are circumstances under which that is not the case; one such occurs in convex quadratic programming.)

Step 3: The above definition of θ gives rise to the *Lagrangian dual problem*:

$$(D) \qquad \begin{array}{ll} \text{maximize} & \theta(u, v) \\ \text{subject to} & u \geq 0, \ v \text{ free.} \end{array} \qquad (11.26)$$

The property of dual feasibility is a particularly simple condition in this formulation.

We next illustrate the above process with examples.

Example 11.9: LAGRANGIAN DUAL OF AN LP. Here we illustrate how to find the dual of the following primal linear program

$$\begin{array}{ll} \text{minimize} & c^{\mathrm{T}}x \\ \text{subject to} & Ax - b \geq 0. \end{array}$$

We start by writing down the Lagrangian of this problem.

$$L(x, u) = c^{\mathrm{T}}x - u^{\mathrm{T}}(Ax - b) = (c - A^{\mathrm{T}}u)^{\mathrm{T}}x + b^{\mathrm{T}}u.$$

The next step is to minimize this with respect to x, that is find

$$\theta(u) = \inf_{x \in R^n} (c - A^{\mathrm{T}}u)^{\mathrm{T}}x + b^{\mathrm{T}}u.$$

It is straightforward to see that the solution to this problem is

$$\theta(u) = \begin{cases} b^{\mathrm{T}}u & \text{if } A^{\mathrm{T}}u = c \\ -\infty & \text{if } A^{\mathrm{T}}u \neq c. \end{cases}$$

Thus, from Step 3 of the process, the Lagrangain dual problem is the familiar one:

$$\begin{array}{ll} \text{maximize} & b^{\mathrm{T}}u \\ \text{subject to} & A^{\mathrm{T}}u = c \\ & u \geq 0. \end{array}$$

Example 11.10: LAGRANGIAN DUAL OF A NONCONVEX PROBLEM. Let us consider the problem

$$\begin{array}{ll} \text{minimize} & x^{\mathrm{T}}Hx \\ \text{subject to} & x_j^2 - 1 = 0 \quad \text{for } i = 1, \ldots, n. \end{array}$$

As in the previous example, we start by writing down the Lagrangian of this problem.

$$L(x, v) = x^{\mathrm{T}}Hx - \sum_{j=1}^{n} v_j(x_j^2 - 1) = x^{\mathrm{T}}(H - \mathrm{Diag}(v))x + \sum_{j=1}^{n} v_j.$$

The next step is to minimize this with respect to x, that is find

$$\theta(v) = \inf_{x \in R^n} x^{\mathrm{T}}(H - \mathrm{Diag}(v))x + \sum_{j=1}^{n} v_j.$$

The solution to this problem is

$$\theta(u) = \begin{cases} \displaystyle\sum_{j=1}^{n} v_j & \text{if } (H - \mathrm{Diag}(v)) \text{ positive semidefinite} \\ -\infty & \text{otherwise.} \end{cases}$$

Thus, from Step 3 of the process, the Lagrangian dual problem is

$$\begin{array}{ll} \text{maximize} & \displaystyle\sum_{j=1}^{n} v_j \\ \text{subject to} & (H - \mathrm{Diag}(v)) \text{ positive semidefinite.} \\ & v \text{ free.} \end{array}$$

Example 11.11: LAGRANGIAN DUAL OF A CONVEX QP. Consider the following problem

$$\begin{array}{ll} \text{minimize} & \frac{1}{2}x_1^2 + \frac{1}{2}x_2^2 \\ \text{subject to} & x_1 + x_2 = 2 \\ & x_1 - 4x_2 \leq 4. \end{array}$$

We rewrite this as

$$\begin{array}{ll} \text{minimize} & \frac{1}{2}x_1^2 + \frac{1}{2}x_2^2 \\ \text{subject to} & x_1 + x_2 - 2 = 0 \\ & -x_1 + 4x_2 + 4 \geq 0. \end{array}$$

The Lagrangian of this problem is

$$L(x, u, v) = \frac{1}{2}x_1^2 + \frac{1}{2}x_2^2 - v(x_1 + x_2 - 2) - u(-x_1 + 4x_2 + 4).$$

The next step is to minimize this with respect to x, that is find

$$\theta(u, v) = \inf_{x \in R^n} L(x, u, v).$$

To do this we take the first partials of $L(x, u, v)$ with respect to x and set them equal to zero. That is, we obtain

$$x_1 - v + u = 0$$
$$x_2 - v - 4u = 0.$$

Solving for x_1 and x_2 we obtain

$$x_1 = v - u$$
$$x_2 = v + 4u.$$

Next substituting these values in $L(x, u, v)$, and simplifying, we obtain

$$\theta(u, v) = -v^2 - \frac{17u^2}{2} - 2vu + 2v - 4u.$$

The dual problem is

$$\text{maximize} \quad -v^2 - \frac{17u^2}{2} - 2vu + 2v - 4u$$

$$\text{subject to} \quad u \geq 0, \ v \text{ free.}$$

Theorem 11.12 (Weak Duality) *If x is primal feasible and (u, v) is dual feasible, then*

$$\theta(u, v) \leq f(x). \tag{11.27}$$

Furthermore, if $f(x) = \theta(u, v)$ where x is feasible for (P) and (u, v) is feasible for (D), then x and (u, v) are optimal for their respective problems.

This is not hard to verify. By definition $\theta(u, v) \leq L(x, u, v)$. For primal feasible x and any $u \geq 0$, it follows that $L(x, u, v) \leq f(x)$, so $\theta(u, v) \leq L(x, u, v) \leq f(x)$. The remaining assertion follows immediately. See Exercise 11.27. If it should happen that

$$\sup \{\theta(u, v) : (u, v) \text{ is dual feasible}\} < \inf \{f(x) : x \in R^n \text{ is primal feasible}\},$$

then the problem (P) is said to have a *duality gap*.

Example 11.12: FEASIBILITY. The Lagrangian dual process and the Weak Duality Theorem can be used to determine if a region is nonempty. Consider the set of points defined by:

$$X = \{x \in R^3 : x_1^2 + x_2^2 + x_3^2 = 1 \text{ and } x_1 + x_2 + x_3 \geq 2\}.$$

In order to apply the Lagrangian dual process we create a primal problem where we minimize a zero objective subject to $x \in X$. That is

$$
\begin{aligned}
\text{minimize} \quad & 0x_1 + 0x_2 + 0x_3 \\
\text{subject to} \quad & x_1 + x_2 + x_3 - 2 \geq 0 \\
& x_1^2 + x_2^2 + x_3^2 - 1 = 0.
\end{aligned}
$$

We form the Lagrangian:

$$L(x, u, v) = 0 - u(x_1 + x_2 + x_3 - 2) - v(x_1^2 + x_2^2 + x_3^2 - 1).$$

We next find

$$\theta(u) = \inf_{x \in R^3} -u(x_1 + x_2 + x_3 - 2) - v(x_1^2 + x_2^2 + x_3^2 - 1).$$

To do this we set the partial derivatives with respect to x to zero and solve for x.

$$
\begin{aligned}
-2x_1 v - u = 0 \quad & \text{or} \quad x_1 = u/2v \\
-2x_2 v - u = 0 \quad & \text{or} \quad x_2 = u/2v \\
-2x_3 v - u = 0 \quad & \text{or} \quad x_3 = u/2v.
\end{aligned}
$$

Substituting in $L(x, u, v)$, and simplifying, we obtain $\theta(u, v)$ as

$$\theta(u, v) = -\frac{9u^2}{4v} + 3u + v.$$

The dual problem is then

$$
\begin{aligned}
\text{maximize} \quad & -\frac{9u^2}{4v} + 3u + v \\
\text{subject to} \quad & u \geq 0, \ v \text{ free.}
\end{aligned}
$$

If we pick $v = -1$, then we find that the dual is unbounded above. Hence the primal problem is infeasible; that is, the set X is empty.

Example 11.13: DUALITY GAP. Consider the following univariate nonlinear program

$$\begin{array}{ll} \text{minimize} & -x^2 \\ \text{subject to} & 0 \le x \le 2 \\ & x \in R. \end{array}$$

It is clear that this trivial optimization problem has a unique optimal solution: $x^* = 2$ at which the objective function $f(x) = -x^2$ has the value -4. In this instance, there being no equality constraints in the problem, the function θ is given by

$$\theta(u) = \inf_{x \in R} \left\{ -x^2 - u_1 x - u_2(-x+2) \right\} = -\infty \text{ for all } u \in R_+^2.$$

This implies that

$$\max_{u \in R_+^2} \theta(u) < \min_{0 \le x \le 2} f(x)$$

which reveals a duality gap. Removing the minus sign in the objective function changes the outcome dramatically.

Notice that in the Weak Duality Theorem nothing was assumed about special properties (such as convexity or differentiability) of the functions through which the problem (P) was defined. Under such general conditions, duality gaps can arise. As we shall see below, however, duality gaps do not occur in problems possessing the right combination of properties.

As foreshadowed in the introduction to this chapter, we now impose a restriction on the functions that specify the primal problem.

Wolfe-type duality for convex programming

The following formulation of duality theory is named for Philip Wolfe who introduced it in his paper [193]. Consider a primal problem (WP) having the form

$$\begin{array}{lll} & \text{minimize} & f(x) \\ \text{(WP)} & \text{subject to} & c_i(x) \ge 0, \quad i = 1, \dots, m \\ & & x \in R^n. \end{array}$$

$$(11.28)$$

where f is differentiable and convex on R^n and $c_1, \dots, c_m$ are differentiable and concave on R^n. As observed for (11.20), here too the concavity of the constraint functions c_i implies that the feasible region of (WP) will be a convex set, albeit possibly empty. Accordingly, such a problem is called a *differentiable convex program*. The Lagrangian for (WP) is expressible as

$$L(x, u) = f(x) - u^{\mathrm{T}}c(x).$$

In this setting, the behavior of $L(x, u)$ on $R^n \times R^m_+$ is of interest. Notice that for fixed $u \in R^m_+$, the Lagrangian $L(x, u)$ is a convex function of $x \in R^n$. The *Wolfe dual* (WD) of the differentiable convex program (WP) is defined as the problem

$$
\begin{array}{lll}
 & \text{maximize} & L(x, u) \\
\text{(WD)} & \text{subject to} & \nabla_x L(x, u) = 0 \\
 & & x \in R^n, \ u \in R^m_+.
\end{array}
\tag{11.29}
$$

The problem (WD) is formally different from the Lagrangian dual (D). There being no equality constraints in (WP), the Lagrangian dual objective function would be

$$\theta(u) := \inf_{x \in R^n} L(x, u). \tag{11.30}$$

Nevertheless, we point out that the constraints $\nabla_x L(x, u) = 0$ with $x \in R^n$ are tantamount to the minimization process through which $\theta(u)$ is defined.

Looking at the Wolfe dual (WD), and particularly its constraints, one might be concerned about just what sort of problem this really is. In some cases such as linear programming presented below in Example 11.14, the statement of the Wolfe dual can be reduced to the conventional LP dual. Yet there still could be problems whose (WD) constraints are not compatible with convex programming. (Nonlinear equations would be a case in point.) However, for primal programs of the form (11.28), even without convexity or differentiability being present, the *Lagrangian* dual objective function (11.30) turns out to be a concave maximization problem over the nonnegative orthant (and hence is obviously feasible).

Theorem 11.13   *Relative to a primal problem (11.28), the objective function of the Lagrangian dual problem is concave.*

Indeed, for fixed $x = \bar{x}$, $L(\bar{x}, u) = f(\bar{x}) - u^{\mathrm{T}}c(\bar{x})$ is an affine function of u. Since each value of the Lagrangian dual objective function is given as the infimum of a family of affine functions, it follows from Theorem A.8 that θ is a concave function on R^m_+.

Example 11.14: APPLICATION TO LINEAR PROGRAMMING. As (WP) take the linear programming problem

$$\begin{array}{ll} \text{minimize} & c^{\mathrm{T}}x \\ \text{subject to} & Ax \geq b. \end{array}$$

Then we have

$$f(x) = c^{\mathrm{T}}x \quad \text{and} \quad C(x) = Ax - b.$$

For the stated primal problem (WP), the objective in the Wolfe Dual would be

$$c^{\mathrm{T}}x - u^{\mathrm{T}}(Ax - b) = (c - A^{\mathrm{T}}u)^{\mathrm{T}}x + b^{\mathrm{T}}u$$

so that the problem (WD) can be written as

$$\begin{array}{ll} \text{maximize} & (c - A^{\mathrm{T}}u)^{\mathrm{T}}x + b^{\mathrm{T}}u \\ \text{subject to} & c - A^{\mathrm{T}}u = 0 \\ & x \in R^n, \ u \geq 0. \end{array}$$

This, of course, is not quite the dual problem we are accustomed to, but when u satisfies $c - A^{\mathrm{T}}u = 0$, the objective function of the Wolfe dual reduces to $b^{\mathrm{T}}u$, and $x \in R^n$ disappears from the formulation. Accordingly, the Wolfe dual and the corresponding LP dual are the same problem.

We have already seen the Weak Duality Theorem in the case of the Lagrangian dual. The following is the same sort of theorem for Wolfe's primal-dual pair.

Theorem 11.14 (Weak Duality (Wolfe)) *Let x' be a feasible vector for the (Wolfe) primal problem (WP), and let (x'', u) be a feasible vector for the (Wolfe) dual (WD). Then*

$$L(x'', u) := f(x'') - u^{\mathrm{T}}c(x'') \leq f(x').$$

Furthermore, if $f(x') = L(x'', u)$ where x' is feasible for (WP) and (x'', u) is feasible for (WD), then x' and (x'', u) are optimal for their respective problems.

Indeed, since f and the functions $-c_i$ involved in (WP) are convex and differentiable, we can apply the *gradient inequality* (see page 569). Using this and (11.29) we have

$$L(x'', u) = f(x'') - u^{\mathrm{T}}c(x'')$$
$$\leq f(x'') - u^{\mathrm{T}}[c(x'') - c(x')]$$
$$\leq f(x'') + \sum_{i=1}^{m} u_i \nabla \left(c_i(x'')\right)^{\mathrm{T}}(x' - x'')$$
$$= f(x'') + \nabla f(x'')^{\mathrm{T}}(x' - x'')$$
$$\leq f(x').$$

That this Weak Duality Theorem does not hold for arbitrary differentiable functions can be seen from the following example.

Example 11.15: FAILURE OF THE WEAK DUALITY INEQUALITY. In the case of the problem

$$\begin{array}{ll} \text{minimize} & x \\ \text{subject to} & e^x \geq 0 \\ & x \in R, \end{array}$$

the constraint function c is not concave (but instead is strictly convex), hence the hypothesis of the (Wolfe) Weak Duality Theorem does not hold. (The inequality $c(x) \geq 0$ does not really impose a constraint at all because every real number x is feasible.) Under these circumstances, the infimum of the objective function is $-\infty$. The (Wolfe) Weak Duality Theorem does not apply here, and its conclusion does not hold either. This is illustrated by taking $x' = -1$ as a primal feasible solution and $(x'', u) = (1, e^{-1})$ as a dual feasible solution. We then get

$$L(1, e^{-1}) = 0 > -1 = f(-1).$$

Theorem 11.15 (Strong Duality Theorem (Wolfe)). *If x^* is an optimal solution of (WP) at which a first-order constraint qualification is satisfied, then there exists a vector u^* such that (x^*, u^*) solves (WD) and*

$$f(x^*) = L(x^*, u^*).$$

The KKT Theorem can be applied in this case. Choose the vector $\bar{u}$ to be the vector of Lagrange multipliers guaranteed by the KKT Theorem.

From the complementary slackness conditions, we have $f(\bar{x}) = L(\bar{x}, \bar{u})$. We also have $\nabla_x L(\bar{x}, \bar{u}) = 0$, $\bar{x} \in R^n$. Hence $(\bar{x}, \bar{u})$ is feasible for (WD), and it is optimal by virtue of the Weak Duality Theorem.

What can be said when we have an optimal solution of the Wolfe dual (WD)? The following result [129], [130, Chapter 8] gives an indication.

Theorem 11.16 (Strict Converse Duality Theorem (Mangasarian)). *Let (WP) have an optimal solution x^* at which the constraints satisfy a first-order constraint qualification. If $(\hat{x}, \hat{u})$ solves (WD) and $L(\cdot, \hat{u})$ is strictly convex[13] at $\hat{x}$, then $\hat{x} = x^*$ and $L(\hat{x}, \hat{u}) = f(x^*)$.*

To see why this is so, apply Wolfe's Duality Theorem to (WP). Then there exists a vector $\bar{u}$ such that $(x^*, \bar{u})$ solves (WD). Hence

$$f(x^*) = L(x^*, \bar{u}) = L(\hat{x}, \hat{u}).$$

Suppose $x^* \neq \hat{x}$. By the strict convexity at $\hat{x}$

$$L(x^*, \hat{u}) - L(\hat{x}, \hat{u}) > (x^* - \hat{x})^T \nabla_x L(\hat{x}, \hat{u}) = 0.$$

This implies
$$L(x^*, \hat{u}) > L(\hat{x}, \hat{u}) = L(x^*, \bar{u}),$$

and from this it follows that

$$-\hat{u}^T c(x^*) > -\bar{u}^T c(x^*) = 0,$$

which is to say $\hat{u}^T c(x^*) < 0$. But $\hat{u}^T c(x^*) \geq 0$ since $\hat{u} \geq 0$ and $c(\bar{x}) \geq 0$. This is a contradiction, so we must have $x^* = \hat{x}$.

What is the essence of duality?

In Chapter 5 we saw that for every given linear program (the primal) there is a second linear program (the dual) that is formally related to it. These two LPs form what is called a *primal-dual pair*. The formal correspondence among the various components of the paired LPs is spelled out in Table 5.1. With a little bit of artifice, we showed that (for an LP of a particular form),

---

[13]This means that the gradient inequality holds with strict inequality at that point.

"the dual of the dual is the primal." The numerous theorems about primal-dual pairs of linear programs are not—in and of themselves—what we mean by duality. Rather, from the mathematical perspective, it is the *involutory property* of the correspondence between the paired problems that makes the correspondence a duality. That is, applying the correspondence twice in succession gives back the original object. Here are some familiar examples of operations where an involutory property is present:

Object	Operation	Object	Operation	Object
x	negate	$-x$	negate	x
A (nonsingular)	invert	A^{-1}	invert	A
A (matrix)	transpose	A^{T}	transpose	A
$x + iy$ (complex no.)	conjugate	$x - iy$	conjugate	$x + iy$
S (subset of T)	complement	$T \setminus S$	complement	S
$\leq$ (inequality sign)	reverse	$\geq$	reverse	$\leq$

Obviously, these examples are all much simpler than the process by which one converts a linear program to its dual and then finds the dual of the dual. Other, less elementary, examples will be found in Section A.8.

It must also be said that there is more to duality in mathematical programming than just the involutory property. It typically involves issues of feasibility and the corresponding relationships between primal and dual objective function values.

Symmetric dual quadratic programs

There is a symmetric duality theory in convex quadratic programming as well as in linear programming. We take the primal problem to be

$$\begin{array}{lll} & \text{minimize} & \tfrac{1}{2}y^{\mathrm{T}}Qy + \tfrac{1}{2}x^{\mathrm{T}}Hx + c^{\mathrm{T}}x \\ (\mathrm{P}) & \text{subject to} & Qy + Ax \geq b \\ & & y \text{ free}, \ x \geq 0. \end{array} \qquad (11.31)$$

In this problem, $H \in R^{n \times n}$ and $Q \in R^{m \times m}$ are assumed to be symmetric and positive semidefinite. The other data are $A \in R^{m \times n}$, $b \in R^m$, and $c \in R^n$.

For the dual problem we have

$$\text{maximize} \quad -\tfrac{1}{2}v^\mathsf{T}Qv - \tfrac{1}{2}u^\mathsf{T}Hu + b^\mathsf{T}v$$

$$(\text{D}) \qquad \text{subject to} \qquad A^\mathsf{T}v - Hu \le c \qquad\qquad (11.32)$$

$$v \ge 0, \; u \text{ free.}$$

Two comments are in order here. First, it is obvious that when both H and Q are zero matrices, problems (P) and (D) reduce to the familiar symmetric dual pair of linear programs. When only Q is a zero matrix, we recover what might be called a conventional (convex) quadratic program and its dual. (See Dorn [57] and [58].) Second, the names of the variables in (D) are not the same as in (P). This has been done to avoid biasing the proof of weak duality (below). It turns out, though, that if either problem has an optimal solution, there exists a single pair of nonnegative vectors that solves *both* (P) and (D).

One can think of the relationship between (P) and (D) in a formal way, much as we ordinarily do when we "mechanically" form the dual of a primal linear program. Proceeding in this manner, we can demonstrate that *the dual of the dual* (D) *is the primal* (P). The steps of the process go as follows. (i) Write the dual (D) in the form of a minimization problem with $\ge$ constraints. (ii) Reverse the order of the variables, putting the free variables u first. (iii) Write the dual of the problem obtained in the previous step. (iv) Write the aforementioned dual as a minimization problem. (v) Reverse the order of the variables, putting the free variables y first. The resulting problem will be (P). This line of reasoning demonstrates[14] that (P) and (D) are a symmetric dual pair.

The weak duality theorem is not difficult to establish. To that end, we let F and G denote the objective functions of (P) and (D), respectively, and let Z and W denote the feasible regions of (P) and (D), respectively. Using the standard convention that

$$Z = \emptyset \implies \inf_{(x,y)\in Z} F(x,y) = +\infty,$$

$$W = \emptyset \implies \sup_{(u,v)\in W} G(u,v) = -\infty,$$

we are left with the task of proving

$$\sup_{(u,v)\in W} G(u,v) \le \inf_{(x,y)\in Z} F(x,y), \qquad\qquad (11.33)$$

---

[14]Steps (i) and (iv) involve changing the direction of optimization; steps (ii) and (v) involve reversing the order of the variables. All of these produce equivalent problems.

under the assumption that the feasible regions are nonempty. Let $(x, y) \in Z$
and $(u, v) \in W$ be arbitrary. It follows from the nonnegativity of x and v
that

$$-v^{\mathrm{T}}Qy + b^{\mathrm{T}}v \leq v^{\mathrm{T}}Ax \leq x^{\mathrm{T}}Hu + c^{\mathrm{T}}x. \tag{11.34}$$

Since H and Q are symmetric and positive semidefinite, we have

$$2x^{\mathrm{T}}Hu \leq x^{\mathrm{T}}Hx + u^{\mathrm{T}}Hu \quad \text{and} \quad -2v^{\mathrm{T}}Qy \geq -v^{\mathrm{T}}Qv - y^{\mathrm{T}}Qy. \tag{11.35}$$

The required inequality (11.33) now follows by appropriate substitution from
(11.35) to (11.34).

This brings us to the Strong Symmetric Duality Theorem for Quadratic
Programming, see [31].

Theorem 11.17 (Strong duality) *If (x^*, y^*) is an optimal solution of the
primal quadratic program (P), then there exists a vector $v^* \geq 0$ such that
(x^*, v^*) is feasible for (P) and (D), $F(x^*, v^*) = G(x^*, v^*)$, and $Qy^* = Qv^*$.*

One proof of this theorem is based on the following ideas. First, the
hypotheses of Exercise 11.25 are satisfied because the objective function of
(P) is convex and differentiable and the pair (x^*, y^*) is optimal for (P).
Hence (x^*, y^*) solves the linear program

$$\begin{aligned}
\text{minimize} \quad & \nabla F(x^*, y^*) \cdot (x, y) \\
\text{subject to} \quad & Qy + Ax \geq b \\
& x \geq 0.
\end{aligned} \tag{11.36}$$

By the duality theorem of linear programming, there exists an optimal
solution $v^* \geq 0$ to the dual of the LP (11.36). From the feasibility and
complementary slackness conditions of the LP, it is not hard to show that
$(y^* - v^*)^{\mathrm{T}}Q(y^* - v^*) = 0$. Because the matrix Q is symmetric and positive semidefinite it follows that $Qy^* = Qv^*$. From this it follows that
(x^*, v^*) is feasible for both the primal and the dual of the QP and yields
$F(x^*, v^*) = G(x^*, v^*)$. Accordingly, (x^*, v^*) is optimal for both the primal
and the dual QP.

11.6   Exercises

11.1   An appliance manufacturer would like to design a carton whose surface
area is the maximum possible subject to the sum of the dimensions being

240 inches.

(a) Formulate this as an optimization problem.

(b) Use the first order optimality conditions to solve it.

(c) Suppose the objective function is the same but the manufacturer would like the length plus the girth (twice the width plus the height) to be equal to 240. Formulate the revised problem, and use the first order optimality conditions to solve it.

11.2   Given an angle of less than π radians in R^2 and a point P in its interior, find a line L passing through P that cuts off from the angle a triangle of minimum area.

11.3   Consider a plane in R^3 that passes through a point $(1, 2, 3)$. Find the intercepts (a, b, c) in the positive orthant such that the volume enclosed by the plane in the positive orthant is minimized. (Hint: The enclosed volume is that of a tetrahedron, and is given by $abc/6$. The intercept equation of a plane is $x/a + y/b + z/c = 1$.)

11.4   Consider a rectangle with side lengths a and b. Assume that points P and Q are selected to lie on the sides of length a so that PQ is parallel to the side of length b. Find points X and Y, one on each of the sides of length b, such that the length of the piecewise linear line connecting P, X, Y, Q is minimized. Show that the minimum possible value is $\sqrt{4a^2 + b^2}$. See [101].

11.5   Let $f \in C^2$. Using notation introduced in Section 11.2, write a formula relating the Hessian of f to the Jacobian of the gradient of f.

11.6   Consider the Lagrange Multiplier Rule, Theorem 11.2, in the case where the constraint functions $c_i(x)$, $i = 1, \ldots, \ell$ are all affine in the equality-constrained problem (11.2). Show that the LICQ is not required to obtain the same first-order optimality conditions. (Hint: This can be done by applying an alternative theorem to the system $A^T\lambda^* = \nabla f(x^*)$.)

11.7   Let $x^* \in R^n$ be an optimal solution of the nonlinear optimization problem

$$\begin{aligned} \text{minimize} \quad & f(x) \\ \text{subject to} \quad & c(x) \geq 0 \\ & x \geq 0. \end{aligned}$$

Assume the KKT conditions hold at x^*. What are they in this case?

11.8   Even though the first-order optimality conditions were intended for use on nonlinear programming problems, they apply to linear programs as well. Consider the LP

$$\begin{array}{ll} \text{minimize} & c^{\mathrm{T}}x \\ \text{subject to} & Ax \geq b \\ & x \geq 0. \end{array}$$

Write the necessary and sufficient conditions of optimality for this optimization problem and compare them with what you learned in Chapter 5.

11.9   Write down the KKT conditions for the following quadratic program:

$$\text{minimize} \quad x_1^2 + x_2^2 + x_3^2 - x_1 x_2 - x_2 x_3 - 3x_1 - 2x_2 + x_3$$

$$\text{subject to} \quad x_1 \geq 0, \ x_2 \geq 0, \ x_3 \geq 0.$$

11.10   Consider the linearly constrained nonlinear program

$$\begin{array}{ll} \text{maximize} & x_1 x_2 \cdots x_n \\ \text{subject to} & x_1 + x_2 + \cdots + x_n = 1 \\ & x_j \geq 0, \quad j = 1, 2, \ldots, n. \end{array}$$

(a) Write the KKT conditions for the above nonlinear program.
(b) Use the KKT conditions to show that the inequality

$$(x_1 x_2 \cdots x_n)^{1/n} \leq \frac{1}{n}(x_1 + x_2 + \cdots + x_n)$$

holds for all $x_j \geq 0$, $j = 1, 2, \ldots, n$.

11.11   Let $Q \in R^{n \times n}$ be a symmetric positive definite matrix. The level set

$$\mathcal{X} = \left\{ x \in R^n : x^{\mathrm{T}}Qx \leq 1 \right\}$$

is an *ellipsoid*. Let c be a nonzero vector in R^n. What point of $\mathcal{X}$ minimizes the linear function $c^{\mathrm{T}}x$? (Hint: Apply the KKT conditions.)

11.12   Consider a point $\bar{y} \in R^n$ and a hyperplane

$$\mathcal{H} = \{ x \in R^n : a^{\mathrm{T}}x = b \}, \quad b \in R.$$

The hypothesis on $\mathcal{H}$ implies that $a \neq 0$.

(a) Formulate a minimization problem to find a point on $\mathcal{H}$ nearest to $\bar{y}$ under the Euclidean metric.
(b) Using the optimality conditions for the minimization problem in part (a), derive a formula for a point in $\mathcal{H}$ that is nearest to $\bar{y}$.
(c) Let $\bar{x}$ be a solution to a linear inequality system $Ax \leq b$, with $A \in R^{m \times n}$, and $b \in R^m$. Assuming that the rows of A all have Euclidean norm 1 (that is, $\|A_{i\bullet}\| = 1$), give an interpretation of the slack variables $\bar{y}$ corresponding to the vector $\bar{x}$.

11.13   A function $f : R^n \longrightarrow R$ is said to be *separable* if it can be written as the sum of functions of a single real variable, that is

$$f(x_1, x_2, \ldots, x_n) = \sum_{j=1}^{n} f_j(x_j).$$

(a) Let f be a differentiable separable function and consider the nonlinear program

minimize $f(x_1, \ldots, x_n)$

subject to $\displaystyle\sum_{j=1}^{n} x_j = \sigma$

$x_j \geq 0, \quad j = 1, \ldots, n,$

where σ is a positive scalar. Write the first-order conditions for a local minimizer $\bar{x}$ in this nonlinear program.

(b) Recall that if $x \in R^n$, the set $\{j : x_j \neq 0\}$ is called the *support* of x, denoted by $\mathrm{supp}(x)$. Show that every feasible solution in part (a) has a nonempty support.

(c) Show that there is a real number κ such that if x^* is a local minimizer of the nonlinear program in part (a), then for all $j \in \mathrm{supp}(x^*)$

$$\frac{\partial f_j(x_j^*)}{\partial x_j} = \kappa.$$

11.14   Let f be a differentiable convex function on R^n. Suppose we seek a point in the nonnegative orthant that minimizes f. That is, we want to solve the nonlinear program

minimize $f(x)$

subject to $x \geq 0.$

Show that x^* solves this problem if and only if

$$\nabla f(x^*) \geq 0$$
$$x^* \geq 0$$
$$x^* \cdot \nabla f(x^*) = 0.$$

11.15   Suppose you have a strictly convex quadratic function in n variables, say

$$f(x) = x^{\mathrm{T}} H x + c^{\mathrm{T}} x,$$

and wish to minimize it over the nonnegative orthant R^n_+.

(a) What vector is the global minimizer of f on R^n?

(b) To find the minimizer of f on R_+^n, would it be correct just to find the vector asked for in part (a) and then set all of its negative components (if any) to zero? If so, explain why. If not, give a counterexample.

11.16 By a *domed cylinder* we mean a convex body composed of a right-circular cylinder with a hemisphere of the same radius at one end. Suppose the volume $V > 0$ of such a body is given. Sometimes it is required to determine the height h of the cylinder and the radius r of the cylinder and its dome such that the total surface area of the convex body is as small as possible. It matters whether the area of the end opposite the dome is to be included. Let us first assume it is *not* included. Since the volume of a sphere of radius r is given by $\frac{4}{3}\pi r^3$, we can express the volume of the domed cylinder as

$$V = \pi r^2 h + \tfrac{2}{3}\pi r^3.$$

The assumption $V > 0$ and this formula imply that both of the unknowns h and r must be positive. The surface area of a sphere of radius r is given by $4\pi r^2$ and that of an open-ended cylinder is given by $2\pi rh$. Thus the total surface area of the domed cylinder is

$$A = 2\pi rh + 2\pi r^2.$$

Suppose the height h is also required to be no less than the diameter $2r$.

(a) Find the values of h and r that minimize the total surface area A of the domed cylinder of volume V as given above.

(b) Do as in part (a) but under the assumption that the total area includes both ends, one of which is a just a circle and the other is a hemisphere.

(c) Give examples of real-world situations where the problems solved in parts (a) and (b) can be applied.

11.17 (Based on Cottle and Olkin [36].) Let a_1, a_2 and r be given scalars satisfying the conditions

$$a_1 > a_2 > 0 \quad \text{and} \quad 0 < r < 1.$$

Define the diagonal positive definite matrix $A = \operatorname{diag}(a_1, a_2)$. Suppose we seek two vectors u and v that solve the equality-constrained problem

$$
(P) \quad
\begin{aligned}
\text{maximize} \quad & u^{\mathsf{T}}Au + vA^{\mathsf{T}}v \\
\text{subject to} \quad & u^{\mathsf{T}}u = 1 \\
& v^{\mathsf{T}}v = 1 \\
& u^{\mathsf{T}}v = r.
\end{aligned}
$$

Note that (P) calls for the *maximum* of a strictly convex function subject to a set of nonlinear *equations*. As a rule, such problems are very difficult, but as (we hope) you will see, that is not so in the present case.

(a) Write down the first-order necessary conditions of optimality for (P).

(b) Use the first-order optimality conditions found in part (a) to obtain the values of the Lagrange multipliers and an optimal solution to (P).

(c) Does (P) have a unique optimal solution?

11.18   Fritz John's Theorem was mentioned in this chapter on page 380; it states that if x^* is a local minimizer for problem (11.13) where f and the c_i are differentiable at x^*, there exist nonnegative scalars $\lambda_0^*, \lambda_1^*, \ldots, \lambda_m^*$ *not all of which are zero* such that

$$\lambda_i^* c_i(x^*) = 0, \quad i = 1, \ldots, m$$

and

$$\lambda_0^* \nabla f(x^*) - \sum_{i=1}^{m} \lambda_i^* \nabla c_i(x^*) = 0.$$

(a) What do you get from John's Theorem when you know that λ_0^* must be positive?

(b) Let x^* be a local minimizer for problem (11.13). Show that if the LICQ is satisfied at x^*, then λ_0^* (from John's Theorem) must be positive.

11.19   (Based on Abadie [1, p. 35].) This exercise explores the question of whether the KKTCQ is indispensable. That is to say, if the KKT conditions hold at a local minimum, does the KKTCQ hold there as well?

Define the functions $s(t)$, and $c(t)$ of the real variable t:

$$s(t) = \begin{cases} t^4 \sin \frac{1}{t} & \text{if } t \neq 0 \\ 0 & \text{if } t = 0 \end{cases}$$

$$c(t) = \begin{cases} t^4 \cos \frac{1}{t} & \text{if } t \neq 0 \\ 0 & \text{if } t = 0 \end{cases}$$

These functions are continuously differentiable. The functions and their derivatives vanish at 0.

(a) Consider the nonlinear program

$$\text{minimize} \quad f(x) = x_2$$
$$\text{subject to} \quad c_1(x) = s(x_1) - x_2 + x_1^2 \geq 0$$
$$c_2(x) = x_2 - x_1^2 - c(x_1) \geq 0$$
$$c_3(x) = 1 - x_1^2 \qquad \geq 0$$

Show that the feasible region lies between the curves $x_2 = x_1^2 - x_1^4$ and $x_2 = x_1^2 + x_1^4$. Indeed, show that

$$x_1^2 - x_1^4 \leq x_1^2 + c(x_1) \leq x_2 \leq x_1^2 + s(x_1) \leq x_1^2 + x_1^4,$$

and that the feasible region also lies between the lines $x_1 = -1$ and $x_1 = +1$.

(b) Verify that the unique optimal solution to the problem is $\bar{x} = 0$ and that $\mathcal{A}(\bar{x}) = \mathcal{A}(0) = \{1, 2\}$.

(c) Show that

$$\nabla c_1(0) = (0, -1),$$
$$\nabla c_2(0) = (0, 1).$$

(d) Describe the set of all v such that

$$v^T \nabla c_i(0) \geq 0, \quad i = 1, 2.$$

(e) Show that the only curve lying in the feasible region is identically 0. [Hint: Compute

$$x_1^2 + s(x_1) - (x_1^2 + c(x_1)) = x_1^4 \left(\sin \frac{1}{x_1} - \cos \frac{1}{x_1} \right).$$

Observe that this is not always nonnegative.]

(f) Could the curve in part (e) satisfy the requirements of the KKTCQ? Explain.

(g) Verify that the KKT conditions for this problem reduce to

$$u_1 \geq 0, \quad u_2 \geq 0, \quad 1 + u_1 - u_2 = 0.$$

(h) Write down a solution of the KKT conditions for this problem.

(i) What is the answer to the question posed at the beginning of this exercise?

11.20 (Based on Kelly and Kupferschmid [109, p. 113].) The first-order (KKT) conditions for the nonlinear program

$$\text{minimize} \quad 40x_1 x_2 + 20x_2 x_3$$
$$\text{subject to} \quad -x_1^{-1} x_2^{-1/2} - 3x_2^{-1} x_3^{-2/3} + 5 \geq 0$$
$$x_1 - 1 \geq 0$$
$$x_2 - 1 \geq 0$$
$$x_3 - 1 \geq 0$$

are satisfied at the point $\bar{x} = (\frac{1}{2}, 1, 1)$ with Lagrange multiplier vector $\bar{\lambda} = (250, 0, 0, 0)$. [Note that the first constraint function is not defined at the origin, but the function approaches $+\infty$ as $x > 0$ approaches 0. Such points are not feasible.] Use the second-order optimality criteria to decide whether $\bar{x}$ is a strict local minimizer for this problem.

11.21   Consider the optimization problem

$$\begin{aligned}
\text{maximize} \quad & x_1 \\
\text{subject to} \quad & (x_1 - 1)^3 + x_2 \leq 0 \\
& x_1 \qquad\qquad \leq 0 \\
& \qquad\quad x_2 \leq 0.
\end{aligned}$$

(a) Find an optimal solution for this problem. Is it unique?
(b) What is the feasible region of this problem?
(c) Is the Slater condition satisfied in this case?
(d) Does the Lagrangian function for this problem have a saddle point over $R_+^2 \times R_+^3$?

11.22   In Chapter 8 we spoke of an approach to the portfolio selection problem

$$\begin{aligned}
\text{minimize} \quad & -\gamma r^T x + x^T D x \\
\text{subject to} \quad & e^T x = 1, \ x \geq 0
\end{aligned}$$

in which one chooses γ from a finite set of values. Suppose you wished to solve this problem for a fixed value of γ. Write down the KKT conditions for this problem. Are these conditions sufficient as well as necessary for this problem? Why?

11.23   Other variants of the portfolio selection problem are given as follows:

(a) minimize $x^T D x$ subject to $e^T x = 1, \ x \geq 0$.
(b) maximize $r^T x$ subject to $x^T D x \leq \alpha, \ e^T x = 1, \ x \geq 0$.
(c) minimize $x^T D x$ subject to $e^T x = 1, \ r^T x \geq k, \ x \geq 0$.

Write down the KKT conditions for each of these optimization problems.

11.24   Use the second-order optimality conditions of quadratic programming (Theorem 11.10) to decide whether the point $\bar{x} = (\frac{105}{142}, \frac{529}{142})$ is a local minimizer for the optimization problem

$$\begin{aligned}
\text{minimize} \quad & 3x_1 - 10x_2 + \tfrac{1}{2}(x_1^2 + 4x_1 x_2 + x_2^2) \\
\text{subject to} \quad & 7x_1 - 3x_2 \geq -6 \\
& 5x_1 + 3x_2 \geq 9 \\
& x_1 + 7x_2 \geq 7 \\
& x_j \geq 0, \ j = 1, 2.
\end{aligned}$$

11.25   Let X be a nonempty convex subset of R^n and let $f : X \to R$ be a differentiable convex function. Show that the convex programming problem

$$(P) \qquad \text{minimize} \quad f(x) \quad \text{subject to} \quad x \in X$$

and the *variational inequality problem*

$$(VI) \qquad \text{find } x^* \in X \text{ such that } \nabla f(x^*) \cdot (x - x^*) \geq 0 \qquad \text{for all } x \in X$$

have the same solutions (if any).

11.26   Suppose $f : S \to R$ is a convex function having a local minimum at $\bar{x} \in S$. Show that $f(\bar{x}) \leq f(x)$ for all $x \in S$. Show further that if f is strictly convex, then $\bar{x}$ is a unique minimizer of f on S.

11.27   Let S and T be nonempty sets and let $f : S \to R$ and $g : T \to R$ be given functions. Suppose $\bar{s} \in S$, $\bar{t} \in T$, and $f(\bar{s}) = g(\bar{t})$. Show that if

$$f(s) \geq g(t) \qquad \text{for all } s \in S \text{ and } t \in T,$$

then

$$f(\bar{s}) = \min_{s \in S} f(s) \quad \text{and} \quad g(\bar{t}) = \max_{t \in T} g(t).$$

11.28   Exactly where is the gradient inequality used in the demonstration of Theorem 11.14?

11.29   The Lagrangian dual problem given in Section 11.5 makes use of the objective function

$$\theta(u, v) := \inf_{x \in R^n} L(x, u, v).$$

As noted there, this function is sometimes defined as

$$\theta(u, v) := \inf_{x \in X} L(x, u, v)$$

where X is a *subset* of R^n which perhaps could otherwise be represented among the constraints of the primal problem. Because the evaluation of $\theta(u, v)$ depends only on the primal objective and the set X, rather than any other constraints of the primal problem, the outcome of this evaluation can affect the corresponding duality gap. Illustrate this statement for the primal problem

$$\text{minimize} \ \{-x^2 : x^3 \geq 8, \ x \in X\}$$

first using $X = \{x : 0 \leq x \leq 2\}$ and then $X = R^n$. Find the duality gap in each case.

11.30   For $c \in R^n$ and symmetric positive semidefinite $H \in R^{n \times n}$ consider the quadratic programming problem

$$\text{minimize} \ \{c^T x + \tfrac{1}{2} x^T H x : \ x \geq 0\}.$$

(a) Write down the (Wolfe) dual corresponding to this primal QP.

(b) Show that if $\bar{x}$ is any optimal solution of the given primal QP, then it is also an optimal solution of the (Wolfe) dual QP.

11.31   Find all extremal values of z, and the corresponding values of x_1 and x_2, using the method of Lagrange multipliers in the following two instances.

(a)

$$x_1 x_2 + 4 = z$$
$$x_1^2 + x_2^2 = 8.$$

(b)

$$x_1 + 2x_2 = z$$
$$x_1^2 + x_2^2 = 13.$$

11.32   Consider the following univariate optimization problem

$$\begin{array}{ll} \text{minimize} & x^2 \\ \text{subject to} & x^2 - 5x + 6 \le 0. \end{array}$$

(a) Form the Lagrangian.

(b) Find the dual of the problem.

(c) Verify the Weak Duality Theorem for this problem.

(d) Does Strong Duality hold, or is there a duality gap?

11.33   Find the Lagrangian dual of a linear program in standard form.

11.34   To find the least 2-norm solution of a consistent system of linear equations ne can solve the following optimization problem.

$$\begin{array}{ll} \text{minimize} & x^{\mathrm{T}} x \\ \text{subject to} & Ax = b. \end{array}$$

(a) Why is it permissible to use the square of the 2-norm as the objective in this optimization problem?

(b) Find the Lagrangian dual.

(c) Is the dual objective function concave?

(d) Is there a duality gap in this instance?

11.35   What is the Lagrangian dual of the convex quadratic program

$$\begin{array}{ll} \text{minimize} & x^{\mathrm{T}} H x + c^{\mathrm{T}} x + d \\ \text{subject to} & Ax \ge b. \end{array}$$

11.36   Use the Lagrangian dual process to show that

$$X = \{x \in R^3 : x_1^2 + x_2^2 + x_3^2 \le 1 \text{ and } x_1 + x_2 + x_3 = 2\}$$

is an empty set.

11.37   (Based S.G. Krantz [114].) Let $\mathcal{P}_n$ denote an n-cube centered at the origin in R^n, having edge length 4 and sides parallel to the coordinate axes; in short, $\mathcal{P}_n = [-2, 2]^n$. Consider the set $\mathcal{C}_n$ of all points $c = (c_1, \ldots, c_n)$ for which $c_i = \pm 1$, $i = 1, \ldots, n$. For purposes of this exercise, the vector $e = (1, \ldots, 1)$ can be regarded as a representative element of the set $\mathcal{C}_n$. Let $\mathcal{B}_n(c, 1)$ denote the closed ball of radius 1 with center at $c \in \mathcal{C}_n$.

(a)   Make a graph depicting these definitions for the case $n = 2$.
(b)   In general, how many elements does $\mathcal{C}_n$ contain?
(c)   Write down a constraint that characterizes the points $x \in R^n$ belonging to $\mathcal{B}_n(e, 1)$.
(d)   Formulate a nonlinear convex program with a differentiable objective function and one differentiable constraint whose unique solution gives the least distance from the origin in R^n to the closed ball $\mathcal{B}_n(e, 1)$.
(e)   Explain why the KKT optimality conditions for this problem must have a solution.
(f)   Write the KKT optimality conditions for the above optimization problem.
(g)   Use the optimality conditions to show that the Lagrange multiplier, say λ, on the single constraint of your convex program has the value $\sqrt{n} - 1$ and that the optimal solution vector is

$$\bar{x} = \frac{\sqrt{n} - 1}{\sqrt{n}} e.$$

Notice that $\bar{x} \longrightarrow e$ as $n \longrightarrow \infty$.

(h)   The closed ball $\mathcal{B}_n(0, \sqrt{n} - 1)$ is centered at the origin and has radius equal to the distance from the origin to the nearest point of $\mathcal{B}_n(c, 1)$ for any $c \in \mathcal{C}_n$. Now let $\hat{x}$ be a point lying on one of the coordinate axes of R^n, say $\hat{x} = (\xi, 0, \ldots, 0)$. Then $\xi \in \mathcal{P}_n$ if and only if $\xi^2 \le 4$. On the other hand, if $\hat{x}$ lies on the boundary of $\mathcal{B}_n(0, \sqrt{n} - 1)$, then $\xi^2 = n - 2\sqrt{n} + 1$. Now, taking all dimensions n into consideration, is it always possible for $\hat{x}$ to belong to both $\mathcal{P}_n$ and the boundary of $\mathcal{B}_n(0, \sqrt{n} - 1)$? If not, for what dimensions is it possible?

12. PROBLEMS WITH LINEAR CONSTRAINTS

Overview

The optimization methods presented in this chapter are for solving the important class of nonlinear programs with linear constraints, that is, linear equations and/or linear inequalities. The algorithms covered here are based on ones designed for unconstrained optimization, but they are modified to take account of the constraints.

We have already seen (for example on page 377) that in problems with linear inequality constraints there is a notion of *active constraints* and that for local minimizers, the nonactive constraints are pretty much out of the picture. In the case of such problems, the active constraints give rise to an associated linear equality-constrained problem. We start with a discussion of solving such problems and then move on to active set algorithms for solving linear inequality-constrained problems. Active set algorithms raise some issues having to do with the adding or dropping of inequalities from the current "working set" of constraints.

As a part of the discussion, we present three methods for obtaining the matrix Z whose columns form a basis for the null space of the coefficient matrix corresponding to the active constraints. Actually, commercial solvers update the factorizations of Z as the working set changes; a discussion of such techniques is outside the scope of this book.

The chapter ends with brief treatments of bound-constraint problems, linear programs, and quadratic programs. Among other things, we show the well-known result that the Revised Simplex Method is exactly the same as an Active Set Method applied to a linear program.

12.1 Linear equality constraints

Because of their intrinsic importance as models for real-world problems (see Chapter 8) and, just as importantly, because of their role as *subproblems* in algorithms with more general constraints, we begin with a discussion of the minimization of a smooth nonlinear function subject to linear equality

© Springer Science+Business Media LLC 2017
R.W. Cottle and M.N. Thapa, *Linear and Nonlinear Optimization*,
International Series in Operations Research & Management Science 253,
DOI 10.1007/978-1-4939-7055-1_12

constraints. We write this as

$$\text{(LEP)} \qquad \begin{array}{ll} \text{minimize} & f(x) \\ \text{subject to} & \hat{A}x = \hat{b} \end{array} \qquad (12.1)$$

where $f(x)$ is a nonlinear function and $\hat{A}$ is an $\ell \times n$ matrix. We do not consider the trivial case where $f(x)$ is linear; it is not a linear programming problem (see Exercise 12.1). The reason for using the notation $\hat{A}$ instead of A will become clear in our discussion of solution methods for problems containing linear inequality constraints.

As noted in Chapter 11 (see page 375 and the footnote at the end of Example 11.2), we can use the Method of Lagrange Multipliers to write first-order optimality conditions for (12.1): If x^* is a local minimizer for (12.1), there exists a vector of multipliers λ such that $g(x^*) = \nabla f(x^*) - \hat{A}^\mathrm{T}\lambda = 0$. In some cases, these equations and those of the constraints $\hat{A}x = \hat{b}$ enable us to develop a closed-form solution of (12.1), but this is not to be expected in all instances of the problem. Much depends on the behavior of the objective function f restricted to the set of points x satisfying $\hat{A}x = \hat{b}$.

One approach to solving (12.1) is to find a basis in $\hat{A}$ and then obtain the *restriction* of f to the set of solutions of the system of linear equations; this yields a smooth *unconstrained* function F of the nonbasic variables alone. (See Section 11.2 and Exercise 12.2.) Another approach to solving (12.1), followed here, is to avoid the explicit elimination of basic variables and remain within the framework of constrained optimization.

Optimality conditions

In this section we derive optimality conditions for the linear equality problem (12.1) by considering feasible moves from a feasible starting point. The derivation of these conditions in this manner helps in the algorithm design.

It would be well to recall a matter brought up in Chapter 2, namely, that given the matrix $\hat{A} \in R^{\ell \times n}$, we can decompose R^n into its corresponding complementary orthogonal subspaces. Indeed, let $\mathcal{N}(\hat{A})$ denote the *null space* of $\hat{A}$ and let $\mathcal{R}(\hat{A}^\mathrm{T})$ denote the *range space* of $\hat{A}^\mathrm{T}$:

$$\begin{aligned} \mathcal{N}(\hat{A}) &= \{z \in R^n : \hat{A}z = 0\} \\ \mathcal{R}(\hat{A}^\mathrm{T}) &= \{w \in R^n : w = \hat{A}^\mathrm{T}y, \ y \in R^\ell\}. \end{aligned} \qquad (12.2)$$

There are two key properties of these subspaces. The first is that

$$R^n = \mathcal{R}(\hat{A}^{\mathrm{T}}) + \mathcal{N}(\hat{A}). \tag{12.3}$$

This means that every point of R^n can be expressed as the sum of two vectors, one belonging to the null space of $\hat{A}$, the other belonging to the range space of $\hat{A}^{\mathrm{T}}$. Accordingly, any point $x \in R^n$ can be written as

$$x = \hat{A}^{\mathrm{T}}x_{\hat{A}} + Zx_Z. \tag{12.4}$$

The second key property is that these two subspaces are orthogonal to each other and have only the zero vector in common. Indeed, if $z \in \mathcal{N}(\hat{A})$ and $x \in \mathcal{R}(\hat{A}^{\mathrm{T}})$ we have

$$\hat{A}z = 0 \quad \text{and} \quad x = \hat{A}^{\mathrm{T}}y$$

so

$$x^{\mathrm{T}}z = y^{\mathrm{T}}\hat{A}z = 0. \tag{12.5}$$

If $x \in \mathcal{R}(\hat{A}^{\mathrm{T}}) \cap \mathcal{N}(\hat{A})$ we may take $x = z$ in (12.5) and immediately deduce that $x = 0$.

Why does $R^n = \mathcal{N}(\hat{A}) + \mathcal{R}(\hat{A}^{\mathrm{T}})$? The matrix $\hat{A}$ has rank l, so

$$\dim \mathcal{N}(\hat{A}) = l$$
$$\dim \mathcal{R}(\hat{A}^{\mathrm{T}}) = n - l.$$

Since $\mathcal{R}(\hat{A}^{\mathrm{T}}) \cap \mathcal{N}(\hat{A}) = \{0\}$, it follows that the vectors of $\mathcal{N}(\hat{A})$ and $\mathcal{R}(\hat{A}^{\mathrm{T}})$ span an n-dimensional space, namely R^n.

Note: *This notation Z, as in (12.4), will be used many times in our discussion of nonlinear programming in this book; it will usually mean a basis for the null space of a specified matrix, namely the coefficient matrix $\hat{A}$ of the constraints satisfied as equalities. However, if the columns of $\hat{A}$ are linearly independent, then the null space includes exactly one vector: the zero vector. In this case we will express Z as a zero matrix of dimension $n \times 1$.*

Example 12.1: NULL SPACE AND RANGE SPACE. Consider the following trivial system of equations of the form $\hat{A}x = \hat{b}$:

$$\begin{bmatrix} 1 & 0 & 0 \\ 0 & 0 & 1 \end{bmatrix} \begin{bmatrix} x_1 \\ x_2 \\ x_3 \end{bmatrix} = \begin{bmatrix} 4 \\ 2 \end{bmatrix}.$$

The range space of $\hat{A}^{\mathrm{T}}$ and null space of $\hat{A}$ are spanned by the columns of $\hat{A}^{\mathrm{T}}$ and Z, respectively, where

$$\hat{A}^{\mathrm{T}} = \begin{bmatrix} 1 & 0 \\ 0 & 0 \\ 0 & 1 \end{bmatrix} \quad \text{and} \quad Z = \begin{bmatrix} 0 \\ 1 \\ 0 \end{bmatrix}.$$

Clearly the columns of $\hat{A}^{\mathrm{T}}$ and Z can be used to represent any vector in R^3. Now consider the system of equations

$$\begin{array}{rcl} x_1 & = & 4 \\ 2x_1 + 4x_2 + x_3 & = & 6 \end{array} \quad \text{or, equivalently,} \quad \begin{bmatrix} 1 & 0 & 0 \\ 2 & 4 & 1 \end{bmatrix} \begin{bmatrix} x_1 \\ x_2 \\ x_3 \end{bmatrix} = \begin{bmatrix} 4 \\ 6 \end{bmatrix}.$$

The columns of the matrices

$$\hat{A}^{\mathrm{T}} = \begin{bmatrix} 1 & 2 \\ 0 & 4 \\ 0 & 1 \end{bmatrix} \quad \text{and} \quad Z = \begin{bmatrix} 0 \\ -1 \\ 4 \end{bmatrix}.$$

span the range space of $\hat{A}^{\mathrm{T}}$ and null space of Z respectively. In this case too, any vector in R^3 can be represented by a linear combination of the columns of $\hat{A}$ and Z. For example, the vector $(4, 3, 5)$ can be represented using $x_{\hat{A}} = (2, 1)$ and $x_Z = (1)$ and forming $\hat{A}^{\mathrm{T}} x_{\hat{A}} + Z x_Z$. It is easy to verify that the product $\hat{A}Z = 0$.

Let us assume we have a feasible point $x = \bar{x}$ for the constraints of (12.1). Any other feasible point will differ from $\bar{x}$ by a direction p such that $\bar{x} + p$ satisfies the constraints $\hat{A}x = \hat{b}$. That is,

$$\hat{A}(\bar{x} + p) = \hat{A}\bar{x} + \hat{A}p = \hat{b}.$$

This clearly implies that $\hat{A}p = 0$. Thus, every feasible direction *must* belong to the null space of $\hat{A}$. For this reason, the null space of $\hat{A}$ plays a large role in the approach to be developed.

Let x^* be a local minimizer for (LEP), and let p denote an arbitrary *feasible direction* at x^*. Then p must be an element of $\mathcal{N}(\hat{A})$. If Z be a matrix whose columns form a basis for the null space of $\hat{A}$, every feasible direction p can be written as

$$p = Z p_z$$

for some vector p_Z of dimension $n - \ell$. For any $\alpha \in R$, Taylor's Theorem gives us

$$
\begin{aligned}
f(x^* + \alpha p) &= f(x^* + \alpha Z p_Z) \\
&= f(x^*) + \alpha p_Z^T Z^T g(x^*) + \tfrac{\alpha^2}{2} p_Z^T Z^T G(x^*) Z p_Z + \cdots
\end{aligned}
$$

where, as before, $g(x^*)$ denotes $\nabla f(x^*)$, and $G(x^*)$ denotes $\nabla^2 f(x^*)$. If the linear term is nonzero for any p_Z, then there exists a neighborhood of x^* in which $f(x^* + \varepsilon Z p_Z) < f(x^*)$, and $x^* + \varepsilon Z p_Z$ is feasible. Therefore, in order for x^* to be a local minimizer, it is *necessary* that

$$
p_Z^T Z^T g(x^*) = 0 \quad \text{for all } p_Z \in R^{n-\ell}.
$$

It follows that

$$
Z^T g(x^*) = 0. \tag{12.6}
$$

The condition (12.6) at the solution implies that $g(x^*)$ *must lie entirely in the range space of* $\hat{A}^T$. Indeed, if

$$
g(x^*) = \hat{A}^T g_{\hat{A}} + Z g_Z,
$$

then,

$$
Z^T g(x^*) = Z^T \hat{A}^T g_{\hat{A}} + Z^T Z g_Z = Z^T Z g_Z.
$$

Since Z is a basis for $\mathcal{N}(\hat{A})$, the matrix $Z^T Z$ is positive definite, and thus (12.6) implies

$$
Z^T Z g_Z = 0,
$$

that is, $g_Z = 0$, and

$$
g(x^*) = \hat{A}^T g_{\hat{A}}.
$$

The components of $g_{\hat{A}}$ corresponding to the columns of $\hat{A}^T$ in the representation[1] of $g(x^*)$ above are usually denoted by $\lambda_1^*, \lambda_2^*, \ldots, \lambda_l^*$ and are termed the *Lagrange multipliers* at x^*.

First-order conditions for a minimum

In light of the above, the necessary conditions for a solution to the problem with linear equality constraints can take two equivalent forms:

---

[1] This equation should come as no surprise. It is just another expression of the first-order optimality conditions seen in Chapter 11.

1. $Z^{\mathrm{T}}g(x^*) = 0$; that is, the *projected gradient* at x^* is zero.

2. $g(x^*) = \hat{A}^{\mathrm{T}}\lambda^*$; that is, the gradient at x^* is a linear combination of the columns of $\hat{A}^{\mathrm{T}}$, or equivalently, $g(x^*)$ lies in the range space of $\hat{A}^{\mathrm{T}}$.

Second-order conditions for a minimum

We next wish to derive some *second-order optimality conditions*. Since $Z^{\mathrm{T}}g(x^*) = 0$, the expansion of f along Zp_z can be written as:

$$f(x^* + \varepsilon Zp_z) = f(x^*) + \frac{\varepsilon^2}{2}p_z^{\mathrm{T}}Z^{\mathrm{T}}G(x^*)Zp_z + \cdots$$

Again, as in the unconstrained case, we see that the matrix $Z^{\mathrm{T}}G(x^*)Z$ must be positive semidefinite; otherwise we could find a neighborhood of x^* in which the value of f is less than $f(x^*)$ at feasible points. The matrix $Z^{\mathrm{T}}G(x^*)Z$ is termed the *projected Hessian* of f at x^*.

Thus, in summary, the *necessary conditions* for x^* to be a local minimizer for (LEP) are:

1. $\hat{A}x^* = \hat{b}$.

2. $Z^{\mathrm{T}}g(x^*) = 0$, (or $g(x^*) = \hat{A}^{\mathrm{T}}\lambda^*$).

3. $Z^{\mathrm{T}}G(x^*)Z$ is positive semidefinite.

It can be shown that the second-order *sufficient conditions* for x^* to be an isolated local minimizer for (LEP) are:

1. $\hat{A}x^* = \hat{b}$.

2. $Z^{\mathrm{T}}g(x^*) = 0$, (or $g(x^*) = \hat{A}^{\mathrm{T}}\lambda^*$).

3. $Z^{\mathrm{T}}G(x^*)Z$ is positive definite. This may alternatively be stated as the requirement that $y^{\mathrm{T}}G(x^*)y > 0$ for all $y \neq 0$ such that $\hat{A}y = 0$ (or $y \in \mathcal{N}(\hat{A})$).

Some observations

There are three essential points to note about these results:

- At a local minimizer x^*, the gradient of f might, in general, not be zero, but the projected gradient of f *must* be zero.

- The necessary conditions for x^* to be a local minimizer require that the projected Hessian $Z^{\mathrm{T}}G(x^*)Z$ be positive semidefinite. The Hessian matrix $G(x^*)$ may in fact be indefinite or singular.

- When the constraints are linear equalities, every feasible move can be characterized in terms of the matrix Z, the columns of which represent the null space of $\hat{A}$. Furthermore, *all* points of the form $x + \alpha Z p_z$ are feasible if x is, so that we can guarantee feasibility by moving only along directions of the form $Z p_z$.

The following two examples illustrate the above observations.

Example 12.2: OPTIMALITY CONDITIONS FOR LEP. In the two-variable optimization problem

$$\begin{array}{ll} \text{minimize} & 2x_1^2 - x_2^2 - 12x_1 \\ \text{subject to} & x_1 + x_2 = 2 \end{array}$$

the objective function f is minimized at $x^* = (4, -2)$ (as can be easily verified by reducing the problem to a one-variable unconstrained optimization problem). The gradient at x^* is $g(x^*) = (4x_1^* - 12, -2x_2^*) = (4, 4) \neq (0,0)$. The Hessian at x^*, namely

$$G(x^*) = \begin{bmatrix} 4 & 0 \\ 0 & -2 \end{bmatrix},$$

is not positive semidefinite. We next show that $Z^{\mathrm{T}}g(x^*) = 0$ and $Z^{\mathrm{T}}G(x^*)Z$ is positive definite as expected. For the above LEP it is easy to see that for $\hat{A} = [1\ 1]$, the corresponding matriz Z is given by

$$Z = \begin{bmatrix} 1 \\ -1 \end{bmatrix}.$$

The projected gradient is zero:

$$Z^{\mathrm{T}}g(x^*) = \begin{bmatrix} 1 & -1 \end{bmatrix} \begin{bmatrix} 4 \\ 4 \end{bmatrix} = 0,$$

and the projected Hessian is positive definite:

$$Z^{\mathrm{T}}G(x^*)Z = \begin{bmatrix} 1 & -1 \end{bmatrix} \begin{bmatrix} 4 & 0 \\ 0 & -2 \end{bmatrix} \begin{bmatrix} 1 \\ -1 \end{bmatrix} = 2.$$

Example 12.3: FEASIBLE DIRECTIONS FOR LEP. For an LEP, all vectors in the null space of the constraint matrix can be used to construct feasible directions independent of the current feasible point. For example, consider the two-variable optimization problem in Example 12.2. Let $p_Z = \beta$ for any scalar β (the null space basis has one column, and so p_Z is a scalar). Then it is straightforward to verify that every feasible direction is given by $Z\beta$; that is,

$$p = Zp_Z = Z\beta = \begin{bmatrix} 1 \\ -1 \end{bmatrix}\beta = \begin{bmatrix} \beta \\ -\beta \end{bmatrix}.$$

Algorithms for solving (LEP)

Based on the discussion above, one can create an algorithm to solve (12.1). Given an initial feasible point, the idea is to move to a better feasible point by choosing at every step a suitable feasible search direction p of the form

$$p = Zp_Z. \tag{12.7}$$

The problem, at each iteration, is to find a p (as in the unconstrained case) such that the function decreases when a step is taken along it. When the current iterate is $x = \bar{x}$, one approach would be to choose p so that $\bar{x} + Zp_Z$ is a local minimizer of $f(\bar{x} + Zp_Z)$, or an approximation of such a minimizer. The function $f(x + Zp_Z)$, which is a restriction of the objective function to the feasible region, is sometimes referred to as the *projected function*.

It may appear a bit odd that all the computations are done within the null space of $\hat{A}$. We elaborate further on this. It turns out that the reason for this is that the linear constraints completely determine $x^*_{\hat{A}}$. As we know, all feasible points may be decomposed as $\hat{A}^Tx^*_{\hat{A}} + Zx_Z$. Then

$$\hat{A}x^* = \hat{A}(\hat{A}^Tx^*_{\hat{A}} + Zx^*_Z) = \hat{A}\hat{A}^Tx^*_{\hat{A}} = \hat{b}. \tag{12.8}$$

(See Exercise 12.3 for further details.) Thus (12.8) can be solved for $x^*_{\hat{A}}$; we do not, however, solve for $x^*_{\hat{A}}$ explicitly. We still need to determine the null space component x_Z. The vector x^*_Z is determined not by the constraints, but by the *unconstrained minimization* of $f(\hat{A}^Tx^*_{\hat{A}} + Zx_Z)$ with respect to x_Z. Note that since $x_Z \in R^{n-\ell}$, where $\ell = \hat{A}$, we can reduce the dimensionality of the minimization problem, so our initial intuition was correct.

Obtaining search directions for (LEP)

As in the unconstrained case, the search directions have to be descent directions. The search directions take the form $p = Zp_z$, hence p is a descent direction for f at $x = \bar{x}$ if and only if

$$p^{\mathrm{T}}g(\bar{x}) = p_z^{\mathrm{T}}Z^{\mathrm{T}}g(\bar{x}) < 0$$

or

$$p_z^{\mathrm{T}}g_z < 0;$$

that is, p_z is a direction of descent for the projected function $f(\bar{x} + Zp_z)$.

Steepest descent

As we have seen in Section 10.1, the steepest-descent direction at $x = x_k$ is the direction p that solves the problem

$$\text{minimize} \quad \frac{g_k^{\mathrm{T}}p}{\|p\|_2} \quad \text{where } g_k = g(x_k).$$

Since $p = Zp_z$, this problem becomes

$$\text{minimize} \quad \frac{p_z^{\mathrm{T}}Z^{\mathrm{T}}g_k}{\|p_z\|_{Z^{\mathrm{T}}Z}} \quad \text{where } \|p_z\|^2_{Z^{\mathrm{T}}Z} = p_z^{\mathrm{T}}Z^{\mathrm{T}}Zp_z. \tag{12.9}$$

The solution to this problem is given by equation (10.3) with $C = Z^{\mathrm{T}}Z$; i.e.,

$$p_z = -(Z^{\mathrm{T}}Z)^{-1}Z^{\mathrm{T}}g_k,$$

so that p, the steepest-descent direction, is given by

$$p = Zp_z = -Z(Z^{\mathrm{T}}Z)^{-1}Z^{\mathrm{T}}g_k. \tag{12.10}$$

This search direction $p = -Z(Z^{\mathrm{T}}Z)^{-1}Z^{\mathrm{T}}g_k$ is the direct generalization of the steepest-descent direction in the unconstrained case to the linear equality-constrained case; this generalization was first suggested by Rosen [168], who termed the related algorithm the *gradient-projection algorithm*.

Wolfe [195] suggested using the direction

$$p = -ZZ^{\mathrm{T}}g_k \tag{12.11}$$

which he termed the *reduced-gradient* direction. This choice of search direction may also be viewed as a direction of steepest descent if the problem (12.9) is stated as

$$\text{minimize} \quad \frac{p_z^T Z^T g_k}{\|p_z\|_2},$$

where we have replaced the $Z^T Z$-norm by the 2-norm. The solution is then given by $p_z = -Z^T g_k$; thus p is given by (12.11).

Note that the reduced-gradient direction is the same as the gradient projection direction only if Z is an orthogonal matrix (i.e., $Z^T Z = I$), so that its columns form an *orthogonal basis* for the null space. However, it is clear from this formulation that the two techniques have the same motivation, and the distinction is not substantial. They are sometimes seen as different because of the particular form of Z chosen by Wolfe.

Example 12.4: STEEPEST-DESCENT DIRECTION FOR LEP.

Consider the three-variable optimization problem

$$\begin{aligned} \text{minimize} \quad & 2x_1^2 - x_2^2 + x_3^2 - 12x_1 - x_1 x_3 \\ \text{subject to} \quad & x_1 + x_2 = 2. \end{aligned}$$

At a feasible point $\bar{x} = (1, 1, 1)$, the gradient is

$$g(\bar{x}) = (4\bar{x}_1 - 12 - \bar{x}_3, -2\bar{x}_2, 2\bar{x}_3 - \bar{x}_1) = (-9, -2, 1).$$

For the above LEP we have $\hat{A} = [1\ 1\ 0]$, and a corresponding representation of the null space is given by

$$Z = \begin{bmatrix} 1 & 0 \\ -1 & 0 \\ 0 & 1 \end{bmatrix}.$$

The projected gradient at $\bar{x}$ is

$$Z^T g(\bar{x}) = \begin{bmatrix} 1 & -1 & 0 \\ 0 & 0 & 1 \end{bmatrix} \begin{bmatrix} -9 \\ -2 \\ 1 \end{bmatrix} = \begin{bmatrix} -7 \\ 1 \end{bmatrix}.$$

To compute Rosen's steepest-descent direction we first compute

$$(Z^T Z)^{-1} = \left(\begin{bmatrix} 1 & -1 & 0 \\ 0 & 0 & 1 \end{bmatrix} \begin{bmatrix} 1 & 0 \\ -1 & 0 \\ 0 & 1 \end{bmatrix} \right)^{-1} = \begin{bmatrix} 2 & 0 \\ 0 & 1 \end{bmatrix}^{-1} = \begin{bmatrix} \frac{1}{2} & 0 \\ 0 & 1 \end{bmatrix}.$$

Then Rosen's steepest-descent direction is given by

$$p = -Z(Z^\mathrm{T}Z)^{-1}Z^\mathrm{T}g(\bar{x}) = - \begin{bmatrix} 1 & 0 \\ -1 & 0 \\ 0 & 1 \end{bmatrix} \begin{bmatrix} \frac{1}{2} & 0 \\ 0 & 1 \end{bmatrix} \begin{bmatrix} -7 \\ 1 \end{bmatrix} = \begin{bmatrix} 7/2 \\ -7/2 \\ -1 \end{bmatrix}.$$

Wolfe's reduced-gradient descent direction is given by

$$p = -ZZ^\mathrm{T}g(\bar{x}) = - \begin{bmatrix} 1 & 0 \\ -1 & 0 \\ 0 & 1 \end{bmatrix} \begin{bmatrix} -7 \\ 1 \end{bmatrix} = \begin{bmatrix} 7 \\ -7 \\ -1 \end{bmatrix}.$$

Note that if the columns of Z had been chosen to be orthonormal, for example

$$Z = \begin{bmatrix} 1/\sqrt{2} & 0 \\ -1/\sqrt{2} & 0 \\ 0 & 1 \end{bmatrix},$$

then Rosen's direction and Wolfe's direction would indeed be the same.

Both the gradient projection and reduced gradient directions can be combined with a steplength procedure that ensures a "sufficient" decrease in f at every iteration. However, both these methods would then resemble the Steepest-descent Method (in the unconstrained case) in terms of extremely slow linear convergence. Therefore, we consider algorithms that take account of curvature (second-order information) in the choice of search direction, precisely as in the unconstrained case.

Newton-type Methods

If the Hessian matrix $G(x)$ is available at each iteration, and a feasible point has been found, then as seen above, the subproblem under consideration at $x = x_k$ is to minimize $f(x_k + Zp_z)$. Using the Taylor series expansion along such a direction Zp_z, we have

$$f(x_k + Zp_z) = f(x_k) + p_z^\mathrm{T}Z^\mathrm{T}g_k + \tfrac{1}{2}p_z^\mathrm{T}Z^\mathrm{T}G_kZp_z + \cdots$$

As in the unconstrained case, a Newton-type Method is based on ignoring terms beyond the quadratic and choosing p_z as the step to the minimizer for the local quadratic model. If $Z^\mathrm{T}G_kZ$ is positive definite, p_z is given by the solution of the nonsingular linear system

$$Z^\mathrm{T}G_kZp_z = -Z^\mathrm{T}g_k. \tag{12.12}$$

This is exactly analogous to the unconstrained Newton-type Method except that the linear system involves the projected Hessian on the left-hand side and the projected gradient on the right-hand side. The search direction $p = Zp_z$ is clearly a direction of descent for f. Indeed, since Z^TG_kZ is assumed positive definite, so is its inverse, hence

$$p_z^T Z^T g_k = -g_k^T Z \left(Z^T G_k Z\right)^{-1} Z^T g_k < 0.$$

If $Z^T G_k Z$ is not positive definite, the local quadratic approximation is either unbounded below, or does not have a unique minimum. In such a case, it is necessary to modify the choice of p_z. To do this, techniques similar to those applied for unconstrained optimization can be utilized; the modified Cholesky factorization is such a technique; see Chapter 10. Likewise, there are techniques that cover the case where the Hessian matrix exists but is not available in analytic form. Quasi-Newton Methods, Finite-Difference Newton Methods, and Conjugate-gradient Methods are examples of algorithms that can be used for this purpose; again, see Chapter 10.

Example 12.5: NEWTON DIRECTION FOR LEP. Consider the 3-variable optimization problem in Example 12.4. At a feasible point $\bar{x} = (1, 1, 1)$, the gradient $g(\bar{x}) = (4\bar{x}_1 - 12 - \bar{x}_3, -2\bar{x}_2, 2\bar{x}_3 - \bar{x}_1) = (-9, -2, 1)$ and the Hessian matrix is

$$G(\bar{x}) = \begin{bmatrix} 4 & 0 & -1 \\ 0 & -2 & 0 \\ -1 & 0 & 2 \end{bmatrix}.$$

The projected gradient at $\bar{x} = (1, 1, 1)$ is, as shown before, $Z^T g(\bar{x}) = (-7, 1)$, and the projected Hessian is

$$Z^T G(\bar{x}) Z = \begin{bmatrix} 1 & -1 & 0 \\ 0 & 0 & 1 \end{bmatrix} \begin{bmatrix} 4 & 0 & -1 \\ 0 & -2 & 0 \\ -1 & 0 & 2 \end{bmatrix} \begin{bmatrix} 1 & 0 \\ -1 & 0 \\ 0 & 1 \end{bmatrix} = \begin{bmatrix} 2 & -1 \\ -1 & 2 \end{bmatrix}.$$

Solving $Z^T G(\bar{x}) Z p_z = -Z^T g(\bar{x})$, that is, solving

$$\begin{bmatrix} 2 & -1 \\ -1 & 2 \end{bmatrix} p_z = -\begin{bmatrix} -7 \\ 1 \end{bmatrix},$$

we obtain

$$p_z = \begin{bmatrix} 13/3 \\ 5/3 \end{bmatrix}.$$

Then Newton's direction is given by

$$
Z p_z = \begin{bmatrix} 1 & 0 \\ -1 & 0 \\ 0 & 1 \end{bmatrix} \begin{bmatrix} 13/3 \\ 5/3 \end{bmatrix} = \begin{bmatrix} 13/3 \\ -13/3 \\ 5/3 \end{bmatrix}.
$$

12.2   Methods for computing Z

Until now, we have assumed that the matrix Z is available for use in the algorithms to solve the linear equality problem (12.1). We have done this intentionally to make it clear that the same underlying ideas are present regardless of the choice of Z. For example, the idea of "moving in the null space" is independent of the particular Z used as a basis.

For a robust computer implementation of such an algorithm, it is necessary to ensure that the chosen Z has satisfactory numerical properties. Furthermore, when solving inequality-constrained problems, for reasons of computational efficiency, it will be necessary to consider how to update a given representation of Z, rather than compute it afresh. In discussing the following three methods, we have assumed for simplicity of exposition that $\hat{A} \in R^{\ell \times n}$ and that $\operatorname{rank}(\hat{A}) = \ell$.

Z derived from a basis in $\hat{A}$

To construct such a Z we need to find a nonsingular $\ell \times \ell$ submatrix, say B, a basis, in $\hat{A}$. One method to do this is to start by creating the LU factors, one column at a time, starting with the first column of $\hat{A}$ that has a nonzero element in row 1 and reordering to make that column 1. If the first ℓ columns are linearly independent, then the factorization process terminates after processing ℓ columns because a basis is comprised of the first ℓ columns of $\hat{A}$. On the other hand, suppose that, when eliminations are performed for some column $k > 1$, we discover that fewer than the first k elements are nonzero. In this case, column k is dependent on the first $k - 1$ columns and so cannot be in the basis; we next work with column $k + 1$, and so on until a basis is found. Recall that we have assumed that $\hat{A}$ is of rank ℓ, so a basis exists.

For simplicity of exposition we assume that the columns of $\hat{A}$ have been reordered so that B occupies the first ℓ columns of $\hat{A}$. That is, $\hat{A} = [B\ N]$, where B is a basis in $\hat{A}$ and N denotes the remaining $n - \ell$ columns in $\hat{A}$, called *nonbasic columns*. Then the matrix

$$Z = \begin{bmatrix} -B^{-1}N \\ I \end{bmatrix}$$

has independent columns and satisfies

$$\hat{A}Z = [B\ N] \begin{bmatrix} -B^{-1}N \\ I \end{bmatrix} = 0.$$

We never need to compute Z explicitly. Recall that the search direction is given by $p = Zp_Z$, where the computation of p_Z involves Z^{T}. For example, in the Steepest-descent Method, letting $g = (g_A, g_Z)$, we have

$$p_Z = -Z^{\mathrm{T}}g = -\begin{bmatrix} -B^{-1}N \\ I \end{bmatrix}^{\mathrm{T}} \begin{bmatrix} g_A \\ g_Z \end{bmatrix} = N^{\mathrm{T}}(B^{-1})^{\mathrm{T}}g_A - g_Z.$$

The vector $u = (B^{-1})^{\mathrm{T}}g_A$ can be computed as the solution to $B^{\mathrm{T}}u = g_A$. Analogously, in order to compute

$$p = Zp_Z = \begin{bmatrix} -B^{-1}N \\ I \end{bmatrix} p_Z = \begin{bmatrix} -B^{-1}Np_Z \\ p_Z \end{bmatrix},$$

we need to determine $v = -B^{-1}Np_Z$. This is obtained as the solution to $Bv = -Np_Z$. By maintaining a factorization (for example, LU) of B we can determine all such matrix-vector products involving Z.

Example 12.6: Z DERIVED FROM A BASIS. Consider an LEP with

$$\hat{A} = \begin{bmatrix} -1 & 1 & 1 & 2 \\ 1 & 0 & 1 & 2 \end{bmatrix}.$$

A basis can be identified through the use of an LU factorization; in this simple example, we choose columns 1 and 2 as our basis. Then the basis B and its inverse are

$$B = \begin{bmatrix} -1 & 1 \\ 1 & 0 \end{bmatrix} \quad \text{and} \quad B^{-1} = \begin{bmatrix} 0 & 1 \\ 1 & 1 \end{bmatrix}.$$

Note that in practice we do not compute B^{-1}; we have done so here for ease of illustration. Next we compute $B^{-1}N$ as

$$B^{-1}N = \begin{bmatrix} 0 & 1 \\ 1 & 1 \end{bmatrix} \begin{bmatrix} 1 & 2 \\ 1 & 2 \end{bmatrix} = \begin{bmatrix} 1 & 2 \\ 2 & 4 \end{bmatrix}.$$

Then Z is given by

$$Z = \begin{bmatrix} -B^{-1}N \\ I \end{bmatrix} = \begin{bmatrix} 1 & 2 \\ 2 & 4 \\ 1 & 0 \\ 0 & 1 \end{bmatrix}.$$

Z as the last $n - \ell$ columns of the inverse of a matrix

Another method for obtaining a Z is to first adjoin a matrix V consisting of $n - \ell$ linearly independent rows to the matrix $\hat{A}$ (once again assumed of full rank), so as to obtain a nonsingular $n \times n$ matrix E given by

$$E = \begin{bmatrix} \hat{A} \\ V \end{bmatrix}.$$

Then we can choose Z to be the last $n - \ell$ columns of the inverse of E. This approach is not used in practice because it requires a portion of the inverse of E. It should be noted however that E^{-1} need not be computed. Instead the last $n - \ell$ columns of the inverse of E can be obtained by solving systems of equations of the form $EZ_{\bullet j} = I_{\bullet \ell+j}$.

Example 12.7: Z DERIVED FROM AN INVERSE. To the matrix $\hat{A}$ from Example 12.6, we adjoin rows of the identity matrix to obtain a nonsingular matrix E, that is

$$E = \begin{bmatrix} \hat{A} \\ V \end{bmatrix} = \begin{bmatrix} -1 & 1 & 1 & 2 \\ 1 & 0 & 1 & 2 \\ 0 & 0 & 1 & 0 \\ 0 & 0 & 0 & 1 \end{bmatrix}.$$

Next we solve $EZ_{\bullet 3} = I_{\bullet 3}$ and $EZ_{\bullet 4} = I_{\bullet 4}$ by Gaussian elimination as follows:

$$\begin{bmatrix} -1 & 1 & 1 & 2 \\ 1 & 0 & 1 & 2 \\ 0 & 0 & 1 & 0 \\ 0 & 0 & 0 & 1 \end{bmatrix} [Z_{\bullet 3} \ Z_{\bullet 4}] = \begin{bmatrix} 0 & 0 \\ 0 & 0 \\ 1 & 0 \\ 0 & 1 \end{bmatrix}.$$

Adding the first row to the second row we obtain

$$
\begin{bmatrix} -1 & 1 & 1 & 2 \\ 0 & 1 & 2 & 4 \\ 0 & 0 & 1 & 0 \\ 0 & 0 & 0 & 1 \end{bmatrix} [Z_{\bullet 3} \ Z_{\bullet 4}] = \begin{bmatrix} 0 & 0 \\ 0 & 0 \\ 1 & 0 \\ 0 & 1 \end{bmatrix}.
$$

This system is upper-triangular and can be solved to give

$$
Z_{\bullet 3} = \begin{bmatrix} -1 \\ -2 \\ 1 \\ 0 \end{bmatrix} \quad \text{and} \quad Z_{\bullet 4} = \begin{bmatrix} -2 \\ -4 \\ 0 \\ 1 \end{bmatrix}.
$$

It is easy to verify that $\hat{A}Z_{\bullet 3} = \hat{A}Z_{\bullet 4} = 0$ as would be expected.

Z derived from the TQ factorization

A well-known approach to solving least-squares problems is to create a QR factorization of the coefficient matrix (which has more rows than columns). See [88]. A similar idea can be used here. In this case we construct an orthogonal matrix Q such that

$$
\hat{A}Q = [0 \ \ T]
$$

where, in this equation, 0 represents an $\ell \times (n-\ell)$ matrix of zeros and T is an $\ell \times \ell$ *reverse lower-triangular matrix* (an ordinary lower-triangular matrix with its columns in reverse order). The reason for using such a form has to do with handling *inequality* constraints; see Section 12.3.

In order to define a suitable matrix Z, consider the partition of the orthogonal matrix Q mentioned above:

$$
Q = [Z \ \ Y].
$$

Observe that

$$
\hat{A}Q = \hat{A}[Z \ \ Y] = [\hat{A}Z \ \ \hat{A}Y] = [0 \ \ T] \tag{12.13}
$$

so that the first $n - \ell$ columns of Q are orthogonal to the rows of $\hat{A}$. The columns of Q are linearly independent since they are columns of an orthogonal matrix. Thus the first $n - \ell$ columns of Q form an orthogonal basis for the desired null space. In addition, the columns of Y (the last ℓ columns of Q) form an orthogonal basis for the range space of $\hat{A}$. The discussion of how to obtain and update such a factorization is outside the scope of this book; see [84] for details of the procedure.

12.3   Linear inequality constraints

In the previous sections we examined the linear equality problem and solution techniques for such problems. We next consider the linear inequality problem:

$$\text{(LIP)} \qquad \begin{aligned} &\text{minimize} && f(x) \\ &\text{subject to} && Ax \geq b \end{aligned} \qquad \text{where } A \text{ is } m \times n. \qquad (12.14)$$

For simplicity of exposition, we shall not consider problems with a mixture of equalities and inequalities. The results and algorithms, when there is a mixture of constraints, are actually quite straightforward once the theory has been developed separately for the pure equality-constrained case and for the pure inequality-constrained case.

Without loss of generality we have assumed that all constraints are of the form

$$A_{i\bullet}x \geq b_i, \quad i \in \mathcal{I}$$

Note that this does not preclude any bounds of the form $x_j \geq 0$. In Section 12.5 we will show how to treat these simple bounds in a special manner. Considerable computational advantage may be gained when there are such simple bound constraints, but conceptually the approach is the same as for (12.14).

We shall see that (LIP) is more complicated than (LEP), since, in general, we do not know beforehand which constraints, if any, will be exactly satisfied as equalities at the solution.

Optimality conditions revisited

In this section, as for the equality-constrained case, we derive optimality conditions for the linear inequality problem (12.14) by considering feasible moves from a feasible starting point. The derivation of these conditions in the manner described here is useful for algorithm design. We start by defining a few terms.

A constraint $A_{i\bullet}x - b_i \geq 0$ evaluated at the point $x = \bar{x}$ is said to be:

1. *active (binding, tight)* if $A_{i\bullet}\bar{x} - b_i = 0$;

2. *inactive* if $A_{i\bullet}\bar{x} - b_i > 0$;

3. *violated* if $A_{i\bullet}\bar{x} - b_i < 0$.

If the constraints are of the form $A_{i\bullet}x - b_i \leq 0$, there are analogous definitions. A point $x = \bar{x}$ is said to be *feasible* with respect to a set of constraints $Ax - b \geq 0$ if no constraints are violated at $x = \bar{x}$.

As defined in Chapter 11, the *active set* of constraints $\mathcal{A}(\bar{x})$ at a feasible point $\bar{x}$ is the set of all indices of the constraints active at $\bar{x}$. Thus,

$$\mathcal{A}(\bar{x}) = \{i : A_{i\bullet}\bar{x} - b_i = 0\}.$$

It is important to recognize the active constraints, because feasible directions p from a feasible point $x = \bar{x}$ are *restricted* with respect to active constraints, and *unrestricted* with respect to inactive constraints as we shall see shortly. Because of this, only the active constraints play a key role in the optimality conditions. Hence, a key part of any algorithm for (12.14) is to find an active set of constraints of full rank corresponding to an optimal solution.

If a constraint $A_{i\bullet}x - b_i \geq 0$ is inactive at $\bar{x}$, then, within some neighborhood of $\bar{x}$, it is possible to move in *any* direction and still remain feasible with respect to that constraint. That is, given any inactive constraint $A_{i\bullet}x > b_i$ at $x = \bar{x}$ and any direction (vector) p, there exists an $\varepsilon > 0$ such that $A_{i\bullet}(\bar{x} + \varepsilon p) > b_i$. Since $d_i = A_{i\bullet}\bar{x} - b_i > 0$ by assumption, we have

$$A_{i\bullet}(\bar{x} + \varepsilon p) - b_i = d_i + \varepsilon A_{i\bullet}p \geq 0 \quad \text{for all } \varepsilon \leq \bar{\varepsilon} \text{ where } \bar{\varepsilon} > 0.$$

If $A_{i\bullet}p \geq 0$, all such moves remain feasible for any $\varepsilon > 0$, that is $\bar{\varepsilon} = \infty$. However, if $A_{i\bullet}p < 0$, then $\bar{\varepsilon} = -d_i/A_{i\bullet}p > 0$ is the maximum step we can move from $\bar{x}$ in the direction p and stay feasible with respect to the i-th constraint.

On the other hand, if a constraint $A_{j\bullet}x - b_j \geq 0$ is active at $x = \bar{x}$, then, for $\varepsilon > 0$,

$$A_{j\bullet}(\bar{x} + \varepsilon p) - b_j \geq 0 \quad \text{only if } A_{j\bullet}p \geq 0.$$

A direction p satisfying $Ap \geq 0$ allows a wider class of feasible directions than for the equality-constrained case. If $A_{j\bullet}p = 0$, no change will occur in the value of the j-th constraint, because $A_{j\bullet}(\bar{x} + \varepsilon p) - b_j = 0$. However, it is also possible to remain feasible by moving along a direction p satisfying $A_{j\bullet}p > 0$. Should such a direction p be chosen, then, of course, the j-th constraint will *no longer be active* at the point $\bar{x} + \varepsilon p$ for any $\varepsilon > 0$.

With this in mind, we derive conditions for a minimum for the linear inequality-constrained problem. A point $x = x^*$ will be considered a local minimizer for (12.14) if x^* is feasible, and $f(x^*) \leq f(x)$ for all *feasible* x in some neighborhood of x^*. To characterize feasible moves in a neighborhood of $x = x^*$, we need to consider only the constraints that are *active* at x^* (as seen above). Suppose that ℓ constraints are active at x^*. We then let the vectors $A_{i\bullet}$ corresponding to these active constraints be the rows of the $\ell \times n$ matrix $\hat{A}$ so that $\hat{A}_{1\bullet}$ corresponds to the first active constraint, and so on. The vector b is similarly partitioned. Thus, at x^*,

$$\hat{A}x^* = \hat{b}, \quad \text{or equivalently} \quad \hat{A}x^* - \hat{b} = 0.$$

From the above, it may seem that the conditions for optimality are the same as for the equality-constrained problem whose set of constraints is $\hat{A}x = \hat{b}$. But this will not be the case because here we have greater freedom in choosing feasible directions; we can choose a direction that makes one or more active constraints inactive while remaining feasible. So, to ensure that x^* is locally optimal, no such direction should lower the objective function value in the neighborhood of x^*.

Given x^* such that $\hat{A}x^* = \hat{b}$, we find a matrix Z whose columns represent the null space of $\hat{A}$. As we have seen earlier, for any p_z, the search direction $p = Zp_z$ is such that the $\hat{A}(x^* + \varepsilon Zp_z) = \hat{b}$ for all ε. This condition is often spoken of as "staying on" the ℓ active constraints. We first consider the behavior of $f(x)$ along such a direction:

$$f(x^* + \varepsilon Zp_z) = f(x^*) + \varepsilon p_z Z^{\mathrm{T}} g(x^*) + \cdots$$

Exactly as in the equality-constrained case, we see that $Z^{\mathrm{T}}g(x^*)$ must be zero if $x = x^*$ is a minimizer. As we have seen earlier (e.g., page 424), this condition is equivalent to stating that $g(x^*)$ must be a linear combination of the columns of $\hat{A}^{\mathrm{T}}$.

So far these conditions are the same first-order necessary conditions as for the equality-constrained problem. However, we know that in this case we are allowed to construct feasible directions that make active constraints inactive. In fact, the set $\mathcal{F}$ of feasible directions at $x = x^*$ with respect to active inequality constraints includes all vectors p that satisfy

$$\hat{A}p \geq 0. \tag{12.15}$$

Thus we also need to consider whether f can decrease along any direction p satisfying (12.15) above. To simplify the analysis, consider a direction p

such that

$$\hat{A}_{j\bullet}p = 1 \text{ for some } j, \text{ and } \hat{A}_{i\bullet}p = 0 \text{ for } i \neq j. \qquad (12.16)$$

(In matrix notation we can write this as $\hat{A}p = I_{\bullet j}$.) That is, p is chosen so that a very small movement along such a direction p will maintain the status of all the active constraints, except the j-th active constraint, as equalities. Any movement along such a direction p will cause the j-th constraint to become inactive while maintaining feasibility. This is referred to as "moving off" the j-th constraint. It is easy to show that such a direction p always exists if $\hat{A}$ has full rank.

Since we are interested in determining whether a movement along such a direction p will decrease the function value, we examine under what conditions p will be a descent direction. Clearly the direction p will be a direction of descent for the function f if

$$p^{\mathrm{T}}g(x^*) < 0. \qquad (12.17)$$

However, from the necessary condition of optimality derived earlier, we have

$$g(x^*) = \hat{A}^{\mathrm{T}}\lambda^*. \qquad (12.18)$$

Combining (12.16), (12.17), and (12.18), we obtain

$$p^{\mathrm{T}}g(x^*) = \lambda_j^* \hat{A}_{j\bullet}p < 0$$

as the requirement for descent in that direction. By construction, $\hat{A}_{j\bullet}p > 0$, so if $\lambda_j^* < 0$, it would then be possible to decrease $f(x)$ from its value at $x = x^*$ and remain feasible, in which case x^* would not be a minimizer for the linear-inequality problem (12.14). Examining each active constraint j in this manner, we see that an *additional* necessary condition for optimality is

$$\lambda_i^* \geq 0 \qquad \text{for } i = 1, \dots, \ell \qquad (12.19)$$

or, in vector form,

$$\lambda^* \geq 0. \qquad (12.20)$$

The above result does not require $\hat{A}$ to be of full rank, but for our purposes we will continue to assume that $\hat{A}$ is of full rank to simplify the exposition.

We could include *all* the vectors $A_{i\bullet}$ in the relationship between $g(x^*)$ and $\{A_{i\bullet}\}$ but require that the multiplier corresponding to an *inactive* constraint be zero. This would be equivalent to imposing the *complementarity condition* that

$$\lambda_i^*(A_{i\bullet}x - b_i) = 0 \quad \text{for all } i = 1, \ldots, m,$$

which implies that if a constraint is inactive, its multiplier must be zero. For implementation purposes, there is no need to compute the Lagrange multipliers for the inactive constraints because they are known to be zero. Instead, we only need to compute the Lagrange multipliers that correspond to the *active* constraints.

As for the equality constrained case, we note that the matrix $Z^{\mathrm{T}}G(x^*)Z$ must be positive semidefinite.

Necessary and sufficient conditions for a minimum

In summary, the necessary conditions for x^* to be a local minimizer of (12.14) are:

1. $Ax^* \geq b$ and $\hat{A}x^* = \hat{b}$.

2. $Z^{\mathrm{T}}g(x^*) = 0$ (or equivalently $g(x^*) = \hat{A}^{\mathrm{T}}\lambda^*$) where $\hat{A}Z = 0$.

3. $\lambda^* \geq 0$.

4. $Z^{\mathrm{T}}G(x^*)Z$ is positive semidefinite.

It can be shown that the second-order sufficient conditions for x^* to be an isolated local minimizer of (12.14) are:

1. $Ax^* \geq b$ and $\hat{A}x^* = \hat{b}$.

2. $Z^{\mathrm{T}}g(x^*) = 0$ (or equivalently $g(x^*) = \hat{A}^{\mathrm{T}}\lambda^*$ where $\hat{A}Z = 0$).

3. $\lambda^* \geq 0$.

4. $Z^{\mathrm{T}}G(x^*)Z$ is positive definite.

5. If $\lambda_i^* = 0$, then $p^{\mathrm{T}}G(x^*)p > 0$ for all p such that $A_{i\bullet}p > 0$.

Example 12.8: LINEAR INEQUALITY PROBLEM. Consider the 3-variable optimization problem

$$\begin{array}{ll} \text{minimize} & 2x_1^2 + x_2^2 + x_3^2 - 12x_1 \\ \text{subject to} & x_1 + x_2 \qquad \leq 2 \\ & \qquad\qquad x_3 \leq 1. \end{array}$$

Rewriting this in "$\geq$" form, we obtain

$$\begin{array}{ll} \text{minimize} & 2x_1^2 + x_2^2 + x_3^2 - 12x_1 \\ \text{subject to} & -x_1 - x_2 \qquad \geq -2 \\ & \qquad\qquad - x_3 \geq -1. \end{array}$$

At a feasible point $\bar{x} = (0,\ 0,\ 1)$, the first constraint is inactive and the second constraint is active. The coefficient matrix of the active constraint is

$$\hat{A} = [0 \quad 0 \; - 1]$$

and a corresponding basis of the null space is given by

$$Z = \begin{bmatrix} 1 & 0 \\ 0 & 1 \\ 0 & 0 \end{bmatrix}.$$

The gradient at $\bar{x}$ is $g(\bar{x}) = (4\bar{x}_1 - 12, 2\bar{x}_2, 2\bar{x}_3) = (0, 0, 2)$. The projected gradient at $\bar{x}$ is

$$Z^{\mathrm{T}} g(\bar{x}) = \begin{bmatrix} 1 & 0 & 0 \\ 0 & 1 & 0 \end{bmatrix} \begin{bmatrix} 0 \\ 0 \\ 2 \end{bmatrix} = \begin{bmatrix} 0 \\ 0 \end{bmatrix}.$$

The Hessian matrix is

$$G(\bar{x}) = \begin{bmatrix} 4 & 0 & 0 \\ 0 & 2 & 0 \\ 0 & 0 & 2 \end{bmatrix}.$$

The projected Hessian is

$$Z^{\mathrm{T}} G(\bar{x}) Z = \begin{bmatrix} 1 & 0 & 0 \\ 0 & 1 & 0 \end{bmatrix} \begin{bmatrix} 4 & 0 & 0 \\ 0 & 2 & 0 \\ 0 & 0 & 2 \end{bmatrix} \begin{bmatrix} 1 & 0 \\ 0 & 1 \\ 0 & 0 \end{bmatrix} = \begin{bmatrix} 4 & 0 \\ 0 & 2 \end{bmatrix}.$$

The projected gradient is zero and the projected Hessian is positive definite. If all the constraints were linear equalities, these conditions would be sufficient to declare $\bar{x}$ as an optimal solution. Here we need to check the sign on the multiplier by solving $\hat{A}^T\lambda = g(\bar{x})$, that is, by solving

$$\begin{bmatrix} 0 \\ 0 \\ -1 \end{bmatrix} \lambda = \begin{bmatrix} 0 \\ 0 \\ 2 \end{bmatrix}$$

results in $\lambda = -2$. This means that the objective function can be decreased by moving off the active constraint at $\bar{x}$, that is $\bar{x}_3 = 1$. A search direction to do this is computed by solving $\hat{A}p = 1$, so we solve

$$\begin{bmatrix} 0 & 0 & -1 \end{bmatrix} p = 1$$

to obtain $p = (0, 0, -1)$. This is clearly a feasible descent direction.

12.4   Active Set Methods

If the active set at an optimal solution $x = x^*$ were known for

$$\text{(LIP)} \qquad \begin{array}{ll} \text{minimize} & f(x) \\ \text{subject to} & Ax \geq b, \end{array}$$

then $x = x^*$ would also solve the corresponding equality-constrained problem

$$\begin{array}{ll} \text{minimize} & f(x) \\ \text{subject to} & \hat{A}x = \hat{b}. \end{array}$$

We have already seen that there are a variety of methods for solving problems with linear equality constraints. Consequently, we shall consider a family of algorithms for solving (LIP) that are based directly on algorithms for (LEP). To do this, we select a *working set* of constraints to be treated as equality constraints.[2]

---

[2]The language used in the concept of the working set is often a bit troublesome. For one thing, there is the question of how a working set differs from an active set. This is not so hard to resolve: an active set may have linearly dependent constraints whereas a working set may not. Constraints belonging to a working set are active, but not all active constraints belong to a working set. Indeed a working set is the set of indices of a maximal set of linearly independent active constraints. The trouble (in our view) comes from the fact that an active set is defined as the set of *indices* (subscripts) of active constraints, i.e., those satisfying $c_i(\bar{x}) = 0$. Now when the constraints are linear inequalities of the form $Ax \geq b$ (leading to $c(x) = Ax - b \geq 0$), one has to wonder if it is correct to refer to A or some subset of its rows as a working set. There is, of course, an obvious correspondence between the indices and the rows. This terminology problem is somewhat similar to the one in linear programming wherein basic variables are sometimes spoken of as elements of the basis. A basis in the sense of linear programming is a matrix, and variables are not matrices.

We next describe an active set algorithm that uses the following notation. As before, the gradient at iteration k will be denoted by $g_k = g(x_k)$. The coefficient matrix of the active constraints in the working set at iteration k will be denoted by $\hat{A}_k$, and the null space corresponding to $\hat{A}_k$ will be denoted by Z_k.

Algorithm 12.1 (An Active Set Algorithm)   For simplicity of exposition, we will assume that the set of active constraints at every iteration are linearly independent; i.e., the working set and active set are the same. Set $k = 0$ and let x_0, a feasible point, be given. Construct $\mathcal{A}(x_0)$, the set of indices of all constraints that are active at x_0, and let $\hat{A}_0$ be the associated coefficient matrix of the working set of active constraints. Compute the gradient $g_0 = g(x_0)$.

1. *Test for convergence.* There are two possibilities:

    (a) $\mathcal{A}(x_k) = \emptyset$ (no constraints are active). If the projected gradient satisfies
    $$||(Z_k)^T g_k|| \leq \sqrt{\varepsilon_M},$$
    report $x^* = x_k$ as the solution and stop.

    (b) $\mathcal{A}(x_k) \neq \emptyset$ (at least one constraint is active). If the system $\hat{A}_k^T \lambda = g_k$ has a solution, say $\lambda = \bar{\lambda}$, and $\bar{\lambda} \geq 0$, then report $x^* = x_k$ as the solution and stop.

2. *Compute a search direction.* There are three cases to consider here: no constraints active, and, one or more constraints active when x_k is a solution of the current equality-constrained problem and when it is not a solution.

    (a) $\mathcal{A}(x_k) = \emptyset$ (no constraints are active). In this case the search direction p_k is unconstrained.

    (b) $\mathcal{A}(x_k) \neq \emptyset$ (at least one constraint is active).

       i. $\hat{A}_k^T \lambda = g_k$. Pick $\bar{\lambda}_s < 0$, for some $j = s$. In this case the search direction will be such that the constraint $j = s$ will leave the active set $\mathcal{A}(x_k)$. That is, find a descent direction $p = p_k$ that solves $\hat{A}_k p = I_{\bullet s}$.

ii. $\hat{A}_k^T \lambda \neq g_k$ for any λ. In this case, construct a search direction that continues to satisfy the active constraints. That is, given $g_z = Z_k^T g_k$, find a p_z such that $(g_z)^T p_z < 0$ and construct $p_k = Z_k p_z$. (Any of the search directions—steepest-descent direction, Newton direction, quasi-Newton direction, conjugate-gradient direction—discussed earlier can be used here.)

3. *Compute a steplength.* A steplength is computed so that the objective function decreases and the next iterate is feasible.

   (a) Compute the maximum positive step $\alpha = \Delta$ that can be taken before an inactive constraint becomes active. The bound Δ is given by

   $$\Delta = \min \begin{cases} \dfrac{b_i - A_{i\bullet} x_k}{A_{i\bullet} p} & \text{for all } i \notin \mathcal{A}(x_k), \ A_{i\bullet} p < 0, \\ \infty. \end{cases}$$

   (b) Compute a steplength $\alpha = \alpha_k \leq \Delta$ using one of the algorithms described in Section 9.8.

4. *Update the iterate.* Using p_k and α_k, determine the next iterate $x_{k+1} = x_k + \alpha_k p_k$. Depending on the search direction used, other quantities may also need to be updated. In addition, update the working set if necessary, as follows:

   (a) If Step 2(a) was executed, then constraint s needs to be *deleted* from the working active set; i.e. $\mathcal{A}(x_{k+1}) = \mathcal{A}(x_k) \backslash \{s\}$; otherwise $\mathcal{A}(x_{k+1}) = \mathcal{A}(x_k)$.

   (b) If $\alpha_k = \Delta < \infty$, then a constraint r becomes active and should be *added* to the working active set. That is $\mathcal{A}(x_{k+1}) = \mathcal{A}(x_k) \cup \{r\}$. If $\alpha_k = \Delta = \infty$, then the objective function is unbounded below.

As a part of updating the active set, also update (or compute) the representation of Z_{k+1}. Assign $k \leftarrow k + 1$. Return to Step 1.

Further discussion of this algorithm

When $\alpha = \Delta$, several constraints may be satisfied exactly at the new point, so that there is a tie for the nearest constraint. When this happens, only one of these is added at this step (it may be necessary to add some of the other constraints at other steps). The effect of adding a constraint depends on the representation of Z and the method used to compute p. According to the Active Set Algorithm described above, a constraint is added to the working set as soon as $\alpha_k = \Delta$. This strategy ensures feasibility at the next iteration. However, other strategies can also be used to retain feasibility as well as prevent a change to the active set at the very next iteration. For example, a constraint with index i that is not in the working set at iteration k, but is exactly satisfied at x_{k+1}, can be added to the working set only if $A_{i\bullet} p_{k+1} < 0$.

In Algorithm 12.1, constraints can be added to the active set as the iterations proceed. However, the decision to delete a constraint is only made when the underlying equality-constrained subproblem is solved exactly. In practice, it is inefficient to solve such problems exactly; instead decisions are made to delete constraints even when $\hat{A}_k^T \lambda \neq g_k$. In this case, the system is solved in the least-squares sense, and the components of the solution vector λ are referred to as *Lagrange multiplier estimates*.

When a subproblem is solved inaccurately, the decision to delete a constraint is difficult because the exact behavior cannot be predicted from the Lagrange multiplier estimates. Due to this, a constraint may be repeatedly deleted and added to the working set (a phenomenon known as *zig-zagging*). Various strategies have been proposed to avoid zig-zagging. See Rosen [168] for one such strategy.

As stated above, the search direction for a given active set may be computed using any of the methods (for example, steepest descent, Newton, or quasi-Newton) considered for the linear equality-constrained problem.

Finding an initial solution

So far, in discussing the Active Set Method for linear inequality problems, we have assumed that an initial feasible solution is available. Because the constraints are linear, we can apply the same techniques that we did for LPs. In Section 4.1, we discussed approaches to finding an initial feasible solution; one such approach, called Phase I, was that of applying the Simplex Algorithm to a modified LP. In Phase I an alternative objective function was defined as the sum of artificial variables with the intent of driving the values of such variables to zero to obtain a feasible solution of the original problem if one exists. The same approach can be used for the linear inequality problem. Then in Section 7.2 we described a computationally better approach, called *minimizing the sum of infeasibilities*, for finding an initial feasible solution. The same approach can also be used here.

In Active Set Methods, we can also use the approach of minimizing the sum of infeasibilities (in modified form), whereby we work with the inequality constraints without introducing any slack variables. We describe this next.

Once again consider the inequality-constrained problem (12.14), repeated here for convenience:

$$\text{(LIP)} \qquad \begin{aligned} &\text{minimize} & f(x) \\ &\text{subject to} & Ax \geq b, \quad \text{where } A \text{ is } m \times n. \end{aligned}$$

Pick any point $x_0 \in R^n$ and evaluate each inequality constraint to determine whether it is violated at x_0. If none of the inequality constraints are violated at x_0, then we have a feasible starting solution. If x_0 violates one or more constraints, then create

$$\mathcal{V}(x_0) = \{j \ : \ A_{j\bullet}x_0 < b_j\},$$

the set of indices of the violated constraints at x_0. For $j \in \mathcal{V}(x_0)$, define the corresponding *infeasibility* at x to be $b_j - A_{j\bullet}x$. Next construct the sum of the infeasibilities:

$$\sum_{j\in\mathcal{V}(x_0)} (b_j - A_{j\bullet}x) = - \sum_{j\in\mathcal{V}(x_0)} A_{j\bullet}x + \sum_{j\in\mathcal{V}(x_0)} b_j.$$

The additive constant on the right-hand side above can be dropped in the corresponding Phase I problem which is

$$\text{minimize} \quad - \sum_{j\in\mathcal{V}(x_0)} A_{j\bullet}x$$

$$\text{subject to} \quad A_{k\bullet}x \geq b_k \quad \text{for } k \notin \mathcal{V}(x_0).$$

This Phase I problem is solved by the Active Set Algorithm 12.1, with starting point $x = x_0$.

There are several comments worth making about this.

1. At each subsequent iteration the index set of violated constraints will depend on the iterate x_k and may change. Thus the Phase I problem may require reformulation at every iteration.

2. As a consequence of the set of violated constraints changing at each iteration, the objective function is piecewise linear.

3. Because the objective function is piecewise linear, the search direction at each iteration is the steepest-descent direction.

4. When moving along a search direction, care must be taken to ensure that the step is not unbounded. This is done by ensuring that the step is taken to the final constraint to become feasible. This maximizes the number of constraints that are no longer violated.

12.5   Special cases

In this section we shall consider three special cases of the linear inequality problem (12.14):

1. Bound-constrained minimization.

2. Quadratic programming.

3. Linear programming.

The special case of bound-constrained minimization

A special case of (LIP) occurs when the constraints are all simple bounds on the variables[3]. That is, the problem (12.14) is stated as

$$\begin{aligned} \text{minimize} \quad & f(x) \\ \text{subject to} \quad & l \leq x \leq u. \end{aligned} \tag{12.21}$$

Both l and u are in R^n, hence their components are finite. We will see that such bounds do not impact the computations by much.

Bound-constrained problems are important in practice for at least a couple of reasons:

1. In practical problems bounds are imposed on the variables, even if such variables are, in principle, free.

---

[3]These are also called *box-constrained problems*.

2. Even if the problem at hand is an unconstrained optimization problem, it may be useful to introduce bounds to ensure that the function is evaluated at reasonable points during the solution process. In such cases, if an artificially bounded variable is at one of its bounds in the optimal solution, it maybe that the bound is incorrectly set or that there is an error in the formulation.

For ease of exposition, we rewrite (12.21) in the form of (12.14), that is,

$$
\begin{aligned}
\text{minimize} \quad & f(x) \\
\text{subject to} \quad & x \geq l \\
& -x \geq -u.
\end{aligned}
\tag{12.22}
$$

The problem has $2n$ inequality constraints. In this case, considerable simplifications occur, allowing us, for example, to construct Z *without any computation* as a subset of columns of the identity matrix.

A bound constraint for x_j is *active* if x_j is equal to one of its bounds. A variable x_j corresponding to an active bound constraint is said to be a *fixed* component; other variables are called *free*. A typical iterate x is partitioned into its fixed components x_{FX} and its free components x_{FR}. The working set is defined by the indices of the fixed components.

Because the rows of the inequality constraints are simple bounds, it follows that the coefficient matrix, $\hat{A}$, of the active constraints is comprised of signed rows of an $n \times n$ identity matrix corresponding to the fixed variables. (For such a problem, the working set is the same as the active set.) In this case a basis, Z, for the null space can be constructed as the complementary set of columns of the corresponding identity matrix.

Example 12.9: SPECIAL FORM OF $\hat{A}$ AND Z. For the bound-constrained problem

$$
\begin{aligned}
\text{minimize} \quad & \tfrac{1}{2}(x_1^2 + 2x_2^2 + 4x_3^2 + 2x_4^2) - 6x_2 \\
\text{subject to} \quad & 1 \leq x_1 \leq 5 \\
& 2 \leq x_2 \leq 4 \\
& 0 \leq x_3 \leq 8 \\
& -2 \leq x_4 \leq 4,
\end{aligned}
$$

suppose the current iterate is $(x_1, x_2, x_3, x_4) = (5, 3, 3, -2)$. In this case, two constraints are active: $x_1 = 5$ and $x_4 = -2$; thus $\hat{A}$ and Z are given by

$$\hat{A} = \begin{bmatrix} -1 & 0 & 0 & 0 \\ 0 & 0 & 0 & 1 \end{bmatrix} \quad \text{and} \quad Z = \begin{bmatrix} 0 & 0 \\ 1 & 0 \\ 0 & 1 \\ 0 & 0 \end{bmatrix}.$$

The simple form of the working set of constraints and the corresponding null space results in a considerable simplification of all the algorithms described earlier. For the search direction p, only the components p_{FR} corresponding to the $n - \ell$ free variables need to be specified; the other components p_{FX} will be zero. For example, Newton's Method requires the solution of $Z^{\mathrm{T}}GZp_z = -Z^{\mathrm{T}}g$ at each iteration. Since the columns of Z are columns of the identity matrix, this reduces to solving

$$G_{\mathrm{FR}}p_{\mathrm{FR}} = -g_{\mathrm{FR}}.$$

If the Hessian matrix is not readily available, then we can obtain a finite-difference approximation of the the Hessian and apply the discrete Newton Method (see Section 10.2) to (12.22). In this case only the elements of the submatrix G_{FR} will need to be computed by the finite-difference operations. The quasi-Newton Methods can be simplified in a similar manner.

Example 12.10: NEWTON SEARCH DIRECTION. For the problem defined in Example 12.9, the Hessian matrix $G = \mathrm{Diag}(1, 2, 4, 2)$ is constant. At the specified iterate, the gradient is

$$g(x) = (x_1, 2x_2 - 6, 4x_3, 2x_4) = (5, 0, 12, -4).$$

We note that

$$Z^{\mathrm{T}}g = \begin{bmatrix} 0 & 1 & 0 & 0 \\ 0 & 0 & 1 & 0 \end{bmatrix} \begin{bmatrix} 5 \\ 0 \\ 12 \\ -4 \end{bmatrix} = \begin{bmatrix} 0 \\ 12 \end{bmatrix} \neq 0.$$

The Newton direction is computed by solving $Z^{\mathrm{T}}GZp_z = -Z^{\mathrm{T}}g$ which reduces to the equation $G_{\mathrm{FR}}p_{\mathrm{FR}} = -g_{\mathrm{FR}}$, that is,

$$\begin{bmatrix} 0 & 1 & 0 & 0 \\ 0 & 0 & 1 & 0 \end{bmatrix} \begin{bmatrix} 1 & 0 & 0 & 0 \\ 0 & 2 & 0 & 0 \\ 0 & 0 & 4 & 0 \\ 0 & 0 & 0 & 2 \end{bmatrix} \begin{bmatrix} 0 & 0 \\ 1 & 0 \\ 0 & 1 \\ 0 & 0 \end{bmatrix} p_z = \begin{bmatrix} 2 & 0 \\ 0 & 4 \end{bmatrix} p_z = -Z^{\mathrm{T}}g = -\begin{bmatrix} 0 \\ 12 \end{bmatrix}.$$

Hence $p_z = (0, -3)$ and the Newton direction is $p = Zp_z = (0, 0, -3, 0)$. The steplength α is chosen to ensure a decrease in the objective function value while ensuring that no constraints are violated. That is, α is chosen so that $f(x + \alpha p) < f(x)$ and $l \le x + \alpha p \le u$. This results in $\alpha = 1$, whereby x_3 reaches its lower bound, thus the iterate is $(5, 3, 0, -2)$.

The coefficient matrix $\hat{A}$ of the new active constraints and the matrix Z are

$$\hat{A} = \begin{bmatrix} -1 & 0 & 0 & 0 \\ 0 & 0 & 1 & 0 \\ 0 & 0 & 0 & 1 \end{bmatrix} \quad \text{and} \quad Z = \begin{bmatrix} 0 \\ 1 \\ 0 \\ 0 \end{bmatrix}.$$

Because the coefficient matrix of the active constraints consists of signed rows of the identity matrix, it is easy to see that the first-order Lagrange multipliers are signed elements of the gradient corresponding to the variables that are fixed at their bounds; in particular,

if $x_i = l_i$, then $\lambda_i = g_i$; and, if $x_i = u_i$, then $\lambda_i = -g_i$.

Suppose that we are at an intermediate solution, where $Z^T g = g_{FR} = 0$, but the solution is not optimal for (12.21) as determined by the signs on the Lagrange multipliers. The search direction will be computed so as to move off one of the constraints in the working set. To do this, we pick a variable at its lower bound with a negative multiplier or a variable at its upper bound with a positive multiplier. At this iteration, one constraint will become inactive. At this, or a subsequent iteration, another inactive constraint may become active. Due to the simple nature of the bound constraints, such changes in the working set involve fixing a variable to its bound and/or freeing a variable from its bound.

Example 12.11: LAGRANGE MULIPLIERS. For the problem defined in Example 12.9, and the iterate determined at the end of Example 12.10, the gradient at the iterate is

$$g(x) = (x_1, 2x_2 - 6, 4x_3, 2x_4) = (5, 0, 0, -4).$$

We note that $Z^T g = 0$ and so we are at a local minimum with respect to the current working set, and hence the Lagrange multipliers exist. The Lagrange multipliers are computed by solving the consistent set of equations[4] $\hat{A}^{\mathrm{T}}\lambda = g$, that is,

$$
\begin{bmatrix} -1 & 0 & 0 \\ 0 & 0 & 0 \\ 0 & 1 & 0 \\ 0 & 0 & 1 \end{bmatrix} \lambda = \begin{bmatrix} 5 \\ 0 \\ 0 \\ -4 \end{bmatrix} \quad \text{or} \quad \lambda = \begin{bmatrix} -5 \\ 0 \\ -4 \end{bmatrix} = \begin{bmatrix} -g_1 \\ g_3 \\ g_4 \end{bmatrix}.
$$

Because x_4 is at its lower bound and $\lambda_4 < 0$, the solution is not optimal and we can reduce the objective function value by moving off this bound.

The special case of quadratic programming

Recall that a *quadratic programming problem* is one of the form (12.14) in which $f(x)$ is a quadratic function, i.e., $f(x) = \frac{1}{2}x^{\mathrm{T}}Hx + c^{\mathrm{T}}x$, where H is a symmetric matrix.

One aspect of nonlinear programming—and quadratic programming in particular—is that differences in notation and presentation can disguise the fact that some algorithms are actually *equivalent*. For example, one algorithm may be presented in terms of a tableau, another may be in terms of a particular representation of Z, and so forth. They may in fact be equivalent in the sense that, when applied to the same problem, they generate exactly the same sequence of iterates.

If $Z^{\mathrm{T}}HZ$ is positive definite for Z corresponding to every active set, (i.e., f is bounded below in every null space), then the step to the minimum of f in a null space spanned by the columns of Z is given by Zp_z, where p_z solves

$$
Z^{\mathrm{T}}HZp_z = -Z^{\mathrm{T}}g(x) = -Z^{\mathrm{T}}(c + Hx).
$$

That is, a step of unity along such a p_z is a step to the minimum of f restricted to the null space specified by Z.

The search direction Zp_z is computed at the current iterate $x = \bar{x}$. If a step of unity can be taken without reaching another constraint boundary, we must have reached the minimum with respect to the current working set. The Lagrange multipliers can then be evaluated to determine whether

---

[4]This system may look improper because the number of equations exceeds the number of variables. Nevertheless, the system has a solution because the gradient is orthogonal to the columns of Z and so must lie in the range space of $\hat{A}^{\mathrm{T}}$.

a minimum has been found or whether the index of a constraint should be deleted from the current working set. Otherwise, if an inactive constraint becomes active at some steplength $\alpha < 1$, then the index of this newly satisfied active constraint is added to the working set, and the process is continued.

In the paragraphs below, we present two closely related examples. Both have the same constraints and same objective function, except that in one example we minimize the objective and in the other we minimize its negative, i.e., we maximize the objective. These examples illustrate the steps of the Active Set Method and reveal some properties that distinguish these quadratic programs from linear programs and from each other.

Example 12.12: QP WITH A STRICTLY CONVEX MINIMAND. The vector $x_0 = (0,0)$ is clearly feasible for the quadratic program

$$
\begin{array}{ll}
\text{minimize} & \frac{1}{2}(2x_1^2 - 4x_1x_2 + 4x_2^2) - 2x_1 - 6x_2 \\
\text{subject to} & -x_1 - x_2 \geq -2 \\
& x_1 - 2x_2 \geq -2 \\
& x_1 \geq 0 \\
& x_2 \geq 0.
\end{array}
$$

As mentioned above, the feasible region of this program is a polytope, so there is no question about the existence of minimizer of its objective function. In this case, the objective function is strictly convex, so the problem has a unique (global) minimum. Our aim here is to find it using the Active Set Method.

In this discussion, we include the nonnegativity constraints with the other linear inequalities. In all, then, we have $m = 4$ linear inequality constraints. For this problem the data are

$$
A = \begin{bmatrix} -1 & -1 \\ 1 & -2 \\ 1 & 0 \\ 0 & 1 \end{bmatrix}, \quad b = \begin{bmatrix} -2 \\ -2 \\ 0 \\ 0 \end{bmatrix}, \quad c = \begin{bmatrix} -2 \\ -6 \end{bmatrix}, \quad H = \begin{bmatrix} 2 & -2 \\ -2 & 4 \end{bmatrix}.
$$

The objective function $f(x) = c^{\mathsf{T}}x + \frac{1}{2}x^{\mathsf{T}}Hx$ has the gradient $g(x) = \nabla f(x) = c + Hx$ and constant Hessian H. The gradient at any point x is given by

$$
g(x) = \nabla f(x) = \begin{bmatrix} 2x_1 - 2x_2 - 2 \\ -2x_1 + 4x_2 - 6 \end{bmatrix}.
$$

We note that the *unconstrained* minimizer of f is $-H^{-1}c = (5, 4)$ which does *not* satisfy the constraints of the problem, hence we use an algorithm to find the solution of the constrained optimization problem.

At the starting point $x_0 = (0, 0)$, the third and fourth constraints are active, so we have $\mathcal{A}(x_0) = \{3, 4\} = \mathcal{W}(x_0)$. Then we have

$$\hat{A} = \begin{bmatrix} 1 & 0 \\ 0 & 1 \end{bmatrix}, \quad \hat{b} = \begin{bmatrix} 0 \\ 0 \end{bmatrix}.$$

Since $\hat{A}$ is a nonsingular matrix, its null space is the zero vector. At this point, we have $g(x_0) = (-2, \ -6)$, $Z^{\mathrm{T}}g(x_0) = 0$, and $\mathcal{A}(x_0) \neq \emptyset$, so we compute

$$\bar{\lambda} = (\hat{A}^{-1})^{\mathrm{T}}g(x_0) = \begin{bmatrix} 1 & 0 \\ 0 & 1 \end{bmatrix} \begin{bmatrix} -2 \\ -6 \end{bmatrix} = \begin{bmatrix} -2 \\ -6 \end{bmatrix}.$$

Since this is not a nonnegative vector, we are going to drop one of the constraints from the working set, namely constraint 4. (This will allow the variable x_2 to become positive.) Accordingly, we revise some definitions as follows:

$$\hat{A} = \begin{bmatrix} 1 & 0 \end{bmatrix}, \quad \hat{b} = \begin{bmatrix} 0 \end{bmatrix}, \quad Z = \begin{bmatrix} 0 \\ 1 \end{bmatrix}.$$

What all this says is that we are presently enforcing the constraint $x_1 = 0$. In this simple situation, there can't be much mystery about what the search direction will be, but we compute it anyway. The reduced Newton direction vector is

$$p = -Z(Z^{\mathrm{T}}HZ)^{-1}Z^{\mathrm{T}}g(x_0) = \begin{bmatrix} 0 \\ 3/2 \end{bmatrix}.$$

The idea now is to move from x_0 to $x_0 + \alpha p$ for some $\alpha > 0$. The value $\alpha = 1$ corresponds to the global minimizer of f restricted to the line $x_1 = 0$. It is easy to check that this point is infeasible. We need a smaller α. We find it by enforcing the feasibility condition $A(0 + \alpha p) \geq b$:

$$\begin{bmatrix} -1 & -1 \\ 1 & -2 \\ 1 & 0 \\ 0 & 1 \end{bmatrix} \begin{bmatrix} 0 \\ (3/2)\alpha \end{bmatrix} \geq \begin{bmatrix} -2 \\ -2 \\ 0 \\ 0 \end{bmatrix}.$$

This turns out to imply that $\alpha \leq 2/3$. Thus, the steplength at the current iterate is $2/3$. We define the next iterate to be $x_1 = x_0 + (2/3)p = (0, 1)$.

For this iterate, we have $\mathcal{A}(x_1) = \{2, 3\} = \mathcal{W}(x_1)$; the other corresponding data are

$$\hat{A} = \begin{bmatrix} 1 & -2 \\ 1 & 0 \end{bmatrix}, \quad \hat{b} = \begin{bmatrix} -2 \\ 0 \end{bmatrix}, \quad Z = \begin{bmatrix} 0 \\ 0 \end{bmatrix}, \quad \hat{A}^{-1} = \tfrac{1}{2}\begin{bmatrix} 0 & 2 \\ -1 & 1 \end{bmatrix}.$$

At the beginning of the next iteration, we have $g(x_1) = (-4, \ -2)$, $Z^{T}g(x_1) = 0$, and $\mathcal{A}(x_1) \neq \emptyset$, so we define the Lagrange multiplier vector

$$\bar{\lambda} = (\hat{A}^{-1})^{T}g(x_1) = \tfrac{1}{2}\begin{bmatrix} 0 & -1 \\ 2 & 1 \end{bmatrix}\begin{bmatrix} -4 \\ -2 \end{bmatrix} = \begin{bmatrix} 1 \\ -5 \end{bmatrix}.$$

Since this vector is not nonnegative, we decide to drop constraint 3 from the working set. This means we are going to enforce constraint 2 but allow the variable x_1 to become positive. Updating, we get

$$\hat{A} = \begin{bmatrix} 1 & -2 \end{bmatrix}, \quad \hat{b} = \begin{bmatrix} -2 \end{bmatrix}, \quad Z = \begin{bmatrix} 2 \\ 1 \end{bmatrix}.$$

Again we compute a reduced Newton search direction. It is given by the formula

$$p = -Z(Z^{T}HZ)^{-1}Z^{T}g(x_1) = \begin{bmatrix} 5 \\ 5/2 \end{bmatrix}.$$

To compute the steplength, we consider the system

$$\begin{bmatrix} -1 & -1 \\ 1 & -2 \\ 1 & 0 \\ 0 & 1 \end{bmatrix}\begin{bmatrix} 5\alpha \\ 1 + (5/2)\alpha \end{bmatrix} \geq \begin{bmatrix} -2 \\ -2 \\ 0 \\ 0 \end{bmatrix}.$$

The maximum value of α for which this holds is $2/15$. From this we conclude that the steplengh is $\alpha = \min\{1, 2/15\} = 2/15$. Using this, we define a new iterate $x_2 = x_1 + (2/15)p = (2/3\,, 4/3)$.

Relative to the new iterate x_2, we have $\mathcal{A}(x_2) = \{1,\, 2\} = \mathcal{W}(x_2)$. We then have

$$\hat{A} = \begin{bmatrix} -1 & -1 \\ 1 & -2 \end{bmatrix}, \quad \hat{b} = \begin{bmatrix} -2 \\ -2 \end{bmatrix}, \quad Z = \begin{bmatrix} 0 \\ 0 \end{bmatrix}, \quad \hat{A}^{-1} = \tfrac{1}{3}\begin{bmatrix} -2 & 1 \\ -1 & -1 \end{bmatrix}.$$

We have $g(x_2) = (-10/3, \ -2)$, $Z^{T}g(x_2) = 0$, and $\mathcal{A}(x_2) \neq \emptyset$. Thus, we need to compute

$$\bar{\lambda} = (\hat{A}^{-1})^{T}g(x_2) = \tfrac{1}{3}\begin{bmatrix} -2 & -1 \\ 1 & -1 \end{bmatrix}\begin{bmatrix} -10/3 \\ -2 \end{bmatrix} = \begin{bmatrix} 26/9 \\ -4/9 \end{bmatrix}.$$

We drop the constraint corresponding to the negative component in $\bar{\lambda}$, namely 2. Thus we get a new data set

$$\hat{A} = \begin{bmatrix} -1 & -1 \end{bmatrix}, \quad \hat{b} = \begin{bmatrix} -2 \end{bmatrix}, \quad Z = \begin{bmatrix} -1 \\ 1 \end{bmatrix}.$$

The corresponding reduced Newton direction is

$$p = -Z(Z^{\mathrm{T}}HZ)^{-1}Z^{\mathrm{T}}g(x_2) = \begin{bmatrix} 2/15 \\ -2/15 \end{bmatrix}.$$

We see that $x_2 + \alpha p$ is feasible for $0 \le \alpha \le 10$. But since 10 is greater than 1, the latter is our steplength. Thus, we obtain the new iterate $x_3 = x_2 + 1p = (4/5, 6/5)$.

At x_3 we have $\mathcal{A}(x_3) = \{1\} = \mathcal{W}(x_3)$. With

$$\hat{A} = \begin{bmatrix} -1 & -1 \end{bmatrix}, \hat{b} = \begin{bmatrix} -2 \end{bmatrix}, Z = \begin{bmatrix} -1 \\ 1 \end{bmatrix},$$

we find that $g(x_3) = (-14/5, -14/5)$, $Z^{\mathrm{T}}g(x_1) = 0$, and the corresponding $\bar{\lambda} = 14/5 > 0$. This tells us that x_3 is a local minimizer. Since the objective function is strictly convex on R^2, x_3 is the unique global minimizer for this little quadratic program.

It can be checked that $x^* = x_3 = (4/5, 6/5)$ and $\lambda^* = (14/5, 0, 0, 0)$ satisfy the KKT conditions for this problem.

An important observation

This example illustrates a difference between linear and nonlinear programming, and in particular, quadratic programming. A minimizer of a linearly constrained nonlinear program with a convex objective function can lie anywhere in the feasible region. In the above example the unique optimal solution is not at an extreme point of the feasible region.

Example 12.13: QP WITH A STRICTLY CONCAVE MINIMAND. This example has the same constraints as in Example 12.12, but its objective function is strictly concave rather than strictly convex. In such a case, we expect to find the local minimizers at extreme points of the feasible region.[5] If the

---

[5]This assertion is based on a theorem due to W.M. Hirsch and A.J. Hoffman [97].

constraints of the quadratic program

$$
\begin{array}{ll}
\text{minimize} & -2x_1 - 6x_2 - \frac{1}{2}(2x_1^2 - 4x_1x_2 + 4x_2^2) \\
\text{subject to} & -x_1 - x_2 \geq -2 \\
& x_1 - 2x_2 \geq -2 \\
& x_1 \geq 0 \\
& x_2 \geq 0
\end{array}
$$

are ignored, its objective function has a global maximizer at $(-5, -4)$ and has no local minimizers. (In the unconstrained case, the objective function would be unbounded below.) Let us see what the Active Set Method will do with this problem when we start it at $x_0 = (0, 0)$.

This time, the problem data are

$$
A = \begin{bmatrix} -1 & -1 \\ 1 & -2 \\ 1 & 0 \\ 0 & 1 \end{bmatrix}, \quad b = \begin{bmatrix} -2 \\ -2 \\ 0 \\ 0 \end{bmatrix}, \quad c = \begin{bmatrix} -2 \\ -6 \end{bmatrix}, \quad H = \begin{bmatrix} -2 & 2 \\ 2 & -4 \end{bmatrix}.
$$

The gradient at any point x is given by:

$$
g(x) = \nabla f(x) = \begin{bmatrix} -2x_1 + 2x_2 - 2 \\ 2x_1 - 4x_2 - 6 \end{bmatrix}.
$$

At the starting point $x_0 = (0, 0)$, the third and fourth constraints are active, so we have $\mathcal{A}(x_0) = \{3, 4\} = \mathcal{W}(x_0)$. Then we have

$$
\hat{A} = \begin{bmatrix} 1 & 0 \\ 0 & 1 \end{bmatrix}, \quad \hat{b} = \begin{bmatrix} 0 \\ 0 \end{bmatrix}.
$$

Because the matrix $\hat{A}$ is nonsingular, its null space is $\{0\}$, so for the null space matrix we take

$$
Z = \begin{bmatrix} 0 \\ 0 \end{bmatrix}.
$$

At this point, we have $g(x_0) = (-2, -6)$, $Z^{\mathrm{T}} g(x_0) = 0$, and $\mathcal{A}(x_0) \neq \emptyset$, so we compute

$$
\bar{\lambda} = (\hat{A}^{-1})^{\mathrm{T}} g(x_0) = \begin{bmatrix} 1 & 0 \\ 0 & 1 \end{bmatrix} \begin{bmatrix} -2 \\ -6 \end{bmatrix} = \begin{bmatrix} -2 \\ -6 \end{bmatrix}.
$$

Since $\bar{\lambda}$ is not nonnegative, the point x_0 cannot be a local minimizer. Continuing as in Example 12.12, we drop constraint 4 from the working set. The new definitions of interest are

$$\hat{A} = \begin{bmatrix} 1 & 0 \end{bmatrix}, \ \hat{b} = \begin{bmatrix} 0 \end{bmatrix}, \ Z = \begin{bmatrix} 0 \\ 1 \end{bmatrix}.$$

As before, we seek a feasible descent direction. Computing the reduced Hessian matrix we find that

$$Z^{\mathrm{T}} H Z = -4.$$

This is not positive, and hence we know that using it will not result in a descent direction. In fact the reduced Newton direction is given by

$$p = -Z(Z^{\mathrm{T}} H Z)^{-1} Z^{\mathrm{T}} g(x_0) = \begin{bmatrix} 0 \\ -3/2 \end{bmatrix}.$$

This not a descent direction and it is also clearly not a feasible direction since it leads to a steplength of $\alpha = 0$. In such situations the Modified Cholesky factorization (see Algorithm 10.3) would result in a feasible descent direction. For the purpose of this example, we can simply choose our diretion to go the other way; that is

$$p = \begin{bmatrix} 0 \\ 3/2 \end{bmatrix}.$$

The steplength computation leads to $\alpha = 2/3$; using it we arrive at the point $x_1 = (0, 1)$. The corresponding data are now

$$W(x_1) = \{2, 3\}, \quad \hat{A} = \begin{bmatrix} 1 & -2 \\ 1 & 0 \end{bmatrix}, \ \hat{b} = \begin{bmatrix} -2 \\ 0 \end{bmatrix}, \ Z = \begin{bmatrix} 0 \\ 0 \end{bmatrix}, \ \hat{A}^{-1} = \frac{1}{2}\begin{bmatrix} 0 & 2 \\ -1 & 1 \end{bmatrix}.$$

The gradient is $g(x_1) = (0, -10)$. Here, too, we have $Z^{\mathrm{T}} g(x_1) = 0$ and $\mathcal{A}(x_1) \neq \emptyset$, so we define the Lagrange multiplier vector

$$\bar{\lambda} = (\hat{A}^{-1})^{\mathrm{T}} g(x_1) = \frac{1}{2}\begin{bmatrix} 0 & -1 \\ 2 & 1 \end{bmatrix}\begin{bmatrix} 0 \\ -10 \end{bmatrix} = \begin{bmatrix} 5 \\ -5 \end{bmatrix}.$$

This vector is not nonnegative. We drop the constraint corresponding to the negative component of $\bar{\lambda}$. This allows us to make the variable x_1 positive. The new working set is $W = \{2\}$, and we now have

$$\hat{A} = \begin{bmatrix} 1 & -2 \end{bmatrix}, \ \hat{b} = [-2], \ Z = \begin{bmatrix} 2 \\ 1 \end{bmatrix}.$$

Taking the search direction to be the negative of the projected gradient, we get $p = (4/5, 1/5)$. The next task is to find the steplength. This turns out to be $1/6$, and the new point reached is $x_2 = (2/3, 4/3)$.

At x_2 we have $\mathcal{A}(x_2) = \mathcal{W} = \{1, 2\}$ and

$$\hat{A} = \begin{bmatrix} -1 & -1 \\ 1 & -2 \end{bmatrix}, \quad \hat{b} = \begin{bmatrix} -2 \\ -2 \end{bmatrix}, \quad Z = \begin{bmatrix} 0 \\ 0 \end{bmatrix}, \quad \hat{A}^{-1} = \frac{1}{3}\begin{bmatrix} -2 & 1 \\ -1 & -1 \end{bmatrix}.$$

We have $g(x_2) = (-2/3, -10)$ and $z^{\mathrm{T}} g(x_2) = 0$. We compute the Lagrange multipliers

$$\bar{\lambda} = (\hat{A}^{-1})^{\mathrm{T}} g(x_2) = \begin{bmatrix} 34/9 \\ 28/9 \end{bmatrix}.$$

Since this vector is positive, we stop. The iterate $x^* = x_2$ satisfies the first-order necessary conditions of optimality.

Is x^* a local minimizer? Let's first check it with the second-order necessary and sufficient conditions for optimality stated in Theorem 11.10, namely

$$g(x^*)^{\mathrm{T}} p \geq 0 \text{ for all } p \in \mathcal{F}; \tag{12.23}$$

$$p^{\mathrm{T}} H p \geq 0 \text{ for all } p \in \mathcal{F} \text{ such that } g(x^*)^{\mathrm{T}} p = 0. \tag{12.24}$$

Here $\mathcal{F} = \mathcal{F}(x^*)$ is the set of feasible directions at x^*. It is given as the solution set of the homogeneous linear inequality system

$$\begin{aligned} -p_1 - p_2 &\geq 0 \\ p_1 - 2p_2 &\geq 0. \end{aligned} \tag{12.25}$$

It follows (by Farkas's Lemma) from the nonnegativity of the vector $\bar{\lambda}$ that (12.23) holds.[6] What about (12.24)? Note that the above inequalities (12.25) imply $p_2 \leq 0$. Suppose a vector $p \in \mathcal{F}$ satisfies $g(x^*)^{\mathrm{T}} p = 0$, that is,

$$\begin{bmatrix} -2/3 & -10 \end{bmatrix} \begin{bmatrix} p_1 \\ p_2 \end{bmatrix} = 0 \quad \text{or} \quad p_1 + 15p_2 = 0. \tag{12.26}$$

The above condition together with the first inequality of (12.25) implies $p_2 \geq 0$. Consequently $p_2 = 0$ and hence $p_1 = 0$, implying that, in this case, there are no nonzero vectors $p \in \mathcal{F}$ that satisfy $g(x^*)^{\mathrm{T}} p = 0$. Therefore (12.24) holds at x^*.

---

[6]Of course, we already know this because the algorithm terminated with a solution of the first-order necessary condition.

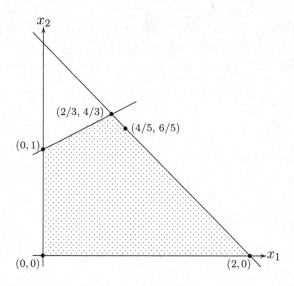

Figure 12.1: Feasible region for Examples 12.12 and 12.13.

Figure 12.1 pertains to the above two examples. As we have seen, the solution found in Example 12.12 is $(4/5, 6/5)$ which lies on the boundary of the feasible region, but not at an extreme point. The local minimizer found in Example 12.13 is at the extreme point $(2/3, 4/3)$.

The special case of linear programming

When f is a linear function, (LIP) reduces to a linear programming problem

$$\begin{array}{ll} \text{minimize} & c^{\mathrm{T}}x \\ \text{subject to} & Ax \geq b. \end{array} \qquad (12.27)$$

If, instead of inequalities, the constraints were all equalities (in which case it would, of course, not be a linear program), the objective function would be unbounded below unless there were exactly n independent equality constraints (see Exercise 12.1). For the above case (12.27) of all inequality constraints, it follows that the objective function (assumed nonconstant) can always be decreased at any iteration where there are fewer than n active constraints. As a result, *a working set of n independent constraints is always maintained when solving such a linear programming problem.*

Suppose that we have an iterate $\bar{x}$ where n constraints (assumed to be linearly independent) are in the working set. We would then have the problem of minimizing $c^{\mathrm{T}}x$ subject to $\hat{A}x = \hat{b}$, where $\hat{A}$ is of rank n. The Lagrange multipliers for this equality-constrained problem are given by $\bar{\lambda}$, the solution of

$$\hat{A}^{\mathrm{T}}\bar{\lambda} = c.$$

If $\bar{\lambda}_i \geq 0$ for all i, then $\bar{x}$ is a local minimizer. Otherwise, if $\bar{\lambda}_j < 0$ for some active constraint $\hat{A}_{j\bullet}$, we can decrease the objective function by moving "off" that constraint, but staying "on" all the others. Because n constraints are active at $\bar{x}$, this gives only *one possible direction* in which to move, since the search direction must be orthogonal to $(n-1)$ vectors. The direction we seek is defined by

$$\begin{aligned} \hat{A}_{i\bullet}p &= 0 \quad (\text{for } i \neq j) \\ \hat{A}_{i\bullet}p &= 1. \end{aligned}$$

The above relationships imply that the direction p is the solution of the equations

$$\hat{A}p = I_{\bullet j}.$$

Hence p is the j-th column of the inverse of $\hat{A}$. Moving along such a direction p will cause the linear objective function to decrease without bound unless we reach an inactive constraint. If we move past the bound specified by the inactive constraint, then the inactive constraint would become violated. Consequently, to preserve feasibility of the iterate, the first constraint to be hit is added to the working set, and the process is repeated.

The algorithm described above is, in fact, a version of the *Simplex Algorithm* for linear programming; it involves moving from vertex to vertex of the polyhedron defined by the set of linear constraints.

Algorithm 12.2 (Simplex Algorithm for LP as an Active Set Algorithm)

Let $k = 0$. Let $x = x_0$ be a feasible point for (12.27) with corresponding working set $\mathcal{W}_k$ of n linearly independent constraints $\hat{A}_0 x = b_0$.

1. *Test for optimality.* Compute the Lagrange multipliers by solving

$$\hat{A}_k^{\mathrm{T}}\bar{\lambda} = c.$$

If $\bar{\lambda}_i \geq 0$ for all active constraints i, report x_k as a solution and stop.

2. *Compute a search direction.* If $\bar{\lambda}_s < 0$, for some active constraint s, find p such that

$$\hat{A}_k p = I_{\bullet s}.$$

Moving a positive amount along such a p will result in constraint s being deleted from the working set.

3. *Add a constraint to the working set.* Find the index of a constraint, that is not in the working set, that will be reached by a move along p i.e., find r such that

$$\alpha_r = \min \begin{cases} \dfrac{b_i - \hat{A}_{i\bullet} x_k}{\hat{A}_{i\bullet} p} & \text{for all } i \notin \mathcal{W}_k, \ \hat{A}_{i\bullet} p < 0, \\ \infty. \end{cases}$$

(a) If $\alpha = 0$, then no constraint is added to the working set. In this case the constraint s stays in the working set which remains unchanged.

(b) If $0 < \alpha_r < \infty$, add constraint r to the working set, increment k, update x_k and go to Step 1.

(c) If $\alpha_r = \infty$, report the problem as unbounded and stop.

Equivalence to the Revised Simplex Algorithm

To see the equivalence of this algorithm to the Revised Simplex Algorithm, we consider an LP problem in standard form

$$\begin{array}{ll} \text{minimize} & c^{\mathrm{T}} x \\ \text{subject to} & Ax = b \\ & x \geq 0, \end{array} \qquad (12.28)$$

where A is $m \times n$ with $m < n$. Note that the general constraints are all equalities; the only inequalities are simple bounds, and thus *only the bound constraints may leave or enter the working set*; the corresponding system of equations is of the form

$$\hat{A} x = \begin{bmatrix} B & N \\ 0 & I_{n-m} \end{bmatrix} \begin{bmatrix} x_B \\ x_N \end{bmatrix} = \begin{bmatrix} b \\ 0 \end{bmatrix} = \hat{b},$$

where B contains the basic columns an N contains the nonbasic columns. As described earlier, the variables in x corresponding to B are termed *basic variables* are denoted by x_B. Similarly, columns in N and their associated variables x_N are termed *nonbasic*. Note that the equation above says that $x = (x_B, \ x_N) = (B^{-1}b, \ 0)$.

Next, partition the vector of Lagrange multipliers into an m-vector λ (the multipliers corresponding to the linear equality constraints) and an $(n-m)$-vector σ (corresponding to the variables fixed at their lower bound). The equations $\hat{A}^T \mu = c$, where $\mu = (\lambda, \sigma)$, can be written as

$$\hat{A}^T \mu = \begin{bmatrix} B^T & 0 \\ N^T & I_{n-m} \end{bmatrix} \begin{bmatrix} \lambda \\ \sigma \end{bmatrix} = \begin{bmatrix} B^T \lambda \\ N^T \lambda + \sigma \end{bmatrix} = \begin{bmatrix} c_B \\ c_N \end{bmatrix},$$

where c_B denotes the components of c corresponding to the basic variables and c_N denotes the components corresponding to the nonbasic variables. Clearly, λ is the solution of the linear system

$$B^T \lambda = c_B$$

and

$$\sigma = c_N - N^T \lambda.$$

The components of σ are the *reduced costs* that we have seen earlier. Similarly, the solution of $\hat{A}p = I_{\bullet s}$ can be written as

$$\begin{bmatrix} B & N \\ 0 & I_{n-m} \end{bmatrix} \begin{bmatrix} p_B \\ p_N \end{bmatrix} = I_{\bullet s}$$

so that p_B is the solution of

$$Bp_B = -A_{\bullet s}.$$

Movement along p to decrease the objective function will cause a basic variable to encounter one of its bounds (i.e., a new bound constraint will be added to the working set). This is determined by finding the largest steplength α such that the inactive constraints remain satisfied, that is,

$$x_B + \alpha p_B \geq 0.$$

Hence we find $\alpha = \alpha_r$, where r is the r-th column in the basis, such that

$$\alpha_r = \left\{ \frac{-(x_B)_r}{(p_B)_r} \right\} = \max \left\{ \frac{-(x_B)_i}{(p_B)_i} \ : \ (p_B)_i < 0, \ i = 1, \ldots, m \right\}$$

$$= \max \left\{ \frac{-(B^{-1}b)_i}{-(B^{-1}A_{\bullet s})_i} \ : \ (p_B)_i < 0, \ i = 1, \ldots, m \right\}$$

$$= \max \left\{ \frac{\bar{b}_i}{-\bar{A}_{is}} \ : \ \bar{A}_{is} > 0, \ i = 1, \ldots, m \right\}.$$

Each complete iteration of the Revised Simplex Algorithm defined in this way has the effect of exchanging the status of a basic and a nonbasic variable. Thus the Revised Simplex Algorithm is equivalent to Algorithm 12.2 applied to the same problem (12.28).

Example 12.14: SOLUTION OF AN LP BY AN ACTIVE SET METHOD.
We compare the Revised Simplex Method illustrated in Example 4.7 with the Active Set Method described here. Recall that the LP in standard form was

$$
\begin{array}{ll}
\text{minimize} & 4x_1 + 6x_2 + 3x_3 + x_4 \\
\text{subject to} & x_1 + 4x_2 + 8x_3 + 6x_4 + x_5   = 11 \\
& 4x_1 + x_2 + 2x_3 + x_4 + x_6 = 7 \\
& 2x_1 + 3x_2 + x_3 + 2x_4 + x_7 = 2 \\
& x_1, \ x_2, \ x_3, \ x_4, \ x_5, \ x_6, \ x_7 \geq 0.
\end{array}
$$

An initial feasible basis B is the identity matrix corresponding to the slack variables. Corresponding to this B, we have a basic feasible solution with $x_B = (x_5, x_6, x_7) = (11, 7, 2)$; and the vector of costs associated with the basic variables is $c_B = (0, 0, 0)$. The coefficient matrix corresponding to the active constraints is

$$
\hat{A} = \begin{bmatrix} B & N \\ 0 & I_3 \end{bmatrix} = \begin{bmatrix} 1 & 4 & 8 & 6 & 1 & 0 & 0 \\ 4 & 1 & 2 & 1 & 0 & 1 & 0 \\ 2 & 3 & 1 & 2 & 0 & 0 & 1 \\ & & & & & & \\ 0 & 0 & 0 & 0 & 1 & 0 & 0 \\ 0 & 0 & 0 & 0 & 0 & 1 & 0 \\ 0 & 0 & 0 & 0 & 0 & 0 & 1 \end{bmatrix}.
$$

The inactive constraints correspond to the basic variables x_B.

Iteration 1:

We start by testing for optimality. To do this, we solve $\hat{A}^T \bar{\lambda} = c$. As we noted earlier, this amounts to solving $B^T \lambda = c_B$; this is the same set of equations as in Example 4.7, but with λ instead of y, and we obtain $\lambda = y = 0$. We next compute $z_j = \lambda^T N_{\bullet j}$ for $j = 1, 2, 3, 4$ to obtain

$$
(z_1, z_2, z_3, z_4) = (0, 0, 0, 0).
$$

These are the same quantities obtained earlier in Example 4.7. Instead of looking at the difference $z_j - c_j$, we reverse the sign and find an index s such that

$$c_s - z_s = \min\{c_s - z_s\} = \min\{-4, -6, -4, -1\} = -6,$$

hence $s = 2$. The condition for optimality is not satisfied (because of the sign reversal, we now require $c_s - z_s \geq 0$), and the nonbasic variable x_2 is a candidate for improving the objective function value.

To determine the level at which we can bring x_2 into the basic set, we first solve $\hat{A}p = I_{\bullet s}$; this is equivalent to solving $Bp_B = -A_{\bullet 2}$ to obtain $p_B = (-4, -1, -3)$. The vector $p = (p_B, p_N) = (-4, -1, -3, 0, 1, 0)$. Notice that $p_B = -\bar{A}_{\bullet 2}$. Next we determine the largest steplength α_r along p_B.

$$\alpha_r = \left\{ \frac{-(x_B)_r}{(p_B)_r} \right\} = \max\left\{ \frac{11}{-4}, \frac{7}{-1}, \frac{2}{-3} \right\} = \frac{2}{-3} = \frac{\bar{b}_3}{-\bar{a}_{32}}.$$

For the Revised Simplex Method, the denominator was the negative of the values above, and we picked the minimum fraction; other than that, the operation is identical. Hence we have $r = 3$. This means that the value of the index j_3 will change from 7 to 2. The new iterate is obtained by setting $x \leftarrow x + \alpha p$; hence $x_B = x_B + \alpha p_B$, $x_2 = \alpha$ and the new $(x_B)_3 = x_7 = 0$. The new basis and associated x_B will be

$$B = [A_{\bullet 5} \; A_{\bullet 6} \; A_{\bullet 2}] = \begin{bmatrix} 1 & 0 & 4 \\ 0 & 1 & 1 \\ 0 & 0 & 3 \end{bmatrix} \quad \text{and} \quad x_B = (x_5, x_6, x_2) = \left(\frac{25}{3}, \frac{19}{3}, \frac{2}{3} \right).$$

In the Revised Simplex Method, we update the basis B and its factorization, and then solve $Bx_B = b$.

The rest of the iterations proceed in a similar manner.

12.6 Exercises

12.1   In this book we have asserted that a problem of the form

$$\text{(LEP)} \quad \begin{array}{ll} \text{minimize} & c^{\mathrm{T}}x \\ \text{subject to} & Ax = b. \end{array}$$

is not a proper linear program. This exercise is meant to justify this statement. To that end, let $X = \{x : Ax = b\}$. Show that for any such problem, exactly one of the following three statements must be true.

(a) The set X is empty, i.e., the constraints have no solution.
(b) The set X is nonempty, and the objective function $c^{\mathrm{T}}x$ is constant on X.
(c) The set X is nonempty, and the objective function $c^{\mathrm{T}}x$ is unbounded on X.

12.2   Suppose we are given a set

$$X = \{x \in R^n : Ax = b\} \neq \emptyset$$

and a point $p \in R^n$ not belonging to X. We seek the point of X that is closest to p as measured by the Euclidean norm. Assume it is known that $A \in R^{m \times n}$ and rank $(A) = m$. The problem can be written as

$$\begin{aligned} \text{minimize} \quad & \tfrac{1}{2}\|x - p\|^2 \\ \text{subject to} \quad & Ax = b. \end{aligned}$$

The objective function $f(x)$ is just

$$\tfrac{1}{2}x^{\mathrm{T}}x - p^{\mathrm{T}}x + \tfrac{1}{2}p^{\mathrm{T}}p.$$

Let B be a basis in A. Let x_B and x_N denote the corresponding basic and nonbasic variables.

(a) Write f as a function of x_N.
(b) Use this to find a closed-form solution.

12.3   Given $\hat{A} \in R^{\ell \times n}$, any vector $x \in R^n$ can be decomposed as

$$x = \hat{A}^{\mathrm{T}}x_{\hat{A}} + Zx_Z.$$

For the linear equality problem (12.1), the linear constraints completely determine x_A^* as shown in equation (12.8).

(a) Show that when $\hat{A}$ is of full rank, x_A^* is the unique solution to (12.8).
(b) Show that when $\hat{A}$ is rank-deficient, (12.8) may have no solution, and, if it does have a solution, then there are infinitely many solutions.

12.4   Consider the $(m + n) \times (m + n)$ matrix

$$H = \begin{bmatrix} B & A \\ -A^{\mathrm{T}} & C \end{bmatrix},$$

where B is an $m \times m$ symmetric positive definite matrix and C is an $n \times n$ symmetric positive definite matrix. (A matrix having the form H above is said to be *bisymmetric*.) Explain why H is positive definite in the sense that the associated quadratic form $z^{\mathrm{T}}Hz$ is positive for all $z \neq 0$.

12.5   Consider the linear equality problem (LEP):

$$\text{minimize} \quad \tfrac{1}{2}x^{\mathrm{T}}x + c^{\mathrm{T}}x$$
$$\text{subject to} \quad \hat{A}x = \hat{b},$$

where
$$c = (1,\ 1,\ 1), \qquad \hat{A} = \begin{bmatrix} 1 & -1 & 1 \end{bmatrix}, \qquad \hat{b} = 1.$$

(a) Write down a null space basis Z for the matrix $\hat{A}$.

(b) Let $x_0 = (1,1,1)$ be an initial point, Write down the projected Hessian $Z^{\mathrm{T}}GZ$ and projected gradient $Z^{\mathrm{T}}g$ at x_0.

(c) Compute a search direction p using x_0 as an initial starting point. (First find p_z and then determine $p = Zp_z$.)

(d) Compute the next approximation (iterate) x_1.

(e) Show that x_1 is a minimizer.

(f) Justify why it took only one step to find a minimum in this case and why α was set to 1 instead of using a steplength method.

12.6   Consider the problem:

$$\text{minimize} \quad f(x) = 2x_1x_2 + x_2x_3 + x_1x_3$$
$$\text{subject to} \quad 3x_1 + 3x_2 + 2x_3 = 3.$$

(a) Write down the gradient of f.

(b) Express the first-order necessary conditions in the form $g(x) = \hat{A}^{\mathrm{T}}\lambda$.

(c) Use these equations together with the equality constraint to solve for x and λ.

(d) Is the solution x a local minimizer? Explain.

12.7   Consider the problem:

$$\text{minimize } f(x) = \tfrac{1}{2}x^{\mathrm{T}}Hx + c^{\mathrm{T}}x$$

where
$$H = \begin{bmatrix} 1 & 0 & 0 \\ 0 & 1 & 0 \\ 0 & 0 & -1 \end{bmatrix}, \qquad c = \begin{bmatrix} 1 \\ 1 \\ 1 \end{bmatrix},$$

(a) Show that the Hessian is indefinite (i.e., neither positive semidefinite nor negative semidefinite).

(b) Demonstrate that minimizing $f(x)$ results in an unbounded solution (i.e., $f(x)$ is unbounded below on R^3).

(c) If we add the constraint $x_3 = 4$, show that $x^* = (-1, -1, 4)$ is a local minimizer of the problem minimize $f(x)$, subject to $x_3 = 4$.

(d) Compute the Lagrange multiplier λ^* at x^*.

    (i) Is x^* still a minimizer if we change the constraint to $x_3 \geq 4$?

    (ii) Is x^* still a minimizer if we change the constraint to $x_3 \leq 4$?

Justify your answer using the necessary conditions for a linear inequality problem.

12.8   Minimize the function

$$f(x_1, x_2) = x_1 - x_2 - x_1^2 + x_2^2$$

subject to the constraints given in Examples 12.12 and 12.13.

12.9   The feasible region shown in Figure 12.1 has four extreme points.

(a) Are all of them local minimizers in Example 12.13?

(b) Where is the global minimum located?

12.10   (Based on an example by P.B. Zwart [202].) Consider the quadratic program

$$\begin{aligned}
\text{maximize} \quad & Q(x) = 2x_1^2 + x_1 x_2 + 2x_2 \\
\text{subject to} \quad & 1.5x_1 + x_2 \leq 1.4 \\
& x_1 + x_2 \leq 1 \\
& x_1 \geq 0 \\
& x_2 \geq -10
\end{aligned}$$

(a) Graph the feasible region of this optimization problem.

(b) Is the point $\bar{x} = (0, 1)$ a local maximum for this problem? Justify your answer.

12.11   Use the well-known *Farkas's Lemma* described on page 125 to establish the necessary condition (12.20).

12.12   Consider the problem

$$\begin{aligned}
\text{minimize} \quad & c^{\mathrm{T}} x \\
\text{subject to} \quad & \hat{A} x = \hat{b}, \quad \text{where} \quad \hat{A} \in R^{\ell \times n}, \; \ell < n.
\end{aligned}$$

Prove that if x^* is a solution to the above problem, then every $\bar{x}$ satisfying $\hat{A}\bar{x} = \hat{b}$ is also a minimizer. (Hint: Use the first-order necessary conditions to show that $\hat{Z}^{\mathrm{T}} c = 0$, where the columns of the matrix $\hat{Z}$ are a basis for the null space of $\hat{A}$, and examine such points $\bar{x}$ in terms of feasible moves from x^*.)

12.13   Consider the three-variable optimization problem

$$\text{minimize} \quad 2x_1^2 - x_2^2 + x_3^2 - 12x_1$$
$$\text{subject to} \quad x_1 + x_2 = 2.$$

(a) Write down the gradient and Hessian of the objective function.

(b) Write down two representations $\widehat{Z}$ and Z of the null space of the constraint matrix: $\widehat{Z}$ that has orthogonal columns and Z that does not.

(c) Compute the following three feasible descent search directions at the point $\bar{x} = (1,\ 1,\ 1)$ using $\widehat{Z}$ and Z in turn.

   (i)   Rosen's steepest-descent direction.
   (ii) Wolfe's reduced-gradient direction.
   (iii) A Newton search direction.

(d) Why are the Newton search direction and Wolfe's reduced-gradient direction multiples of each other for this optimization problem?

12.14   Modify the BFGS Algorithm to specialize it for the bound-constrained problem (12.21).

12.15   The Barrier-function Method is an interior-point method (one in which the iterates stay strictly inside the feasible region) that has been used in solving nonlinear inequality constrained problems. The idea behind the method is to create a "barrier" at the constraint boundary and relax the barrier at "appropriate" intervals. The transformation of the objective function and inequality constraints is such that the problem is converted to an unconstrained minimization problem.

Consider the linear programming problem:

$$\text{minimize} \quad h^{\mathrm{T}}x$$
$$\text{subject to} \quad Ax \geq b, \quad \text{where} \quad A \in R^{m \times n}.$$

A logarithmic barrier function for the above LP problem is

$$B(x, \varepsilon) = h^{\mathrm{T}}x - \varepsilon \sum_{i=1}^{m} \ln\left(c_i(x)\right)$$

where

$$c_i(x) = A_{i.}x - b_i = \sum_{j=1}^{n} a_{ij}x_j - b_i,$$

and $\varepsilon > 0$ is a small scalar.

(a) Write down the Newton equations by determining the expression for the first and second derivatives of $B(x, \varepsilon)$.

(b) What can you say about the Hessian matrix at interior points (that is, where $Ax > b$)?

12.16   Let A be an $\ell \times n$ matrix of full row rank and let Z be a basis matrix for $\mathcal{N}(A)$ satisfying $AZ = 0$. Prove the following assertions:

(a) For any vector v,

$$\left(I - A^{\mathrm{T}}\left(AA^{\mathrm{T}}\right)^{-1}A\right)v = Z(Z^{\mathrm{T}}Z)^{-1}Z^{\mathrm{T}}v.$$

(b) If g is any vector in R^n,

$$g = A^{\mathrm{T}}\lambda_L + Z(Z^{\mathrm{T}}Z)^{-1}g_z$$

where $g_z = Z^{\mathrm{T}}g$ and λ_L is the least-squares solution of $A^{\mathrm{T}}\lambda \approx g$.

(c) If $\bar{\lambda}$ is any multiplier estimate with residual $r = g - A^{\mathrm{T}}\bar{\lambda}$, then

$$r = A^{\mathrm{T}}(\lambda_L - \bar{\lambda}) + Z(Z^{\mathrm{T}}Z)^{-1}g_z$$

and

$$\|Z^{\mathrm{T}}g\|_{(Z^{\mathrm{T}}Z)^{-1}} = \|g - A^{\mathrm{T}}\lambda_L\|_2.$$

13. PROBLEMS WITH NONLINEAR CONSTRAINTS

Overview

In Chapter 12, we presented an Active Set Method for solving optimization problems with a nonlinear objective and linear constraints. There we saw that when moving in a linear direction from one feasible point to another, we could keep a subset of the constraints active (or satisfied). This, in general, is not possible for nonlinear constraints, and the development of appropriate algorithms must take this into account.

As in Chapter 12, we start by deriving optimality conditions for non-linearly constrained problems by considering feasible moves from a local minimizer. After this, we present the Penalty-function Method. This transforms a nonlinearly constrained problem into a sequence of unconstrained problems, the solutions of which converge to an optimal solution of the original problem. Penalty-function methods operate from outside of the feasible region, which means that the iterates are infeasible until an optimal solution is reached. Next we describe reduced-gradient methods which are based on techniques for solving nonlinear problems with linear constraints. After this we describe an Augmented Lagrangian Method (also known as a Multiplier Method) where the objective function is the Lagrangian of the problem plus a penalty term of the violated constraints. This is followed by methods that involve a modification of the Lagrangian as an objective function and linearization of nonlinear constraints to form subproblems solved by Active Set Methods. Finally we describe Sequential Quadratic Programming (SQP) Methods whereby the original problem is approximated by a quadratic program at each iteration. The solution of each approximation results in a search direction for the original problen. Such methods are currently the most widely used for solving nonlinear programs.

13.1   Nonlinear equality constraints

We begin our discussion with optimization problems of the form

$$\text{(NEP)} \quad \begin{array}{ll} \text{minimize} & f(x) \\ \text{subject to} & \hat{c}(x) = 0 \\ & x \in R^n, \end{array} \quad (13.1)$$

© Springer Science+Business Media LLC 2017
R.W. Cottle and M.N. Thapa, *Linear and Nonlinear Optimization*,
International Series in Operations Research & Management Science 253,
DOI 10.1007/978-1-4939-7055-1_13

where f is a smooth nonlinear function and $\hat{c}$ is a vector of ℓ smooth nonlinear functions[1]. The reason for using the notation $\hat{c}$ instead of c will become clear in our discussion of solution methods for problems containing nonlinear inequality constraints. In practice, problems of the form (13.1) could have a linear objective and one or more linear constraints.

In Section 12.1 we saw that one possible approach to solving a problem with linear equations (12.1) is to find a basis in $\hat{A}$ and then obtain the *restriction* of f to the set of solutions of the system of linear equations. This is not always possible when the equations are nonlinear as assumed in (13.1).

Optimality conditions revisited

In this subsection we derive optimality conditions for the nonlinear equality problem (13.1) by considering feasible moves from a feasible starting point. As usual, the derivation of these conditions in this manner helps in algorithmic design.

With linear equality constraints, it is, in general, possible to specify a set of feasible directions from a starting feasible point $x = \bar{x}$, along which unlimited movement always retains feasibility with respect to these constraints. However, with nonlinear equality constraints, there will not always be such feasible directions. For example, consider the nonlinear constraint

$$x_1^2 + x_2^2 = 4.$$

In this case, the set of all feasible points (x_1, x_2) must lie on the circle of radius 2 centered at the origin. Clearly, given a feasible point $(\bar{x}_1, \bar{x}_2)$, there is no p such that $(\bar{x}_1 + \varepsilon p_1, \bar{x}_2 + \varepsilon p_2)$ is feasible for all $\varepsilon > 0$. Instead we need to consider nonlinear movements. In order to do this we look at *feasible arcs* starting at $x = \bar{x}$. An *arc*[2] is defined to be a directed curve γ in R^n parameterized by a single parameter, say θ, and consisting of more than one point. A feasible arc starting at $x = \bar{x}$ is one for which $\gamma(0) = \bar{x}$ and $\hat{c}(\gamma(\theta)) = 0$ for all $\theta \in [0, \bar{\theta}]$ with $\bar{\theta} > 0$.

---

[1] In the literature and in this book, the individual functions which make up the vector $\hat{c}(x)$ are called *constraints*, even though by themselves they are not constraints but *constraint functions*.

[2] A more precise definition can be found in Jost's book, [103, pp. 145–149].

Let $\gamma : [0, \bar{\theta}] \rightarrow R^n$ denote a differentiable arc[3] in R^n. The *tangent* to the arc at $\bar{x}$ is given by the vector

$$p = \frac{d\gamma(\theta)}{d\theta}\bigg|_{\theta=0} = \left(\frac{d\gamma_1(\theta)}{d\theta}, \frac{d\gamma_2(\theta)}{d\theta}, \ldots, \frac{d\gamma_n(\theta)}{d\theta}\right) = \dot{\gamma}(0).$$

If $\gamma(\theta)$ is a feasible arc, then for each $i = 1, 2, \ldots, \ell$, the chain rule of differentiation yields

$$\frac{d\hat{c}_i(\gamma(\theta))}{d\theta}\bigg|_{\theta=0} = \left(\nabla\hat{c}_i(\gamma(\theta))\right)^{\mathrm{T}} p\bigg|_{\theta=0} = \hat{A}_{i\bullet}(\bar{x})p = 0.$$

Hence

$$\hat{A}(\bar{x})p = 0, \tag{13.2}$$

where the rows of $\hat{A}(\bar{x})$ are the rows of the Jacobian matrix of $\hat{c}(x)$ at $\bar{x}$ (i.e., they are the transposes of the gradients of the constraints $1, 2, \ldots, \ell$ evaluated at $\bar{x}$).

The fact that the tangent p to a given a feasible arc starting at $\bar{x}$ satisfies (13.2) reveals a similarity to the linear equality-constrained case, where every feasible direction p satisfies $\hat{A}p = 0$. Furthermore, in the linear case, any nonzero p satisfying $\hat{A}p = 0$ must be a feasible search direction. Can we draw a similar conclusion in the nonlinear case? That is, if we have a p that satisifies (13.2) can we conclude that p is tangent to a feasible arc? Unfortunately the answer is no: not unless other conditions are satisfied. The following example illustrates a case where we can find many vectors p satisfying (13.2), none of which are tangent to a feasible arc; in this case, because no feasible arc exists.

Example 13.1: NO FEASIBLE ARC. The system of constraints

$$\hat{c}_1(x) = (1 - x_1)^3 - x_2 = 0$$
$$\hat{c}_2(x) = \qquad\qquad x_2 = 0$$

has a unique solution: $\bar{x} = (1, 0)$. Hence, obviously there is no feasible arc emanating from $\bar{x}$. The Jacobian (constraint gradient) matrix evaluated at $\bar{x} = (1, 0)$ is:

$$\hat{A}(\bar{x}) = \begin{bmatrix} -3(1 - \bar{x}_1)^2 & -1 \\ 0 & 1 \end{bmatrix} = \begin{bmatrix} 0 & -1 \\ 0 & 1 \end{bmatrix}.$$

---

[3]If $\gamma(\theta)$ is an arc and its components are differentiable functions, then it is called a *differentiable arc*.

It is easy to see that any vector $p = (\beta, 0)$, where β is a real number, satisfies (13.2). Yet no p is tangent to a feasible arc at $\bar{x} = (1, 0)$ because there is no feasible arc at $\bar{x}$.

First-order necessary conditions

Given a p that satisfies (13.2) at a feasible point $\bar{x}$, we would like to determine when it is tangent to a feasible arc emanating at $\bar{x}$. It turns out that if the constraints satisfy certain conditions, called *constraint qualifications*, then such a p will in fact be tangent to a feasible arc. Here we will be concerned only with the practical linear independence constraint qualification (LICQ) introduced earlier on page 373.

Theorem 11.2 specifies the first-order necessary conditions for optimality in (13.1) with the LICQ as a key hypothesis. Here, analogous to our treatment of linear equality constraints, we develop these conditions by examining feasible moves from a minimizer x^*. Clearly such an x^* must be feasible, that is, $\hat{c}(x^*) = 0$. If the LICQ holds at x^*, every vector p satisfying $\hat{A}(x^*)p = 0$ is tangent to a feasible arc emanating from x^*. Let $\gamma(\theta)$ denote an arc with $\gamma(0) = x^*$, and let p denote a tangent to $\gamma(\theta)$ at $\theta = 0$. For $f(x)$ to have a minimum at x^*, it is necessary for the first derivative of $f(x)$ along the arc to be zero, i.e.,

$$\frac{df\left(\gamma(\theta)\right)}{d\theta}\bigg|_{\theta=0} = 0.$$

By the chain rule of differentiation, this becomes

$$p^{\mathrm{T}}g(x^*) = 0, \tag{13.3}$$

where, as before, $g(x^*)$ is the gradient of $f(x)$ evaluated at $x = x^*$. The vector p is tangent to a feasible arc and satisfies

$$\hat{A}(x^*)p = 0. \tag{13.4}$$

As in the linear constraint case, we let $Z(x^*)$ denote a basis for the null space of $\hat{A}(x^*)$; that is, the columns of $Z(x^*)$ are a basis for the set of vectors orthogonal to the columns of $\hat{A}(x^*)$. Note that in the nonlinear case, the null space ordinarily changes with each iterate. By (13.4), p belongs to the null space of $\hat{A}(x^*)$ and hence can be written in the form

$$p = Z(x^*)p_z \tag{13.5}$$

for some p_z. Substituting (13.5) in (13.3), we obtain

$$p_z^T Z(x^*)^T g(x^*) = 0. \tag{13.6}$$

Clearly the above will be true for every p that satisfies (13.4), and every such p is tangent to a feasible arc by the LICQ. Thus, from (13.6) it follows that the projected gradient is zero at x^*, that is to say

$$Z(x^*)^T g(x^*) = 0. \tag{13.7}$$

This implies that $g(x^*)$ is orthogonal to the columns of $Z(x^*)$. Thus, a statement equivalent to (13.7) is

$$g(x^*) = \hat{A}(x^*)^T \lambda^*. \tag{13.8}$$

The components of λ^* are the Lagrange multipliers. Equation (13.8) says that if the LICQ holds at x^*, then x^* is a *stationary point* of the Lagrangian function

$$L(x, \lambda^*) = f(x) - \sum_{i=1}^{\ell} \lambda_i^* \hat{c}_i(x) = f(x) - (\lambda^*)^T \hat{c}(x). \tag{13.9}$$

As noted earlier, if x^* is feasible and the LICQ holds, then (13.8) expresses the Lagrangian first-order necessary conditions of optimality at that point.

Second-order necessary conditions

The derivation of second-order necessary conditions for (NEP) is a lot more complicated than it is for the linear equality problem, hence we just state these conditions here. If the LICQ holds at x^*, the second-order necessary conditions for x^* to be a local minimizer of (NEP) are that

$$g(x^*) = \hat{A}(x^*)^T \lambda^* \tag{13.10}$$

and

$$Z(x^*)^T W(x^*, \lambda^*) Z(x^*) \tag{13.11}$$

must be positive semidefinite, where $W(x, \lambda)$ is the Hessian of the Lagrangian function, i.e.,

$$W(x, \lambda) = G(x) - \sum_{i=1}^{\ell} \lambda_i \hat{G}_i(x), \tag{13.12}$$

with $G(x)$ and $\hat{G}_i(x)$ being the Hessian matrices of the objective function f and the constraint function $\hat{c}_i$, respectively.

Second-order sufficient conditions

The second-order sufficient conditions for x^* to be an *isolated* local minimizer of (NEP) are that all the following hold:

1. The constraints are satisfied at x^*, that is $\hat{c}(x^*) = 0$.

2. The projected gradient is zero: $Z(x^*)^{\mathrm{T}}g(x^*) = 0$, where $Z(x^*)$ denote a basis for the null space with respect to the constraint gradients which are the columns of $\hat{A}(x^*)$. Equivalently the objective function gradient lies in the range space of the constraint gradients: $g(x^*) = \hat{A}(x^*)^{\mathrm{T}}\lambda^*$.

3. The projected Lagrangian Hessian $Z(x^*)^{\mathrm{T}}W(x^*, \lambda^*)Z(x^*)$ is positive definite, where $W(x^*, \lambda^*)Z(x^*)$ is the Hessian of the Lagrangian function.

Some observations

1. As its notation indicates, the matrix $Z(x)$, the columns of which are a basis for the null space with of $\hat{A}(x)$, is a function of the point x at which the constraints are evaluated.

2. Feasible moves from a point $\bar{x}$ are along a feasible arc. This is important for analysis, however, algorithms construct linear search directions.

3. At a local minimizer x^*, the gradient of f might, in general, not be zero, but the projected gradient *must* be zero. Furthermore x^* is a stationary point of the Lagrangian function $L(x, \lambda^*)$, and not necessarily an unconstrained minimizer of f.

4. The necessary conditions for x^* to be a local minimizer require only that the projected Hessian of the Lagrangian $Z(x^*)^{\mathrm{T}}W(x^*, \lambda^*)Z(x^*)$ be positive semidefinite. The Hessian of the Lagrangian $W(x^*, \lambda^*)$ need not be positive definite.

Example 13.2: SECOND-ORDER CONDITIONS. Here we illustrate how the second-order conditions are used to determine if a point satisfying the first-order conditions is a minimizer (see Figure 13.1). Consider the following problem of minimizing a linear function subject to the constraint that all feasible points lie on a circle:

$$\begin{array}{ll} \text{minimize} & x_1 + x_2 \\ \text{subject to} & x_1^2 + x_2^2 - 4 = 0. \end{array}$$

The constraint gradient is given by $\hat{A}(x) = (2x_1, 2x_2)$. The objective function gradient is $g(x) = (1, 1)$ for all x. The Hessian matrix of the constraint is

$$\hat{G}(x) = \begin{bmatrix} 2 & 0 \\ 0 & 2 \end{bmatrix}.$$

A null space representation of the Jacobian as a function of x is

$$Z(x) = \begin{bmatrix} 1 \\ -1 \end{bmatrix}.$$

The Lagrangian and its Hessian are

$$L(x, \lambda) = x_1 + x_2 - \lambda(x_1^2 + x_2^2 - 4)$$
$$W(x, \lambda) = G(x) - \lambda\hat{G}(x) = -\lambda\hat{G}(x) = -\lambda \begin{bmatrix} 2 & 0 \\ 0 & 2 \end{bmatrix}.$$

We examine the optimality conditions at two points: $\bar{x} = (\sqrt{2}, \sqrt{2})$ and $\hat{x} = (-\sqrt{2}, -\sqrt{2})$. At $\bar{x}$, the first-order conditions $g(\bar{x}) = \hat{A}(\bar{x})^{\mathrm{T}}\bar{\lambda}$ are satisfied with $\bar{\lambda} = 1/(2\sqrt{2})$ because

$$\begin{bmatrix} 1 \\ 1 \end{bmatrix} = \begin{bmatrix} 2\sqrt{2} \\ 2\sqrt{2} \end{bmatrix} \bar{\lambda}.$$

The Hessian of the Lagrangian at $\bar{x}$ is

$$W(\bar{x}, \bar{\lambda}) = -\frac{1}{2\sqrt{2}} \begin{bmatrix} 2 & 0 \\ 0 & 2 \end{bmatrix},$$

which is not positive definite. However, recall that we are interested in the projected Hessian of the Lagrangian, that is,

$$Z(\bar{x})^{\mathrm{T}}W(\bar{x}, \bar{\lambda})Z(\bar{x}) = -\frac{1}{2\sqrt{2}} \begin{bmatrix} 1 \\ -1 \end{bmatrix}^{\mathrm{T}} \begin{bmatrix} 2 & 0 \\ 0 & 2 \end{bmatrix} \begin{bmatrix} 1 \\ -1 \end{bmatrix} = -\sqrt{2},$$

which is also not positive definite. Thus, the second-order sufficient conditions do not hold at $\bar{x}$, and we cannot conclude that it is a strict local

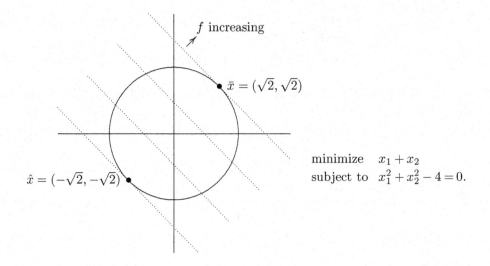

$$\begin{aligned} \text{minimize} \quad & x_1 + x_2 \\ \text{subject to} \quad & x_1^2 + x_2^2 - 4 = 0. \end{aligned}$$

Figure 13.1: Minimizing a linear function over a circle.

minimizer. In fact, in this case, the second-order necessary conditions also do not hold.

Next consider $\hat{x}$. The first-order condition

$$\begin{bmatrix} 1 \\ 1 \end{bmatrix} = \begin{bmatrix} 2\sqrt{2} \\ 2\sqrt{2} \end{bmatrix} \hat{\lambda}$$

holds with $\hat{\lambda} = -1/(2\sqrt{2})$. The sign of $\hat{\lambda}$ is not relevant because the constraint is an equation. The Hessian of the Lagrangian at $\hat{x}$ is

$$W(\hat{x}, \hat{\lambda}) = \frac{1}{2\sqrt{2}} \begin{bmatrix} 2 & 0 \\ 0 & 2 \end{bmatrix},$$

which is positive definite and, hence, so is the projected Hessian of the Lagrangian:

$$Z(\hat{x})^{\mathrm{T}} W(\hat{x}, \hat{\lambda}) Z(\hat{x}) = \frac{1}{2\sqrt{2}} \begin{bmatrix} 1 \\ -1 \end{bmatrix}^{\mathrm{T}} \begin{bmatrix} 2 & 0 \\ 0 & 2 \end{bmatrix} \begin{bmatrix} 1 \\ -1 \end{bmatrix} = \sqrt{2}.$$

Thus, the second-order sufficient conditions show that $\hat{x}$ is an isolated local minimizer for the problem.

13.2 Nonlinear inequality constraints

The nonlinear inequality-constrained problem is stated as

$$\text{(NIP)} \qquad \begin{array}{ll} \text{minimize} & f(x) \\ \text{subject to} & c(x) \geq 0 \\ & x \in R^n, \end{array} \qquad (13.13)$$

where f is a smooth nonlinear function and c is a vector of m smooth nonlinear functions. Without loss of generality we have assumed that the constraints are of the form $c(x) \geq 0$. For the present discussion we assume (as we did in the previous section) that f and *all* the constraint functions of (13.13) are nonlinear.

Optimality conditions revisited

As before we will examine feasible moves from a feasible point. Here the definitions of active, inactive, violated, and feasible are analogous to those for the linear inequality case, (see page 435).

The *active set* of constraints $\mathcal{A}(\bar{x})$ at a point $x = \bar{x}$

$$\mathcal{A}(\bar{x}) = \{i : c_i(\bar{x}) = 0\}.$$

As in the linear case, it is important to recognize the active constraints because, as we shall see shortly, feasible moves from a feasible point $x = \bar{x}$ are *restricted* with respect to active constraints, and *unrestricted* with respect to inactive constraints. For this reasons, the active constraints play a major role in the optimality conditions. Indeed, a key part of any algorithm for (13.13) will be to find the active set corresponding to an optimal solution.

As in the linear-inequality constrained case, if $c_i(x)$ is *inactive* at $\bar{x}$, then

$$c_i(\bar{x} + \varepsilon p) \geq 0$$

for sufficiently small ε and all p. For *active* constraints, however, feasible moves are restricted as noted above. As we have seen, in general, there is no direction such that a move along it will maintain the status of an active constraint remaining active. For the active constraint $\hat{c}_j(\bar{x}) = 0$, any move from $\bar{x}$ that remains feasible with respect to the j-th constraint must satisfy $\hat{c}_j(\bar{x} + \varepsilon p) \geq 0$. For any $\varepsilon > 0$, Taylor's Theorem gives us

$$\hat{c}_j(\bar{x} + \varepsilon p) = \hat{c}_j(\bar{x}) + \varepsilon \hat{A}_{j\bullet}(\bar{x})^{\mathrm{T}} p + \frac{\varepsilon^2}{2} p^{\mathrm{T}} \hat{G}_j(\bar{x}) p + \cdots$$

where $\hat{A}_{j\bullet}(\bar{x})$ denotes the transpose of the gradient of the j-th constraint, and $\hat{G}_j(\bar{x})$ is Hessian of the j-th constraint. Then, to the first order, the direction p should satisfy

$$\hat{A}_{j\bullet}(\bar{x}) p \geq 0. \tag{13.14}$$

First-order necessary conditions

Let $x = x^*$ be a local minimizer of (13.13) and let $\hat{c}(x^*)$ denote the vector of active constraints of which there are, say, ℓ. If the constraint qualification LICQ holds, then x^* must satisfy the necessary conditions corresponding to the equality constrained problem (13.1). That is,

$$Z(x^*)^{\mathrm{T}} g(x^*) = 0,$$

where the columns of $Z(x^*)$ represent the null space of $\hat{A}(x^*)$ as noted in Section 13.1. This is equivalent to saying

$$g(x^*) = \hat{A}(x^*)^{\mathrm{T}} \lambda^*.$$

Note that if $\ell = 0$, the null space can be represented by an identity matrix, and we have a stationary point of the objective function.

However here, in contrast to the equality-constrained case, we have additional freedom in choosing p; we can move off one or more of the active constraints and stay feasible. Thus, there will be additional conditions satisfied if $x = x^*$ is a minimizer of (13.13). As in the linear inequality-constrained case, consider a direction p such that

$$\hat{A}_{j\bullet}(x^*) p > 0$$
$$\hat{A}_{i\bullet}(x^*) p = 0 \quad \text{for } i \neq j. \tag{13.15}$$

If the objective function f is decreasing along such a feasible direction p, then x^* is not a local minimizer. Thus, there can be no such direction p that satisfies the above. This implies that

$$g(x^*)^{\mathrm{T}} p \geq 0. \tag{13.16}$$

Substituting $g(x^*) = \hat{A}(x^*)^T \lambda^*$, we obtain

$$\lambda^{*T} \hat{A}(x^*) p \geq 0.$$

Combining this with (13.15) we get

$$\lambda_j^* \hat{A}_{j\bullet}(x^*)^T p \geq 0,$$

which implies that $\lambda_j^* \geq 0$. If in fact $\lambda_j < 0$, then there would exist a feasible descent direction along which we can move off the j-th constraint and decrease the value of the objective function f.

Second-order necessary conditions

If the LICQ holds at $x = x^*$, the *necessary conditions* for $x = x^*$ to be a minimizer of (13.13) are:

1. The constraints are feasible at x^*, that is, $c(x^*) \geq 0$, with $\hat{c}(x^*) = 0$.

2. The projected gradient is zero: $Z(x^*)^T g(x^*) = 0$, where $Z(x^*)$ denotes a basis for the null space with respect to the constraint gradients which are the columns of $\hat{A}(x^*)$. Equivalently the objective function gradient lies in the range space of the constraint gradients: $g(x^*) = \hat{A}(x^*)^T \lambda^*$.

3. The Lagrange multipliers are nonnegative, that is, $\lambda^* \geq 0$.

4. The projected Hessian of the Lagrangian $Z(x^*)^T W(x^*, \lambda^*) Z(x^*)$ is positive semidefinite, where W is the Hessian of the Lagrangian function and is defined by (13.4).

Second-order sufficient conditions

Sufficient conditions for x^* to be a solution of (13.13) are that all second-order necessary conditions hold and that the projected Hessian of the Lagrangian mentioned in condition 4 is *positive definite* instead of positive semidefinite.

Example 13.3: OPTIMALITY—NONLINEAR INEQUALITY CONSTRAINTS.

We show how the first- and second-order conditions are used to determine if a point is a minimizer. Consider a modification of the problem discussed in Example 13.2 to minimize a linear function subject to the inequality constraint that all feasible points lie inside a circle of radius 2 centered at $(0,0)$.

$$\begin{aligned} \text{minimize} \quad & x_1 + x_2 \\ \text{subject to} \quad & x_1^2 + x_2^2 - 4 \le 0. \end{aligned}$$

Suppose that the current iterate is $\bar{x} = (\sqrt{2}, \sqrt{2})$ where the constraint is active; the first-order conditions $g(\bar{x}) = \hat{A}(\bar{x})^{\mathrm{T}}\bar{\lambda}$ are satisfied with $\bar{\lambda} = 1/(2\sqrt{2}) > 0$ because

$$\begin{bmatrix} 1 \\ 1 \end{bmatrix} = \begin{bmatrix} 2\sqrt{2} \\ 2\sqrt{2} \end{bmatrix} \bar{\lambda}.$$

At a minimizer, for $\ge$ constraints, the first-order optimality conditions require $\bar{\lambda} \ge 0$. Since our inequality constraint is $\le$, we require $\bar{\lambda} \le 0$; hence $\bar{x}$ is not a minimizer.

Next consider $\hat{x} = (-\sqrt{2}, -\sqrt{2})$. The first-order condition

$$\begin{bmatrix} 1 \\ 1 \end{bmatrix} = \begin{bmatrix} 2\sqrt{2} \\ 2\sqrt{2} \end{bmatrix} \hat{\lambda}$$

holds with $\hat{\lambda} = -1/(2\sqrt{2})$. Because $\hat{\lambda} < 0$, the point $\hat{x}$ satisfies the first-order conditions of optimality. The projected Hessian of the Lagrangian,

$$Z(\bar{x})^{\mathrm{T}}W(\bar{x}, \bar{\lambda})Z(\bar{x}) = \frac{1}{2\sqrt{2}} \begin{bmatrix} 1 \\ -1 \end{bmatrix}^{\mathrm{T}} \begin{bmatrix} 2 & 0 \\ 0 & 2 \end{bmatrix} \begin{bmatrix} 1 \\ -1 \end{bmatrix} = \sqrt{2},$$

is positive definite. Thus, the second-order sufficient conditions show that $\hat{x}$ is an isolated local minimizer.

13.3   Overview of algorithm design

When we considered algorithms for unconstrained nonlinear functions, the approach consisted of attempting to reduce the function value at each iteration from an initial starting point. This involved solving two subproblems:

- Determining a search direction p.

- Calculating a steplength α to obtain a decrease in the value of the objective function by moving in the search direction p.

The algorithms for linear constraints consisted of the same two subproblems with the added restriction of starting with a feasible point and then maintaining feasibility during the solution of the subproblems. The subproblems, slightly modified, were:

- Determining a *feasible* search direction p.

- Calculating a steplength α to obtain a decrease in the value of the objective function while maintaining feasibility.

In both the unconstrained and linearly constrained case, the determination of the search direction, as discussed in this book, was *deterministic* because the amount of computation needed was known. However, the computation of the steplength was *adaptive*; that is, it was not possible to determine in advance how many iterations would be required to compute it. In both these cases, the progress of the algorithm was measured by observing the reduction of the objective function value at each iteration.

For nonlinearly constrained optimization problems, similar ideas apply. In this case, both the subproblems are often solved adaptively. Typically the search direction is obtained by solving a linearly constrained optimization problem. However, complications may arise during the solution of a subproblem to determine a search direction p; the subproblem may have no solution or may be unbounded, even though the full optimization problem is well defined and has a solution. When the subproblem has a solution, we call it a *valid* subproblem, otherwise we call it a *defective* subproblem; see Example 13.7. When the subproblem is defective, a well-designed algorithm should have the ability to recover, for example, by providing an alternate subproblem.

In the nonlinearly constrained case, movement along a search direction does not maintain feasibility, in general. Hence, the progress of the execution of the algorithms should not be measured only by observing the decrease in value of the objective function value. Instead, we need to find a way to balance the goals of maintaining feasibility and reducing the objective function value at each iteration. We do this by employing a *merit function* defined so as to balance these two goals. In the unconstrained case and the linearly constrained case, we also implicitly used a simple merit function: the given objective function.

13.4   Penalty-function Methods

The idea behind a penalty-function method is to create an unconstrained subproblem such that constraint violations result in a large value ("penalty") being added to the objective function. This, in effect, is a merit function. The solution of such a subproblem is then examined. If the constraints are not satisfied, then a new subproblem is created that increases the impact of constraint violations; this is done by increasing the *penalty parameter*. The process is repeated until termination conditions are satisfied. Such a method has the advantage of being simple, but it has some disadvantages as we shall see. Probably the biggest disadvantage is that it is not a feasible-point method; hence, if the algorithm is terminated prematurely, the solution obtained might not satisfy the constraints of the original problem. Such an approach was first proposed in 1943 by Courant [38]. Penalty functions are also called *exterior* penalty functions because the iterates of the unconstrained subproblem start *outside* the feasible region and converge to an optimal solution satisfying the feasibility conditions in the limit.

Algorithm outline

Many different forms exist for specifying penalty functions. They can be created to be differentiable or nondifferentiable; we will look at both such types of penalty functions. A nonlinearly constrained problem (13.13) is converted to an unconstrained problem with minimand

$$F(x, \rho) = f(x) + \rho P(x), \tag{13.17}$$

where $P(x)$ is a function that measures constraint violations and $\rho \geq 1$ is termed the *penalty parameter*. A measure of the constraint violations can be, for example, the sum of the squares of the violations or the sum of the absolute value of the violations. The penalty parameter ρ magnifies the level of constraint violations. Thus as ρ is increased, the expectation is that the algorithm will attempt to drive the constraint violations to zero.

Independent of the particular penalty function used, a penalty function algorithm has the following form.

Algorithm 13.1 (A Penalty Function Algorithm Outline) Given a nonlinear program with nonlinear constraints, create an unconstrained subproblem with objective function $F(x, \rho)$ as in (13.17). Set $k = 0$. Set $\rho_0 = 1$, for example. Pick a constant $\beta > 1$ for updating the penalty parameter. Let $x_0(\rho_0)$ be a starting point and include only the constraints that are violated into the penalty part of the objective function. (Note that the starting point does not depend on ρ_0; we use this convention to distinguish between the starting points of the various subproblems.) Pick a tolerance τ_c such that constraints satisfied to within this tolerance are considered to be feasible.

1. *Test for convergence.* If the first-order optimality conditions are met and the constraints are satisfied to the pre-specified tolerance τ_c, stop and report the solution.

2. *Solve subproblem k.* Let the solution to the unconstrained subproblem be $x^*(\rho_k)$. (We assume for this discussion that the subproblem is *not defective.*) Any of the techniques for solving an unconstrained problem can be used here. Some care must be exercised in case (contrary to our assumption) the subproblem actually is *defective.* The iterations for the subproblem are referred to as *minor* iterations. During the solution process, constraints that become feasible have their contributions removed from the penalty term; similarly constraints that were feasible but become infeasible at the new iterate have their contributions added back to the penalty term.

3. *Update for the next major iteration.* Set the initial solution for the next subproblem to the optimal solution of the current subproblem: $x_0(\rho_{k+1}) = x^*(\rho_k)$. Let $k \leftarrow k + 1$. Update the penalty parameter, $\rho_{k+1} = \beta \rho_k$. Go to Step 1.

Properties of successive $x^*(\rho)$

Let $x^*(\rho)$ denote a local minimizer of the penalty function $F(x, \rho)$. Given that $\rho_{k+1} > \rho_k$, it can be shown (see, for example, Fiacco and McCormick, [64], and Luenberger and Ye [125, p. 403]) that successive minima $x^*(\rho_k)$ of the penalty function subproblems (13.17) have the following properties:

1. The optimal objective function values of the subproblems are nondecreasing:
$$F(x^*(\rho_{k+1}), \rho_{k+1}) \geq F(x^*(\rho_k), \rho_k).$$

2. The objective function values of the original problem is nondecreasing when computed at a minimizer of each successive subproblem:

$$f\left(x^*(\rho_{k+1})\right) \geq f\left(x^*(\rho_k)\right).$$

3. The measure of constraint infeasibilities is nonincreasing with the solution of each subproblem:

$$P(x^*(\rho_{k+1})) \leq P(x^*(\rho_k)).$$

4. The $x^*(\rho)$ approach a local minimizer of the original problem:

$$\lim_{\rho \to \infty} x^*(\rho) = x^*. \tag{13.18}$$

The quadratic penalty function

A commonly used differentiable penalty function is the *quadratic penalty function*. The penalty term consists of the sum of squares of the violated constraints multiplied by a penalty parameter ρ.

Equality constraints are treated a bit differently than inequality constraints. If the nonlinear optimization problem has equality constraints, then *all* of them are included in the penalty term. Equality constraints are always considered to be violated, whether or not they actually are; if they are satisfied, then their contribution to the quadratic penalty function will be zero, which is what we desire.

Example 13.4: PENALTY FUNCTION FOR EQUALITY CONSTRAINTS.
Consider the following two-variable nonlinear program with one equality constraint.

$$
\begin{aligned}
\text{minimize} \quad & x_1^2 + x_2^2 \\
\text{subject to} \quad & x_1^2 - x_2^2 - 2 = 0.
\end{aligned}
$$

The quadratic penalty function associated with the above for a fixed ρ is:

$$\text{minimize} \ F(x, \rho) = (x_1^2 + x_2^2) + \frac{\rho}{2}(x_1^2 - x_2^2 - 2)^2.$$

Notice that the penalty parameter is divided by 2; the reason for doing this is to avoid a multiplication by 2 when the gradient is computed.

When inequality constraints are present, the form is a bit different. The contribution of the inequality constraints evaluated at any point $\bar{x}$ is zero if they are satisfied and positive if they are violated.

Example 13.5: PENALTY FUNCTION FOR INEQUALITY CONSTRAINTS.

Consider the following two-variable nonlinear program with one inequality constraint:

$$\text{minimize} \quad x_1^2 + x_2^2$$
$$\text{subject to} \quad x_1^2 - x_2^2 - 4 \geq 0.$$

For a fixed ρ, the quadratic penalty function associated with the above inequality-constrained problem is

$$\text{minimize} \ F(x,\rho) = (x_1^2 + x_2^2) + \frac{\rho}{2}\left(\max\{0, (-x_1^2 + x_2^2 + 4)\}\right)^2.$$

In this case, the function F is continuous, but its gradient is not.

In general, the quadratic penalty function for an equality-constrained nonlinear program (13.1) is

$$F(x,\rho) = f(x) + \frac{\rho}{2}\sum_{i=1}^{\ell}(\hat{c}_i(x))^2 = f(x) + \frac{\rho}{2}\hat{c}(x)^{\mathrm{T}}\hat{c}(x). \qquad (13.19)$$

For the nonlinearly constrained problem (11.1), it is

$$F(x,\rho) = f(x) + \frac{\rho}{2}\sum_{i=1}^{m}(\max\{0, -c_i(x)\})^2 = f(x) + \frac{\rho}{2}\tilde{c}(x)^{\mathrm{T}}\tilde{c}(x), \qquad (13.20)$$

where $\tilde{c}(x)$ contains all the constraints violated at the current iterate. Recall that equality constraints are considered to be violated. Thus $\tilde{c}(x)$ contains all the violated inequality constraints and all the equality constraints. The term $\tilde{c}(x)^{\mathrm{T}}\tilde{c}(x)$ is called the *penalty term*. Note that the penalty term has continuous first derivatives but has discontinuous second derivatives wherever an inequality constraint changes from satisfied to violated, or vice versa.

Increasing the penalty parameter ρ in successive subproblems

As before, let x^* denote a solution to the original nonlinear optimization problem and let $x^*(\rho)$ denote the minimum of the penalty function subproblem $\min_x F(x,\rho)$. As we have noted earlier in (13.18), $\lim_{\rho\to\infty} x^*(\rho) = x^*$. We illustrate this with a simple example.

Example 13.6: INCREASING ρ.   Consider the following univariate optimization problem with one linear inequality constraint:

$$\begin{array}{ll} \text{minimize} & x^2 \\ \text{subject to} & x \geq 1. \end{array}$$

In this case, the constraint is a linear-inequality, and we could use the Active Set Method described in Chapter 12. Here we set up a penalty function problem to illustrate the limiting behavior of solutions to penalty-function subproblems. The quadratic penalty-function subproblem associated with the above problem is

$$F(x, \rho) = \begin{cases} x^2 + (\rho/2)(x-1)^2 & \text{if } x < 1 \\ x^2 & \text{otherwise.} \end{cases}$$

When we start at an infeasible point, the first-order conditions of optimality for the unconstrained penalty function result in $x^*(\rho)$ as the minimizer. In this case,

$$2x^*(\rho) + \rho(x^*(\rho) - 1) = 0 \quad \text{or} \quad x^*(\rho) = \frac{\rho}{\rho+2}.$$

It is easy to see that

$$\lim_{\rho \to \infty} x^*(\rho) = 1 = x^*.$$

In Figure 13.2, we can observe that $x^*(\rho)$ is approaching x^* as a sequence of quadratic problems is solved with increasing ρ.

Why do we solve a sequence of subproblems?

At the k-th major iteration, the solution found is denoted by $x^*(\rho_k)$. We have stated that the sequence $\{x^*(\rho_k)\}$ converges to a solution of the original problem as we increase ρ_k for each successive major iteration. So it would be reasonable to ask: why not just choose ρ to be a very large number at the start, for example $\rho = 10^{30}$, and either find a solution right away or cut down on the number of subproblems that need to be solved? The reason we do not start with a very large ρ has to do with numerical stability, in particular ill-conditioning. It turns out that the condition number of the Hessian of the quadratic penalty function increases with the size of ρ as the solution of the subproblem is reached. It can be shown that the Hessian becomes singular in the limit; see Murray [142]. This results in numerical difficulties for any unconstrained minimization algorithm. To avoid them, we solve a sequence of unconstrained problems, with ρ being increased for

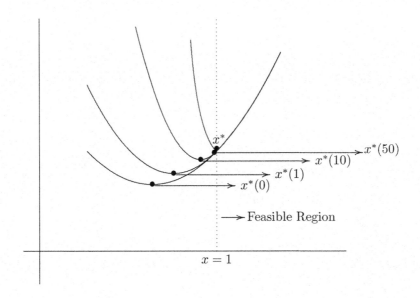

Figure 13.2: Penalty-function Method.

each successive subproblem, and then stop when a reasonable approximation to the solution is reached.

Defective subproblem

The purpose of constructing a penalty function subproblem is to obtain a local minimizer of the penalty function which, for sufficiently large ρ, is near a minimizer x^* of the original problem. When the problems are well-behaved this works fine; however, the penalty function subproblem may be defective in the sense that the penalty function may be unbounded below as illustrated in the following example due to Powell [160].

Example 13.7: DEFECTIVE SUBPROBLEM. Consider the univariate optimization problem

$$\begin{aligned} \text{minimize} \quad & x^3 \\ \text{subject to} \quad & x - 1 = 0. \end{aligned}$$

The quadratic penalty function associated with this problem is

$$F(x, \rho) = x^3 + \frac{\rho}{2}(x - 1)^2.$$

Clearly this function is unbounded below for all $\rho > 0$, thereby making the subproblem defective.

Lagrange multipliers

We next discuss how each penalty function subproblem provides an estimate of the Lagrange multipliers of the original problem. Let $x^*(\rho)$ be an unconstrained minimizer of the penalty function, let g be the gradient of the original objective function, let $\hat{A}$ be a matrix whose rows are the gradients of the violated constraints, and let $\tilde{c}$ be the vector of violated constraints. The first-order necessary optimality conditions for the penalty function subproblem are

$$0 = \nabla F\left(x^*(\rho), \rho\right) = g\left(x^*(\rho)\right) + \rho \hat{A}\left(x^*(\rho)\right)^T \tilde{c}\left(x^*(\rho)\right)$$

or

$$g\left(x^*(\rho)\right) = \hat{A}\left(x^*(\rho)\right)^T \left(-\rho \tilde{c}\left(x^*(\rho)\right)\right) = -\rho \hat{A}\left(x^*(\rho)\right)^T \tilde{c}\left(x^*(\rho)\right).$$

This says that g^T is a linear combination of the rows of $\hat{A}$. Thus, estimates of the Lagrange multipliers at a minimizer x^* of the original problem are given by

$$\lambda^*(\rho) = -\rho \tilde{c}\left(x^*(\rho)\right). \tag{13.21}$$

It can be shown that

$$\lim_{\rho \to \infty} \lambda^*(\rho) = \lambda^*; \tag{13.22}$$

see, for example, Nocedal and Wright [152, p. 497].

An exact penalty function

We mentioned earlier that as ρ is increased during the solution process, the quadratic penalty function becomes more and more ill-conditioned. We can obtain an alternative well-conditioned penalty function by allowing it to be nondifferentiable. A popular nondifferentiable penalty function is the *absolute-value penalty function* given by

$$F_E(x, \rho) = f(x) + \rho \sum_{i \in \mathcal{V}(\bar{x})} \left|c_i(x)\right| = f(x) + \rho \|\tilde{c}(x)\|_1,$$

where $\mathcal{V}(\bar{x})$ is the set of indices of all the violated inequality constraints at $\bar{x}$ and all the equality constraints; as before $\tilde{c}(x)$ is the vector of all the violated inequality constraints at $\bar{x}$ and all the equality constraints.

It can be shown that there exists a $\bar{\rho}$ such that a minimizer x^* of the original problem is also an unconstrained minimizer of $F_E(x, \rho)$ for all $\rho > \bar{\rho}$, see [84]. Because of this property, $F_E(x, \rho)$ is sometimes called an *exact penalty function*.

Example 13.8: EXACT PENALTY FUNCTION. Consider the univariate optimization problem

$$\begin{array}{ll} \text{minimize} & x^2 \\ \text{subject to} & x - 1 = 0. \end{array}$$

The absolute-value penalty function associated with it is

$$F_E(x, \rho) = x^2 + \rho|x - 1|.$$

It is straightforward to see that for any $\rho > 2$, the unconstrained minimizer x^* of $F_E(x, \rho)$ is 1.

While an exact penalty function is better conditioned than the quadratic penalty function, it is nondifferentiable. Furthermore, the other issues associated with penalty functions persist.

13.5   Reduced-gradient and Gradient-projection Methods

The Reduced-gradient and Gradient-projection Methods for nonlinearly constrained optimization problems are based on extending the techniques used for solving nonlinear optimization problems with linear constraints. In the linear equality-constrained case, the search directions were chosen to preserve the active constraints. When nonlinear constraints are present, it is not possible to satisfy the active nonlinear constraints by moving in a linear search direction, and so the algorithms have to include a method for correcting the next iterate to satisfy the active constraints.

Recall that one idea to solve the linearly constrained nonlinear problem (12.1) was to find a basis, and then eliminate the corresponding basic variables so as to obtain a restriction of the objective function to nonbasic variables and solve for these nonbasic variables. It was noted there that the explicit elimination of the basic variables was not necessary; instead we could find search directions that were restricted to the null space of the constraint matrix. Using this approach, we showed on page 427 that the steepest-descent direction is

$$p_k = -Z p_z = -Z(Z^T Z)^{-1} Z^T g_k, \tag{13.23}$$

where g_k is the gradient of the objective function, and Z is a matrix whose columns are a basis for the null space of the constraint matrix $\hat{A}$. This direction was first suggested by Rosen [168], who termed the related algorithm the *Gradient-projection Algorithm*. Later Wolfe [195] suggested using a modified version of the direction

$$p_k = -ZZ^{\mathrm{T}}g_k \qquad (13.24)$$

which he termed the *reduced-gradient* direction. Notice that (13.24) and (13.23) are of the same form when the columns of Z are orthonormal vectors.

When applying this approach to the nonlinear equality-constrained problem (13.1), the constraint gradients are represented by $\hat{A}_k = \hat{A}(x_k)$ and the null space corresponding to this matrix is denoted by Z_k. At each iteration, the constraint gradients change and so does the null space representation. As a result, a search direction p_k computed to be in the null space Z_k is not typically in the null space at an updated point $x_k + \alpha p_k$. A movement along a linear direction such as p_k almost surely causes the next iterate to violate the nonlinear equality constraints $\hat{c}(x) = 0$. Thus, as a part of this algorithm, the next step is to modify the iterate so that it satisfies the constraints $\hat{c}(x) = 0$.

One approach to restoring feasibility is to define

$$s_k = \alpha p_k + Y_k p_Y,$$

where α is a scalar. The columns of Y_k form a basis for the range space of the constraints at x_k, and the vector p_Y is to be determined so that $x_k + s_k$ is feasible. Hence we attempt to solve the nonlinear equations

$$\hat{c}(x_k + s_k) = \hat{c}(x_k + \alpha p_k + Y_k p_Y) = 0 \qquad (13.25)$$

by varying α. This is not an easy problem to solve because sufficient decrease of the objective function is also required; we want $f(x_k + s_k) \leq f(x_k) - \delta$ for some positive scalar $\delta > 0$. It is not enough to just find any α and p_Y. Even after a solution is obtained, a further adjustment may be required of α and the system re-solved. It should be noted that for (13.25) it is possible that no such p_Y exists or the solution process encounters difficulties.

A widely used method of this type is the GRG (Generalized Reduced-Gradient) Method which was developed by Abadie and Carpentier [2] (the original paper appeared in French in 1965). The Microsoft Excel spreadsheet software includes a solver which employs a GRG-type method developed by Lasdon and Waren [121].

The original GRG method used the representation of Z_k (as described on page 431):

$$\widehat{Z}_k = \begin{bmatrix} -V^{-1}U \\ I \end{bmatrix},$$

where $\widehat{A}_k = [V \; U]$ and V is an $\ell \times \ell$ matrix. Now if we partition p_k as (p_V, p_U), we can see that $p_Z = p_U$.

When the constraints are "nearly linear," Reduced-gradient Algorithms can work quite well. When this is not so, then in general, they can be inefficient due to the fact that they can take many tiny steps along a nonlinear constraint boundary.

As for the linearly constrained case, we can define a Newton direction by

$$Z_k^{\mathrm{T}} W_k Z_k p_Z = -Z_k^{\mathrm{T}} g_k,$$

where W_k is the Hessian of the *Lagrangian function* at iteration k.

When nonlinear inequality constraints are present, the method can be extended along the lines of the Active Set Methods for linear inequality constraints. But there are complications, since now it is essential to make sure that the previously satisfied constraints are not violated when adjusting α to solve (13.25). In the linear-inequality case this was straightforward to ensure, but it is not necessarily so in the nonlinear-inequality case. We note that some Reduced-gradient Methods convert all nonlinear inequality constraints to equalities by the addition of slack variables that are subject to simple bounds, and then implement an active set strategy where only the bounds are the inequalities.

13.6   Augmented Lagrangian Methods

Augmented Lagrangian Methods are also known as *Multiplier Methods*. In such methods, the objective function is the usual Lagrangian plus a penalty term. In contrast to Penalty-function Methods, Augmented Lagrangian Methods construct an unconstrained optimization subproblem such that ill-conditioning does not have to occur as the penalty parameter is increased, and, furthermore, the objective function of the subproblem is continuously differentiable. The reduction in the possibility of ill-conditioning is done through the introduction of explicit multipliers in the objective function. As a consequence, any of the approaches for unconstrained nonlinear optimization can be applied to the augmented Lagrangian function. These methods can be applied to both equality and inequality constraints.

We start by considering the nonlinear equality-constrained problem (13.1) and defining the equivalent optimization problem where the objective function is the Lagrangian of (13.1).

$$
\begin{array}{ll}
\text{minimize} & f(x) - \lambda^{\mathrm{T}} \hat{c}(x) \\
\text{subject to} & \hat{c}(x) = 0 \\
& x \in R^n.
\end{array}
\tag{13.26}
$$

The equivalence of the above to (13.1) follows from the observation that adding multiples of its *equality* constraints to the objective function cannot affect the optimal solution.

Next we construct a penalty function of the form

$$
L_A(x, \lambda, \rho) = f(x) - \lambda^{\mathrm{T}} \hat{c}(x) + \frac{\rho}{2} \hat{c}(x)^{\mathrm{T}} \hat{c}(x).
\tag{13.27}
$$

This function is called the *augmented Lagrangian* because it is the Lagrangian of (13.1) with a penalty term added to it. If we knew the optimal multipliers $\lambda = \lambda^*$ for (13.1), then the gradient with respect to x of the function $L_A(x, \lambda, \rho)$ would be zero at x^*, that is,

$$
\begin{aligned}
\nabla L_A(x^*, \lambda^*, \rho) &= \nabla f(x^*) - (\lambda^*)^{\mathrm{T}} \nabla \hat{c}(x^*) + \rho \nabla \hat{c}(x^*)^{\mathrm{T}} \hat{c}(x^*) \\
&= g(x^*) - (\lambda^*)^{\mathrm{T}} \hat{A}(x^*) = 0,
\end{aligned}
$$

where $\hat{A}(x^*)$ is the matrix whose rows are the gradients of the constraints at x^*. It can be shown (see Nocedal and Wright [152]) that under mild conditions, given $\lambda = \lambda^*$, there exists a $\bar{\rho}$ such that for all $\rho > \bar{\rho}$, x^* is a minimizer of $L_A(x, \lambda^*, \rho)$. In this sense the augmented Lagrangian function is an exact penalty function for (13.26); see Section 13.4.

Example 13.9: AUGMENTED LAGRANGIAN. Consider the following problem of minimizing a quadratic function subject to a linear constraint:

$$
\begin{array}{ll}
\text{minimize} & \frac{1}{2}(x_1^2 + x_2^2) \\
\text{subject to} & x_1 + x_2 - 2 = 0.
\end{array}
\tag{13.28}
$$

In this case, the constraint gradient, the gradient of the objective function, and the Hessian of the objective function are as follows:

$$\nabla \hat{c}(x) = \begin{pmatrix} 1 \\ 1 \end{pmatrix}, \quad g(x) = \begin{pmatrix} x_1 \\ x_2 \end{pmatrix}, \quad \text{and} \quad G(x) = \begin{bmatrix} 1 & 0 \\ 0 & 1 \end{bmatrix},$$

which means that every stationary point of the objective function that satisfies the constraint is optimal. A null space representation of the constraint Jacobian as a function of x is

$$Z(x) = \begin{bmatrix} 1 \\ -1 \end{bmatrix}.$$

For a feasible x to be optimal it must satisfy $Z(x)^{\mathrm{T}} g(x) = 0$; thus

$$x_1 = x_2.$$

Hence $x_1^* = x_2^* = 1$. Solving $\hat{A}^{\mathrm{T}} \lambda = g$ results in $\lambda^* = 1$.

The augmented Lagrangian for (13.28) is

$$L_A(x, \lambda, \rho) = \tfrac{1}{2}(x_1^2 + x_2^2) - \lambda(x_1 + x_2 - 2) + \tfrac{\rho}{2}(x_1 + x_2 - 2)^2.$$

The gradient of the augmented Lagrangian function with respect to x is given by

$$\nabla L_A(x, \lambda, \rho) = \begin{bmatrix} (1 + \rho)x_1 + \rho x_2 - \lambda - 2\rho \\ \rho x_1 + (1 + \rho)x_2 - \lambda - 2\rho \end{bmatrix}.$$

Setting the gradient equal to zero, we obtain the stationary point

$$x(\lambda, \rho) = \begin{bmatrix} \dfrac{2\rho + \lambda}{2\rho + 1} \\ \dfrac{2\rho + \lambda}{2\rho + 1} \end{bmatrix}.$$

In this case, if we substitute $\lambda = \lambda^* = 1$, thereby making $x(1, \rho) = (1, 1)$, the solution is independent of the value of $\rho > \bar{\rho} = 0$ as noted earlier. This implies that as λ is chosen closer to the true optimal value, the iterates approach the optimal solution of the original problem.

It might appear that we should be able to use the Lagrangian function $L(x, \lambda) = f(x) - \lambda^{\mathrm{T}} \hat{c}(x)$ as the unconstrained objective function. However this is not so: x^* is *not necessarily* a minimizer of $L(x, \lambda)$ even if λ^* is known.

Example 13.10: LAGRANGIAN AS THE UNCONSTRAINED OBJECTIVE.
Consider the problem of minimizing $f(x_1, x_2) = x_2$ subject to the constraints

$$
\begin{aligned}
c_1(x) &= -x_1^9 + x_2^3 \geq 0 \\
c_2(x) &= x_1^9 + x_2^3 \geq 0 \\
c_3(x) &= x_1^2 + (x_2 + 1)^2 - 1 \geq 0.
\end{aligned}
\tag{13.29}
$$

The first two constraints imply that $x_2^3 \geq 0$ and hence that $x_2 \geq 0$. The
solutions of the third constraint are the points *outside* the open disc of radius
1 centered at $(0, -1)$. It follows that the feasible region is a proper subset of
the upper half plane in R^2. Indeed, it is clear that $x_2^* = 0$ is the minimum
value of the objective function, but no feasible point other than $x^* = (0, 0)$
has its second coordinate equal to zero. Consequently, the optimal solution
of this problem is $x^* = (0, 0)$. All three constraints are active at this point.

We have $\nabla f(x^*) = (0, 1)$. As for the constraints, $\nabla c_i(x^*) = (0, 0)$ for
$i = 1, 2$ and $\nabla c_3(x^*) = (0, 2)$. The rows of the Jacobian matrix at x^* are
clearly not linearly independent, so the LICQ does *not* hold in this case.[4]
Nevertheless, the first-order optimality conditions are satisfied at x^*. To see
this, let
$$L(x, \lambda) = f(x) - \lambda_1 c_1(x) - \lambda_2 c_2(x) - \lambda_3 c_3(x).$$

The stationarity condition $\nabla_x L(x, \lambda) = 0$ has the solution (x^*, λ^*) where
x^* is as above, and $\lambda^* = (\lambda_1^*, \lambda_2^*, \lambda_3^*) = (\lambda_1^*, \lambda_2^*, 1/2)$ for arbitrary values of
$\lambda_1^*, \lambda_2^* \geq 0$. The vector $\lambda^* = (0, 0, 1/2)$ can be used as a vector of Lagrange
multipliers in the first-order optimality conditions.

Now fix $\lambda = \lambda^*$ as above and consider the problem

$$\min_x \; M(x) = L(x, \lambda^*) = x_2 - \tfrac{1}{2}[x_1^2 + (x_2 + 1)^2 - 1].$$

It is clear that
$$\nabla^2 M(x^*) = \begin{bmatrix} -1 & 0 \\ 0 & -1 \end{bmatrix},$$

which is negative definite. In fact, the minimand $M(x)$ in this problem is
strictly concave on R^2 and has no local minimum there. Thus, even when
an optimal set of multipliers λ^* is known in advance, it not always the case
that the original constrained minimization problem can be reduced to the
unconstrained minimization of $L(x, \lambda^*)$.

---
[4]It can be shown that the Karush-Kuhn-Tucker constraint qualification *does* hold here.

The penalty parameter ρ in the augmented Lagrangian function plays an important role. In fact, the augmented Lagrangian function has x^* as its minimizer when λ^* is known and ρ is sufficiently large. We illustrate this in the next example.

Example 13.11: CHOICE OF ρ. The augmented Lagrangian for (13.29) in Example 13.10 is

$$L(x, \lambda, \rho) = x_1^3 + x_2^3 - \lambda(x_1 + x_2 + 2) + \frac{\rho}{2}(x_1 + x_2 + 2)^2. \qquad (13.30)$$

Its Hessian matrix is given by

$$\nabla^2 L(x, \lambda, \rho) = \begin{bmatrix} 6x_1 + \rho & 0 \\ 0 & 6x_2 + \rho \end{bmatrix}.$$

As long as $\rho > 6$, the Hessian will be positive definite, and $x^* = (-1,\ -1)$ will be a minimizer of the augmented Lagrangian.

In general, the multipliers λ^* will not be known. Thus, it makes sense to estimate them, solve the subproblem and then update the multiplier estimate; this is why such an approach is sometimes called the *Method of Multipliers*.

We have already noted that the penalty term is required in order to successfully find a minimizer. The choice of ρ is quite important; if ρ is chosen to be too small, the augmented Lagrangian function may be defective, whereas, if ρ is chosen to be too large, then the Hessian of the augmented Lagrangian function may become ill-conditioned.

Example 13.12: ILL CONDITIONING OF $\nabla^2 L_A(x, \lambda, \rho)$. The Hessian matrix of the augmented Lagrangian of Example 13.9 is given by

$$W(x, \lambda, \rho) = \nabla^2 L_A(x, \lambda, \rho) = \begin{bmatrix} 1 + \rho & \rho \\ \rho & 1 + \rho \end{bmatrix}.$$

As ρ becomes large, all the terms become almost equal implying that the matrix is close to being singular.

Next we show how the multiplier estimates are updated. Let λ_k be an estimate of the Lagrange multipliers during iteration k, and let ρ_k be the current penalty factor. The subproblem to be solved requires minimizing $L(x, \lambda_k, \rho_k)$. At a minimizer x_k of this subproblem, we must have

$$g(x) - \hat{A}(x_k)^{\mathrm{T}}\lambda_k + \rho\hat{A}(x_k)^{\mathrm{T}}\hat{c}(x_k) = 0,$$

or

$$g(x) - \hat{A}(x_k)^{\mathrm{T}}(\lambda_k - \rho\hat{c}(x_k)) = g(x) - \hat{A}(x_k)^{\mathrm{T}}\lambda_{k+1} = 0.$$

This implies that the Lagrange multipliers can be updated by the formula

$$\lambda_{k+1} = \lambda_k - \rho\hat{c}(x_k). \tag{13.31}$$

Example 13.13: UPDATING MULTIPLIER ESTIMATES. Let λ_k be the Lagrange multiplier estimate at iteration k, and let ρ_k be the corresponding penalty parameter. Then a stationary point for the augmented Lagrangian of Example 13.9 is given by

$$x_k(\lambda_k, \rho_k) = \begin{bmatrix} \dfrac{2\rho_k + \lambda_k}{2\rho_k + 1} \\ \dfrac{2\rho_k + \lambda_k}{2\rho_k + 1} \end{bmatrix}.$$

For the next iteration, the Lagrange multiplier estimate is updated taking (13.31) to be:

$$\begin{aligned}
\lambda_{k+1} &= \lambda_k - \rho_k\left(\frac{2\rho_k + \lambda_k}{2\rho_k + 1} + \frac{2\rho_k + \lambda_k}{2\rho_k + 1} - 2\right) \\
&= \lambda_k - \rho_k\left(\frac{2\lambda_k - 2}{2\rho_k + 1}\right) \\
&= \frac{\lambda_k + 2\rho_k}{2\rho_k + 1}.
\end{aligned}$$

The optimizer x^* is a stationary point of the Lagrangian only if $\lambda = \lambda^*$. Thus, Augmented Lagrangian Methods will converge only if λ_k converges to λ^*. In fact, it can be shown that the rate of convergence of x_k to x^* will be no better than the rate of convergence of the multiplier estimates λ_k to λ^*; see [84]. This result implies that even an otherwise quadratically convergent algorithm used to solve the subproblems will not converge quadratically unless a second-order estimate of the multipliers is used. In the exposition above, we have described how to obtain first-order estimates of the Lagrange multipliers; for a discussion on second-order estimates of the multipliers, see [84].

An outline of an Augmented Lagrangian Method is shown below.

Algorithm 13.2 (Augmented Lagrangian Method) Let K be the maximum number iterations allowed. Set $k = 0$ and pick a $\rho_0 > 0$, say $\rho_0 = 10$, and a λ_0. Set $\tau_c > 0$ to be the tolerance for constraint satisfaction and $\tau_g > 0$, the tolerance for checking that the gradient is zero. In addition, define a tolerance $\tau_v > 0$ to decide if the constraint violation at an iteration k is too large and thus requires an increase in the penalty parameter.

1. *Solve the Augmented Lagrangian subproblem.* Find a minimizer x_k for $L_A(x, \lambda_k, \rho_k)$. If the solution process terminates with an indication of unboundedness, try again after increasing the penalty parameter

$$\rho_k \leftarrow \beta \rho_k$$

   where, for example, $\beta = 10$.

2. *Test for convergence.*

   (a) If x_k satisfies the termination criteria

   $$\|\nabla L_A(x_k, \lambda_k, \rho_k)\| \leq \tau_g$$
   $$\|\hat{c}(x_k)\| \leq \tau_c,$$

   set $x^* = x_k$, $\lambda^* = \lambda_k$ and exit with success.

   (b) If $k = K$ exit with failure.

3. *Constraints not satisfied.* If the unconstrained minimizer is found but the constraints are violated, that is

   $$\|\nabla L_A(x_k, \lambda_k, \rho_k)\| \leq \tau_g$$
   $$\|\hat{c}(x_k)\| > \tau_c,$$

   then first update the multipliers

   $$\lambda_{k+1} = \lambda_k - \rho_k \hat{c}(x_k).$$

   Next update other quantities based on the constraint violations.

(a) If the constraint violations are not sufficiently decreased (that is, $\|\hat{c}(x_k)\| > \tau_v$), increase the penalty parameter:

$$\rho_{k+1} \leftarrow \beta \rho_k$$

where, for example, $\beta = 10$.

(b) If the constraint violations are sufficiently decreased (that is, $\|\hat{c}(x_k)\| \leq \tau_v$), decrease the constraint violation tolerance:

$$\tau_v \leftarrow \max\{\tau_c, \gamma \tau_v\}$$

where, for example, $\gamma = 0.1$.

4. *Update the iteration counter.* Set $k \leftarrow k + 1$ and go to Step 1.

Typically the augmented Lagrangian subproblem is not solved accurately at the start of the algorithm, and a tolerance $\bar{\tau}_g > \tau_g$ is used. As the algorithm proceeds the tolerance $\bar{\tau}_g$ is reduced systematically until a tolerance of τ_g is used before declaring the current solution as optimal.

Example 13.14: AUGMENTED LAGRANGIAN METHOD. Once again consider the problem in Example 13.9 which is repeated here for convenience:

$$\begin{aligned} \text{minimize} \quad & \tfrac{1}{2}(x_1^2 + x_2^2) \\ \text{subject to} \quad & x_1 + x_2 - 2 = 0. \end{aligned} \tag{13.32}$$

Choose $\rho = 10$, $\beta = 10$, $\lambda = 0$, $\tau_g = 10^{-7}$, and $\tau_v = 10^{-7}$. Let the initial starting point be $x_0(\lambda_0, \rho_0) = x_0(0, 10) = (0, 0)$. The augmented Lagrangian for (13.32) is

$$L_A(x, \lambda_0, \rho_0) = \frac{1}{2}(x_1^2 + x_2^2) + \frac{10}{2}(x_1 + x_2 - 2)^2.$$

The gradient at $x_0(\lambda_0, \rho_0)$ is

$$\nabla L_A(x, \lambda_0, \rho_0) = \begin{bmatrix} x_1 + 10(x_1 + x_2 - 2) \\ x_2 + 10(x_1 + x_2 - 2) \end{bmatrix} = \begin{bmatrix} 11x_1 + 10x_2 - 20 \\ 10x_1 + 11x_2 - 20 \end{bmatrix} = \begin{bmatrix} -20 \\ -20 \end{bmatrix}.$$

Clearly

$$\|\nabla L_A(x, \lambda_0, \rho_0)\| > \tau_g$$

and also

$$\|x_1 + x_2 - 2\| > \tau_c,$$

so the solution fails to satisfy the termination criteria. Hence we solve this unconstrained problem by applying Newton's Method. To do this, we compute the Hessian at $x_0(\lambda_0, \rho_0)$:

$$\nabla^2 L_A(x, \lambda_0, \rho_0) = \begin{bmatrix} 11 & 10 \\ 10 & 11 \end{bmatrix}.$$

On solving

$$\nabla^2 L_A(x, \lambda_0, \rho_0)\, p = \nabla L_A(x, \lambda_0, \rho_0),$$

we obtain

$$p = \left(\frac{20}{21}, \frac{20}{21} \right).$$

We have seen that under these circumstances the Newton step for an unconstrained quadratic form results in an optimal solution

$$x_0^*(0, 10) = x_0(0, 10) + p = \left(\frac{20}{21}, \frac{20}{21} \right).$$

Observe that $x_0^*(0, 10)$ is close to the optimal solution as we saw in Example 13.9. It is easy to verify that this does not satisfy the termination criteria, so we update the penalty parameter and the Lagrange multiplier obtaining

$$\rho_1 = \beta \rho_0 = 100$$
$$\lambda_1 = \lambda_0 - \rho_0(x_1 + x_2 - 2) = \frac{20}{21}.$$

We see that the Lagrange multiplier is already close to its optimal value. The new augmented Lagrangian is

$$L_A(x, \lambda_1, \rho_1) = \frac{1}{2}(x_1^2 + x_2^2) - \frac{20}{2}(x_1 + x_2 - 2) + \frac{100}{2}(x_1 + x_2 - 2)^2,$$

and the process continues.

Algorithm 13.2 can also be used when nonlinear inequalities are present. There are two simple approaches to doing this. One approach is to subtract slack variables and then solve the resulting bound-constrained problem. Thus, for example, we can modify (13.13) to be

$$\begin{array}{ll} \text{minimize} & f(x) \\ \text{subject to} & c(x) - s = 0 \\ & x \in R^n, \ s \in R^m, \ s \geq 0. \end{array} \qquad (13.33)$$

Most real-world problems have bounds on the variables, so we can simply consider problems of the form

$$
\begin{aligned}
\text{minimize} \quad & f(x) \\
\text{subject to} \quad & \hat{c}(x) = 0, \\
& x \in R^n, \ l \le x \le u.
\end{aligned} \tag{13.34}
$$

Rather than solve an unconstrained problem at Step 1 of Algorithm 13.2, we now solve the bound-constrained problem

$$
\begin{aligned}
\text{minimize} \quad & L(x, \lambda_k, \rho_k) \\
\text{subject to} \quad & l \le x \le u.
\end{aligned}
$$

This problem can be solved by any of the methods described in Chapter 12.

An alternate approach to adding slack variables is to subtract the square of slack variables, that is

$$
\begin{aligned}
\text{minimize} \quad & f(x) \\
\text{subject to} \quad & c_i(x) - s_i^2 = 0 \ \text{ for } i = 1, \dots, m \\
& x \in R^n.
\end{aligned} \tag{13.35}
$$

Whereas this has the advantage of solving an unconstrained problem at Step 1 of Algorithm 13.2, it does so at the expense of introducing additional nonlinearity. However, we note that if the original problem had bounds to start with, then we would still need to solve a bound-constrained subproblem; in this case, there would be no advantage to introducing squared slack variables.

Practical problems typically have a mixture of nonlinear constraints, linear constraints, and bounds on the variables. One approach is to convert all the inequality constraints (except the bounds) into equality constraints as described above for nonlinear constraints. Once this has been done, the optimization problem will be of the form

$$
\begin{aligned}
\text{minimize} \quad & f(x) \\
\text{subject to} \quad & \hat{c}(x) = 0 \\
& Ax = b \\
& l \le x \le u \\
& x \in R^n.
\end{aligned} \tag{13.36}
$$

An augmented Lagrangian function can be created by only using the non-linear constraints, thereby resulting in the linearly constrained problem

$$
\begin{aligned}
\text{minimize} \quad & L_A(x, \lambda, \rho) = f(x) - \lambda^{\mathrm{T}}\hat{c}(x) + \tfrac{\rho}{2}\hat{c}(x)^{\mathrm{T}}\hat{c}(x) \\
\text{subject to} \quad & Ax = b \\
& l \le x \le u \\
& x \in R^n .
\end{aligned}
\tag{13.37}
$$

To solve (13.36) by the Augmented Lagrangian Method, we modify Step 1 of Algorithm 13.2 to find the minimizer for (13.37). The test for convergence is also suitably modified to take into account the active linear constraints. Here too, such linearly constrained subproblems can be solved by any of the methods described in Chapter 12.

13.7   Projected Lagrangian Methods

Here we describe a method that does not make use of a penalty parameter. Instead, an extension of the Lagrangian function is minimized subject to a linearization of the constraints. Such methods, sometimes referred to as *Projected Lagrangian Methods*, are based on solving a sequence of linearly constrained subproblems. The methods were first proposed in 1972 by Robinson [166] and Rosen and Kreuser [170].

Once again, the purpose is to solve the nonlinearly constrained problem (13.1). Let x_k be the current, not-necessarily-feasible approximation to a solution at iteration k. The Taylor series expansion of $\hat{c}(\cdot)$ around such an x_k is

$$
\hat{c}\left(x_k + (x - x_k)\right) = \hat{c}_k + \hat{A}_k(x - x_k) + O\left(||x - x_k||^2\right),
\tag{13.38}
$$

where $\hat{c}_k = \hat{c}(x_k)$ and $\hat{A}_k = \hat{A}(x_k)$. As in Newton's Method for solving equations, we assume that the x in the expression above satisfies $\hat{c}(x) = 0$. Dropping the higher-order terms, we then obtain

$$
\hat{A}_k x = -\hat{c}_k + \hat{A}_k x_k.
\tag{13.39}
$$

Using this we can define a sequence of linearly constrained subproblems.

As before, we cannot just use the Lagrangian function. To see this, consider the subproblem

$$
\begin{aligned}
\text{minimize} \quad & L(x, \lambda_k) = f(x) - \lambda_k^{\mathrm{T}}\hat{c}(x) \\
\text{subject to} \quad & \hat{A}_k x = -\hat{c}_k + \hat{A}_k x_k.
\end{aligned}
\tag{13.40}
$$

The first-order optimality conditions for (13.40) are:

$$\hat{A}_k x_k^{\star} = -\hat{c}_k + \hat{A}_k x_k$$
$$\hat{A}_k^{\mathrm{T}} \lambda_k^{*} = \nabla L(x_k^{*}, \lambda_k). \qquad (13.41)$$

The multipliers λ_k^{*} can be used as an estimate of λ_{k+1}. For an optimal solution, a second-order necessary condition is that the projected Hessian of $L(x, \lambda_k)$ evaluated at (x_k^{*}, λ_k) be positive semidefinite. If only the first-order necessary conditions are satisfied, we cannot be sure that we have an optimal solution at (x^{*}, λ_k). However, if $\nabla L(x^{*}, \lambda_k) = 0$ and $\hat{A}_k$ is of full rank, we see that $\lambda_{k+1} = \lambda_k^{*} = 0$. This is undesirable since we would no longer have an estimate of the Lagrange multipliers for the next iteration. Hence, instead of (13.40), we solve the following subproblem

$$\begin{aligned} \text{minimize} \quad & L_P(x, \lambda_k) = f(x) - \lambda_k^{\mathrm{T}} \hat{c}(x) + \lambda_k^{\mathrm{T}} \hat{A}_k x \\ \text{subject to} \quad & \hat{A}_k x = -\hat{c}_k + \hat{A}_k x_k. \end{aligned} \qquad (13.42)$$

Note that we have assumed that $\hat{A}_k$ is of full rank; if $\hat{A}_k$ is rank deficient or if there are more than n constraints, the above subproblem can be defective. It can be shown (see Robinson [166]) that with this method the iterates (x_k, λ_k) converge at a quadratic rate provided the initial starting pair (x_0, λ_0) is "close" to an optimal solution (x^{*}, λ^{*}).

An outline of a Projected Lagrangian Method is shown below.

Algorithm 13.3 (Projected Lagrangian Method) Let K be the maximum number of iterations allowed. Set $k = 0$.

1. *Test for convergence.*

    (a) If x_k satisfies the termination criteria, set $x^{*} = x_k$, $\lambda^{*} = \lambda_k$ and exit with success.

    (b) If $k = K$ exit with failure.

2. *Solve the Projected Lagrangian subproblem.* Find a minimizer x_{k+1} for (13.42) and the associated optimal multipliers λ_{k+1}.

3. *Update the iteration counter.* Set $k \leftarrow k + 1$ and go to Step 1.

Given the excellent local convergence properties of this method, it is desirable, if possible, to find a starting point near an optimal solution. One possibility, as suggested by Rosen [169], is to use a two-phase approach, where the first phase is to use a few iterations of a penalty function method to obtain a good initial starting point.

Alternatively, the phases can be combined as suggested by Murtagh and Saunders [143]. In this approach, they define

$$d_k(x) = \hat{c}(x) - \hat{c}_k - \hat{A}_k(x - x_k).$$

Notice that we can equivalently use $L_P(x, \lambda_k) = f(x) - \lambda_k^T d_k(x)$ in (13.42). In order to combine the phases, Murtagh and Saunders propose solving the subproblem

$$
\begin{aligned}
\text{minimize} \quad & L_P(x, \lambda_k) = f(x) - \lambda_k^T d_k(x) + \tfrac{\rho}{2} d_k(x)^T d_k(x) \\
\text{subject to} \quad & \hat{A}_k x = -\hat{c}_k + \hat{A}_k x_k,
\end{aligned}
\tag{13.43}
$$

where ρ is a penalty parameter. The purpose of defining such a d_k is to be able to penalize the error in satisfying the constraints at each iteration k. With this approach, a quadratic rate of convergence will be achieved, provided the iterates get close to the solution and ρ is adjusted appropriately. For details, see [143].

Some argue that the above approach requires more computational effort than the Augmented Lagrangian Method because a linearly constrained subproblem is solved at each iteration. However, it should be noted that practical problems typically contain a mixture of linear and nonlinear constraints, and so the Augmented Lagrangian Method formulations also include linearly constrained subproblems as shown earlier.

The multiplier estimates are true multipliers only at a solution of the original problem. At all intermediate points, the multipliers obtained are only an approximation to the true multipliers. Often the subproblem is not solved accurately to a minimum and, furthermore, the linearization may result in a rank deficient system. In such cases, it may not be possible to obtain the multipliers by solving $\hat{A}_k^T \lambda = -\nabla L_P(x_k, \lambda_k)$. Hence, when the multipliers cannot be obtained, they can be estimated by solving such a system in a least-squares sense.

13.8 Sequential Quadratic Programming (SQP) Methods

As the name implies, Sequential Quadratic Programming (SQP) Methods approximate the nonlinear programming problem by a quadratic programming problem at each iteration.

SQP Methods were first proposed by Wilson [192] in 1963. He developed his approach for the special case of convex programming. The Hessian of this quadratic program is the exact Hessian of the Lagrangian of the original convex program, and the solution of the quadratic program is a search direction p for the original program. The next iterate is then obtained as $x_{k+1} \leftarrow x_k + p$. The method Wilson proposed did not perform a linesearch. Later in 1969, Murray [141] described an SQP algorithm for general nonlinear programs. He utilized a quasi-Newton approximation of the Hessian of the Lagrangian and defined a merit function to measure progress of the algorithm for finding a solution to the nonlinear program. The next iterate was determined by invoking a steplength algorithm aimed at reducing the value of the merit function. Since then much work has been done on SQP methods, and they have become widely used in the solution of nonlinear programs.

Nonlinear equality constraints

As before, we start by considering the nonlinear equality-constrained problem (13.1), repeated below.

$$\begin{array}{ll} \text{minimize} & f(x) \\ \text{(NEP)} \quad \text{subject to} & \hat{c}(x) = 0 \\ & x \in R^n. \end{array} \tag{13.44}$$

Let x_k be the current—not necessarily feasible—approximation to a solution at iteration k. In this discussion we assume the following holds at every iterate x_k:

1. The projected Hessian of the Lagrangian function is positive definite.

2. The Jacobian matrix of the constraints has full rank.

As a reminder, for the discussion to follow, we let $g_k = \nabla f(x_k)$, $G_k = \nabla^2 f(x_k)$, $\hat{c}_k = \hat{c}(x_k)$, and $\hat{A}_k = \hat{A}(x_k)$. Furthermore, let $Z_k = Z(x_k)$ be a matrix whose columns, as usual, form a basis for the null space of $\hat{A}_k$.

Recall Theorem 11.2 gives the first-order necessary conditions for this problem, namely

$$\nabla f(x) - \hat{A}(x)^T\lambda = 0$$
$$\hat{c}(x) = 0.$$

With $x = x_k$, we can use Newton's Method to solve this system of nonlinear equations. To see this, define

$$\mathcal{K}(x, \lambda) = \begin{bmatrix} \nabla f(x) - \hat{A}(x)^T\lambda \\ \hat{c}(x) \end{bmatrix}.$$

Then we can use Newton's Method to solve $\mathcal{K}(x, \lambda) = 0$, that is, we solve

$$\mathcal{K}(x_k + p, \lambda_k + \gamma) = 0$$

and iteratively set $x_{k+1} = x_k + p_k$ and $\lambda_{k+1} = \lambda_k + \gamma_k$. From the Taylor series expansion of $\mathcal{K}(x_k + p, \lambda_k + \gamma)$ to first order we have

$$0 = \mathcal{K}(x_k + p, \lambda_k + \gamma) = \nabla\mathcal{K}(x_k, \lambda_k)^T(p, \gamma) + \mathcal{K}(x_k, \lambda_k).$$

That is, we solve

$$\begin{bmatrix} G_k - \nabla\hat{A}_k^T\lambda_k & -\hat{A}_k^T \\ \hat{A}_k & 0 \end{bmatrix} \begin{bmatrix} p \\ \gamma \end{bmatrix} = - \begin{bmatrix} g_k - \hat{A}_k^T\lambda_k \\ \hat{c}_k \end{bmatrix}.$$

In terms of the Lagrangian, the above can be written as

$$\begin{bmatrix} \nabla^2 L(x_k, \lambda_k) & -\hat{A}_k^T \\ \hat{A}_k & 0 \end{bmatrix} \begin{bmatrix} p \\ \gamma \end{bmatrix} = - \begin{bmatrix} \nabla L(x_k, \lambda_k) \\ \hat{c}_k \end{bmatrix}. \tag{13.45}$$

The solution of the above system of equations gives p_k and γ_k, and the next iterate is obtained as $x_{k+1} = x_k + p_k$ and $\lambda_{k+1} = \lambda_k + \gamma_k$.

It turns out, as we shall now show, that solving the above system of equations is equivalent to solving a quadratic program. To set up the quadratic program, we start by creating a linear approximation to the nonlinear constraints. Recall that in (13.38) we used the Taylor series expansion of the nonlinear equality constraints around x_k to obtain a linear approximation. The Taylor series expansion of $\hat{c}(x_k + p)$ is

$$\hat{c}(x_k + p) = \hat{c}_k + \hat{A}_k p + O\left(||p||^2\right).$$

As in Newton's Method for solving equations, we assume that $\hat{c}(x_k + p) = 0$ in the expression above. Thus, dropping the higher-order terms, we then obtain

$$\hat{A}_k p = -\hat{c}_k \qquad (13.46)$$

as a linearization of the nonlinear constraints.

Next we need to decide on an appropriate quadratic objective function for each subproblem, such that the sequence of solutions to the subproblems will converge to an optimal solution to the original problem (13.44). We will use a quadratic function, denoted by $L_Q(p, \lambda_k)$, such that at the solution of the quadratic subproblem, we will have a search direction p and also an estimate of the Lagrange multipliers for the next iterate. The subproblem we are interested in is of the form

$$\begin{array}{ll} \text{minimize} & L_Q(p, \lambda_k) \\ \text{subject to} & \hat{A}_k p = -\hat{c}_k, \end{array} \qquad (13.47)$$

where $L_Q(p, \lambda_k)$ is a quadratic function of p.

The function $L_Q(p, \lambda_k)$ is typically taken to be a quadratic approximation to the Lagrangian function $L(x, \lambda) = f(x) - \lambda^{\mathrm{T}} c(x)$ around x_k, that is,

$$\begin{aligned} L_Q(p, \lambda_k) &\approx L(x_k, \lambda_k) + \nabla L(x_k, \lambda_k)^{\mathrm{T}} p + \tfrac{1}{2} p^{\mathrm{T}} \nabla^2 L(x_k, \lambda_k) p \\ &= L(x_k, \lambda_k) + g_k^{\mathrm{T}} p - \lambda_k^{\mathrm{T}} \hat{A}_k p + \tfrac{1}{2} p^{\mathrm{T}} \nabla^2 L(x_k, \lambda_k) p. \end{aligned}$$

The term $L(x_k, \lambda_k)$ is a constant, and, because $\hat{A}_k p = -\hat{c}_k$, so is the term $\lambda_k^{\mathrm{T}} \hat{A}_k p$. These two terms can be dropped from the objective function without affecting an optimal solution. Hence the quadratic subproblem can be written as

$$\begin{array}{ll} \text{minimize} & L_Q(p, \lambda_k) = g_k^{\mathrm{T}} p + \tfrac{1}{2} p^{\mathrm{T}} \nabla^2 L(x_k, \lambda_k) p \\ \text{subject to} & \hat{A}_k p = -\hat{c}_k. \end{array} \qquad (13.48)$$

Next we show the equivalence of (13.48) to (13.45). The first-order necessary conditions for (13.48) are:

$$\begin{aligned} \nabla^2 L(x_k, \lambda_k) p_k + g_k &= \hat{A}_k^{\mathrm{T}} \lambda_{k+1} \\ \hat{A}_k p_k &= -\hat{c}_k \end{aligned}$$

Note that the multipliers are λ_{k+1} since we expect these to be an estimate of the multipliers for the next subproblem. Substituting $\lambda_{k+1} = \lambda_k + \gamma$ in the first equation and rearranging terms, we get

$$\nabla^2 L(x_k, \lambda_k)p_k - \hat{A}_k^T\gamma = \hat{A}_k^T\lambda_k - g_k$$
$$\hat{A}_k p_k = -\hat{c}_k,$$

which is the same as (13.45).

An outline of an SQP Method for solving (13.44) is shown below.

Algorithm 13.4 (Sequential Quadratic Programming) Let K be the maximum number of iterations allowed. Set $k = 0$.

1. *Test for convergence.*

   (a) If x_k satisfies the termination criteria, set $x^* = x_k$, $\lambda^* = \lambda_k$, and exit with success.

   (b) If $k = K$, exit with failure.

2. *Solve the quadratic programming subproblem.* Find a minimizer p_k for (13.48) and the associated multiplier vector λ_{k+1}.

3. *Update for the next major iteration.* Compute $x_{k+1} = x_k + p_k$. Set $k \leftarrow k + 1$ and go to Step 1.

It can be shown (see Nocedal and Wright [152, p. 564]), under mild conditions, that provided (x_0, λ_0) is close to (x^*, λ^*), the SQP Method described above converges quadratically.

Use of a merit function

In the event that $Z_k^T\nabla^2 L_Q(p, \lambda_k)Z_k$ is not positive definite, p_k obtained as described above might not be a minimizer. In this case, a steplength $\alpha_k = 1$ may not provide a step to a minimizer, and hence a steplength algorithm would have to be used. The question is: to what function should such a steplength algorithm be applied? Clearly we would like the next iterate to be "better" than the current iterate in some sense. Unlike the linear case, where ensuring feasibility is straightforward, here, at each iteration, we need to balance the goals of decreasing the objective function and reducing the infeasibility. To try to accomplish this, we use a *merit function* and apply a linesearch to this merit function. Much research has been done on the choice of such functions.

One approach (first proposed by Murray [141]) is to choose the merit function as a quadratic penalty function:

$$\mathcal{M}_Q(x, \rho) = f(x) + \frac{\rho}{2}\hat{c}(x)^{\mathrm{T}}\hat{c}(x).$$

Han [91] proposed an exact penalty function

$$\mathcal{M}_E(x, \rho) = f(x) + \frac{\rho}{2}\|\hat{c}(x)\|_1$$

as a merit function. Another approach, first proposed by Wright [196] in the context of SQP methods, is to use the augmented Lagrangian function as the merit function, that is

$$\mathcal{M}_A(x, \rho, \lambda) = f(x) - \lambda^{\mathrm{T}}\hat{c}(x) + \frac{\rho}{2}\hat{c}(x)^{\mathrm{T}}\hat{c}(x).$$

This merit function includes one more parameter, namely λ, than the quadratic penalty function. Since the exact multipliers of the solution of the original problem are not known, an estimate is used. As noted on page 495, the Augmented Lagrangian function has the property that x^* is its minimizer when λ^* is known and ρ is sufficiently large.

An obvious choice for a multiplier estimate is to use the multipliers obtained at the solution of the quadratic program (13.48). An SQP algorithm that incorporates this is as follows.

Algorithm 13.5 (SQP with a Merit Function) Let K be the maximum number of iterations allowed. Set $k = 0$.

1. *Test for convergence.*

   (a) If x_k satisfies the termination criteria, set $x^* = x_k$, $\lambda^* = \lambda_k$ and exit with success.

   (b) If $k = K$, exit with failure.

2. *Solve the quadratic programming subproblem.* Find a minimizer p_k for (13.48) and the associated multipliers λ_{k+1}.

3. *Compute a steplength.* Compute a steplength using one of the algorithms described in Section 9.8 on the merit function,

$$\mathcal{M}_A(x_k, \rho, \lambda_k) = f(x_k + \alpha p_k) - \lambda_k^{\mathrm{T}} \hat{c}(x_k + \alpha p_k) + \frac{\rho}{2} \hat{c}(x_k + \alpha p_k)^{\mathrm{T}} \hat{c}(x_k + \alpha p_k).$$

4. *Update for the next major iteration.* Compute $x_{k+1} = x_k + \alpha_k p_k$. Set $k \leftarrow k + 1$ and go to Step 1.

Quasi-Newton approximations

The SQP algorithm described so far requires computation of the Hessian of the Lagrangian, and this can be computationally expensive. Furthermore, the projected Hessian $Z_k^{\mathrm{T}} \nabla^2 L(x_k, \lambda_k) Z_k$ may not be positive definite at each iteration. A quasi-Newton approximation to the projected Hessian can be used to reduce the computational effort and also guarantee positive definiteness at each iteration.

Solving for p_k

The search direction p_k is obtained by computing the range space and null space components separately and then combining them. Let Y_k be a basis for the range space of $\hat{A}_k$, and let Z_k be a basis for the null space of $\hat{A}_k$. If $Z_k^{\mathrm{T}} \nabla^2 L_Q(x_k, \lambda_k) Z_k$ is positive definite, the solution p_k can be obtained by solving two systems of equations. To see this, we write

$$p_k = Y_k p_Y + Z_k p_Z \tag{13.49}$$

and then compute the components p_Y and p_Z as follows.

To obtain the range-space component of p_k, we multiply both sides of (13.49) by $\hat{A}_k$ to obtain

$$\hat{A}_k p_k = \hat{A}_k Y_k p_Y + \hat{A}_k Z_k p_Z = \hat{A}_k Y_k p_Y.$$

Noting that $\hat{A}_k p_k = -\hat{c}_k$, we can solve $\hat{A}_k Y_k p_Y = -\hat{c}_k$ for p_Y.

From the first-order conditions of optimality for the linearly constrained problem, we have $Z_k^{\mathrm{T}} \nabla L_Q = 0$, where $\nabla L_Q = g_k + \nabla^2 L(x_k, \lambda_k) p$. Substituting $p = Y_k p_Y + Z_k p_Z$, we obtain

$$Z_k^{\mathrm{T}} \Big(g_k + \nabla^2 L(x_k, \lambda_k) (Y_k p_Y + Z_k p_Z) \Big) = 0.$$

By rearranging terms, we can solve for p_Z using

$$Z_k^T \nabla^2 L_Q Z_k p_Z = -Z_k^T(g_k + \nabla^2 L(x_k, \lambda_k) Y_k p_Y). \qquad (13.50)$$

The above system can be solved by using the Cholesky factorization of $Z_k^T \nabla^2 L_Q Z_k$. As a part of creating the factorization we can also determine if the matrix is positive definite.

Nonlinear inequality constraints

So far we have described the SQP approach for a nonlinear equality problem. For the moment we consider an optimization problem in which all the constraints are nonlinear inequalities:

$$\begin{aligned}
&\text{minimize} && f(x) \\
(\text{NIP}) \quad &\text{subject to} && c(x) \geq 0 \\
& && x \in R^n,
\end{aligned} \qquad (13.51)$$

where $f(x)$ is a smooth nonlinear function and $c(x)$ is a vector of m smooth nonlinear functions.

One approach to solving such a problem would be to identify a working set of active constraints and then apply an active-set method. This is not so straightforward because of the nonlinearities present in the system, hence we shall consider alternative techniques.

We can also apply Algorithm 13.5 when nonlinear inequalities are present. There are two simple approaches to doing this. We could subtract slack variables and then solve the resulting bound-constrained problem. Thus, for example, we can modify (13.51) to be

$$\begin{aligned}
&\text{minimize} && f(x) \\
&\text{subject to} && c(x) - s = 0 \\
& && x \in R^n, \ s \in R^m, \ s \geq 0.
\end{aligned} \qquad (13.52)$$

Most real-world problems have bounds on the variables, so, as discussed for the Augmented Lagrangian Method, we can simply consider solving problems of the form

$$\begin{aligned}
&\text{minimize} && f(x) \\
&\text{subject to} && \hat{c}(x) = 0 \\
& && x \in R^n, \ l \leq x \leq u.
\end{aligned} \qquad (13.53)$$

Now instead of solving (13.48) at Step 2 of Algorithm 13.2, we solve the following bound-constrained problem

$$
\begin{aligned}
\text{minimize} \quad & \tfrac{1}{2} p^{\mathrm{T}} \nabla^2 L(x_k, \lambda_k) p \\
\text{subject to} \quad & \hat{A}_k p = -\hat{c}_k \\
& l \leq x \leq u.
\end{aligned}
$$

When there is a mixture of equality and inequality constraints (with both nonlinear and linear constraints) together with bounds on the variables, we can convert all the inequality constraints (except the bounds) into equality constraints by introducing slack variables. Once this has been done, the optimization problem will be of the form

$$
\begin{aligned}
\text{minimize} \quad & f(x) \\
\text{subject to} \quad & \hat{c}(x) = 0 \\
& Ax = b \\
& l \leq x \leq u \\
& x \in R^n.
\end{aligned}
\tag{13.54}
$$

Instead of introducing slack variables in (13.51), an alternative approach would be to subtract the square of slack variables as was described for the Augmented Lagrangian Method on page 500.

Lagrange multipliers

At a solution to (13.48), the Lagrange multipliers $\lambda = \lambda_{k+1}$ are obtained by solving

$$
\hat{A}_k^{\mathrm{T}} \lambda = \nabla L_Q = \nabla^2 L(x_k, \lambda_k) p_k + g_k.
\tag{13.55}
$$

The multipliers will be the true multipliers only if p_k is a local minimizer for (13.48), which will be true if $Z_k^{\mathrm{T}} \nabla^2 L_Q Z_k$ is positive definite. The solution to such a system may fail to exist if $Z_k^{\mathrm{T}} \nabla^2 L_Q Z_k$ is not positive definite. In this case the system is solved in a least-squares sense for an estimate of the Lagrange multipliers.

Termination Criteria

Typically more than one criterion is used to decide on optimality. Gill, Murray and Wright [84] propose the following set of criteria when nonlinear constraints are present. These criteria are applicable for any algorithm used to solve optimization problems with nonlinear constraints. Let

- t_k be the number of active constraints in the working set at iteration k,

- $\lambda_{\min}$ be the smallest multiplier corresponding to an inequality constraint in the working set,

- $\sigma_{\max}$ be the largest absolute value of *all multipliers*,

- $\tau_c > \varepsilon_M$ denote a user-specified tolerance for the size of infeasibility that will be accepted in a computed solution,

- τ_f be a tolerance on the difference of function values between consecutive iterations.

For successful termination, it is usually required that the constraints be almost exactly satisfied (within ε_M), the gradient be close to zero, and the multipliers on the active inequality constraints be positive. If the constraints are not satisfied to within ε_M but are satisfied to within τ_c, then the iterates must also get closer to each other. This is summarized below, where, for termination either the first three or the last four of the following conditions hold:

1. $||\hat{c}_k|| < \varepsilon_M$.

2. $||g_z|| \leq \sqrt{\tau_F}\, ||g_k||$.

3. If $t_k > 1$, $\sigma_{\min} \geq \sqrt{\tau_f}\, |\lambda_{\max}|$; if $t_k = 1$, $\sigma_{\min} \geq \sqrt{\tau_f}\, (1 + |f_k|)$.

4. $||\hat{c}_k|| \leq \tau_c$.

5. $||x_{k+1} - x_k|| < \sqrt{\tau_f}\, (1 + ||x_k||)$.

13.9   Exercises

13.1   Consider the nonlinear function $c(\cdot)$ defined as follows:

$$
c(x_1, x_2) = \begin{cases} \frac{1}{2}x_1 + x_2 + \frac{1}{2} & \text{if } (x_1, x_2) \in (-\infty, -1] \times R, \\ x_1^2 - x_2 & \text{if } (x_1, x_2) \in [-1, 1] \times R, \\ x_1 - x_2 - \frac{1}{2} & \text{if } (x_1, x_2) \in [1, \infty) \times R. \end{cases}
$$

Let $X = \{x \in R^2 : c(x) = 0\}$ where $c(\cdot)$ is the function defined above.

(a) Plot the the set X defined above.

(b) Are there feasible directions at points in X? If so, what are the points and what are the directions?

13.2 Consider the following problem of minimizing a quadratic function subject to linear inequality constraints:

$$\begin{aligned}
\text{minimize} \quad & x_1^2 + x_2^2 \\
\text{subject to} \quad & x_1 + 3x_2 \leq 6 \\
& x_1 \qquad\quad \geq 1.
\end{aligned}$$

Solve the problem by the Penalty-function Method as follows:

(a) Let the initial starting point be $(2, 2)$.

(b) Write down the unconstrained penalty function objective.

(c) Set the penalty parameter $\rho = 10$.

(d) Apply the Penalty-function algorithm to solve the problem.

13.3 Consider the following problem of minimizing a quadratic function subject to a linear equality constraint:

$$\begin{aligned}
\text{minimize} \quad & \tfrac{1}{2}(x_1^2 + x_2^2) \\
\text{subject to} \quad & 2x_1 + x_2 - 1 = 0.
\end{aligned} \qquad (13.56)$$

(a) Analytically and graphically find the solution x^* and corresponding optimal multiplier λ^* to (13.56).

(b) Write down the augmented Lagrangian function L_A for (13.56).

(c) Find a strict local minimizer $\bar{x}$ of $L_A(x, \lambda, \rho)$, where ρ denotes the penalty parameter.

(d) Examine the behavior of $\bar{x}$ as $\lambda \to \lambda^*$.

(e) Show that the Hessian, $\nabla^2 L_A$, of the augmented Lagrangian becomes ill-condiitonedd as the penalty parameter ρ increases.

(f) Solve (13.56) by the Augmented Lagrangian Method.

13.4 Consider the convex programming problem

$$\begin{aligned}
\text{minimize} \quad & \tfrac{1}{2}(x_1^2 + x_2^2) \\
\text{subject to} \quad & x_1 + x_2 = 4 \\
& x_2^2 \leq 4.
\end{aligned} \qquad (13.57)$$

(a) Write down the necessary and suffficient conditions of optimality.

(b) What is the Lagrangian dual of the problem (13.57)?

(c) What is the Wolfe dual of the problem (13.57)?

(d) Solve the problem by the Augmented Lagrangian Method.

13.5   Consider the following optimization problem

$$\begin{aligned}
\text{minimize} \quad & x_1^2 + x_2^2 \\
\text{subject to} \quad & x_1 + x_2 \leq 5 \\
& x_2 \geq 1
\end{aligned}$$

Solve the above problem by each of the following methods:

(a) Penalty-function Method.
(b) The Augmented-Lagrangian Method.
(c) The Projected-Lagrangian Method.
(d) The SQP Method.

13.6   Consider the optimization problem:

$$\begin{aligned}
\text{minimize} \quad & f(x) \\
\text{subject to} \quad & h_i(x) \leq 0, \ i = 1, \ldots, m,
\end{aligned}$$

where, the functions $f, h_1, \ldots, h_m$ are convex and differentiable.

(a) Write down the necessary and sufficient conditions of optimality.
(b) What is the Lagrangian dual of the problem?
(c) What is the Wolfe dual of the problem?

13.7   Consider the problem of maximizing the volume of a rectangular box, where one side is square, and satisfies the constraints that the sum of the dimensions is 20 and surface area is at most 200. This can be formulated as the following nonlinear programming problem

$$\begin{aligned}
\text{minimize} \quad & -x_1^2 x_2 \\
\text{subject to} \quad & 2x_1 + x_2 = 20 \\
& 2x_1^2 + 4x_1 x_2 \leq 200 \\
& x_1 \geq 0, \ x_2 \geq 0.
\end{aligned}$$

(a) Write down the necessary and sufficient conditions of optimality.
(b) Find optimal x_1 and x_2.

13.8   Consider the nonlinear optimization problem

$$\begin{aligned}
\text{minimize} \quad & x_1 x_2 \\
\text{subject to} \quad & x_1 + x_2^2 \leq 1 \\
& x_1 + x_2 \geq 0.
\end{aligned}$$

(a) Give a graphical illustration of the above problem.

(b) Solve the problem using the Augmented Lagrangian Method.

13.9　　For an integer $n \geq 3$, consider the optimization problem

$$\text{minimize} \quad \sum_{i=1}^{n} x_i^2$$

$$\text{subject to} \quad \sum_{i=1}^{n} \left(\frac{x_i}{a_i}\right)^2 - 1 = 0$$

where $0 < a_1 < a_2 < \cdots < a_n$ are given constants.

(a) Form an appropriate Lagrangian function for this problem.

(b) Write down the first-order necessary conditions of optimality for this problem.

(c) List *all* solutions of the first-order necessary conditions of optimality stated in (b).

(d) How many optimal solutions does this problem possess? Justify your answer. In so doing, apply the second-order necessary conditions of optimality.

13.10　Design a box of maximum volume according to the following specifications. The perimeter of the bottom is a given positive number $p = 50$. The height of the box is the difference between its other two dimensions.

(a) Formulate the problem as a constrained optimization problem and solve it.

(b) Re-formulate the problem as an unconstrained optimization problem and solve it.

(c) Compare the solutions from the two different formulations. Are they the same?

13.11　(Forst and Hoffmann [72, p. 85].) Let D_1 and D_2 be two closed disjoint disks in R^2 whose radii are both 1, but whose centers are arbitrary.

(a) Formulate the nonlinear optimization problem of finding a triangle of smallest area containing the two disks D_1 and D_2.

(b) Write down the augmented Lagrangian for this formulation.

13.12　Formulate the following geometric extremal problems as nonlinear optimization problems:

(a) Find the equilateral triangle of maximum area contained in a square of side 1.

(b) Find the regular tetrahedron having maximum volume contained in a cube of side 1.

13.13   (Heron's Problem.) Given two points p and q in R^2 belonging to an open half-space determined by a given line L. Formulate the following problem as a nonlinear optimization problem: Find the point x on L such that the sum of the distances to p and q is a minimum.

13.14   A generalization of the problem stated in Exercise 8.8 is the following. Let $p_1, p_2, \ldots, p_n$ be a set of $n > 2$ points in the plane R^2. Suppose the coordinates (x_i, y_i) of each point p_i are given. For $i = 1, 2, \ldots, n$, suppose D_i is a closed disk of given radius $r_i > 0$ centered at (x_i, y_i). Thus

$$D_i = \left\{ (x, y) : \sqrt{\sum_{i=1}^{n} ((x - x_i)^2 + (y - y_i)^2)} \leq r_i \right\}.$$

Formulate the problem of finding the smallest closed disk that contains the union of $D_1, D_2, \ldots, D_n$ as a nonlinear program.

14. INTERIOR-POINT METHODS

Overview

This chapter discusses interior-point methods for *linear programming* as promised in Chapter 7. Their placement here is justified by the fact that they rely on the theory and methods of nonlinear optimization for which they were originally developed.

Interior-point methods are algorithms that traverse the (relative) interior of the feasible region. In contrast, as we have previously seen, the Simplex Algorithm for linear programs goes from one feasible point to another along edges of the feasible region. Early in the history of linear programming, attempts were made to develop interior-point algorithms. Perhaps the first discussion of such algorithms appears in the 1951 paper of Brown and Koopmans who gave what they called "suggestions" for solving linear programs by making "big jumps rather than 'crawling along the edges' of the convex set" [24, p. 377]. In chronological order, subsequent publications of this type include von Neumann [150], Frisch [74], Tompkins [183, 184], and Dikin [51]. At the time of their publication, none of these methods were claimed to have been superior to the Simplex Algorithm. Interest in such methods dwindled for a while. Then in 1972, Klee and Minty [112] discovered a class of problems that exhibit the worst-case behavior for the Simplex Algorithm; see Section 7.5. This development regenerated strong interest in the development of interior-point algorithms.

In 1979, Khachiyan [110] developed a polynomial-time algorithm for linear programming. Its execution time was shown to be $O(n^6 L^2)$, where n is the number of variables in the problem and L is the number of bits[1] in the input. The method came to be known as an ellipsoid method and owes its name to the fact that it is based on using a sequence of shrinking ellipsoids that enclose a minimizer. Although Khachiyan's ellipsoid method has the desired theoretical property of polynomial-time execution, in practice it performs poorly. There are two major difficulties associated with this method. First, the number of iterations tends to be very large; second, the amount of computation associated with each iteration is much more than that of the Simplex Algorithm.

---

[1]On a computer, numbers are represented in the binary system. A *bit* is a binary digit and takes one of two values: 0 or 1.

© Springer Science+Business Media LLC 2017

R.W. Cottle and M.N. Thapa, *Linear and Nonlinear Optimization*,
International Series in Operations Research & Management Science 253,
DOI 10.1007/978-1-4939-7055-1_14

In 1984, Karmarkar [107] presented a polynomial-time *interior-point* algorithm for linear programming having execution time $O(n^{3.5}L^2)$, where n and L are as above. Clearly Karmarkar's method has a better polynomial bound than Khachiyan's. Comparing his algorithm with the Simplex Algorithm, Karmarkar claimed that, in some cases, his method is 50 times faster. However, the claim was never independently verified. While the method generally takes much fewer iterations than the Simplex Algorithm, it requires the solution of a linear least-squares problem at each iteration; this is computationally very expensive as compared with a single iteration of the Simplex Algorithm. Nevertheless, Karmarkar's claim stimulated the search for new interior-point approaches and improved solution techniques for solving linear programs; see, for example, M. Wright [198], S. Wright [199], Vanderbei [186], and Ye [200]. where some of these methods are discussed. The newer interior-point methods perform well in comparison to the Simplex Algorithm on very large problems. However, they do not find an extreme point of the feasible region as is usually required. The transition from the interior-point solution found by such an algorithm to an optimal extreme-point solution entails substantial additional computation; for a discussion of this topic see Dantzig and Thapa [45, pp. 137–139].

Interior-point methods typically inscribe an ellipsoid in the feasible region with its center at the current iterate or first transform the feasible space and then inscribe a sphere with the current iterate at its center. Next an improving direction is found by joining the current iterate to the point on the boundary of the ellipsoid or sphere that maximizes (or minimizes) the linear objective function. A point on the improving direction line is then selected as the next iterate.

In this chapter, we focus attention on a class of interior-point methods known as Barrier-Function Methods. We start by describing a Barrier-function Method for *nonlinear programming*, and then apply it to linear programming problems. Finally we describe a Primal-Dual Barrier Method for linear programming.

14.1   Barrier-function methods

In Section 13.4 we looked at penalty function methods which could be applied to problems with equalities as well as inequalities. When inequalities are present, another approach is to start with a feasible interior point and minimize an objective function in such a way that its value increases sharply as the boundary of the feasible region is approached. That is, in effect, we create a "barrier" at the boundary of the feasible region as we describe below. Such methods, developed primarily for nonlinear optimization, appear in two pioneering publications: Frisch [74] and Fiacco and McCormick [64].

Barrier-function Methods have the important property that all their iterates remain feasible; such methods are appropriately classified as *feasible-point methods*. If such an algorithm is terminated prematurely, the iterate available upon termination will be feasible. An algorithm may be terminated early if only approximate solutions are required or if there is a constraint on total computational time, for example, when solving a sequence of problems. Another important reason for creating a feasible-point method is that the objective function f or some of the constraint functions c_i may be undefined or ill-defined outside the feasible region.

Logarithmic barrier function

The *logarithmic barrier-function* approach, which we now describe, was first proposed in 1955 by Frisch [74] who considered the inequality-constrained problem (13.13); for convenience we repeat it below:

$$
\begin{array}{ll}
\text{minimize} & f(x) \\
\text{subject to} & c(x) \geq 0 \\
& x \in R^n,
\end{array}
\tag{14.1}
$$

where $c(x)$ is a vector of m smooth nonlinear constraint functions. Let $x = \bar{x}$ be an initial interior point, i.e., one satisfying $c(\bar{x}) > 0$. Starting at $\bar{x}$, we would like the iterates to traverse the interior of the feasible region by staying away from the boundary until an optimal solution is found. To accomplish this, Frisch created a barrier at the boundary of the constraints by using natural logarithms in the objective function. In particular, he converted the inequality-constrained problem (14.1) to an *unconstrained* problem of the form

$$
\text{minimize } B(x, \mu) = f(x) - \mu \sum_{i=1}^{m} \ln(c_i(x))
\tag{14.2}
$$

where $0 < \mu \leq 1$ is termed the *barrier parameter*, and strict satisfaction of the constraints is ensured by the fact that $-\ln(c_i(x)) \to \infty$ as $c_i(x) \to 0$. As the barrier parameter μ is decreased, the impact of the barrier is reduced, allowing the iterates to approach the constraint boundaries. Thus as μ is decreased, the expectation is that the iterates will approach a solution of the original problem at which there may be several active constraints.

Barrier-function algorithm outline

For given $\mu_k > 0$, let $x_j(\mu_k)$ be the j-th iterate in the solution of the k-th subproblem

$$\text{minimize } B(x, \mu_k) = f(x) - \mu_k \sum_{i=1}^{m} \ln(c_i(x)) \qquad (14.3)$$

and let $x^*(\mu_k)$ denote an unconstrained local minimizer thereof. A barrier-function algorithm has the following form.

Algorithm 14.1 (A Barrier-function Algorithm Outline) Given a nonlinear program (14.1) with inequality constraints, create an unconstrained subproblem with objective function $B(x, \mu)$ as in (14.2). In general, let $x_0(\mu_k)$ denote the starting point for solving (14.3). Pick a starting point $\bar{x}$ such that $c(\bar{x}) > 0$. Set $k = 0$, $\mu_0 = 1.0$, and $x_0(\mu_0) = \bar{x}$.

1. *Test for convergence.* If the first-order optimality conditions are met to a pre-specified tolerance, stop and report the solution as $x_0(\mu_k)$

2. *Solve subproblem k.* Let a solution to the unconstrained subproblem be $x^*(\mu_k)$.

3. *Update for the next major iteration.* Update the barrier parameter: set $\mu_{k+1} = \beta \mu_k$, where $\beta < 1$ is a constant. Set the initial solution for the next subproblem to the optimal solution of the current subproblem; that is, $x_0(\mu_{k+1}) = x^*(\mu_k)$. Let $k \leftarrow k + 1$. Go to Step 1.

Note: In Step 2 we assume for this discussion that subproblem k is not defective. Any of the techniques for solving an unconstrained problem can be used here. Some care must be exercised in case (contrary to our assumption) the subproblem actually is *defective* (see page 523). The iterations for solving each subproblem are referred to as *minor* iterations.

Properties of successive $x^*(\mu)$ Given that $\mu_{k+1} < \mu_k$ for all k, it can be shown that successive unconstrained minimizers $x^*(\mu_k)$ of the barrier-function subproblems (14.2) have the following properties.

1. For sufficiently small μ_k, the optimal objective function values of the subproblems are decreasing:

$$B(x^*(\mu_{k+1}), \mu_{k+1}) < B(x^*(\mu_k), \mu_k).$$

2. The objective function of the original problem is nonincreasing when computed at a minimizer of each successive subproblem:

$$f(x^*(\mu_{k+1})) \le f(x^*(\mu_k)).$$

3. The minimizers $x^*(\mu_k)$ approach the boundary of the constraints that will be in the optimal active set. Consequently

$$-\sum_{i=1}^{m} \ln(c_i(x^*(\mu_{k+1}))) \ge -\sum_{i=1}^{m} \ln(c_i(x^*(\mu_k))).$$

4. The unconstrained minimizers $x^*(\mu_k)$ converge to a solution x^* of the original problem:

$$\lim_{\mu_k \to 0} x^*(\mu_k) = x^*.$$

Example 14.1: A LOGARITHMIC BARRIER-FUNCTION APPROACH. To illustrate this approach we use an extremely simple example due to Gill, Murray, and Wright [84]. Consider the univariate optimization problem

$$\begin{array}{ll}
\text{minimize} & x^2 \\
\text{subject to} & x \ge 1.
\end{array}$$

The logarithmic barrier function for a fixed μ for this problem is

$$B(x, \mu) = x^2 - \mu \ln(x - 1).$$

The first-order conditions of optimality for the unconstrained barrier function result in $x^*(\mu)$ as the minimizer of $B(x, \mu)$. In particular,

$$x^*(\mu) = \frac{1}{2} + \frac{1}{2}\sqrt{1 + 2\mu}.$$

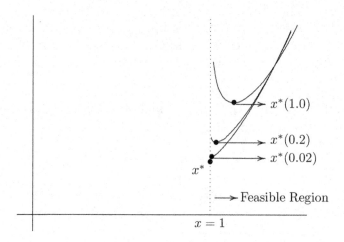

Figure 14.1: Barrier-function Method.

It is easy to see that

$$\lim_{\mu \to 0} x^*(\mu) = 1 = x^*.$$

The approach of the barrier-function minimizers $x^*(\mu)$ to the solution x^* is illustrated in Figure 14.1. Note that the iterates are approaching the boundary of the constraint that will be active at an optimal solution of the optimzation problem.

Example 14.2: BARRIER FUNCTION VS PENALTY FUNCTION. Once again consider the inequality-constrained problem stated in Example 13.5:

$$
\begin{array}{ll}
\text{minimize} & x_1^2 + x_2^2 \\
\text{subject to} & x_1^2 - x_2^2 - 4 \geq 0.
\end{array}
$$

For $0 < \mu \leq 1$, the logarithmic barrier function for this problem is given by

$$B(x, \mu) = x_1^2 + x_2^2 - \mu \ln(x_1^2 - x_2^2 - 4).$$

The function is not defined at points outside the feasible region, and the approach requires that the starting point be strictly in the interior of the feasible region. Subsequently the algorithm will ensure that the iterates stay inside of the feasible region. This is in contrast to the penalty function method which takes *any* starting point, and if an iterate is feasible, the penalty is set to zero.

Properties of barrier-function methods

Unlike penalty-function methods every iterate of a barrier-function method is a feasible solution of the constraints of the optimization problem being solved. Nevertheless, many properties of barrier-function methods are analogous to those of penalty-function methods. We now highlight these properties.

Sequence of subproblems

We have noted that as we decrease μ for each subsequent subproblem, the sequence of solutions obtained by solving each such subproblem approaches a solution of the original problem. So it would be reasonable to ask: why not just choose μ to be a very small number at the start, for example $\mu = 10^{-30}$, and either find a solution right away or cut down on the number of subproblems that need to be solved. The reason we do not start with a very small μ has to do with numerical stability, in particular, ill-conditioning. It turns out that the condition number of the Hessian of the barrier function increases as μ is decreased and as a solution of the subproblem is reached. It can be shown that the Hessian becomes singular in the limit. This results in numerical issues for unconstrained minimization algorithms applied to such problems. To avoid this, we solve a sequence of unconstrained subproblems, with μ being decreased for each subsequent subproblem, and then stop when a reasonable approximation to a solution of the original problem is reached.

Ill-conditioning

As noted earlier, decreasing μ to 0, causes the Hessian matrix at $x^*(\mu)$ to become increasingly *ill-conditioned* with singularity occurring in the limit. Thus, the unconstrained subproblems become increasingly difficult to solve.

Defective subproblems

Similar to the penalty-function case, the solution to a barrier-function subproblem is a local minimizer of the barrier function which, for sufficiently small μ, is near a minimizer x^* of the original problem. When the problems are well-behaved this works fine; however, the subproblem may be defective in the sense that the barrier function may be unbounded below. Of course, this is less likely to occur than in the case of the penalty function, since the feasible region in many problems is closed and bounded (this is especially true if bounds are imposed on each variable).

Linesearch

The barrier-function approach creates singularities[2] in the objective function. This causes difficulties when using a standard unconstrained linesearch algorithm. For example, the barrier function is undefined outside the feasible region, and it is not possible to place an a priori upper bound on the step to the nearest nonlinear constraint along a given direction. Furthermore, polynomial approximations within the linesearch are likely to be poor for small μ because of their closeness to the singularities.

Lagrange multipliers

As in penalty-function methods, each barrier-function subproblem provides an estimate of the Lagrange multipliers of the original problem. Let $x^*(\mu)$ be an unconstrained minimizer of the barrier function and let g be the gradient of the original objective function. The first-order necessary optimality conditions for a minimizer $x^*(\mu)$ of $B(x, \mu)$ are

$$\nabla B(x^*(\mu), \mu) = g(x^*(\mu)) - \mu \sum_{i=1}^{m} \frac{1}{c_i(x^*(\mu))} \nabla c_i(x^*(\mu)) = 0$$

or equivalently,

$$g(x^*(\mu)) = \sum_{i=1}^{m} \nabla c_i(x^*(\mu)) \left(\frac{\mu}{c_i(x^*(\mu))} \right).$$

Thus, estimates of the Lagrange multipliers are given by

$$\lambda_i = \frac{\mu}{c_i(x^*(\mu))}, \quad \text{for } i = 1, \ldots, m.$$

It can be shown that

$$\lim_{\mu \to 0} \frac{\mu}{c_i(x^*(\mu))} = \lambda_i^*;$$

see, for example, M. Wright [197, p. 375].

Notice that *all* the constraints appear for the Lagrange multiplier estimates in contrast to the analogous result for penalty-function methods, in which only the violated constraints appear.

---

[2]A function has a singularity at a point if it is not defined there. In particular, the logarithm function has a singulariy at 0. Accordingly, for the barrier function, a singularity occurs at any point where the inequality constraint is exactly satisfied as an equality.

14.2 Primal barrier-function method for linear programs

In this section we describe a barrier-function approach combined with Newton's Method applied to the primal linear programming problem in standard form. Theoretical details and computational results with the method to be described are covered in Gill, Murray, Saunders, Tomlin, and Wright [82].

Developing the approach

To create a barrier-function subproblem for a linear program in standard form, i.e.,

$$\begin{array}{ll} \text{minimize} & c^{\mathrm{T}}x \\ \text{subject to} & Ax = b \\ & x \geq 0, \end{array} \qquad (14.4)$$

we note that an interior point only exists for the nonnegativity constraints. Applying the barrier construction, we obtain the linear equality-constrained problem

$$\begin{array}{ll} \text{minimize} & c^{\mathrm{T}}x - \mu \sum_{j=1}^{n} \ln(x_j) \\ \text{subject to} & Ax = b. \end{array} \qquad (14.5)$$

We next apply Newton's Method (see Section 12.1) to solve (14.5).

We now define a notation used in the remainder of this chapter. Given $x \in R^n$, we define the matrix D_x to be the the diagonal matrix with d_{jj} equal to the j-th component of x for $j = 1, \ldots, n$.

The gradient and Hessian matrix of the barrier function in (14.5) at points $x > 0$ are given by:

$$g(x) = c - \mu D_x^{-1} e \quad \text{and} \quad G(x) = \mu D_x^{-2}, \qquad (14.6)$$

where $e = (1, 1, \ldots, 1)$ and D_x is the diagonal matrix defined above. Note that because μ is positive, $G(x)$ is positive definite for all $x > 0$.

As seen on page 422, for a linear equality-constrained problem, every feasible search direction must satisfy $Ap = 0$. Note that $Gp = -g$ is the first-order optimality condition for minimizing the quadratic function

$(1/2)p^TGp + g^Tp$. Hence we can compute the Newton search direction for (14.5) by solving

$$\begin{aligned} \text{minimize} \quad & \tfrac{1}{2}p^TGp + g^Tp \\ \text{subject to} \quad & Ap = 0, \end{aligned} \tag{14.7}$$

with g and G defined by (14.6). If we let λ denote the multipliers on the constraints $Ap = 0$, then the optimal p must satisfy the first-order necessary conditions for optimality, namely, $Ap = 0$ and $Gp + g = A^T\lambda$. Rearranging terms and writing this in matrix form, we see that the solution to (14.7) must satisfy the system of equations

$$\begin{bmatrix} \mu D_x^{-2} & A^T \\ A & 0 \end{bmatrix} \begin{bmatrix} -p \\ \lambda \end{bmatrix} = \begin{bmatrix} c - \mu D_x^{-1}e \\ 0 \end{bmatrix}. \tag{14.8}$$

To find the search direction we could solve the above problem directly, but this would be computationally expensive. Instead we show that p can be obtained from the solution of the least-squares problem

$$\min_{\lambda \in R^m} \|D_x c - \mu e - D_x A^T\lambda\|_2^2. \tag{14.9}$$

Let $r = D_x c - \mu e - D_x A^T\lambda$ be the residual vector. At the optimal solution $\lambda = \lambda^*$ to (14.9), the residual vector will be denoted by r^*. We have seen earlier (see page 258) that the optimal solution to (14.9) can be obtained by solving the normal equations

$$AD_x D_x A^T\lambda^* = AD_x(D_x c - \mu e). \tag{14.10}$$

The optimal residual r^* is given by

$$r^* = D_x c - \mu e - D_x A^T\lambda^*. \tag{14.11}$$

This equation can be rewritten as

$$r^* + D_x A^T\lambda^* = D_x c - \mu e. \tag{14.12}$$

Multiplying the left- and right-hand sides of (14.11) by AD_x, and using (14.10), we find that $AD_x r^* = 0$; that is,

$$AD_x r^* = AD_x(D_x c - \mu e) - AD_x D_x A^T\lambda^* = 0. \tag{14.13}$$

Writing equations (14.12) and (14.13) in block-matrix notation, we obtain

$$\begin{bmatrix} I & D_x A^T \\ AD_x & 0 \end{bmatrix} \begin{bmatrix} r^* \\ \lambda^* \end{bmatrix} = \begin{bmatrix} D_x c - \mu e \\ 0 \end{bmatrix}. \tag{14.14}$$

In order to see that (14.14) is equivalent to (14.8), multiply the top n rows of (14.8) by D_x to obtain

$$\begin{bmatrix} \mu D_x^{-1} & D_x A^{\mathrm{T}} \\ A & 0 \end{bmatrix} \begin{bmatrix} -p \\ \lambda \end{bmatrix} = \begin{bmatrix} D_x c - \mu e \\ 0 \end{bmatrix}.$$

Define $r = -\mu D_x^{-1} p$, to obtain

$$\begin{bmatrix} I & D_x A^{\mathrm{T}} \\ A D_x & 0 \end{bmatrix} \begin{bmatrix} r \\ \lambda \end{bmatrix} = \begin{bmatrix} D_x c - \mu e \\ 0 \end{bmatrix}.$$

Once we solve the least-squares problem (14.9) and obtain r^*, we can find the projected-Newton barrier direction p from $D_x r^* = -\mu p$; that is to say,

$$p = -\frac{1}{\mu} D_x r^*. \tag{14.15}$$

The new approximation to the solution of (14.5) is then given by

$$x_{k+1} = x_k + \alpha p.$$

Note that since we will be using a steplength algorithm, we could ignore the scaling factor $\mu > 0$ and define the search direction (14.15) by

$$p = -D_x r^*.$$

In this case, the Newton step would be $\alpha = 1/\mu$.

During the linesearch, care must be taken to ensure that all the components of x stay away from their lower bound of 0. The maximum step allowed will be the value of the smallest component of x. If the steplength algorithm attempts to reach this value, then the step will be reduced to, say, 99% of that value.

Primal barrier-function algorithm for LPs

Algorithm 14.2 (A Primal Barrier-function Algorithm for LPs) Given a linear program in standard form (14.4), create an unconstrained subproblem as in (14.2). Set $k = 0$. Set $\mu_0 = 1.0$ and $\beta = 0.1$, for example. Pick a feasible interior point $x = \bar{x} > 0$. Let $x_0(\mu_0) = \bar{x}$.

1. *Test for convergence.* If the first-order optimality conditions are met to a pre-specified tolerance, stop and report the solution.

2. *Solve subproblem k.* Perform the following steps:

   (a) Let λ^* be the solution of the linear least-squares problem

   $$\underset{\lambda \in R^m}{\text{minimize}} \ ||D_x c - \mu_k e - D_x A^{\mathrm{T}} \lambda||_2^2.$$

   (b) Compute $r^* = D_x c - \mu_k e - D_x A^{\mathrm{T}} \lambda^*$.

   (c) Set $p = -D_x r^*$.

   (d) Use a steplength algorithm with the initial Newton step being set to $\alpha = 1/\mu_k$ to determine $\bar{\alpha}$. If the steplength algorithm causes a bound to be reached, then let $\bar{\alpha} \leftarrow 0.99\bar{\alpha}$.

   (e) Set $x^*(\mu_k) = x_0(\mu_k) + \bar{\alpha} p$.

3. *Update for the next major iteration.* Set the initial solution for the next subproblem to the optimal solution of the current subproblem; that is, $x_{k+1}(\mu_k) = x^*(\mu_k)$. Update the barrier parameter: Set $\mu_{k+1} = \beta\mu_k$. Let $k \leftarrow k + 1$. Go to Step 1.

The solution of the linear least-squares problem in Step 2 of the algorithm is computationally expensive. Since the introduction of interior-point methods, considerable progress has been made to improve the efficiency of solving large least-squares problems; as a result, large degenerate problems and large problems with a block-staircase structure can be solved quite efficiently by this method.

Finding an initial feasible solution

In the discussion above we have assumed that an initial feasible interior-point solution is available; this is usually not the case for practical problems. One way to construct an initial feasible solution is based on a Big-M type of approach (see, for example, [45]). Say we choose $x = \bar{x} > 0$ where $\bar{x} > 0$ is arbitrary. Then, by choosing a sufficiently large cost M, we can modify the linear program in standard form (14.4) to the equivalent linear program

$$
\begin{aligned}
\text{minimize} \quad & c^{\mathrm{T}} x + \qquad M\theta \\
\text{subject to} \quad & Ax + (b - A\bar{x})\theta = b \\
& x \geq 0, \ \theta \geq 0.
\end{aligned}
\tag{14.16}
$$

For (14.16) to have the same solution as (14.4), the artificial variable θ must be 0 at the solution. It is easy to verify that an initial feasible interior-point solution to the above linear program is $x = \bar{x}$ and $\theta = 1$.

14.3   Primal-Dual barrier function for linear programs

In the previous section we discussed the primal barrier-function method for solving linear programs. In this section we develop an approach where we attempt to solve both the primal and dual problems simultaneously by taking steps in both the dual and primal spaces at each iteration. Such primal-dual algorithms have become the most important interior-point algorithms for solving linear programs; see S. Wright [199].

Developing the approach

The dual of a linear program (14.4) in standard form is:

$$
\begin{array}{ll}
\text{maximize} & b^{\mathrm{T}}y \\
\text{subject to} & A^{\mathrm{T}}y \leq c \\
& (y \text{ free}).
\end{array}
\qquad (14.17)
$$

Adding a vector of slack variables s, we obtain

$$
\begin{array}{ll}
\text{maximize} & b^{\mathrm{T}}y \\
\text{subject to} & A^{\mathrm{T}}y + s = c \\
& y \text{ free}, \ s \geq 0.
\end{array}
\qquad (14.18)
$$

Next we apply a barrier transformation to move the inequality (nonnegativity) constraints to the objective function. Thereby obtaining

$$
\begin{array}{ll}
\text{maximize} & b^{\mathrm{T}}y + \mu \sum_{j=1}^{n} \ln(s_j) \\
\text{subject to} & A^{\mathrm{T}}y + s = c.
\end{array}
\qquad (14.19)
$$

Having transformed the problem, we write the Lagrangian of (14.19), with x acting as the vector of Lagrange multipliers,

$$
L(y, s, \mu, x) = b^{\mathrm{T}}y + \mu \sum_{j=1}^{n} \ln(s_j) - x^{\mathrm{T}}(A^{\mathrm{T}}y + s - c).
\qquad (14.20)
$$

To obtain the first-order conditions for a minimum of the Lagrangian we set
the gradient of the Lagrangian with respect to x, y, and s to 0. This results
in the following three sets of equations:

$$A^\mathsf{T}y + s = c \qquad \text{(Dual feasibility)} \tag{14.21}$$

$$Ax = b \qquad \text{(Primal feasibility)} \tag{14.22}$$

$$D_x D_s e = \mu e \qquad \text{(``Complementary slackness'')}, \tag{14.23}$$

where D_x, D_s, and e are as defined on page 525. The last condition above
follows from first setting the gradient of the Lagrangian with respect to s to
0, that is $\mu D_s^{-1}e - D_x e = 0$, multiplying by D_s, and then rearranging the
terms.

As before, we assume that given a $\mu > 0$, an initial feasible interior-point
solution, say $(x, y, s) = (\bar{x}, \bar{y}, \bar{s})$, is available for the combined primal and
dual. This solution does not necessarily satisfy the complementary slackness
condition (14.23). We can improve the solution by finding appropriate search
directions $(\Delta x, \Delta y, \Delta s)$. Setting the new iterate to be $(x+\Delta x, y+\Delta y, s+\Delta s)$
and substituting in (14.21), (14.22), and (14.23), we obtain

$$A^\mathsf{T}\Delta y + \Delta s = 0 \tag{14.24}$$

$$A\Delta x = 0 \tag{14.25}$$

$$(\bar{x}_j + \Delta x_j)(\bar{s}_j + \Delta s_j) = \mu \quad \text{for } j = 1, \ldots, n. \tag{14.26}$$

In matrix form, with the second-order terms $\Delta x_j \Delta s_j$ ignored, (14.26) becomes

$$D_{\bar{x}}\Delta s + D_{\bar{s}}\Delta x = \mu e - D_{\bar{x}}D_{\bar{s}}^{-1}e. \tag{14.27}$$

What we have described here is a variant of Newton's Method to solve a
nonlinear system of equations. See, for example, [84, pp. 139–140].

We next solve the system of equations (14.24), (14.25), and (14.27).
Rewriting (14.27), and multiplying through by $D_{\bar{s}}^{-1}$, we obtain

$$\Delta x = D_{\bar{s}}^{-1}(\mu e - D_{\bar{x}}D_{\bar{s}}^{-1}e) - D_{\bar{s}}^{-1}D_{\bar{x}}\Delta s$$

$$= D_{\bar{s}}^{-1}(\mu e - D_{\bar{x}}D_{\bar{s}}^{-1}e) - D_{\bar{x}}D_{\bar{s}}^{-1}\Delta s. \tag{14.28}$$

Now we multiply through by A and note that $A\Delta x = 0$ from (14.25) and
$\Delta s = -A^\mathsf{T}\Delta y$ from (14.24) to obtain

$$0 = AD_{\bar{s}}^{-1}(\mu e - D_{\bar{x}}D_{\bar{s}}^{-1}e) - AD_{\bar{x}}D_{\bar{s}}^{-1}A^\mathsf{T}\Delta y.$$

Rewriting the above, we obtain

$$AD_{\bar{x}}D_{\bar{s}}^{-1}A^{T}\Delta y = AD_{\bar{s}}^{-1}(\mu e - D_{\bar{x}}D_{\bar{s}}^{-1}e). \qquad (14.29)$$

Solving for Δy, and noting that $D_{\bar{x}} = D_{\bar{s}}^{-1}$, we obtain

$$\begin{aligned}
\Delta y &= (AD_{\bar{x}}D_{\bar{s}}^{-1}A^{T})^{-1}AD_{\bar{s}}^{-1}(\mu e - D_{\bar{x}}D_{\bar{s}}^{-1}e) \\
&= (AD_{\bar{s}}^{-1}D_{\bar{s}}^{-1}A^{T})^{-1}AD_{\bar{s}}^{-1}(\mu e - D_{\bar{x}}D_{\bar{s}}^{-1}e)\mu \\
&= (AD_{\bar{s}}^{-1}D_{\bar{s}}^{-1}A^{T})^{-1}AD_{\bar{s}}^{-1}(\mu^{2}e - D_{\bar{x}}D_{\bar{x}}e).
\end{aligned}$$

These are the normal equations for the least-squares problem

$$\underset{\Delta y \in R^{m}}{\text{minimize}} \ ||(\mu^{2}e - D_{\bar{x}}D_{\bar{x}}e) - D_{\bar{s}}^{-1}A^{T}\Delta y||_{2}^{2}.$$

Commercial computer implementations (solvers) do not solve the above linear least-squares problem; instead they solve (14.29) by using a Cholesky factorization of $AD_{\bar{x}}D_{\bar{s}}^{-1}A^{T}$.

Once $(\Delta y, \Delta s, \Delta x)$ has been obtained, simple linesearches can be performed in the dual and primal spaces to generate the next approximation to the solution. To that end, we determine α_{p} and α_{d} such that

$$\begin{aligned}
\bar{x} + \alpha_{p}\Delta x &> 0 \\
\bar{s} + \alpha_{d}\Delta s &> 0.
\end{aligned} \qquad (14.30)$$

The Newton step is, of course, unity; but we adjust it to ensure that the new approximation is in the interior of the region.

Primal-Dual barrier-function algorithm

In the previous subsection we developed an approach for solving linear programs by a primal-dual interior point algorithm. Here we provide an outline of such an algorithm.

Algorithm 14.3 (A Primal-Dual Barrier-function Algorithm for LPs)

Given a linear program in standard form (14.4), set $k = 0$. Set $\mu_{0} = 1.0$ and $\beta = 0.1$, for example. Pick a feasible primal and dual interior point $(\bar{x}, \bar{s}, \bar{y})$ with $(\bar{x}, \bar{s}) > 0$. Let $x_{0}(\mu_{0}) = \bar{x}$, $s_{0}(\mu_{0}) = \bar{s}$, and $y_{0}(\mu_{0}) = \bar{y}$.

1. *Test for convergence.* If the first-order optimality conditions are met to a pre-specified tolerance, stop and report the solution. Typically the test for convergence will include a check to see if the duality gap $c^T x_k - b^T y_k$ is close to zero and μ is small.

2. *Solve subproblem k.* Perform the following steps:

    (a) Solve the least-squares type system of equations (14.29), that is, solve

    $$AD_{\bar{x}} D_{\bar{s}}^{-1} A^T \Delta y = AD_{\bar{s}}^{-1}(\mu e - D_{\bar{x}} D_{\bar{s}}^{-1} e)$$

    for Δy, using a Cholesky factorization.

    (b) Compute $\Delta s = -A^T \Delta y$.

    (c) Compute $\Delta x = D_{\bar{s}}^{-1}(\mu e - D_{\bar{x}} D_{\bar{s}}^{-1} e) - D_{\bar{x}} D_{\bar{s}}^{-1} \Delta s$.

    (d) Compute steplengths in a manner that ensures that the primal and dual iterates stay in the interior. First compute the steps α_p and α_d in the primal and dual to be the largest values so that

    $$\bar{x} + \alpha_p \Delta x \geq 0$$
    $$\bar{s} + \alpha_d \Delta s \geq 0.$$

    In order to ensure that the new iterate is in the interior do the following:

    - If $\alpha_p \leq 1$, let $\alpha_p \leftarrow \gamma \alpha_p$ and α_d, else if $\alpha_p > 1$, let $\alpha_p \leftarrow 1$.
    - If $\alpha_d \leq 1$, let $\alpha_d \leftarrow \gamma \alpha_d$, else if $\alpha_d > 1$, let $\alpha_d \leftarrow 1$.

    The constant γ is typically chosen to be in the interval $0.9995, 1.0)$

    (e) Set

    $$x_k^*(\mu_k) = x_k(\mu_k) + \alpha_p \Delta x$$
    $$y_k^*(\mu_k) = x_k(\mu_k) + \alpha_d \Delta y$$
    $$s_k^*(\mu_k) = x_k(\mu_k) + \alpha_d \Delta s.$$

3. *Update for the next major iteration.* Set the initial solution for the next subproblem to the optimal solution of the current subproblem; that is, $x_{k+1}(\mu_k) = x^*(\mu_k)$, $y_{k+1}(\mu_k) = y^*(\mu_k)$, and $s_{k+1}(\mu_k) = s^*(\mu_k)$. Update the barrier parameter by setting $\mu_{k+1} = \beta \mu_k$. Let $k \leftarrow k + 1$. Go to Step 1.

Finding an initial feasible solution

Here again, we have assumed that an initial feasible interior-point solution is available. If one is not available, we can use an approach similar to that described on page 528 for the primal log-barrier function method for linear programs.

14.4 Exercises

14.1   Consider the univariate optimization problem in Example 14.1. For $\mu = 1, 0.2, 0.02$, show that the properties listed on page 521 hold for successive $x^*(\mu)$.

14.2   For the linear program:

$$\begin{array}{ll} \text{minimize} & 2x_1 + 3x_2 \\ \text{subject to} & x_1 \geq 1, \; x_2 \geq 2. \end{array}$$

(a) Write down the barrier function $B(x, \mu)$.
(b) Write down the first-order conditions of optimality for $B(x, \mu)$.
(c) From the first-order conditions of optimality find $x^*(\mu)$.
(d) Show that $\lim_{\mu \to 0} x^*(\mu) = x^*$.
(e) What are the Lagrange multipliers at an optimal solution?
(f) Show that

$$\lim_{\mu \to 0} \frac{\mu}{c_i(x^*(\mu))} = \lambda_i^*.$$

14.3   Consider the linear program

$$\begin{array}{ll} \text{minimize} & x_1 + 4x_2 \\ \text{subject to} & -2 \leq x_1 \leq 2 \\ & -1 \leq x_2 \leq 1. \end{array}$$

(a) Write down the corresponding barrier function $B(x, \mu)$.
(b) Write down the first-order conditions of optimality for $B(x, \mu)$.
(c) From the first-order conditions of optimality find $x^*(\mu)$.
(d) Show that $\lim_{\mu \to 0} x^*(\mu) = x^*$.
(e) What are the Lagrange multipliers at the optimal solution?
(f) Show that

$$\lim_{\mu \to 0} \frac{\mu}{c_i(x^*(\mu))} = \lambda_i^*.$$

14.4   Consider the linear program

$$\begin{array}{ll} \text{minimize} & 2x_1 + x_2 \\ \text{subject to} & x_1 + x_2 = 1 \\ & x_1 \geq 0, \ x_2 \geq 0. \end{array}$$

(a) Write down the barrier function $B(x, \mu)$ corresponding to the above linear program.
(b) Write down the first-order conditions of optimality for $B(x, \mu)$.
(c) From the first-order conditions of optimality find $x^*(\mu)$.
(d) Show that $\lim_{\mu \to 0} x^*(\mu) = x^*$.

14.5   Justify the assertion made on page 528 to the effect that the linear programs (14.4) and (14.16) are equivalent.

14.6   In this exercise you are asked to set up an initial feasible interior-point solution as described on page 528. Consider the linear program

$$\begin{array}{ll} \text{minimize} & 2x_1 + x_2 + 4x_3 \\ \text{subject to} & x_1 + x_2 + x_3 = 7 \\ & x_1 + x_2 \qquad\;\; \geq 3 \\ & x_1 \geq 0, \ x_2 \geq 0, \ x_3 \geq 0. \end{array}$$

(a) Choose an initial point $x = \bar{x} > 0$.
(b) Use the description given on page 528, to set up a linear program such that $x = \bar{x} > 0$ is a feasible interior point for the problem above.
(c) Write down the barrier function $B(x, \mu)$ for the problem.

14.7   Consider the following problem of minimizing a quadratic function subject to linear inequality constraints:

$$\begin{array}{ll} \text{minimize} & x_1^2 + x_2^2 \\ \text{subject to} & x_1 + 3x_2 \leq 6 \\ & x_1 \qquad\quad \geq 1. \end{array}$$

Solve the problem by the Barrier-function Method as follows:

(a) Let the initial starting point be $(2, 1)$.
(b) Write down the unconstrained barrier function objective.
(c) Set the barrier parameter $\mu = 0.1$.
(d) Apply the Barrier-function algorithm to solve the problem.

14.8   For the linear program

$$\begin{array}{ll} \text{minimize} & 2x_1 + 3x_2 \\ \text{subject to} & x_1 \geq 1, \ x_2 \geq 2, \end{array}$$

manually perform one or more iterations of the Primal Barrier-function Algorithm 14.2. Use any initial feasible starting point; for example, $x_1 = 2$, $x_2 = 3$.

14.9   For the following linear program, manually perform one or more iterations of the Primal-Dual Barrier-function Algorithm 14.3

$$\begin{array}{ll} \text{minimize} & 2x_1 + 3x_2 \\ \text{subject to} & x_1 + x_2 = 1 \\ & x_1 \geq 0, \ x_2 \geq 0. \end{array}$$

Use any initial feasible starting point; for example, $x_1 = 0.5$, $x_2 = 0.5$.

14.10   Consider the linear program

$$\begin{array}{ll} \text{minimize} & 2x_1 + x_2 + 4x_3 \\ \text{subject to} & x_1 + x_2 + x_3 = 7 \\ & x_1 \geq 0, \ x_2 \geq 0, \ x_3 \geq 0. \end{array}$$

Choose $x = (2, 1, 4)$; an initial feasible interior point solution to the above linear program. Manually perform one iteration of the Primal-Dual Algorithm 14.3 on the above linear program.

14.11   Recall the portfolio selection problem described on page 254. The following is a simple problem of this type given by Markowitz [131]:

$$\begin{array}{ll} \text{minimize} & \frac{1}{2}x^{\mathrm{T}}Hx - \theta r^{\mathrm{T}}x \\ \text{subject to} & e^{\mathrm{T}}x = 1 \\ & x \geq 0, \end{array}$$

where the covariance matrix H and expected returns vector r are

$$H = \begin{bmatrix} .0146 & .0187 & .0145 \\ .0187 & .0854 & .0104 \\ .0145 & .0104 & .0289 \end{bmatrix} \quad \text{and} \quad r = \begin{bmatrix} .062 \\ .146 \\ .128 \end{bmatrix}.$$

Suppose $\theta = 10$.

(a) Set up the problem as an optimization problem so as to minimize a barrier function subject to one equality constraint.

(b) Find an initial feasible interior starting point.

(c) Manually perform one iteration of the Barrier-function Algorithm 14.1 on the above quadratic programming problem.

(d) How would you expect the solution to change if θ were chosen to be a thousand times larger?

14.12   Consider the following optimization problem:

$$\begin{array}{ll} \text{minimize} & x_1^2 + x_2^2 + x_1 x_2 \\ \text{subject to} & x_1 \geq 1. \end{array}$$

(a) Write down the unconstrained barrier function objective.
(b) Write down the Hessian of the barrier function.
(c) Show that for any μ, as the iterates approach the solution, the Hessian becomes increasingly ill-conditioned.

14.13   For most linear programs it is desirable to obtain an extreme point solution. When using interior-point methods, an extreme point solution is typically not obtained. Describe how to construct an optimal extreme point solution from an "optimal" interior-point solution.

A. APPENDIX: SOME MATHEMATICAL CONCEPTS

Overview

This appendix on mathematical concepts is simply intended to aid readers who would benefit from a refresher on material they learned in the past or possibly never learned at all. Included here are basic definitions and examples pertaining to aspects of set theory, numerical linear algebra, metric properties of sets in real n-space, and analytical properties of functions. Among the latter are such matters as continuity, differentiability, and convexity.

A.1   Basic terminology and notation from set theory

This section recalls some general definitions and notation from set theory. In Section A.4 we move on to consider particular kinds of sets; in the process, we necessarily draw certain functions into the discussion.

A *set* is a collection of objects belonging to some universe of discourse. The objects belonging to the collection (i.e., set) are said to be its *members* or *elements*. The symbol $\in$ is used to express the membership of an object in a set. If x is an object and A is a set, the statement that x is an element of A (or x belongs to A) is written $x \in A$. To negate the membership of x in the set A we write $x \notin A$. The set of all objects, in the universe of discourse, not belonging to A is called the *complement* of A. There are many different notations used for the complement of a set A. One of the most popular is A^c.

The *cardinality* of a set is the number of its elements. A set A can have cardinality zero in which case it is said to be *empty*. The empty set is sometimes called the *null set*; in either case it is denoted by the symbol $\emptyset$. The cardinality of a set can be finite or infinite. When the cardinality of a set is one, the set is called a *singleton*. If x is the one element of a singleton A, we write $A = \{x\}$.

If A and B are two sets and every element of A is an element of B, we call A a *subset* of B and write $A \subseteq B$. This notation allows for the possibility that A and B are the same set, but when A and B are not equal, we write

© Springer Science+Business Media LLC 2017
R.W. Cottle and M.N. Thapa, *Linear and Nonlinear Optimization*,
International Series in Operations Research & Management Science 253,
DOI 10.1007/978-1-4939-7055-1

$A \subset B$. There is an analogous notion of containment as well. In particular, when A is a subset of B, we say that B *contains* A and write $B \supseteq A$ or $B \supset A$ if the sets are not equal. The binary relations $\subset$ and $\supset$ are called *proper inclusion* and *proper containment*, respectively. By convention, the empty set is a subset of every set.

A set is customarily described by some criterion that its members must satisfy. The criterion is a statement involving a variable which is made true by each of the members of the set. This gives rise to expressions such as

$$A = \{x : \text{statement about } x \text{ is true}\}.$$

This is called *set-builder notation*.[1] For example, when we take R, the real number field, to be the universe of discourse, the nonnegative real numbers (often associated with the nonnegative part of the real number line) is

$$R_+ = \{x : x \in R \text{ and } x \geq 0\}.$$

As a set, R^n can be described as follows.

$$R^n = \{x : x = (x_1, \ldots, x_n), \text{ where } x_i \in R, \ i = 1, \ldots, n\}.$$

When we take R^n to be the universe of discourse, the corresponding *nonnegative orthant* can be expressed as

$$R^n_+ = \{x : x \in R^n \text{ and } x_i \geq 0, \ i = 1, \ldots, n\}.$$

This can also be put in a slightly different way, namely as

$$R^n_+ = \{x \in R^n : x_i \geq 0, \ i = 1, \ldots, n\}.$$

The set-builder notation is useful in defining the various operations on sets itemized below.

- The *union* of two sets, say A and B, is the set

$$A \cup B = \{x : x \in A \text{ or } x \in B\}.$$

---

[1]Some authors use a vertical bar instead of a colon, and whatever symbol is used, one sometimes sees the universe of discourse mentioned *before* that symbol. Thus, if U is the universe of discourse, a set A might be declared as $\{x \in U : \text{ statement about } x \text{ is true}\}$. But more often than not, the universe of discourse is clear from the context and is not even stated.

If A_ω is a set for each index ω belonging to a set Ω, their union is

$$\bigcup_{\omega\in\Omega} A_\omega = \{x : x \in A_\omega \text{ for some } \omega \in \Omega\}.$$

- The *intersection* of two sets, say A and B, is the set

$$A \cap B = \{x : x \in A \text{ and } x \in B\}.$$

If A_ω is a set for each index ω belonging to a set Ω, the intersection of all these sets is

$$\bigcap_{\omega\in\Omega} A_\omega = \{x : x \in A_\omega \text{ for all } \omega \in \Omega\}.$$

- As mentioned above, the *complement* of a set A is the set of objects in the universe of discourse that do not belong to A, that is

$$A^c = \{x : x \notin A\}.$$

- Given two sets, say A and B with B regarded as the universe of discourse, the *set-theoretic difference* of B and A is the complement of A relative to B. In set-builder notation, we have

$$B \setminus A = \{x : x \in B \text{ and } x \notin A\} = B \cap A^c.$$

- Another construct is the *symmetric difference* of two sets, say A and B. This is defined by

$$A \,\Delta\, B = (A \setminus B) \cup (B \setminus A).$$

- The *Cartesian product* of two (or more) sets, A and B is another important concept. This set consists of all ordered pairs of elements from A and B. Accordingly, we write

$$A \times B = \{(x,y) : x \in A \text{ and } y \in B\}.$$

(Note that when A and B are *distinct sets*, i.e. $A \neq B$, the sets $A \times B$ and $B \times A$ are likewise distinct.) The Cartesian product of sets can be iterated as in the definition of R^n with $n \geq 2$.

$$R^n = R^{n-1} \times R.$$

A.2   Norms and metrics

The importance of distance

The concept of the *distance* between objects arises frequently, both in practice and in theory (see [48]). We are all familiar with the measurement of length and of the ordinary distance between physical objects. Sometimes the "distance" between locations is reckoned in terms of travel time. One can also use the notion of "distance" to measure the "similarity" of objects. Optical character recognition is one of many types of distance measurement done in technological methods aimed at automatically identifying "objects." In this case, the objects would normally be printed or hand-written characters belonging to an alphabet. Such objects can be described by vectors of numbers representing chosen features of the character set. A similar approach can be applied to recognizing faces or fingerprints by various biometric measurements. In optimization, the measurement of distance from one vector to another vector is often needed. The distance between successive iterates in an algorithm, or the distance of a point from the origin are matters that come up repeatedly. A common kind of problem is that of minimizing the distance from a point to a set. For example, the question of whether the system $Ax = b$, $x \geq 0$ has a solution can be settled by determining the distance from the point (vector) b to the finitely generated cone $\text{pos}\,(A) = \{y : y = Ax,\ x \geq 0\}$. If the distance is positive, then $b \notin \text{pos}\,(A)$ and hence $Ax = b$, $x \geq 0$ has no solution. The converse of this statement is also true.

As we saw briefly in Chapter 8, distance can be calculated in various ways. Metrics are the tools that enable us to measure distance. However, because norms[2] are a fundamental concept used in defining metrics, we begin with an introduction to vector norms.

Norms on R^n

In general, vector norms measure the size of vectors in some linear space. For our purposes, the linear space of interest is R^n. A *norm* on R^n is a real-valued function φ with the following properties:

---

[2]There are vector norms and matrix norms.

1. $\varphi(x) \geq 0$ for all $x \in R^n$.

2. $\varphi(\alpha x) = |\alpha|\varphi(x)$ for all $\alpha \in R$, $x \in R^n$, where $|\alpha|$ is the absolute value of α, i.e., $|\alpha| = \max\{\alpha, -\alpha\}$.

3. $\varphi(x + y) \leq \varphi(x) + \varphi(y)$ for all $x, y \in R^n$.

If φ is a norm on R^n, then for any $x \in R^n$, $\varphi(x)$ is called the *norm* of x. Note that property 2 implies $\varphi(0) = 0$. It also implies that for every $x \in R^n$, the vectors x and $-x$ "have the same norm," that is, $\varphi(x) = \varphi(-x)$. Property 3 is called the *triangle inequality*.

The most commonly encountered vector norms are:

1. The 1-norm: $\|x\|_1 = \sum_{j=1}^{n} |x_j|$ also known as the *Manhattan norm*.

2. The 2-norm: $\|x\|_2 = \sqrt{\sum_{j=1}^{n} x_j^2}$ also known as the *Euclidean norm*.

3. The ∞-norm: $\|x\|_\infty = \max_{1 \leq j \leq n} |x_j|$.

Norms can also be defined on $R^{m \times n}$, that is, on real $m \times n$ matrices. In a formal sense, a matrix norm must possess the same three properties that a vector norm possesses. It is customary to use the same notation for both vector norms and matrix norms. Matrix norms are continuous functions $\|\cdot\|$ that possess the following properties:

1. $\|A\| \geq 0$, with equality if and only if $A = 0$.

2. $\|\alpha A\| = |\alpha| \, \|A\|$ for any scalar α.

3. $\|A + B\| \leq \|A\| + \|B\|$. (Triangle inequality.)

In the case of two matrices A and B for which the product AB is well defined, a further—so called *submultiplicative*—property is possessed by *some* matrix norms. The submultiplicative property holds if

$$\|AB\| \leq \|A\| \, \|B\| \qquad \text{for all } A \in R^{m \times q} \text{ and } B \in R^{q \times n}.$$

LINEAR AND NONLINEAR OPTIMIZATION

Golub and Van Loan [88, p. 56] exhibit a matrix norm that does not possess the submultiplicative property.

A matrix norm on $R^{m \times n}$ is said to be *induced* by given vector norms on R^m and R^n if it is defined as follows:

$$||A|| = \sup_{x \neq 0} \frac{||Ax||}{||x||}. \tag{A.1}$$

By virtue of the properties of vector norms, equation (A.1) in this definition is equivalent to

$$||A|| = \max_{||x||=1} ||Ax||. \tag{A.2}$$

The following are the most commonly used examples of matrix norms.

1. The 1-norm: $||A||_1 = \max\limits_{1 \leq j \leq n} \sum\limits_{i=1}^{m} |a_{ij}|.$

2. The 2-norm: $||A||_2 = (\lambda_{\max}[A^T A])^{1/2}$; see Section A.3.

3. The ∞-norm: $||A||_\infty = \max\limits_{1 \leq i \leq m} \sum\limits_{j=1}^{n} |a_{ij}|.$

4. The Frobenius-norm: $||A||_F = \left(\sum\limits_{i=1}^{m} \sum\limits_{j=1}^{n} a_{ij}^2 \right)^{1/2}.$

The 1-norm, 2-norm, and the ∞-norm are vector-induced norms. On the other hand, the Frobenius-norm is not a vector-induced norm (see Exercise A.3).

General definition of a metric

In general, a *metric* defined on a set S is a function $d : S \times S \longrightarrow R_+$ having the following properties:

1. Symmetry: $d(x, y) = d(y, x)$ for all x and $y \in S$.

2. Definiteness: $d(x, y) \geq 0$ with $d(x, y) = 0$ if and only if $x = y$.

3. Triangle inequality: $d(x, z) \leq d(x, y) + d(y, z)$ for all x, y, and $z \in S$.

A set S on which a metric d is defined is called a *metric space*. Note that the set S of a metric space need not be a linear space. In particular, it is not necessary for there to be any notion of addition or scalar multiplication among the elements of S as there is in a linear space. For example, when S is a nonempty set, the function $d : S \times S \rightarrow \{0, 1\}$ given by the rule

$$d(x, y) = \begin{cases} 0 & \text{if } x = y \\ 1 & \text{if } x \neq y \end{cases}$$

is a metric.

Equivalence of norms

Two norms, say $||\cdot||_p$ and $||\cdot||_q$ are said to be *equivalent* if there exist positive constants γ and δ such that

$$\gamma ||\cdot||_p \leq ||\cdot||_q \leq \delta ||\cdot||_p \quad \text{for all } x \in R^n.$$

It can be shown that for all $x \in R^n$

1. $||x||_2 \leq ||x||_1 \leq \sqrt{n} ||x||_2$

2. $||x||_\infty \leq ||x||_2 \leq \sqrt{n} ||x||_\infty$

3. $||x||_\infty \leq ||x||_1 \leq n ||x||_\infty$

These inequalities imply that the 1-norm, the 2-norm, and the ∞-norm are equivalent.

Metrics induced by norms

Using vector norms on R^n, we can define corresponding metrics. A metric on R^n is a function that measures the distance between two vectors, say x and y in R^n. The technique for obtaining the metric *induced* by a norm is the same in every case. If φ is a norm on R^n, the corresponding metric is $d(x, y) = \varphi(x - y)$. The fact that R^n is a linear space makes it meaningful to write the expression $x - y$ which, of course, is a vector in R^n.

The metrics induced by the specific norms defined above are

$$d_1(x, y) = ||x - y||_1,$$
$$d_2(x, y) = ||x - y||_2,$$
$$d_\infty(x, y) = ||x - y||_\infty.$$

Of these, the *Euclidean metric*, d_2, is by far the most often used, and when that meaning is understood, we often drop the subscript 2 from the notation. The linear space R^n "equipped" with the Euclidean metric goes by the name *Euclidean n-space*, often denoted E^n.

In more detail, these metrics look as follows. Given $x = (x_1, x_2, \ldots, x_n)$ and $y = (y_1, y_2, \ldots, y_n)$:

1. The ℓ_1 or "Manhattan" distance from x to y is

$$d_1(x, y) = \sum_{j=1}^{n} |x_j - y_j| = ||x - y||_1.$$

2. The *Euclidean distance* from x to y is

$$d_2(x, y) = \sqrt{\sum_{j=1}^{n} (x_j - y_j)^2} = ||x - y||_2.$$

3. The "max" distance from x to y is

$$d_\infty(x, y) = \max_{1 \le j \le n} |x_j - y_j| = ||x - y||_\infty.$$

In the Chebyshev problem one wants to minimize the ∞-norm of the difference between two vectors. The problem is modeled as a linear program (see page 35).

A.3 Properties of vectors and matrices

Eigenvalues and eigenvectors

The real number field is a subfield of the field of complex numbers (those of the form $a + ib$ where a and b are real numbers and $i^2 = -1$). For each complex number, $c = a + ib$, there is an associated ordered pair (a, b). The first component of this ordered pair is called the *real part* of c, whereas the second component is called the *imaginary part* of c. These components are sometimes denoted by writing $a = \Re(c)$ and $b = \Im(c)$, respectively.

Viewed this way, every real number a can be interpreted as a complex number whose "imaginary part" b equals zero. Unlike the real number field, the complex number field is *algebraically closed.* This means that every polynomial equation of degree n with complex coefficients has n (not necessarily distinct) roots. In fact, every such polynomial p is the product of n polynomials of degree one. Moreover, the complex roots occur in what are called "conjugate pairs," meaning that if $a + ib$ is a root of p, then so is $a - ib$.

Now, given a real (or complex) $n \times n$ matrix A, one can form the equation

$$Ax = \lambda x \tag{A.3}$$

in which the n-vector x and scalar λ are unknowns. If, together, $x \neq 0$ and λ satisfy (A.3), then x is called an *eigenvector* of A, and λ is called an *eigenvalue* of A. Furthermore, there is an associated polynomial equation, called the *characteristic equation* of A, namely

$$\det(A - \lambda I) = 0 \tag{A.4}$$

in which λ plays the role of an indeterminate. Over the complex field, this n-th degree polynomial equation (sometimes called the *secular equation of A*) has n roots, these being the eigenvalues of A. The eigenvalues are also known as *characteristic roots* or *latent roots*. If A is $n \times n$ and n is odd, then A must have at least one real eigenvalue.

It can also be shown that the following properties hold:

$$\sum_{i=1}^{n} \lambda_i = \sum_{i=1}^{n} a_{ii} \tag{A.5}$$

$$\prod_{i=1}^{n} \lambda_i = \det(A). \tag{A.6}$$

For *any* square matrix A, the sum of its diagonal elements is called its *trace.* Equation (A.5) says that the trace of such a matrix also equals the sum of its eigenvalues.

From (A.6) it follows that if the imaginary part of each eigenvalue of A is nonzero, then n is even and $\det(A) > 0$ since the product of nonzero conjugate pairs of complex numbers is positive. But more interesting from our perspective is the following important matrix-theoretic fact.

Theorem A.1   *All eigenvalues of a real symmetric matrix are real numbers.*

It follows that the 2-norm of the matrix A (see page 542) is well defined.

Condition number of a matrix

On a computer not all real numbers are represented exactly. This can be troublesome when solving a system of linear equations. Even though the errors may be small to start with, they can get magnified depending on the numerical properties of the matrix in the system being solved.

Perturbation theory is used to analyze how sensitive the computed solution is to small changes in the data. If it is "very sensitive" we say that the system is *ill-conditioned*. Note that this is not the same as stating that an algorithm is *unstable*, that is, one in which the errors made in the computation cannot be attributed to small errors in the data. In the case of ill-conditioning of the system to be solved, an algorithm may be very stable, yet the computed solution may differ substantially from the true answer. We illustrate the effect of ill-conditioning next.

Example A.1: ILL-CONDITIONING.   Consider the following two matrices.

$$A = \begin{bmatrix} 0.0001 & 1.0 \\ 1.0 & 1.0 \end{bmatrix} \quad \text{and} \quad B = \begin{bmatrix} 1.0 & 1.0 \\ 1.0 & 1.0001 \end{bmatrix}.$$

In this instance, the matrix A is well-conditioned and the matrix B is ill-conditioned (its rows are nearly identical). To see this, select two vectors that are only slightly different from each other: $b^1 = (2, 2.0001)$ and $b^2 = (2, 2.0002)$. Solving $Ax = b^1$ we obtain $x^1 = (0.0001, 2)$ and solving $Ax = b^2$ results in $x^2 = (0.0002, 2)$; there is not much change in the solutions. On the other hand the solution of $By = b^1$ is $y^1 = (1, 1)$, and that of $By = b^2$ is $y^2 = (0, 2)$; these are very different solutions. For the matrix B, small changes in the right-hand side lead to large changes in the solution. The reason for this is that B is ill-conditioned, as we shall see next.

To assess the conditioning of a nonsingular matrix A, we compute its *condition number* which we will denote by cond (A). This quantity has a simple form:

$$\text{cond}(A) = ||A|| \cdot ||A^{-1}||.$$

In order to derive the above number, we examine the impact of perturbations to the data when solving $Ax = b$. Consider the system of equations $Ax = b$, where A is a square matrix. The exact solution of this system is given by

$$x = A^{-1}b. \tag{A.7}$$

Now suppose that we perturb the right-hand side slightly to be $b + \delta b$, where the components of the vector δb are small numbers[3]. The exact solution $x + \delta x$ of the same system with a perturbed right-hand side $b + \delta b$ is

$$x + \delta x = A^{-1}(b + \delta b). \tag{A.8}$$

Thus, it follows that $\delta x = A^{-1}\delta b$. From (A.1) we obtain

$$||\delta x|| \leq ||A^{-1}|| \, ||\delta b|| \tag{A.9}$$

$$||b|| \leq ||A|| \, ||x||. \tag{A.10}$$

Combining these two inequalities, we get

$$\frac{||\delta x||}{||x||} \leq ||A|| \, ||A^{-1}|| \frac{||\delta b||}{||b||} = \text{cond}(A) \frac{||\delta b||}{||b||}. \tag{A.11}$$

From this we see that the condition number is essentially a magnification factor. Indeed, for any square matrix A, cond $(A) = ||A|| \, ||A^{-1}|| \geq ||I|| = 1$ for any vector-induced norm.

Example A.2: CONDITION NUMBERS. The inverses of the matrices A and B in Example A.1 are

$$A^{-1} = \begin{bmatrix} \dfrac{-1}{1 - 10^{-4}} & \dfrac{1}{1 - 10^{-4}} \\ \dfrac{1}{1 - 10^{-4}} & 1 - \dfrac{1}{1 - 10^{-4}} \end{bmatrix} \quad \text{and} \quad B^{-1} = \begin{bmatrix} 1 + 10^4 & -10^4 \\ -10^4 & 10^4 \end{bmatrix}.$$

---

[3]We could, of course, perturb A or both A and b; the results, while more complicated to derive will be similar. Since we are only interested in an estimate, the right magnitude of the condition number will suffice and we choose the simpler perturbation of the right-hand side for our analysis.

In terms of the 1-norm, their condition numbers are computed as

$$\|A\|_1 = 2, \qquad \|A^{-1}\|_1 = \frac{2}{1-10^{-4}}, \qquad \text{cond}(A) = \|A\|_1\|A^{-1}\|_1 \approx 4$$

$$\|B\|_1 = 2.0001, \quad \|B^{-1}\|_1 = 2.0001 \times 10^4, \quad \text{cond}(B) = \|B\|_1\|B^{-1}\|_1 \approx 4 \times 10^4.$$

The condition number of B is of the order 10^4 which is large. We saw the effect of this in Example A.1.

Orthogonality

Two vectors p, $q \in R^n$ are said to be *orthogonal* if $p^T q = 0$. A set of vectors $\{q_1, \ldots, q_n\} \in R^n$ are said to be *mutually orthogonal* if and only if

$$q_i^T q_j = 0 \text{ if } i \neq j \quad \text{and} \quad q_i^T q_j \neq 0 \text{ if } i = j. \tag{A.12}$$

The vectors are said to be *orthonormal* if they are orthogonal and

$$q_i^T q_j = 1 \text{ when } i = j. \tag{A.13}$$

Clearly every set of nonzero orthogonal vectors must be linearly independent.

An *orthogonal matrix* Q is a square matrix such that $Q^T Q = I$. This definition implies that for every orthogonal matrix Q, the equation $Q^{-1} = Q^T$ holds. Note that the columns of Q, regarded as a set of vectors, would then constitute an orthonormal set. Orthonormal matrices define linear transformations called *isometries* because they preserve Euclidean distances between points (see page 544). Orthogonal matrices are important from a numerical point of view because their condition number is 1, and they turn up in many matrix factorization schemes, as presented in [88].

A.4   Properties of sets in R^n

Sets and functions are ever-present and intertwined concepts in mathematical work. The following discussion illustrates this point.

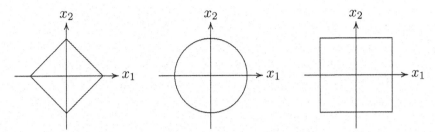

Unit circle in 1-norm   Unit circle in 2-norm   Unit circle in ∞-norm

Figure A.1: Unit circles in various norms.

Balls, spheres, and neighborhoods

Let r be a positive real number and let $c \in R^n$. If $\|\cdot\|$ is a norm on R^n, then an *open ball* of radius r centered at c is

$$\mathcal{B}(c,r) = \{x : \|x - c\| < r\}.$$

A (closed) *ball* of radius r centered at c is

$$\overline{\mathcal{B}}(c,r) = \{x : \|x - c\| \leq r\}.$$

The set

$$\mathcal{S}(c,r) = \{x : \|x - c\| = r\}$$

is the *sphere* of radius r centered at c. In the case of $n = 2$, these three kinds of sets are called *open disks*, *closed disks* and *circles*, respectively.

This terminology is inspired by the everyday language based on the 2-norm; nevertheless, it is readily transferred to the analogous structures related to other norms. Figure A.1 illustrates the idea.

A ball centered at a point $\bar{x}$ is called a *neighborhood* of that point.[4] This concept is particularly important in the case where the ball is open and the radius is small.

Bounded and unbounded sets

A set S that lies within a ball (either open or closed) of some finite radius is said to be *bounded*. If there is no such ball containing the set S, it is said to

---

[4]In topology, one considers neighborhoods of a more general nature.

be *unbounded*. There are countless examples of bounded sets and unbounded sets in the context of convex polyhedra alone. As a reminder, we mention that all nonzero cones (including, in particular R^n and R^n_+) are unbounded whereas the feasible regions of transportation problems are bounded polyhedra (polytopes). Naturally the terms bounded and unbounded apply to many other kinds of sets. Two of these are given in the example below.

Example A.3: MORE BOUNDED AND UNBOUNDED SETS. The set

$$S_1 = \{x \in R^2_+ : x_1 \leq 1, \ x_2 \leq (x_1 - 1)^2\}$$

is bounded. It is contained in the closed ball of radius 1 centered at the origin. On the other hand, the set

$$S_2 = \{x \in R^2_+ : x_2 \leq e^{-x_1}\}$$

is not bounded because it contains points $x = (x_1, e^{-x_1})$ having arbitrarily large coordinate x_1.

Open sets and closed sets in R^n

A set $S \subseteq R^n$ is said to be *open* if for each $x \in S$, there is an open ball centered at x that is entirely contained in S. Notice that R^n is clearly open. A set is said to be *closed* if its complement is open. By convention, the empty set is both open and closed, hence so is R^n.

Example A.4: A CLOSED SET AND ITS COMPLEMENT. The set

$$S = \{x \in R : -1 \leq x \leq 1\}$$

is closed. Its complement is

$$\{x \in R : x < 1\} \ \cup \ \{x \in R : x > 1\},$$

which is the union of two open sets.

More generally, the solution set of a linear inequality $a^{\mathrm{T}}x \leq \beta$ is a closed set, and its complement $\{x : a^{\mathrm{T}}x > \beta\}$ is open. Consequently, all polyhedral sets $\{x : Ax \leq b\}$ are closed.

Note that if $\bar{x}$ is any element of an open set, S, and p is any element of R^n, then $\bar{x} + \theta p$ belongs to S for sufficiently small $\theta \in R$. In other words, it is always possible to move away from $\bar{x}$ in any direction and still remain in S. See Exercise A.9.

It should be clear that *the union of a family of open sets is open and the intersection of a family of closed sets is closed*:

$$S_\alpha \text{ open for all } \alpha \implies \bigcup_\alpha S_\alpha \text{ open}$$

and

$$S_\alpha \text{ closed for all } \alpha \implies \bigcap_\alpha S_\alpha \text{ closed}.$$

In these two statements, the family in question need not be finite, but when it is finite, one can also prove the following assertions.

The intersection of finitely many open sets in R^n is open, and the union of finitely many closed sets in R^n is closed.

See Exercise A.10.

Limit points

Let $S \subseteq R^n$. Then $\bar{x} \in R^n$ is a *limit point* of S if every open ball centered at $\bar{x}$ contains an element of S. Intuitively, this means that there are points of S arbitrarily close to $\bar{x}$. Limit points of S need not belong to S, but when all of them do, the set S must be closed. Conversely, a closed set contains all of its limit points.

In optimization, the sets S referred to in the preceding paragraph are typically *sequences* of points in R^n or possibly real numbers (in which case $n = 1$). By such sequences, we mean sets whose elements are in one-to-one correspondence with the set of *natural numbers* $N = \{k : k = 1, 2, \ldots\}$. A sequence of points in R^n could be denoted $\{x_k : k \in N\}$, but it is more customary to write it as $\{x_k\}$ although it is important to distinguish the sequence from the singleton, x_k. This should always be clear from the context in which the symbol is used.

Closure of a set

For a set $S \subseteq R^n$, let S' denote the set of all limit points of S. Then

$$\overline{S} = S \cup S'$$

is known as the *closure* of S. The notation $\operatorname{cl} S$ is also used for the closure of S.

Our use of the notation $\overline{\mathcal{B}}(c, r)$ for the set $\{x : \|x - c\| \leq r\}$ would suggest that $\overline{\mathcal{B}}(c, r)$ is the closure of $\mathcal{B}(c, r)$, and indeed it is.

A set $S \subseteq R^n$ is closed if and only if it equals its closure.

Boundary of a set

Let $S \subset R^n$. A point $\bar{x} \in S$ is a *boundary point* of S if every neighborhood of $\bar{x}$ contains an element of S and an element of the complement of S. The *boundary* of a set S is the set of all boundary points of S. We denote the boundary of S by $\operatorname{bdy} S$; some authors use ∂S to denote the boundary of S.

As an example of the boundary of a set, consider the closed ball $\overline{\mathcal{B}}(c, r)$ in R^n. In this case $\operatorname{bdy} \overline{\mathcal{B}}(c, r) = \{x : \|x - c\| = r\}$, which is just the sphere of radius r centered at c.

Compact sets

An important property of *some* (but by no means all) sets in R^n is that of compactness. A set $S \subset R^n$ is defined to be *compact* if it is both closed and bounded. For instance, when $n = 1$, the closed, bounded intervals are compact sets. A closed subset of a compact set is clearly compact, hence the intersection of any number of compact sets is compact. There are many situations in which the compactness of a set turns out to be a key element in justifying an assertion. An example of such an assertion is Weierstrass's Theorem found on page 554.

Interior of a set

The *interior* of a set $S \subseteq R^n$ is the set of points $x \in S$ that are not contained in bdy S. Such points belong to open neighborhoods that are entirely contained in S. We denote the interior of a set S by int S. A set S for which int $S \neq \emptyset$ is said to be *solid*.

Relative neighborhoods and related concepts

A set having no interior points may still have something like interior points. A line segment in Euclidean n-space with $n \geq 2$ will serve to illustrate the idea. Consider the line L that contains the given line segment. We may think of this line L as 1-dimensional space R^1. For a point $\bar{x}$ in the given line segment, consider the intersection of L with an open neighborhood containing $\bar{x}$. That intersection is called a *relatively open neighborhood* of $\bar{x}$. The concepts of relative boundary and relative interior can be built around those of boundary and interior. For an example of why the concept of relative interior is of interest, see the discussion of continuity of convex functions on page 568.

In general, the definition of a relatively open neighborhood of a point in a set $S \subset R^n$ relies on the concept of the *carrying plane*[5] of S. This is the affine set of least dimension that contains S. (Such a thing will be a translate of a linear subspace of R^n.)

With the carrying plane of S denoted by $V(S)$, we obtain the relatively open neighborhoods of points in S by considering the intersection of $V(S)$ with all open neighborhoods of the points of S. The collection of all these relatively open neighborhoods that also lie within S constitutes the relative interior of S, denoted rel int S.

In the little example shown in Figure A.2, the points of the line segment that lie strictly between its endpoints a and b constitute the relative interior. The endpoints of the line segment make up its relative boundary. As a second example, consider a closed disk in R^3. The interior of a such a set is empty, but it does have a nonempty relative interior.

---

[5]The carrying plane of S is also called the *affine hull* of S which is defined as the intersection of all affine sets containing S.

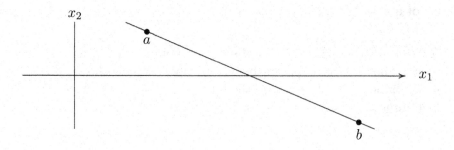

Figure A.2: Relative interior and relative boundary of a line segment.

A.5   Properties of functions on R^n

In this section we briefly review of some concepts from multivariate differential calculus. Most of this material should be familiar.

Continuity of functions

A function $f : S \rightarrow R$ is *continuous* at the point $\bar{x} \in S$ if for every $\varepsilon > 0$ there exists a $\delta > 0$ such that

$$\|x - \bar{x}\| < \delta \Longrightarrow |f(x) - f(\bar{x})| < \varepsilon.$$

The function f is continuous on S if it is continuous at every point $\bar{x} \in S$. The function f is continuous on S if and only if for every convergent sequence $\{x^k\}$ in S

$$\lim_{k \to \infty} f(x^k) = f\left(\lim_{k \to \infty} x^k\right).$$

The following much-used theorem of Weierstrass[6] combines the property of continuity with that of compactness to guarantee the existence of maxima and minima.

Theorem A.2 (Weierstrass)   *If $f : S \longrightarrow R$ is a continuous function and $S \subset R^n$ is a compact set, then the function f attains its maximum and minimum values on S.*

---

[6]Karl Theodor Wilhelm Weierstrass (1815–1897) is known as the father of modern analysis. The sketch of Weierstrass's life, including his contributions, [59, p. 599] makes fascinating reading.

This means, there exist elements $\bar{x}$ and $\bar{y} \in S$ such that

$$f(\bar{x}) \geq f(x) \text{ for all } x \in S \quad \text{and} \quad f(\bar{y}) \leq f(y) \text{ for all } y \in S.$$

The power of this theorem is the assertion that the set S *contains* the elements $\bar{x}$ and $\bar{y}$ which maximize and minimize the value of f on S.

Partial derivatives and the gradient vector

Let f be a real-valued function defined on the open set S. If, at $\bar{x} \in S$, the limit

$$\lim_{\delta \to 0} \frac{f(\bar{x} + \delta I_{\bullet i}) - f(\bar{x})}{\delta}$$

exists, we call it the *partial derivative* of f at $\bar{x}$ with respect to x_i and denote the limit as

$$\frac{\partial f(\bar{x})}{\partial x_i}.$$

The vector

$$g(x) = \left(\frac{\partial f(x)}{\partial x_1}, \frac{\partial f(x)}{\partial x_2}, \dots, \frac{\partial f(x)}{\partial x_n} \right)$$

is called the *gradient* of f at $\bar{x}$ and, in addition to the above, is variously denoted $\nabla f(\bar{x})$ or grad $f(\bar{x})$, or $f'(\bar{x})$.

Differentiability of functions

Let $f : S \longrightarrow R$ where S is an open subset of R^n. Then f is *differentiable* at $\bar{x} \in S$ if for every $x \in R^n$ such that $\bar{x} + x \in S$,

$$f(\bar{x} + x) = f(\bar{x}) + g(\bar{x})^{\mathrm{T}} x + \alpha(\bar{x}, x) \|x\| \qquad (A.14)$$

where the gradient $g(\bar{x})$ is bounded and α is a real-valued function such that

$$\lim_{x \to 0} \alpha(\bar{x}, x) = 0. \qquad (A.15)$$

If f is differentiable at every point of an open set, it is said to be differentiable on that set.[7]

---

[7]In 1872, Karl Weierstrass gave an example of a function that is continuous and *nowhere differentiable*. Such things are now called *Weierstrass functions*.

Differentiable functions have gradients. If $f : S \longrightarrow R$ is differentiable at $\bar{x} \in S$, then f is continuous at $\bar{x}$ and $\nabla f(\bar{x})$ exists. The existence of partial derivatives at a point $\bar{x}$ does *not* imply differentiability there.

A sufficient condition for differentiability. A function $f : S \longrightarrow R$ having continuous partial derivatives at the point $\bar{x} \in S$ is differentiable at $\bar{x}$.

Example A.5: NONDIFFERENTIABLE FUNCTION HAVING A GRADIENT.
Consider the real-valued function f defined on R^2 by

$$f(x_1, x_2) = \begin{cases} x_1 & \text{if } |x_1| > |x_2| \\ -x_1 & \text{otherwise} . \end{cases}$$

From the definition of partial derivative, we see that

$$\nabla f(0) = \left(\frac{\partial f(0)}{\partial x_1}, \frac{\partial f(0)}{\partial x_2} \right) = (1, 0).$$

Nevertheless, the function f is not differentiable at $\bar{x} = 0$. The reason has to do with how $x \to 0$ in the limit process (A.15). Indeed, if f were differentiable, and we let x approach the origin along the halfline where $0 < x_1 = x_2$, we would have in (A.14)

$$f(x_1, x_1) = -x_1$$
$$= x_1 + x_1 \cdot 1 + x_1 \cdot 0 + \alpha(x_1, x_1) \cdot x_1.$$

Equating the two right-hand sides here, dividing through by x_1, and then letting x_1 go to zero, we would obtain a contradiction. Consequently, the function f cannot be differentiable, even though it has a gradient. This example is based on [190, Example H, p. 16].

Twice-differentiable functions

Consider $f : S \longrightarrow R$ where S is an open subset of R^n; if $\bar{x} \in S$, then f is said to be *twice-differentiable* at $\bar{x}$ if

$$f(\bar{x} + x) = f(\bar{x}) + \nabla f(\bar{x})^{\mathrm{T}} x + \tfrac{1}{2} x^{\mathrm{T}} \nabla^2 f(\bar{x}) x + \beta(\bar{x}, x)(\|x\|)^2$$

where $\nabla^2 f(\bar{x})$ is a bounded matrix and β is a real-valued function such that

$$\lim_{x \to 0} \beta(\bar{x}, x) = 0.$$

The Hessian matrix of second partial derivatives

The symbol $\nabla^2 f(\bar{x})$ denotes the $n \times n$ (Hessian) matrix of second partial derivatives of f at $\bar{x}$, assuming these derivatives exist.

$$\nabla^2 f(\bar{x}) = \left[\frac{\partial^2 f(\bar{x})}{\partial x_i \partial x_j} \right] \quad \text{(Hessian)}.$$

In line with the common nonlinear programming practice of denoting the Hessian matrix by $G(x)$, we write the ij-th element of the Hessian as

$$G_{ij}(x) = \frac{\partial^2 f(x)}{\partial x_i \, \partial x_j}.$$

Theorem A.3 Hessian symmetry   *When f is a twice-continuously differentiable function, its Hessian matrix is symmetric, that is,*

$$\frac{\partial^2 f(\bar{x})}{\partial x_i \partial x_j} = \frac{\partial^2 f(\bar{x})}{\partial x_j \partial x_i} \quad \text{for all } i, j.$$

Continuity of the second derivatives is essential for the theorem to be true. An example of twice-differentiable function having an *asymmetric* Hessian matrix is given in [103, p. 120].

Example A.6: A GRADIENT AND HESSIAN.   Let $f(x) = e^{x_1 x_2 x_3}$, where e denotes the base of the natural logarithm, that is, $e = 2.7182818284\ldots$. Then

$$g(x) = e^{x_1 x_2 x_3}(x_2 x_3, \quad x_1 x_3, \quad x_1 x_2)$$

and

$$G(x) = e^{x_1 x_2 x_3} \begin{bmatrix} x_2^2 x_3^2 & x_3 + x_1 x_2 x_3^2 & x_2 + x_1 x_2^2 x_3 \\ x_3 + x_1 x_2 x_3^2 & x_1^2 x_3^2 & x_1 + x_1^2 x_2 x_3 \\ x_2 + x_1 x_2^2 x_3 & x_1 + x_1^2 x_2 x_3 & x_1^2 x_2^2 \end{bmatrix}.$$

Note that in writing the gradient and Hessian of f, we have factored out the function $e^{x_1 x_2 x_3}$.

Partial derivatives of higher order

Systematic repeated application of the partial derivative operator $\partial f(\bar{x})/\partial x_j$ allows one to define partial derivatives of higher order, provided, of course, that the required limits exist. We shall not get into this subject here except to note that a k-th order partial derivative of f would be of the form

$$\frac{\partial^k f}{\partial x_{j_1} \partial x_{j_2} \cdots \partial x_{j_k}}$$

where the values of the subscripts of the variables $x_{j_1}, \ldots, x_{j_k}$ need not be distinct.

Just as continuity is a desirable property of functions, so too is the continuity of their derivatives. There is a handy notation C^k often used to express the continuity of a function or its derivatives where k is a nonnegative integer. Thus, the statement $f \in C^0$ at $\bar{x}$ is shorthand for the statement "the function f is continuous at the point $\bar{x}$." For an integer $k \geq 1$, the statement $f \in C^k$ at $\bar{x}$ conveys the statement "the function f has continuous partial derivatives of orders 1 through k at the point $\bar{x}$.

In the literature, one finds the adjective *smooth* used in the context of functions possessing continuous derivatives "of higher order." Just what is meant by higher order is sometimes omitted from the discussion, but normally, it means that the function belongs to C^k, $k \geq 2$.

Quadratic forms and functions

For $H \in R^{n \times n}$, the function

$$f(x) = x^{\mathrm{T}} H x \quad (x \in R^n)$$

is called a *quadratic form* on R^n. Notice that

$$x^{\mathrm{T}} H x = \tfrac{1}{2} x^{\mathrm{T}} (H + H^{\mathrm{T}}) x$$

for all x; moreover the matrix $\frac{1}{2}(H + H^{\mathrm{T}})$ is symmetric, even if H is not. The matrix $\frac{1}{2}(H + H^{\mathrm{T}})$ is called the *symmetric part* of H.

A *quadratic function* is the sum of a quadratic form and an affine function. Thus, a quadratic function has (or can be brought to) the form

$$x^{\mathrm{T}} H x + c^{\mathrm{T}} x + d$$

where $H \in R^{n \times n}$, $c \in R^n$, and $d \in R$.

It is worth noting that when the function f is a quadratic form $x^{\mathrm{T}}Hx$ where H is a symmetric $n \times n$ matrix and $x \in R^n$, then the corresponding gradient and Hessian at x are given by $g(x) = 2Hx$ and $G(x) = 2H$. There are two observations to be made here. The first—and more important—is that in the case of a quadratic form, the Hessian matrix is a *constant*: its entries do not depend on x. The second point is that it is fairly customary to write a quadratic form as

$$f(x) = \tfrac{1}{2}x^{\mathrm{T}}Hx$$

so that when the gradient and Hessian are taken, the factor 2 gets multiplied by $\tfrac{1}{2}$ and does not show up in the expression for the gradient or the Hessian because the product is 1 (one). That is,

$$g(x) = Hx \quad \text{and} \quad G(x) = H.$$

Example A.7: A QUADRATIC FUNCTION. Consider the quadratic function

$$f(x) = 3x_1^2 - 8x_1x_2 + 7x_2^2 + 2x_1 - 3x_2 + 9.$$

This can be written in the form

$$f(x) = \tfrac{1}{2}x^{\mathrm{T}}Hx + c^{\mathrm{T}}x + d$$

by defining H, c, and d as follows:

$$H = \begin{bmatrix} 6 & -8 \\ -8 & 14 \end{bmatrix}, \quad c = \begin{bmatrix} 2 \\ -3 \end{bmatrix}, \quad d = 9.$$

For any $x \in R^2$, the gradient and Hessian matrix are:

$$g(x) = \begin{bmatrix} 6 & -8 \\ -8 & 14 \end{bmatrix} x + \begin{bmatrix} 2 \\ -3 \end{bmatrix} \quad \text{and} \quad G = \begin{bmatrix} 6 & -8 \\ -8 & 14 \end{bmatrix}.$$

Notice that the gradient is $g(x) = Gx + c$.

Positive semidefinite and positive definite matrices and quadratic forms

The matrix $H \in R^{n \times n}$ and the associated quadratic form $x^\mathrm{T}Hx$ are called *positive semidefinite* if

$$x^\mathrm{T}Hx \geq 0 \quad \text{for all } x \in R^n.$$

They are called *positive definite*[8] if

$$x^\mathrm{T}Hx > 0 \quad \text{for all } x \in R^n, \ x \neq 0.$$

The negatives of positive definite and positive semidefinite matrices are called *negative definite* and *negative semidefinite*, respectively.[9]

There are various ways of testing for the positive semidefiniteness or positive definiteness of a real symmetric matrix, H.

1. One of these is through matrix factorization. If $H = XX^\mathrm{T}$ for some matrix X, then H is positive semidefinite. If $H = XX^\mathrm{T}$ and the rows of X are linearly independent, then H is positive definite.[10]

2. Another is through its eigenvalues. If all the eigenvalues of H are non-negative (positive), then H is positive semidefinite (positive definite).

3. Finally, a real symmetric matrix H is positive definite if and only if the *leading* principal minors of H are all positive. The analogous statement for positive semidefiniteness is false as shown by the matrix

$$H = \begin{bmatrix} 0 & 0 & 0 \\ 0 & 0 & 1 \\ 0 & 1 & 0 \end{bmatrix}. \tag{A.16}$$

The leading principal minors of H are

$$\det([0]) = 0, \ \det\left(\begin{bmatrix} 0 & 0 \\ 0 & 0 \end{bmatrix}\right) = 0, \ \det(H) = 0.$$

[8]The definitions of positive semidefinite and positive definite make sense even if the matrix H is not symmetric.

[9]Some writers use the term *indefinite* for a square matrix that is *not* positive definite. But others apply this term to square matrices that are neither positive semidefinite nor negative semidefinite, i.e., whose associated quadratic forms take on both positive and negative values. Thus, in reading the literature, it is definitely important to be clear about the author's definition of the term "indefinite".

[10]The converses of these last two statements are true as well.

There is, however, a criterion for positive semidefiniteness in terms of principal minors, namely that a symmetric matrix is positive semidefinite if and only if *all* of its principal minors—not just leading principal minors—are nonnegative.

For matrices of large order, the computation of eigenvalues or the determination of principal minors is computationally time-consuming and so the latter two approaches are not used in practice. Instead, the determination of positive definiteness and positive semidefiniteness is done through a factorization of the matrix.

A.6 Convex sets

Recall that a set $S \subseteq R^n$ is *convex* if it contains the line segment between every pair of its points. By convention, the empty set is convex. A singleton (i.e., a set consisting of a single point) is also convex. This makes it possible to assert that the intersection of any two convex subsets of R^n is another convex set. As we know, every polyhedral set (i.e., the solution set of a linear inequality system) is convex. Many convex sets are not polyhedral. Balls (either open or closed) are examples of such convex sets.

One may ask: Why is a ball a convex set? This can be seen as a consequence of the *Cauchy-Schwartz inequality*; see [66] and [103], for example. It says that in the Euclidean- or 2-norm

$$|x^{\mathrm{T}}y| \leq \|x\|\|y\| \qquad \text{for all } x,\, y \in R^n \tag{A.17}$$

with equality if and only if x and y are linearly dependent. Incidentally, another relationship of this sort is the formula

$$x^{\mathrm{T}}y = \|x\|\|y\| \cos\theta,$$

where θ denotes the angle between the vectors x and y in R^n.

For any set $S \subseteq R^n$ and vector $a \in R^n$, the set of points

$$S + \{a\} = \{t \,:\, t = s + a,\ s \in S\}$$

is called the *translate* of S. It is quite easy to see that any translate of a convex set is convex, so in particular, if we are given a ball of radius ρ

centered at c, we may translate the ball so that it is centered at the origin. This will simplify the notation to follow. Now let x and y be arbitrary elements of the ball.[11] Our ball is the set of points whose 2-norm squared is less than or equal to ρ^2. Thus, $\|x\|^2 \leq \rho^2$ and $\|y\|^2 \leq \rho^2$. Let $\lambda \in (0,1)$ be arbitrary. We need to verify that the norm of the convex combination $\lambda x + (1-\lambda)y$ is at most ρ or equivalently that $\|\lambda x + (1-\lambda)y\|^2 \leq \rho^2$. Thus,

$$
\begin{aligned}
\|\lambda x + (1-\lambda)y\|^2 &= (\lambda x + (1-\lambda)y)^{\mathrm{T}}(\lambda x + (1-\lambda)y) \\
&= \lambda^2 x^{\mathrm{T}}x + 2\lambda(1-\lambda)x^{\mathrm{T}}y + (1-\lambda)^2 y^{\mathrm{T}}y \\
&\leq \lambda^2 x^{\mathrm{T}}x + 2\lambda(1-\lambda)|x^{\mathrm{T}}y| + (1-\lambda)^2 y^{\mathrm{T}}y \\
&\leq \lambda^2 \|x\|^2 + 2\lambda(1-\lambda)\|x\|\|y\| + (1-\lambda)^2 \|y\|^2 \\
&= (\lambda\|x\| + (1-\lambda)\|y\|)^2 \\
&\leq (\lambda\rho + (1-\lambda)\rho)^2 \\
&= \rho^2.
\end{aligned}
$$

A much simpler argument in terms of functions for the convexity of a ball (and related sets) will be given as Exercise A.8 at the end of the Appendix.

Separation and support of convex sets

A hyperplane $\mathcal{H} = \{x : p^{\mathrm{T}}x = \alpha\}$ is said to *separate* S_1 and S_2, two nonempty convex subsets of R^n, if $p^{\mathrm{T}}x \leq \alpha$ for all $x \in S_1$ and $p^{\mathrm{T}}x \geq \alpha$ for all $x \in S_2$. The intention is to say that S_1 and S_2 lie on opposite sides of $\mathcal{H}$; allowing the inequalities to hold weakly (i.e., as equations) results in a rather weak property. For that reason, it is appropriate to distinguish three more powerful types of separation as defined below:

(i) $\mathcal{H}$ separates S_1 and S_2 *properly* if $S_1 \cup S_2 \not\subseteq \mathcal{H}$. This would still allow *one* of the two sets to lie in $\mathcal{H}$.

(ii) $\mathcal{H}$ separates S_1 and S_2 *strictly* if $p^{\mathrm{T}}x < \alpha$ for all $x \in S_1$ and $p^{\mathrm{T}}x > \alpha$ for all $x \in S_2$. This does not prevent either of the sets S_1 and S_2 from coming arbitrarily close to $\mathcal{H}$.

---

[11] We may assume the ball is closed. The argument for open balls is similar.

(iii) $\mathcal{H}$ separates S_1 and S_2 *strongly* if there exists a number $\varepsilon > 0$ such that $p^{\mathrm{T}}x < \alpha + \varepsilon$ for all $x \in S_1$ and $p^{\mathrm{T}}x > \alpha - \varepsilon$ for all $x \in S_2$. This would allow the *slab* $\mathcal{H}(\varepsilon) = \{x : \alpha - \varepsilon \le p^{\mathrm{T}}x \le \alpha + \varepsilon\}$ to fit between S_1 and S_2.

Example A.8: FARKAS'S LEMMA AS A STRONG SEPARATION THEOREM.
A good example of the strong separation of convex sets can be gleaned from a geometric interpretation of Farkas's Lemma (see page 125). It says, in effect, that if the system $Ax = b$, $x \ge 0$ has no solution, then there exists a vector p such that $p^{\mathrm{T}}A \ge 0$ and $p^{\mathrm{T}}b < 0$. Now $\mathrm{pos}\,(A) = \{Ax : x \ge 0\}$ is a nonempty closed convex cone. Call this S_1. The singleton $\{b\}$ is a compact convex set; call it S_2. The assumption that the given system has no solution means that S_1 and S_2 are disjoint. The conclusion of Farkas's Lemma implies that S_1 and S_2 are strongly separated by a hyperplane $\mathcal{H}$. The vector p gives the normal direction of the (strongly) separating hyperplane. The position of $\mathcal{H}$ is then determined by a constant α which can be chosen as any number in the open interval $(p^{\mathrm{T}}b, 0)$. This observation is a special case of the theorem stated below.

Theorem A.4   *If $S \subset R^n$ is a nonempty closed convex set and $y \in R^n \setminus S$, there exists a unique point $\bar{x} \in S$ such that*

$$\|\bar{x} - y\| \le \|x - y\| \quad \text{for all } x \in S.$$

This can be seen from the following argument. Choose an arbitrary point $\hat{x} \in S$ and define $\delta = \|\hat{x} - y\|$. The ball

$$\bar{B}(y, \delta) = \{x \in R^n : \|x - y\| \le \delta\}$$

is compact and must contain a point $\bar{x}$ that minimizes $\|x - y\|$ over the set $\bar{B}(\bar{x}, \delta) \cap S$ which is compact.[12] The point $\bar{x}$ is unique because it minimizes the strictly convex function $\|x - y\|$. The point $\bar{x}$ is the *projection* of y onto S.

---

[12]The intersection of a closed set and a compact set is compact. The existence of the minimum here follows from Weierstrass's Theorem.

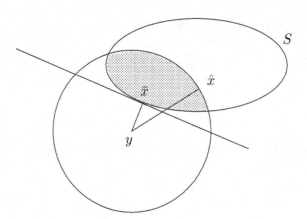

Figure A.3: Projection of y onto closed convex set S.

Theorem A.5 Separating hyperplane   *If S is a nonempty closed convex set and $y \notin S$, then there exists a vector p and a real number α such that $p^{\mathrm{T}}y > \alpha$ and $p^{\mathrm{T}}x \le \alpha$ for all $x \in S$.*

To see this, take $p = y - \bar{x}$ and $\alpha = \sup\{\, p^{\mathrm{T}}x : x \in S \,\}$ as above.

A more general result states that *if S_1 and S_2 are nonempty closed convex sets in R^n, at least one of which is bounded (and hence compact), there exists a hyperplane that strongly separates them.*

The dual of a set

Let S be a nonempty subset of R^n. The *dual*[13] of S is the set

$$S^* = \{y : y \cdot x \ge 0 \ \text{ for all } x \in S\}.$$

Every nonzero element of S^* makes a nonobtuse (i.e., acute or right) angle with every element of S. It is obvious that S^* is a cone. It is not difficult to show that S^* is also closed and convex.

---

[13]The concept defined here is one of the least standardized to be found in the literature of optimization. Most—but not all—authors apply it to cones rather than arbitrary sets. While most authors use a linear inequality system such as $y \cdot x \ge 0$ for all $x \in S$, many state the system as $y \cdot x \le 0$ for all $x \in S$. And, while many refer to the solution set of whatever conditions they impose as the *dual* of the relevant set S, many others call it the *polar* of the set. Some authors make a distinction between the dual and the polar. In addition to this diversity of statements, there is a variety of notations in use; the most common are S^* and S° (usually for the polar).

A question arises: When does S equal S^{**}, the dual of S^*? It is clear from the above discussion that such a thing can be true only if S is a closed convex cone. (Interest in the conditions under which $S = S^{**}$ probably explains why the set S is often taken to be a cone right from the start.) It can be shown that this is also a sufficient condition for S to be the same set as S^{**}. The argument for this assertion makes use of the separating hyperplane theorem.

Supporting hyperplanes

Let S be a nonempty subset of R^n, and let $\bar{x} \in \text{bdy } S$. Then the hyperplane

$$\mathcal{H} = \{\, x : p^{\mathrm{T}}(x - \bar{x}) = 0 \,\}$$

is a *supporting hyperplane* at $\bar{x}$ if $S \subset \mathcal{H}^+$ or $S \subset \mathcal{H}^-$ where

$$\mathcal{H}^+ = \{\, x : p^{\mathrm{T}}(x - \bar{x}) \geq 0 \,\} \quad \text{and} \quad \mathcal{H}^- = \{\, x : p^{\mathrm{T}}(x - \bar{x}) \leq 0 \,\}.$$

Proper support requires that $S \not\subset \mathcal{H}$. If $\mathcal{H}$ properly supports S at $\bar{x} \in \text{bdy } S$ the set $\mathcal{H}^+$ or $\mathcal{H}^-$ that contains S is called a *supporting halfspace*. The following is a much-used result.

Theorem A.6 Supporting hyperplane   *If S is a nonempty convex set in R^n and $\bar{x} \in \text{bdy } S$, then there exists a hyperplane that supports S at $\bar{x}$, i.e., there exists a vector $p \neq 0$ such that*

$$p^{\mathrm{T}}(x - \bar{x}) \leq 0 \quad \text{for all } x \in \text{cl } S.$$

Once p is known, the supporting hyperplane is

$$\mathcal{H} = \{x : p^{\mathrm{T}}x = p^{\mathrm{T}}\bar{x}\}.$$

Convex sets are characterized by their supporting halfspaces. A case in point is the following. (See [167, p. 99].)

Theorem A.7 Closure   *The closure of every proper convex subset of R^n is the intersection of its supporting halfspaces.*

In some cases, there is only one supporting hyperplane at a point $\bar{x}$; it should be noted, however, that in other cases there can be infinitely many hyperplanes that support a convex set at a particular boundary point $\bar{x}$. An easily understood instance of this sort arises at an extreme point of a polyhedron. In Figure A.4, S has unique supporting hyperplanes at b and d but not at a, c, and e.

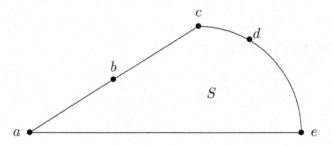

Figure A.4: Points with unique/non-unique supporting hyperplanes.

A.7   Properties of convex functions

In this book, convex functions were defined in Chapter 8 (see, in particular, page 235). Here we collect several useful properties of convex functions. Most of these are quite easily proved; nonetheless, we offer some references where proofs can be found. There are analogous results for concave functions, but we leave most of them unstated.

Convexity of the epigraph

One of the criteria for convexity of functions involves a concept which relates the function to a set in a higher-dimensional space. Let $f : S \to R$ be a real-valued function defined on the nonempty convex set $S \subseteq R^n$. The set

$$\text{epi } f = \{(x, \alpha) : x \in S \text{ and } \alpha \geq f(x)\}$$

is called the *epigraph* of f. It now follows that *the function f is convex on S if and only if its epigraph is a convex set*. The corresponding construct for concave functions is called the *hypograph*. The hypograph of f is the set of points

$$\text{hyp } f = \{(x, \alpha) : x \in S \text{ and } \alpha \leq f(x)\}.$$

Then f *is concave on S if and only if its hypograph is convex.*

Viewing the convexity of functions through their epigraphs is particularly useful in dealing with convex optimization. Indeed, for a problem such as minimize $f(x)$ subject to $x \in S$, the feasibility of a vector can be expressed through the formalism

$$\bar{f}(x) = \begin{cases} f(x) \text{ if } x \in S \\ +\infty \text{ otherwise.} \end{cases}$$

Note that $\bar{f}$ extends the set of points on which f is defined to all of R^n by allowing function values in the *extended reals*, $\bar{R}$. It becomes interesting to distinguish between the points of R^n where f is finite valued, and those where it is not. For this purpose, one defines the *domain* of $\bar{f}$ to be the set

$$\text{dom } \bar{f} = \{x \in R^n : \bar{f}(x) \in R\}.$$

The function $\bar{f}$ is said to be *proper* if $\text{dom } \bar{f} \neq \emptyset$ and $\bar{f}(x) > -\infty$ for all $x \in R^n$. Saying that $\bar{f}$ is proper does not imply that it is bounded below on R^n.

Caution: *Unless stated otherwise, we shall deal here only with proper functions and for simplicity we drop the bar above the symbol.*

Convexity of level sets

The term *level set* was defined on page 305. It is very straightforward to show that *the level sets of a convex function are always convex.* The converse of this fact is not true, however.

Example A.9: NONCONVEX FUNCTION WITH CONVEX LEVEL SETS.

The *negative* of the familiar "bell-shaped curve" associated with the density function of a normally distributed random variable provides an example of a smooth nonconvex function on R having convex level sets. In this case, the level sets are closed bounded intervals on the real line. The nonconvexity of this function can be seen from its graph as well as by several other criteria for convexity. We take this example up in Exercise A.14.

Continuity of convex functions

Let $f : S \to R$ be a convex function. (Saying this is meant to imply that S is a convex set.) If S contains at least two points, then it can be shown (see, for example, [63, p. 75] or [167, p. 82]) that S *has a nonempty relative interior on which f must be continuous.* Of course, any function defined on a one-point set is continuous. So, in particular, if S is an open (or relatively open) set in R^n, then every convex function $f : S \to R$ is continuous there.

Example A.10: A DISCONTINUOUS CONVEX FUNCTION. Let S be the line segment between two distinct points p and q in R^n. Thus,

$$S = \{x : x = \lambda p + (1 - \lambda)q, \ \lambda \in [0, 1]\}.$$

Then

$$\mathrm{rel\,int}\,S = \{x : \lambda p + (1 - \lambda)q, \ \lambda \in (0, 1)\}.$$

When $n = 1$ the relative interior and the interior are the same set, otherwise they are not. Now let f be any convex function on rel int S. Then, as asserted above, f is continuous there. If $f(x)$ has a finite upper bound β on rel int S, we can extend f to all of S by setting $f(p)$ and $f(q)$ equal to any values greater than the upper bound β. This yields a discontinuous convex function on S.

Propagation of convex functions

If f is a convex function on a given set S, then so is αf for all $\alpha \geq 0$. The set of all convex functions on a given set S is a convex cone. That is, if f_1 and f_2 are convex functions on S and α_1 and α_2 are nonnegative real numbers, then $\alpha_1 f_1 + \alpha_2 f_2$ is a convex function on S. This assertion follows immediately from the basic definitions of convex function and cone. This property is useful in the construction (or analysis) of new convex functions from known convex functions. Some of these will be found among the exercises at the end of this appendix.

Theorem A.8　*The pointwise supremum of a family of convex functions on a convex set is convex. Formally, if $f_\gamma : S \to R$ is convex for each $\gamma \in \Gamma$,* *then*

$$f(x) = \sup_{\gamma \in \Gamma} f_\gamma(x), \quad x \in S$$

is a convex function.

Figure A.5 illustrates the pointwise maximum of two convex functions.

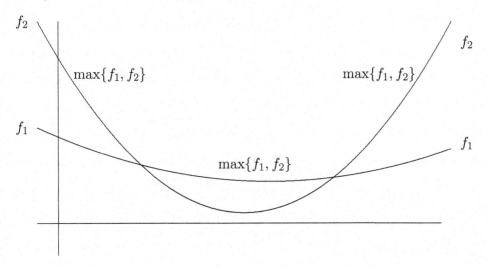

Figure A.5: The pointwise maximum of two convex functions.

Sometimes convex functions arise as *composite functions*. For example, if $f : S \to R$ is convex and g is nondecreasing on the range of f, then the composite function given by $g(f(x))$ for all $x \in S$ is convex. If f is convex on R^n and $x = Az + b$ is an affine mapping on R^m, then the composite function $g(z) = f(Az + b)$ is convex on R^m.

Differential criteria for convexity

Gradient inequality

For differentiable functions, the *gradient inequality* provides a nice characterization of convexity, namely the statement that $f : S \to R$ *is convex if and only if*

$$f(y) \geq f(x) + (y - x) \cdot \nabla f(x) \quad \text{for all } x, y \in S. \qquad (A.18)$$

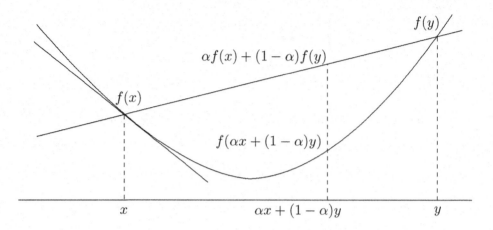

Figure A.6: The gradient inequality on a univariate convex function.

The idea behind this important property is illustrated in Figure A.6 showing the tangent line at a point $(x, f(x))$ on the graph of a convex function and the secant drawn from this point to another point $(y, f(y))$. The feature of interest is the slope of the tangent line (which is given by the first derivative $f'(x)$) and the larger slope of the secant. This form of the gradient inequality says

$$\frac{f(y) - f(x)}{y - x} \geq f'(x);$$

such an inequality is only meaningful in the univariate case.

It should be noted that the differentiable function f is *strictly convex* if and only if strict inequality holds in the gradient inequality when x and y are distinct points.

Directional derivatives and subgradients

The concept of a directional derivative comes into play when one is interested in the rate of change of a function at a given point in a particular direction. To that end, let $f : R^n \to \bar{R}$ be any function. Assume that $\bar{x} \in R^n$ is a vector for which $f(\bar{x}) \in R$, and let $d \in R^n$ be given. When

$$\lim_{\tau \to 0^+} \frac{f(\bar{x} + \tau d) - f(\bar{x})}{\tau}$$

exists, it is denoted $f'_+(\bar{x}; d)$ and called the *positive one-sided directional derivative* of f at $\bar{x}$ in the direction d. There is an analogous *negative one-*

sided directional derivative $f'_-(\bar{x}; d)$ obtained by taking the limit from the negative side of zero.

In a case where the function f is not differentiable at a particular point $\bar{x} \in S$, it can still happen that an inequality resembling (A.18) holds. The kind of inequality we have in mind is

$$f(y) \geq f(\bar{x}) + (y - \bar{x}) \cdot \gamma \quad \text{for all } y \in S. \tag{A.19}$$

Such a vector γ is called a *subgradient* of f at $\bar{x}$. For given $\bar{x}$, (A.19) is called the *subgradient inequality*. The set of all subgradients of a function f at a specific point $\bar{x}$ is called its *subdifferential* there. The subdifferential of f at $\bar{x}$ is denoted $\partial f(\bar{x})$; it is a generalization of the concept of the gradient $\nabla f(\bar{x})$. For this and the following theorem, see [7, Section 4.3] and [63].

Theorem A.9   *For a convex function f, the set $\partial f(\bar{x})$ is closed, convex, and nonempty. When f is differentiable at $\bar{x}$, the gradient $\nabla f(\bar{x})$ is the only element of $\partial f(\bar{x})$.*

First-order convexity criterion

Perhaps the most familiar result involving first-order differential criteria for convexity is the theorem saying that *if $S \subset R$ is an open interval and $f : S \to R$ is differentiable on S, then f is convex on S if and only if f' is a nondecreasing function on S.*

Second-order convexity criterion

Let S be an open convex subset of R^n, and let $f : S \longrightarrow R$ be twice continuously differentiable on S. Then f is convex on S if and only if

$$y^{\mathrm{T}} G(x) y \geq 0 \quad \text{for all } x \in S \text{ and all } y \in R^n.$$

This says that the Hessian matrix $G(x)$ is positive semidefinite for every $x \in S$. This characterization of convexity of twice continuously differentiable functions on open sets implies that a positive semidefinite quadratic form on R^n is convex there as it has a constant Hessian matrix. Accordingly, a quadratic function is convex on an open subset S of R^n if and only its Hessian matrix is positive semidefinite. Moreover, it can be shown that a quadratic function is convex on an open subset of R^n if and only if it is convex on *all* of R^n.

Extrema of convex functions

When a function $f : S \longrightarrow R$ is given, we say that $\bar{x}$ is a *global minimizer* of f on S if

$$f(\bar{x}) \leq f(x) \quad \text{for all } x \in S,$$

and $f(\bar{x})$ is the *minimum value* of f on S. Analogous definitions hold for the terms *global maximizer* and *maximum value*. If for some neighborhood $N(\bar{x})$ of $\bar{x}$

$$f(\bar{x}) \leq f(x) \quad \text{for all } x \in N(\bar{x}) \cap S,$$

then $\bar{x}$ is a *local minimizer* of f on S, and $f(\bar{x})$ is a *local minimum value* of f on S.

One of the reasons that convex functions are of interest in nonlinear optimization is that their local minima are always global minima. That is, if S is a convex set and $f : S \longrightarrow R$ is a convex function, then

$$f(\bar{x}) \leq f(x) \text{ for all } x \in N(\bar{x}) \quad \Longrightarrow \quad f(\bar{x}) \leq f(x) \text{ for all } x \in S.$$

Nonconvex functions can have local minima that are *not* global minima. An example of such a function is depicted in Figure A.7. This function has three local minima that are the only local minima in a neighborhood around them. These are said to be *isolated local minima*. This function also has *nonisolated* local minima. Notice that the (apparent) global minimum shown in the figure occurs at a point where the function is not differentiable.

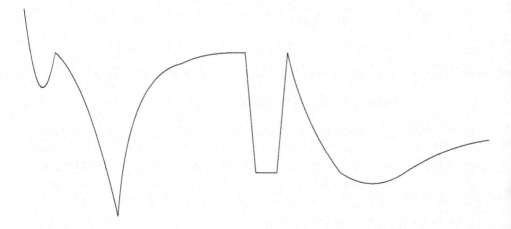

Figure A.7: Graph of a function with infinitely many local minima.

Sufficiency of the KKT conditions in the convex case

As we have noted earlier on page 370, for a convex program the $c_i(x)$ are normally required to be concave. However, for convex programs, it seems natural to depart from the "$\geq$" form (11.13) and instead think of the problem in the form

$$\begin{array}{ll} \text{minimize} & f(x) \\ \text{subject to} & c_i(x) \leq 0, \quad i \in \mathcal{I} \\ & x \in R^n \end{array} \qquad (A.20)$$

in which $\mathcal{I} = \{1, \ldots, m\}$, the functions f and $c_1, \ldots, c_m$ are differentiable and *convex*. For problems of this form, the first-order (KKT) necessary conditions of local optimality (11.15) are also *sufficient* for global optimality. This means that when these differentiability and convexity conditions hold, a KKT stationary point $\bar{x}$ yields a *global minimizer* of (11.13). The proof of this assertion rests on a nice application of the *gradient inequality* applied to the *Lagrangian function*

$$L(x, y) = f(x) + \sum_{i=1}^{m} y_i c_i(x).$$

When the functions $f, c_1, \ldots, c_m$ are all convex (differentiable or not) and the scalars $y_1, \ldots, y_m$ are all nonnegative, $L(\cdot, y)$ is a convex function of x. Hence when all the functions are also differentiable, the gradient inequality applies to L and (by virtue of the complementary slackness) yields the global optimality of a KKT stationary point.

It is important to remember that, in discussing sufficient conditions for optimality, we invoke no regularity condition because the existence of the multiplier vector y is assumed.

Local maxima

At a more advanced level, the following property is of interest. The *local maxima* of a convex function on a nonempty polyhedral convex set occur at extreme points of the set. The analogous statement for a concave function is that its minima on a nonempty polyhedral convex set occur at extreme points of the set. (This is not meant to suggest that all extreme points are local maxima or that these are the only places where such extrema can occur.) Since affine functions are both convex and concave, this pair of statements gives another way to understand why extreme points are of so much interest in linear programming. See [97].

A.8   Conjugate duality for convex programs

Conjugate duality derives its name from the use of conjugate convex functions which were introduced in a paper by Fenchel [62]. The concept of conjugate functions led to an important body of literature on duality which stems largely from another work by Fenchel[14] that influenced many authors, among whom are Karlin [106] and Rockafellar [167]. We give only a cursory review of this extensive subject. Our aim is to present some of its geometric flavor and to indicate how the involutory property shows up in conjugate convex functions and the related duality theory. The discussion given here makes use of a brief account of this topic by Veinott [187] which in turn follows the previously cited work of Rockafellar. (See also Geoffrion [78].)

Closure and closedness of convex functions

Let f be a proper convex function on R^n so that $\operatorname{dom} f \neq \emptyset$ and $f(x) > -\infty$ for all $x \in R^n$. This assumption implies the existence of *affine minorants* of f, that is, functions of the form $h(x) = y^{\mathrm{T}}x - \mu$ such that $h(x) \leq f(x)$ for all $x \in R^n$. We denote the family of all affine minorants of f by $\mathsf{A}(f)$. See Figure A.8 for an example of affine minorants.

The concept of affine minorant lays the foundation for the definitions of closure and closedness of the proper convex function f. The function given by the pointwise supremum of all affine minorants of f is called the *closure* of f, denoted $\operatorname{cl} f$. Symbolically, we have

$$\operatorname{cl} f(x) = \sup \{h(x) : h \in \mathsf{A}(f)\}.$$

The function f is said to be *closed* if $f = \operatorname{cl} f$. It also makes sense to define the function f to be closed at a single point $\bar{x} \in \operatorname{dom} f$ (see Figure A.9) if

$$f(\bar{x}) = \sup \{h(\bar{x}) : h \in \mathsf{A}(f)\}.$$

Let $S = \operatorname{dom} f$ and let $\bar{x} \in \operatorname{rel} \operatorname{bdy} S$. It can be shown that then

$$\liminf_{x \to \bar{x}} f(x) \leq f(\bar{x}).$$

---

[14]The mathematician Werner Fenchel gave a series of lectures at Princeton University in 1951, the notes of which inspired much of the early research on mathematical programming theory. "Fenchel's notes" (as they are called) were never published, but they were circulated in mimeographed form. See [63]. Biographical details are available on various Wikipedia sites.

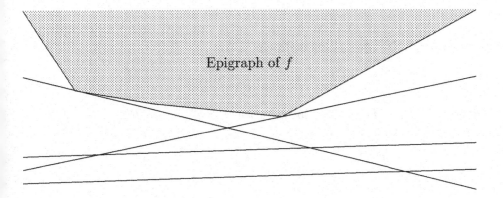

Figure A.8: Four affine minorants of a (piecewise-linear) convex function.

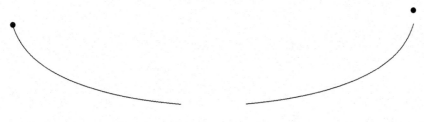

Function is closed   Function is not closed

Figure A.9: Closed and non-closed functions.

When this relationship holds with equality, the function f is said to be *lower semicontinuous* (lsc) at $\bar{x}$. If it holds with equality for every relative boundary point of S, then f is said to be *lower semicontinuous* (lsc) on S.

The following equivalence theorem plays an important part in the development of conjugate duality theory. It is a standard result which appears in Rockafellar [167, p. 51] for instance.

Theorem A.10   *Let f be a proper convex function on R^n. Then the following are equivalent:*

(i) *f is closed.*

(ii) *epi f is a closed subset of R^{n+1}.*

(iii) *for every $\alpha \in R$ the level set $\{x : f(x) \leq \alpha\}$ is closed in R^n.*

(iv) *f is lower semicontinuous.*

Conjugate convex functions

If f is a proper convex function on R^n and $h \in A(f)$, the condition

$$h(x) := y^T x - \mu \leq f(x) \quad \text{for all } x \in R^n$$

is equivalent to saying

$$y^T x - f(x) \leq \mu \quad \text{for all } x \in R^n.$$

This way of expressing the condition for h to be an affine minorant of f leads to the definition of the *conjugate function*

$$f^*(y) := \sup_{x \in R^n} \{y^T x - f(x)\}.$$

Clearly, $h \in A(f)$ if and only if $f^*(y) \leq \mu$. Indeed, the *smallest* value of μ (or equivalently, the *largest* value of $-\mu$) for which the function h (above) is an affine minorant of f is $f^*(y)$.

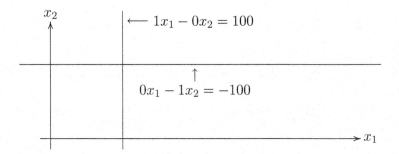

Figure A.10: Vertical and nonvertical lines and their equations.

Note that $f^* : R^n \to [-\infty, +\infty]$. When f is a convex function, f^* is called the *Fenchel (or conjugate) transform* of f. The definition of conjugate function f^* gives what is called *Fenchel's inequality*[15], namely

$$y^{\mathrm{T}}x \leq f(x) + f^*(y) \quad \text{for all } x, y \in R^n. \tag{A.21}$$

Geometric interpretation

The epigraph of a given convex function f is a convex set in R^{n+1}. As applied to this convex set, Theorem A.6 says that epi f is the intersection of the closed halfspaces in R^{n+1} which contain it. These closed halfspaces have associated boundary hyperplanes. A hyperplane in R^{n+1} corresponds to the solution set of a linear equation which in turn is given by a linear function set equal to a constant. Such linear functions are of the form $y^{\mathrm{T}}x + \eta\mu$. The coefficients of this linear form can be scaled so that either $\eta = 0$ or $\eta = -1$. Those with $\eta = 0$ are said to be *vertical*, whereas those with $\eta = -1$ are *nonvertical* (see Figure A.10). Nonvertical hyperplanes in R^{n+1} have the form

$$\{(x, \mu) \in R^{n+1} : y^{\mathrm{T}}x - \mu = \zeta\} \quad \text{where } y \in R^n, \ \zeta \in R.$$

The following theorem states two key properties of convex conjugate functions.

Theorem A.11   *If f is a convex function on R^n, then the following hold:*

---

[15]In 1912 an inequality of this sort for univariate functions was introduced by W.H. Young [201] and later extended (to its present form) by Fenchel [62] in 1949.

1. f^* *is closed and convex.*

2. $f^{**} = \text{cl}\, f$, *the closure of* f.

In the second of these assertions $f^{**} = (f^*)^*$. This is a duality theorem for *closed* convex functions, i.e., when f is a closed convex function, $f = f^{**}$ (and conversely).

The following is a sketch of how this is proved. For fixed $x \in R^n$, the mapping $y \mapsto y^\mathrm{T} x - f(x)$ is affine, so from the definition, and Theorem A.8 on pointwise suprema of convex functions, it follows that f^* is convex.

To show that f^* is closed, we can use the Equivalence Theorem A.10 and Fenchel's inequality (A.21). Let α be an arbitrary real number, and let y^k be a convergent sequence of points belonging to the level set $\{z : f^*(z) \leq \alpha\}$. Assume $y^k \to y$. We want to show that y belongs to the same level set, i.e., that $f^*(y) \leq \alpha$. By Fenchel's inequality, we get

$$(y^k)^\mathrm{T} x \leq f(x) + f^*(y^k) \leq f(x) + \alpha \quad \text{for all } k$$

so that in the limit we have $f^*(y) \leq \alpha$.

For every $x \in R^n$, Fenchel's inequality implies

$$y^\mathrm{T} x - f^*(y) \leq f(x) \quad \text{for all } y \in R^n.$$

Taking the supremum (with respect to y) on the left-hand side, we get f^{**} which turns out to be an affine minorant of f. For this reason

$$\sup_{y \in R^n} \{y^\mathrm{T} x - f^*(y)\} = f^{**}(x) \leq \text{cl}\, f(x). \qquad (A.22)$$

Now consider an arbitrary affine minorant of f, say $a^\mathrm{T} x - \beta$. Thus, suppose $a^\mathrm{T} x - \beta \leq f(x)$ for all $x \in R^n$. Then $a^\mathrm{T} x - f(x) \leq \beta$ for all $x \in R^n$. By definition, this implies $f^*(a) \leq \beta$ or equivalently $-\beta \leq -f^*(a)$. Therefore

$$a^\mathrm{T} x - \beta \leq a^\mathrm{T} x - f^*(a) \leq f^{**}(x),$$

the second inequality coming from the definition of the second conjugate $f^{**}(x)$. This inequality shows that every affine minorant of f is bounded above by f^{**}. This holds as well for the supremum over all affine minorants of f which by definition is just the function $\text{cl}\, f$. That is, $\text{cl}\, f(x) \leq f^{**}(x)$ for all $x \in R^n$. Combining this with (A.22), we obtain $\text{cl}\, f = f^{**}$.

This result implies that a function cannot equal its second conjugate unless it is closed and convex. (There is a similar statement that pertains to the theory of cones and their polars.)

Duality theory

Geometrically, conjugate duality theory rests on a few key facts and definitions. We list them here.

1. The epigraph of every proper closed convex function on R^n is a closed proper convex subset of R^{n+1}. (See Theorem A.10.)

2. Every closed proper convex subset of R^{n+1} is the intersection of its supporting halfspaces. (See Theorem A.7.)

We are interested in a duality theory for mathematical programming problems of the formalism

$$\text{(P(0))} \qquad \begin{array}{ll} \text{minimize} & f(x) \\ \text{subject to} & c_i(x) \leq 0, \quad i = 1, \ldots, m \end{array}$$

where $f(x)$ and c_i $(i = 1, \ldots, m)$ are real-valued and convex. We can redefine f to take values in the extended reals so that $f(x) = +\infty$ if $c_i(x) > 0$ for some i. Thereby, we obtain a function f defined over all of R^n, and (P(0)) becomes a matter of the "unconstrained" minimization of f.

Suppose there exists a proper convex function $F : R^n \times R^m \to (-\infty, +\infty]$ such that

$$F(x, 0) = f(x) \qquad \text{for all } x \in R^n.$$

In a more general sense, we think of $F(x, y)$ as being related to a *perturbation* of (P(0)), for example, the problem

$$\text{(P(y))} \qquad \begin{array}{ll} \text{minimize} & f(x) \\ \text{subject to} & c_i(x) \leq y_i, \quad i = 1, \ldots, m. \end{array}$$

Lemma A.12 *Let* $F : R^n \times R^m \to (-\infty, +\infty]$ *be a proper convex function. The function* $\Phi : R^m \to [-\infty, +\infty]$ *given by*

$$\Phi(y) = \inf_{x \in R^n} F(x, y)$$

is convex.

To see this, it suffices to show that epi Φ is convex. Since F is proper, its epigraph is nonempty. Hence Φ has a nonempty epigraph. Let (y^1, α^1) and (y^2, α^2) belong to epi Φ. Now, for all $\varepsilon > 0$ there exist x^1, x^2 such that

$$F(x^1, y^1) < \inf_{x \in R^n} F(x, y^1) + \varepsilon = \Phi(y^1) + \varepsilon \leq \alpha^1 + \varepsilon,$$

and

$$F(x^2, y^2) < \inf_{x \in R^n} F(x, y^2) + \varepsilon = \Phi(y^2) + \varepsilon \leq \alpha^2 + \varepsilon.$$

Let $\lambda \in (0, 1)$ and define $\lambda_1 = \lambda$ and $\lambda_2 = 1 - \lambda$. Since epi F is convex, the above inequalities imply

$$F(\lambda_1 x^1 + \lambda_2 x^2, \lambda_1 y^1 + \lambda_2 y^2) < \lambda_1 \alpha_1 + \lambda_2 \alpha_2 + \varepsilon.$$

Hence for all $\varepsilon > 0$, we have

$$\begin{aligned}
\Phi(\lambda_1 y^1 + \lambda_2 y^2) &= \inf_x F(x, \lambda_1 y^1 + \lambda_2 y^2) \\
&\leq F(\lambda_1 x^1 + \lambda_2 x^2, \lambda_1 y^1 + \lambda_2 y^2) \\
&< \lambda_1 \alpha_1 + \lambda_2 \alpha_2 + \varepsilon,
\end{aligned}$$

so the conclusion follows.

Now for a fixed perturbation vector, say $z \in R^m$, the optimal value of the corresponding "primal problem" $(P(z))$ is denoted by $\Phi(z)$. To this problem, one can associate the following "dual problem" of finding $(\pi, \mu) \in R^m \times R$ so as to

$$\begin{aligned}
\text{maximize} \quad & \pi^T z - \mu \\
\text{subject to } & \pi^T y - \mu \leq \Phi(y) \quad \text{for all } y \in R^m.
\end{aligned} \tag{A.23}$$

This problem has a linear objective and an infinite number of linear constraints, each of which is "indexed" by a (perturbation) vector y. This problem asks for an affine minorant of Φ whose value at z is maximum.

In the dual problem (A.23), if we fix π, the objective function is maximized by setting

$$\mu = \Phi^*(\pi) = \sup_{y \in R^m} \{\pi^T y - \Phi(y)\},$$

so the dual problem can be recast as

$$\max_{\pi \in R^m} \{\pi^T z - \Phi^*(\pi)\}.$$

The supremum of this objective function (with respect to π) is just $\Phi^{**}(z)$, the conjugate of Φ^* and the second conjugate of Φ. As we know, $\Phi^{**}(z) = \Phi(z)$ if and only if Φ is closed at z.

As a proper convex function, $F(\cdot, \cdot)$ has a convex conjugate, $F^*(\cdot, \cdot)$. Moreover,

$$\Phi^*(\pi) = F^*(0, \pi).$$

To see this, note that

$$\Phi^*(\pi) = \sup_y \{\pi^T y - \Phi(y)\}$$

$$= \sup_{x,y} \{0^T x + \pi^T y - F(x, y)\}$$

$$= F^*(0, \pi).$$

Theorem A.13 (Duality)   *In $(P(z))$, when $z = 0$,*

$$\inf_x F(x, 0) = \sup_\pi -F^*(0, \pi)$$

if and only if Φ is closed at 0.

Let us assume F is closed and proper, and that Φ is closed at 0. Consider the process depicted in the following diagram:

(Primal) $\displaystyle\inf_{x \in R^n} F(x, 0)$ $\longrightarrow$ $\displaystyle\sup_{\pi \in R^m} -F^*(0, \pi)$ (Dual)

$\uparrow$ $\downarrow$

("Dual") $- \displaystyle\sup_{x \in R^n} -F^{**}(x, 0)$ $\longleftarrow$ $- \displaystyle\inf_{\pi \in R^m} F^*(0, \pi)$ ("Primal")

Starting at the upper left-hand corner and first moving to the right, we pass from the primal problem to the dual and then, moving downward, convert the dual into an equivalent "primal" problem. Moving leftward, we pass to the "dual" and complete the cycle by converting it to an equivalent primal problem. In light of the assumption that F is closed and proper, we can replace F^{**} by F, thereby showing that "the dual of the dual is the primal."

A.9 Exercises

A.1 Show that $A \triangle B = (A \cup B) \setminus (B \cap A)$. Illustrate $A \triangle B$ with a Venn diagram.

A.2 Let C be a symmetric positive definite matrix. Show that the function

$$||x||_C^2 = x^T C x$$

satisfies the properties of a vector norm (see page 540).

A.3 Construct an example to show that the Frobenius norm is not a vector-induced norm. (Hint: Think of a very simple square matrix.)

A.4 Show that $||A||_1 = ||A^T||_\infty$

A.5 Let $S \subset T \subset R^n$ with S closed and T compact. Explain why S is compact.

A.6 Give an example of an unbounded level set.

A.7 We have asserted that the convexity of a closed ball is a consequence of the Cauchy-Schwartz inequality. Identify the line where it was used.

A.8 Let $\bar{B}$ denote a closed ball of radius $\rho > 0$ centered at the origin in R^n. Interpreting this set as a level set, explain why $\bar{B}$ is convex.

A.9 Justify the remark on page 551 to the effect that from a point in an open set it is possible to move in any direction and still remain within the set.

A.10 Let S_1 and S_2 be two open sets in R^n.
(a) Explain why their intersection $S_1 \cap S_2$ must also be open.
(b) Generalize your argument to handle the intersection of a finite family of open sets.
(c) Why does your argument not work for the intersection of infinitely many open sets in R^n?
(d) How would you use what you have shown in (b) to establish the analogous result for the union of finitely many closed sets in R^n?

A.11 Are all compact subsets of R intervals? Justify your answer.

A.12 Let

$$A = \begin{bmatrix} 3 & -4 & 9 \\ -5 & 0 & 6 \end{bmatrix}$$

and define the matrix

$$M = \begin{bmatrix} 0 & A^T \\ -A & 0 \end{bmatrix}.$$

(a) Let $z \in R^5$ be an arbitrary vector and show that $z^T M z = 0$.
(b) Is the analogous statement true for every matrix $A \in R^{m \times n}$? Justify your answer.

A.13   Illustrate a case where the closure theorem A.7 does not hold when the convexity assumption is removed.

A.14   Plot the univariate function $f : R \longrightarrow R$ given by $f(x) = -e^{-x^2}$ using graph paper. [To do this, it will help to compute f' and f''.]

(a) Describe the level sets of this function.
(b) Show that f is not a convex function on R.
(c) What do you conclude from your analysis of this function?

A.15   Show that the pointwise minimum of two convex functions has convex level sets but need not be a convex function. See for example, Figure A.5.

A.16   Let $\varphi(t)$ be a monotone nondecreasing convex function of a real variable. [This means $\varphi : R \to R$ is convex and $\varphi(t) \leq \varphi(t')$ for all $t' > t$.] Now consider the composite function $\varphi(f(x))$ where $f : C \to R$ is convex. Show that this composite function is convex on R.

A.17   (a) Use properties of a norm function to show that $||x||$ is a convex function.
(b) Let $Q \in R^{n \times n}$ be a symmetric positive definite matrix. Show that the function $f(x) = \sqrt{x^T Q x}$ is convex on R^n. [Hint: Use a suitable factorization of the matrix Q.]
(c) Using the definition (8.1) of a convex function, illustrate the convexity of the function f above, when Q, x, y, and λ are given by

$$Q = \begin{bmatrix} 4 & -1 \\ -1 & 2 \end{bmatrix}, \quad x = \begin{bmatrix} 3 \\ -4 \end{bmatrix}, \quad y = \begin{bmatrix} -2 \\ 5 \end{bmatrix}, \quad \text{and} \quad \lambda = \frac{1}{5}.$$

A.18   Classify the following functions as convex, concave, or neither, and explain why this is the case

(a) $f(x_1, x_2) = x_1^2 + 2x_1x_2 - 10x_1 + 5x_2$.
(b) $f(x_1, x_2) = x_1 e^{-(x_1+x_2)}$.
(c) $f(x_1, x_2) = -x_1^2 - 5x_2^2 + 2x_1x_2 + 10x_1 - 10x_2$.
(d) $f(x_1, x_2, x_3) = x_1x_2 + 2x_1^2 + x_2^2 + 2x_3^2 - 6x_1x_3$.
(e) $\ln x$ for scalar $x > 0$.

A.19   For $i = 1, 2, \ldots, n$, let

$$f_i(x_i) = \begin{cases} x_i \ln x_i & x > 0 \\ 0 & x \leq 0. \end{cases}$$

(a) Verify that f_i is convex on R.
(b) Verify that for all $x, y \in R^n$

$$\sum_{i=1}^{n} x_i y_i \leq \sum_{i=1}^{n} f_i(x_i) + \sum_{i=1}^{n} e^{y_i - 1}.$$

A.20   Let g denote a real-valued convex function on R^n such that $g(x) \geq \alpha$ for all $x \in R^n$. Verify the convexity of

$$f(x) = (g(x) - \alpha)^2 + e^{g(x)}.$$

is convex on R^n.

A.21   Suppose $f : C \to R$ is a convex function having more than one global minimizer. Verify that the set of all global minimizers of f is convex.

A.22   For $i = 1, \ldots, k$, let $f_i(x)$ be a convex function on the nonempty convex set C.

(a) Verify that the *pointwise maximum*

$$F(x) := \max_i f_i(x)$$

of this collection of convex functions is itself a convex function.
(b) Is $F(x)$ differentiable if all the functions $f_i(x)$ are differentiable? Why?

A.23   Consider the dual, S^*, of a nonempty set, S.

(a) Why must S^* be a cone?
(b) Why must the dual of S^* be closed?
(c) Why must the dual of S^* be convex?
(d) Assume that S is a closed convex cone. Why must $S = S^{**}$?

A.24   Let n be a positive integer. Show that $R_+^n = (R_+^n)^*$.

A.25   Let C_1 and C_2 be two cones in R^n. Verify the following assertions.

(a) $C_1 \subseteq C_2 \Longrightarrow C_1^* \supseteq C_2^*$.
(b) $(C_1 + C_2)^* = C_1^* \cap C_2^*$.
(c) $C_1^* + C_2^* = (C_1 \cap C_2)^*$.
(d) $C \subseteq (C^*)^*$.

A.26   This exercise pertains to a comment (made on page 573) about the local maxima of convex functions on a polyhedral convex set. Is it true that all extreme points of a polyhedron are local maxima of every convex function defined on that set? If so, explain why; if not, give a counterexample.

GLOSSARY OF NOTATION

Spaces, sets, and set-theoretic notation

R^n	Real n-space
E^n	Euclidean n-space
$\mathcal{N}(A)$	Null space of the matrix A
$\mathcal{R}(A)$	Range space of the matrix A
$\mathcal{I}$	Index set of the inequality constraints
$\mathcal{E}$	Index set of the equality constraints
$\mathcal{A}(\bar{x})$	Set of subscripts of all constraints that are active at $\bar{x}$
$\mathcal{A}_k$	$\mathcal{A}(x_k)$
C^1	Class of continuously differentiable functions
C^2	Class of twice-continuously differentiable functions
$[a, b]$	Closed interval
(a, b)	Open interval
$\in$	Belongs to
$\subset$	Is a subset of
$\subseteq$	Is a subset of or equals
$\supset$	Contains
$\supseteq$	Contains or equals
$\cup$	Set-theoretic union
$\cap$	Set-theoretic intersection
$\setminus$	Set-theoretic difference
$\triangle$	Symmetric difference (of sets)
$\emptyset$	The empty set
$\mathrm{supp}(x)$	Support of x

© Springer Science+Business Media LLC 2017
R.W. Cottle and M.N. Thapa, *Linear and Nonlinear Optimization*,
International Series in Operations Research & Management Science 253,
DOI 10.1007/978-1-4939-7055-1

Functions, gradients, and Hessians

x_k	The solution vector at iteration k
f_k	$f(x_k)$
∇f	$\left(\dfrac{\partial f(x)}{\partial x_1}, \ldots, \dfrac{\partial f(x)}{\partial x_n} \right)$
$g(x)$	Gradient of a function $f(x)$ evaluated at x
g_k	$g(x_k)$
$\nabla^2 f$	Hessian matrix of function f
$G(x)$	Hessian matrix of a function $f(x)$ evaluated at x
G_k	$G(x_k)$
$A_{i\bullet}$	Row i of the matrix A
$A_{\bullet j}$	Column j of the matrix A
$\|x\|$	Norm of vector x
$\|x\|_p$	p-norm of vector x; $p = 1, 2, \infty$
$\|A\|_p$	p-norm of matrix A; $p = 1, 2, \infty$
$\|A\|_F$	Frobenius norm of matrix A
$\|A\|_C$	Norm induced by a positive-definite symmetric matrix C
$\sum_i^n x_i$	$x_1 + x_2 + \cdots + x_n$
$\prod_{i=1}^n \lambda_i$	$\lambda_1 \lambda_2 \cdots \lambda_n$

Miscellaneous

$\times$	Multiplication sign (times)
$\ln$	Natural logarithm
$<$	Less than
$\ll$	Much less than
$\leq$	Less than or equal to
$>$	Greater than
$\gg$	Much greater than
$\geq$	Greater than or equal to
$O(n)$	Order n
Z	Matrix of columns representing null space of a specific matrix
$\leftarrow$	Replaces
ε_M	Machine precision (around 10^{-15} for double precision)

Bibliography

[1] J. Abadie, On the Kuhn-Tucker Theorem, in (J. Abadie, ed.) *Nonlinear Programming*. Amsterdam: North-Holland Publishing Company, 1967, pp. 19–36.

[2] J. Abadie and J. Carpentier, Generalization of the Wolfe Reduced Gradient Method, in (R. Fletcher, ed.) *Optimization*. New York: Academic Press, 1969, pp. 37–47.

[3] E.D. Andersen and K.D. Andersen, Presolving in linear programming, *Mathematical Programming* 71 (1995), 221–245.

[4] J.G. Andrews and R.R. McLone, *Mathematical Modelling*. London: Butterworths, 1976.

[5] A. Antoniou and W-S. Lu, *Practical Optimization*. New York: Springer, 2007.

[6] http://www-fp.mcs.anl.gov/otc/Guide/CaseStudies/diet.

[7] M. Avriel, *Nonlinear Programming: Analysis and Methods*. Engelwood Cliffs, N.J.: Prentice-Hall, Inc., 1976.

[8] M. Avriel and R.S. Dembo, *Engineering Optimization*. Amsterdam: North-Holland Publishing Company, 1979. [See *Mathematical Programming Study 11*.]

[9] M. Bazaraa, H. Sherali, and C.M. Shetty, *Nonlinear Programming*. New York: John Wiley & Sons, 1993.

[10] A. Ben-Tal and A.S. Nemirovskii, *Lectures on Modern Convex Optimization: Analysis, Algorithms, and Engineering Optimization*. Philadelphia: SIAM, 2001.

© Springer Science+Business Media LLC 2017
R.W. Cottle and M.N. Thapa, *Linear and Nonlinear Optimization*,
International Series in Operations Research & Management Science 253,
DOI 10.1007/978-1-4939-7055-1

[11] J.F. Benders, Partitioning procedures for solving mixed-variable pro-
     gramming problems, *Numerische Mathematik* 4 (1962), 238–252.

[12] D.P. Bertsekas, *Nonlinear Programming*. Belmont, Mass.: Athena,
     1995.

[13] D.P. Bertsekas, *Convex Optimization Algorithms*. Belmont, Mass.:
     Athena Scientific, 2015.

[14] D. Bertsimas and R.M. Freund, *Data, Models, and Decisions*. Cincin-
     nati, Ohio: South-Western College Publishing, 2000.

[15] D. Bertsimas and J.N. Tsitsiklis, *Introduction to Linear Optimization*.
     Belmont, Mass.: Athena Scientific, 1997.

[16] R.G. Bland, The allocation of resources by linear programming, *Sci-
     entific American*, June 1981, pp. 126-144.

[17] F. Bonnans, J.C. Gilbert, C. Lemaréchal, and C.A. Sagastizábal, *Nu-
     merical Optimization*. Berlin: Springer-Verlag, 2003.

[18] S. Boyd and L. Vandenberghe, *Convex Optimization*. Cambridge:
     Cambridge University Press, 2004.

[19] J. Bracken and G.P. McCormick, *Selected Applications of Nonlinear
     Programming*. New York: John Wiley & Sons, 1968.

[20] S.P. Bradley, A.C. Hax, and T.L. Magnanti, *Applied Mathematical
     Programming*. Reading, Mass.: Addison-Wesley Publishing Company,
     1977.

[21] D.M. Bravata, R.W. Cottle, B.C. Eaves, and I. Olkin, Measuring
     conformability of probabilities, *Statistics and Probability Letters* 52
     (2001), 321–327.

[22] P. Breitkopf and R.F. Coelho, eds., *Multidisciplinary Design Opti-
     mization in Computational Mechanics*. Hoboken, N.J.: John Wiley &
     Sons, 2010.

[23] C. Brezinski, The life and work of André Louis Cholesky, *Numerical
     Algorithms* 43 (2006), 279–288.

[24] G.W. Brown and T.C. Koopmans, Computational suggestions for
     maximizing a linear function subject to linear inequalities in

(T.C. Koopmans, ed.) *Activity Analysis of Production and Allocation*. New York: John Wiley and Sons, 1951, pp. 377–380.

[25] C.G. Broyden, The convergence of a class of double-rank minimization algorithms, *Journal of the Institute of Mathematics and its Applications* 6 (1970), 76–90.

[26] E. Castillo, A.J. Conejo, P. Pedregal, R. Garcia, and N. Alguacil, *Building and Solving Mathematical Programming Models in Engineering and Science*. New York: John Wiley & Sons, 2002.

[27] A. Charnes, Optimality and degeneracy in linear programming, *Econometrica* 20 (1952), 160–170.

[28] V. Chvátal, *Linear Programming*. New York: W.H. Freeman, 1983.

[29] L. Contesse, Une caractérisation complète des minima locaux en programmation quadratique, *Numerishe Mathematik* 34 (1980), 315–332.

[30] R.W. Cottle, A theorem of Fritz John in mathematical programming, RAND Corporation Memo RM-3858-PR, 1963. [Appears in [85].]

[31] R.W. Cottle, Symmetric dual quadratic programs, *Quarterly of Applied Mathematics* 21 (1963), 237–243.

[32] R.W. Cottle, Manifestations of the Schur complement, *Linear Algebra and its Applications* 8 (1974), 189–211.

[33] R.W. Cottle, On dodging degeneracy in linear programming. Unpublished manuscript. Stanford, California (1997).

[34] R.W. Cottle, William Karush and the KKT theorem, in (M. Grötschel, ed.) *Optimization Stories*. Extra volume of DOCUMENTA MATHEMATICA, 2012, pp. 255–269.

[35] R.W. Cottle, S.G. Duvall, and K. Zikan, A Lagrangean relaxation algorithm for the constrained matrix problem, *Naval Research Logistics Quarterly* 33 (1986), 55–76.

[36] R.W. Cottle and I. Olkin, Closed-form solution of a maximization problem, *Journal of Global Optimization* 42 (2008), 609–617.

[37] R.W. Cottle, J.S. Pang, and R.E. Stone, *The Linear Complementarity Problem*. Boston: Academic Press, 1992. [Republished, with corrections, in 2009 by the Society for Industrial and Applied Mathematics, Philadelphia.]

[38] R. Courant, Variational methods for the solution of problems of equilibrium and vibrations, *Bulletin of the American Mathematical Society* 49 (1943), 1–23.

[39] R. Courant and H. Robbins, *What is Mathematics?* London: Oxford University Press, 1941.

[40] G.B. Dantzig, *Linear Programming and Extensions*. Princeton, N.J.: Princeton University Press, August, 1963. Revised edition Fall 1966; fourth printing, 1968.

[41] G.B. Dantzig, Making progress during a stall in the Simplex Algorithm, *Linear Algebra and its Applications* 114/115 (1989), 251259.

[42] G.B. Dantzig, S. Johnson, and W. White, A linear programming approach to the chemical equilibrium problem, *Management Science* 5 (1958), 38-43.

[43] G.B. Dantzig, A. Orden, and P. Wolfe, The generalized simplex method for minimizing a linear form under linear restraints, *Pacific Journal of Mathematics* 5 (1955), 183–195.

[44] G.B. Dantzig and M.N. Thapa, *Linear Programming 1: Introduction*. New York: Springer, 1997.

[45] G.B. Dantzig and M.N. Thapa, *Linear Programming 2: Theory and Extensions*. New York: Springer, 2003.

[46] G.B. Dantzig and P. Wolfe, The decomposition principle for linear programming, *Operations Research* 8 (1960), 101–111.

[47] G.B. Dantzig and P. Wolfe, The decomposition algorithm for linear programming, *Econometrica* 29 (1961), 767–778.

[48] J. Dattorro, *Convex Optimization & Euclidean Distance Geometry*. Palo Alto, Calif.: Meboo Publishing, 2005.

[49] W.C. Davidon, Variable metric methods for minimization, Research and Development Report ANL-5990 (Rev.), Argonne National Laboratory, Argonne, Illinois, 1959. [*Reprinted in SIAM J. Optimization* 1 (1991), 1–17.]

[50] A. Dax, An elementary proof of Farkas' lemma, *SIAM Review* 39 (1997). 503–507.

[51] I.I. Dikin, Iterative solution of problems of linear and quadratic programming, *Doklady Akademiia Nauk USSR* 174 (1967). Translated in *Soviet Mathematics Doklady* 18, 674–675.

[52] L.C.W. Dixon, Quasi-Newton methods generate identical points, *Mathematical Programming* 2 (1972), 383–387.

[53] L.C.W. Dixon, Quasi-Newton methods generate identical points. II. The proof of four new theorems, *Mathematical Programming* 3 (1972), 345–358.

[54] J. Dongarra and F. Sullivan, The top 10 algorithms, *Computing in Science and Engineering* 2 (2000), 22–23.

[55] R. Dorfman, *Application of Linear Programming to the Theory of the Firm.* Berkeley: University of California Press, 1951.

[56] R. Dorfman, P.A. Samuelson, and R.M. Solow, *Linear Programming and Economic Analysis.* New York: McGraw-Hill, 1958.

[57] W.S. Dorn, Duality in quadratic programming, *Quarterly of Applied Mathematics* 18 (1960), 155–162.

[58] W.S. Dorn, A symmetic dual theorem for quadratic programming, *Journal of the Operations Research Society of Japan* 2 (1960), 93–97.

[59] *Encyclopaedia Britannica, Micropaedia, Volume X, Fifteenth Edition.* Chicago: Encyclopaedia Britannica, Inc., 1974.

[60] L. Euler [Leonhardo Eulero], *Methodus Inviendi Lineas Curvas Maximi Minimive Proprietate Gaudentes*, Lausanne & Geneva, 1744, p. 245.

[61] D.P. Feder, Automatic lens design with a high-speed computer, *Journal of the Optical Society of America* 52 (1962), 177–183.

[62] W. Fenchel, On conjugate convex functions, *Canadian Journal of Mathematics* 1 (1949), 73–77.

[63] W. Fenchel, *Convex Cones, Sets, and Functions.* Lecture Notes, Princeton, N.J.: Princeton University, 1953.

[64] A.V. Fiacco and G.P. McCormick, *Nonlinear Programming: Sequential Unconstrained Minimization Techniques.* New York: John Wiley & Sons, 1968.

[65] A.H. Fitch, Synthesis of dispersive delay characterisitics by thickness tapering in ultrasonic strip delay lines, *Journal of the Acoustical Society of America* 35 (1963), 709–714.

[66] W. Fleming, *Functions of Several Variables, Second Edition.* New York: Springer-Verlag, 1977.

[67] R. Fletcher, A new approach to variable metric algorithms, *Computer Journal* 13 (1970), 317–322.

[68] R. Fletcher, *Practical Methods of Optimization, Volume 1 Unconstrained Optimization.* Chichester: John Wiley & Sons, 1980.

[69] R. Fletcher and M.J.D. Powell, A rapidly convergent descent method for minimization, *Computer Journal* 6 (1963), 163–168.

[70] R. Fletcher and C.M. Reeves, Function minimization by conjugate gradients, *Computer Journal* 7 (1964), 149–154.

[71] J.J. Forrest and D. Goldfarb, Steepest-edge simplex algorithms for linear programming, *Mathematical Programming* 57 (1992), 341–374.

[72] W. Forst and D. Hoffmann, *Optimization—Theory and Practice.* New York: Springer, 2010.

[73] C.G. Fraser, Joseph Louis Lagrange, *Théorie des Fonctions Analytiques,* First Edition (1797), in (I. Grattan-Guinness, ed.), *Landmark Writings in Western Mathematics,* Amsterdam: Elsevier, 2005, pp. 258–276.

[74] R.A.K. Frisch, The logarithmic potential method of convex programs, Memorandum of May 13, 1955, University Institute of Economics, Oslo, Norway (1955).

[75] D. Gale, *The Theory of Linear Economic Models.* New York: McGraw-Hill, 1960.

[76] M. Gardner, Mathematical Games: The curious properties of the Gray code and how it can be used to solve puzzles, *Scientific American* 227 (August 1972), pp. 106-109.

[77] S.I. Gass, *Linear Programming, Fifth Edition.* Danvers, Mass.: Boyd & Fraser, 1985.

[78] A.M. Geoffrion, Duality in nonlinear programming: A simplified applications-oriented approach, *SIAM Review* 13 (1971), 1–37.

[79] P.E. Gill and W. Murray, Quasi-Newton methods for unconstrained optimization, *Journal of the Institute of Mathematics and its Applications* 9 (1972), 91–108.

[80] P.E. Gill, G.H. Golub, W. Murray, and M.A. Saunders, Methods for modifying matrix factorizations, *Mathematics of Computation* 28 (1974), 505–535.

[81] P.E. Gill, W. Murray, and S.M. Picken, The implementation of two modified Newton algorithms for unconstrained optimization, *Technical Report NAC* 11, (1972), National Physical Laboratory, England.

[82] P.E. Gill, W. Murray, M.A. Saunders, J.A. Tomlin, and M.H. Wright, On projected Newton barrier methods and an equivalence to Karmarkar's projective method, *Mathematical Programming* 36 (1986), 183–209.

[83] P.E. Gill, W. Murray, M.A. Saunders, and M.H. Wright, A practical anti-cycling procedure for linearly constrained optimization, *Mathematical Programming* 45 (1989), 437–474.

[84] P.E. Gill, W. Murray, and M.H. Wright. *Practical Optimization*, London: Academic Press, 1981.

[85] G. Giorgi and T.H. Kjeldsen, eds., *Traces and Emergence of Nonlinear Programming*. Basel: Birkhäuser, 2014.

[86] D. Goldfarb, A family of variable-metric methods derived by variational means, *Mathematics of Computation* 24 (1970), 23–26.

[87] A.J. Goldman, Resolution and separation theorems for polyhedral convex sets, in (H.W. Kuhn and A.W. Tucker, eds.) *Linear Inequalities and Related Systems*. Princeton, New Jersey: Princeton University Press, 1956, pp. 41–51.

[88] G.H. Golub and C.F. Van Loan, *Matrix Computations, Third Edition*. The Johns Hopkins University Press, Baltimore, Maryland, 1996.

[89] P. Gordan, Über die Auflösung linearer Gleichungen mit reellen Coefficienten, *Mathematische Annalen* 6 (1873), 23–28.

[90] F. Gray, Pulse code communication, U.S. Patent 2,632,058 (March 17, 1953).

[91] S.P. Han, Superlinearly convergent variable-metric algorithms for constrained optimization, *Mathematical Programming* 11 (1976), 263–282.

[92] H. Hancock, *Theory of Maxima and Minima*. New York: Dover Publications, Inc., 1960.

[93] M.R. Hestenes, *Optimization Theory: The Finite Dimensional Case.* New York: John Wiley & Sons, 1975.

[94] M.R. Hestenes and E. Stiefel, Methods of conjugate gradients for solving linear systems, *Journal of Research of the National Bureau of Standards* 49 (1952), 409–436.

[95] J.R. Hicks, *Value and Capital.* Oxford: Oxford University Press, 1939.

[96] F.S. Hillier and G.J. Lieberman, *Introduction to Operations Research, Eighth Edition.* Boston: McGraw-Hill, 2005.

[97] W.M. Hirsch and A.J. Hoffman, Extreme varieties, concave functions, and the fixed charge problem, *Communications on Pure and Applied Mathematics* 14 (1961), 355–369.

[98] F.L. Hitchcock, The distribution of a product from several sources to numerous localities, *Journal of Mathematical Physics* 20 (1941), 224–230.

[99] M. Hubbard, N.J. de Mestre, and J. Scott, Dependence of release variables in the shot put, *Journal of Biomechanics* 34 (2001), 449–456.

[100] H.E. Huntley. *The Divine Proportion: A Study in Mathematical Beauty.* New York: Dover Publications, 1970.

[101] I.M. Isaacs, *Geometry for College Students.* American Mathematical Society, Providence, R.I. 2009

[102] F. John, Extremum problems with inequalities as subsidiary conditions, in (K.O. Friedrichs, O.E. Neugebauer, and J.J. Stoker, eds.) *Studies and Essays, Courant Anniversary Volume.* New York: Wiley-Interscience, 1948, pp. 187–204.

[103] J. Jost, *Postmodern Analysis, Second Edition.* Berlin: Springer-Verlag, 2003.

[104] L.V. Kantorovich, On the translocation of masses, *Comptes Rendus (Doklady) de l'Academie des Sciences de l'URSS* XXXVII, (1942), 199–201. Translated in *Management Science* 5 (1958), 1–4.

[105] L.V. Kantorovich, Mathematical methods in the organization and planning of production, Publication House of the Leningrad State University, 1939. Translated in *Management Science* 6 (1960), 366–422.

[106] S. Karlin, *Mathematical Methods and Theory in Games, Programming and Economics, Vol. 1.* Reading, Mass.: Addison-Wesley Publishing Company, Inc., 1959.

[107] N. Karmarkar, A new polynomial-time algorithm for linear programming, *Combinatorica* 4 (1984) 373–395,

[108] W. Karush, *Minima of Functions of Several Variables with Inequalities as Side Conditions*, Master's thesis, Department of Mathematics, University of Chicago, 1939.

[109] T.K. Kelly and M. Kupferschmid, Numerical verification of second-order sufficiency conditions for nonlinear programming, *SIAM Review* 40 (1998), 310–414.

[110] L.G. Khachiyan, A polynomial algorithm in linear programming, *Soviet Mathematics Doklady* 20 (1979), 191–194.

[111] S-J. Kim, K. Koh, M. Lustig, S. Boyd, D. Gorinevsky, An interior-point method for large-scale l_1-regularized least squares, *IEEE Journal of Selected Topics in Signal Processing* 1, No. 4 (2007), 606–617.

[112] V. Klee and G.J. Minty, How good is the simplex algorithm? in (O. Shisha, ed.) *Inequalities–III.* New York: Academic Press, 1972, pp. 159–175.

[113] F.H. Knight, A note on Professor Clark's illustration of marginal productivity. *Journal of Political Economy* 33 (1925), 550–553.

[114] S.G. Krantz, Higher dimensional conundra, *The Journal of Online Mathematics and its Application* 7 (2007), Article ID 1612.

[115] S.G. Krantz and H.R. Parks, *The Implicit Function Theorem.* Boston: Birkhäuser, 2002.

[116] H.W. Kuhn, Solvability and consistency for linear equations and inequalities, *American Mathematical Monthly* 63 (1956), 217–232.

[117] H.W. Kuhn, Nonlinear programming: A historical view, in (R.W. Cottle and C.E. Lemke, eds.) *Nonlinear Programming* [SIAM-AMS Proceedings, Volume IX]. Providence, R.I.: American Mathematical Society, 1976, pp. 1–26.

[118] H.W. Kuhn and A.W. Tucker, Nonlinear programming, in (J. Neyman, ed.) *Proceedings of the Second Berkeley Symposium on Mathematical Statistics and Probability*. Berkeley: University of California Press, 1951, pp. 481–492.

[119] M. Laikin, *Lens Design, Third Edition*. Boca Raton, Fl.: CRC Press, 2001.

[120] L.S. Lasdon, *Optimization Theory for Large Systems*. New York: Macmillan, 1972.

[121] L.A. Lasdon and A.D. Waren, Generalized Reduced Gradient Method for linearly and nonlinearly constrained programming, in (H.J. Greenberg, ed.) *Design and Implementation of Optimization Software*. Alphen aan den Rijn, The Netherlands: Sijthoff and Noordhoff, 1978, pp. 363–396.

[122] L.A. Lasdon and A.D. Waren, Survey of nonlinear programming applications, *Operations Research* 5 (1980), 1029–1073.

[123] C.L. Lawson and R.J. Hanson, *Solving Least Squares Problems*. Englewood Cliffs, N.J.: Prentice-Hall, 1974.

[124] C.E. Lemke, The dual method of solving the linear programming problem, *Naval Research Logistics Quarterly* 1 (1954), 36–47.

[125] D.G. Luenberger and Y. Ye, *Linear and Nonlinear Programming, Third Edition*. New York: Springer, 2008.

[126] T.L. Magnanti, Fenchel and Lagrange duality are equivalent, *Mathematical Programming* 7 (1974), 253–258.

[127] H.J. Maindonald, *Statistical Computation*. New York: John Wiley & Sons, 1989.

[128] A. Majthay, Optimality conditions for quadratic programming, *Mathematical Programming* 1 (1971), 359–365.

[129] O.L. Mangasarian, Duality in nonlinear programming, *Quarterly of Applied Mathematics* 20 (1962), 300–302.

[130] O.L. Mangasarian, *Nonlinear Programming*. SIAM Classics in Applied Mathematics (CL10), Philadelphia: SIAM, 1994. [Same as the 1969 edition published by McGraw-Hill.]

[131] H.M. Markowitz, Portfolio selection, *Journal of Finance* 7 (1952), 77–91.

[132] H.M. Markowitz, *Portfolio Selection*. New Haven: Yale University Press, 1959.

[133] H.M. Markowitz, *Mean-Variance Analysis in Portfolio Choice and Capital Markets*. Oxford: Basil Blackwell, 1987.

[134] G.P. McCormick, Second-order conditions for constrained minima, *SIAM Journal on Applied Mathematics* 15 (1967), 641–652.

[135] H. Minkowski, *Geometrie der Zahlen*. Leipzig: Teubner, 1896.

[136] L. Mirsky, *An Introduction to Linear Algebra*. London: Oxford University Press, 1955. [Reprinted in Oxford, 1961, 1963, 1972.]

[137] G. Monge, Mémoire sur la théorie des déblais et remblais, *Histoire de l'Académie Royale des Sciences, avec les Mémoires de Mathématique et de Physique* [année 1781] (2e partie) (1784) [*Histoire*: 34–38, *Mémoire*: 666–704.]

[138] J. Moré, B. Garbow, and K. Hillstrom, Testing unconstrained optimization software, *ACM Transactions on Mathematical Software* 7 (1981), 17–41.

[139] T.S. Motzkin, *Beiträge zur Theorie der Linearen Ungleichungen*. Jerusalem: Azriel, 1936.

[140] T.S. Motzkin and E.G. Straus, Maxima for graphs and a new proof of a theorem of Turán. *Canadian Journal of Mathematics* 17 (1965), 533–550.

[141] W. Murray, An algorithm for constrained minimization, in (R. Fletcher, ed.) *Optimization*. New York and London: Academic Press, 1976, pp. 247–258.

[142] W. Murray, Constrained optimization, in (L.C.W. Dixon, ed.) *Optimization in Action.* New York and London: Academic Press, 1976, pp. 217–251.

[143] B.A. Murtagh and M.A. Saunders, A projected Lagrangian algorithm and its implementation for sparse nonlinear constraints, *Mathematical Programming Study* 16 (1982), 84–117.

[144] K.G. Murty, *Linear Programming.* New York: John Wiley & Sons, 1983.

[145] K.G. Murty, *Operations Research: Deterministic Optimization Models.* Englewood Cliffs, N.J.: Prentice-Hall, 1995.

[146] P.J. Nahin, *When Least is Best.* Princeton, N.J.: Princeton University Press, 2007. [Fifth printing, first paperback; with revised Preface].

[147] S.G. Nash and A. Sofer, *Linear and Nonlinear Programming.* New York: McGraw-Hill, 1996.

[148] G.L. Nemhauser, A.H.G. Rinnooy Kan, and M.J. Todd, eds., *Optimization*, (Handbooks in Operations Research and Management Science). Amsterdam: North-Holland Publishing Company, 1989.

[149] E.D. Nering and A.W. Tucker, *Linear Programs and Related Problems.* Boston: Academic Press, 1993.

[150] J. von Neumann, On a maximization problem. Manuscript, Institute for Advanced Study, Princeton University, Princeton, N.J., 1947.

[151] B. Noble, *Applications of Undergraduate Mathematics in Engineering.* New York: The Macmillan Company [for the Mathematical Association of America], 1967.

[152] J. Nocedal and S.J. Wright, *Numerical Optimization.* New York: Springer, 1999.

[153] W.A. Oldfather, C.A. Ellis, and D.M. Brown, Leonhard Euler's elastic curves, *ISIS* 20 (1933), 72–160. [English translation of [60].]

[154] W. Orchard-Hays, *Advanced Linear-Programming Computing Techniques.* New York: McGraw-Hill, 1968.

[155] J.M. Ortega and W.C. Rheinboldt, *Iterative Solution of Nonlinear Equations in Several Variables.* New York: Academic Press, 1970.

[156] J. O'Rourke, *Computational Geometry, Second Edition*. Cambridge: Cambridge University Press, 1998.

[157] A.S. Posamentier and I. Lehrman, *The (Fabulous) Fibonacci Numbers*. Amherst, Mass: Prometheus Books, 2007.

[158] M.J.D. Powell, A hybrid method for nonlinear equations, in (P. Rabinowitz, ed.) *Numerical Methods for Algebraic Equations*. London: Gordon and Breach 1970, pp. 87–114.

[159] M.J.D. Powell, On the convergence of the variable metric algorithm, *Journal of the Institute of Mathematics and its Applications* 7 (1971), 21–36.

[160] M.J.D. Powell, Problems relating to unconstrained optimization, in (W. Murray, ed.) *Numerical Methods for Unconstrained Optimization*. New York and London: Academic Press 1972, pp. 29–55.

[161] M.J.D. Powell, Some global convergence properties of a variable variable metric algorithm without exact line searches, in (R.W. Cottle and C.E. Lemke, eds.) *SIAM-AMS Proceedings, Volume IX, Mathematical Programming* (1976), American Mathematical Society, Providence, R.I., pp. 53–72.

[162] M.J.D. Powell and Ph.L. Toint, On the estimation of sparse Hessian matrices, *SIAM Journal of Numerical Analysis* 16 (1979), 1060–1073.

[163] H. Pulte, Joseph Louis Lagrange, Méchanique Analitique, First Edition (1788), in (I. Grattan-Guinness, ed.), *Landmark Writings in Western Mathematics*, Amsterdam: Elsevier, 2005, pp. 208–224.

[164] G.V. Reklaitis, A. Ravindran, and K.M. Ragsdell, *Engineering Optimization: Methods and Applications*. New York: John Wiley & Sons, 1983.

[165] K. Ritter, Stationary points of quadratic maximum-problems, *Zeitschrift für Wahrscheinlichkeitstheorie und verwandte Gebiete* 4 (1965), 149–158.

[166] S.M. Robinson, A quadratically convergent algorithm for general nonliear programming problems, *Mathematical Programming* 3 (1972), 145–156.

[167] R.T. Rockafellar, *Convex Analysis*. Princeton, N.J.: Princeton University Press, 1970.

[168] J.B. Rosen, The gradient projection method for nonlinear programming. Part I. Linear constraints, *SIAM Journal on Applied Mathematics* 8 (1960), 181–217.

[169] J.B. Rosen, Two-phase algorithm for nonlinear constraint problems, in (O.L. Mangasarian, R.R. Meyer, and S.M. Robinson, eds.) *Nonlinear Programming 3*. New York and London: Academic Press, 1978, pp. 97–124.

[170] J.B. Rosen and J. Kreuser, A gradient projection algorithm for nonlinear constraints, in (F.A. Lootsma, ed.) *Numerical Methods for Nonlinear Optimization*. New York and London: Academic Press, 1972, pp. 297-300.

[171] S.M. Ross, *An Introduction to Mathematical Finance*. Cambridge: Cambridge University Press, 1999.

[172] W. Rudin, *Principles of Mathematical Analysis, Second Edition*. New York: McGraw-Hill Book Company, 1964.

[173] Y. Saad, *Iterative Methods for Sparse Linear Systems, Second Edition*. Philadelphia: SIAM, 2003.

[174] P.A. Samuelson, Market mechanisms and maximization, in (J. Stiglitz, ed.) *The Collected Scientific Papers of Paul A. Samuelson, Vol. 1*. Cambridge, Mass.: M.I.T. Press, 1966. [This paper (Chapter 33 of the cited book) was originally published in 1949 as two research memoranda at the RAND Corporation, Santa Monica, Calif.]

[175] P.A. Samuelson, Frank Knight's theorem in linear programmning. *Zeitschrift für Nationalökonomie* 18(3) (1958), 310–317. [This appears as Chapter 35 of (J. Stiglitz, ed.) *The Collected Scientific Papers of Paul A. Samuelson, Vol. 1*. Cambridge, Mass.: M.I.T. Press, 1966.]

[176] A. Schrijver, *Theory of Linear and Integer Programs*. Chichester: John Wiley & Sons, 1986.

[177] J.T. Schwartz, M. Sharir, and J. Hopcroft (eds.), *Planning, Geometry, and Complexity*. New Jersey: Ablex Publishing Corporation, 1987.

[178] D.F. Shanno, Conditioning of quasi-Newton methods for function min-imization, *Mathematics of Computation* 24 (1970), 647–657.

[179] M.L. Slater, Lagrange Multipliers Revisited: A Contribution to Non-linear Programming, *Cowles Commission Discussion Paper, Math-ematics*, 403, 1950. [Released as RM-676, The RAND Corporation, Santa Monica, Calif., August, 1951; also in [85].]

[180] G.W. Stewart, *Afternote on Numerical Analysis.* Philadelphia: SIAM, 1996.

[181] G.J. Stigler, The cost of subsistence, *Journal of Farm Economics* 27 (1945), 303–314.

[182] G. Strang, *Linear Algebra and its Applications, 4th Edition.* Belmont, Calif: Thomson Brooks/Cole, 2006.

[183] C.B. Tompkins, Projection methods in calculation, in (H.A. An-tosiewicz, ed.) *Proceedings of the Second Symposium in Linear pro-gramming.* United States Air Force, Washington, D. C., 1955, pp. 425–448.

[184] C.B. Tompkins, Some methods of computational attack on program-ming problems, other than the Simplex Method, *Naval Research Lo-gistics Quarterly* 4 (1957), 95–96.

[185] M.S. Townend, *Mathematics in Sport.* Chichester: Ellis Horwood, Ltd., 1984.

[186] R.J. Vanderbei, *Linear Programming: Foundations and Extensions.* Boston: Kluwer Academic Publishers, 1996.

[187] A.F. Veinott, Jr., Conjugate duality in convex programs: A geomet-ric development, *Linear Algebra and its Applications* 114/115 (1989), 663–667.

[188] A. Weber, *The Location of Industries.* Chicago: University of Chicago Press, 1929 [English translation by C.J. Friedrichs of *Reine Theorie des Standorts*].

[189] P. Whittle, *Optimization under Constraints.* London: Wiley-Inter-science, 1971.

[190] D.V. Widder, *Advanced Calculus, Second Edition.* New York: Dover Publications, Inc., 1989.

[191] H.P. Williams, *Model Building in Mathematical Programming, Third Edition*. Chichester: John Wiley & Sons, 1990.

[192] R.B. Wilson, *A simplicial algorithm for concave programming*, Ph.D. thesis, Graduate School of Business Administration, Harvard University, 1963.

[193] P. Wolfe, A duality theorem for nonlinear programming, *Quarterly of Applied Mathematics* 19 (1961), 239–244.

[194] P. Wolfe, A technique for resolving degeneracy in linear programming, *Journal of the Society for Industrial and Applied Mathematics* 11 (1963), 205–211.

[195] P. Wolfe, Methods for nonlinear programming, in (J. Abadie, ed.) *Nonlinear Programming*. Amsterdam: North-Holland Publishing Company, 1967, pp. 67–86.

[196] M.H. Wright, *Numerical methods for nonlinearly constrained optimization*, Ph.D. thesis, Department of Computer Science, Stanford University, 1976.

[197] M.H. Wright, Interior methods for constrained optimization, *Acta Numerica* 1 (1992), 341–407.

[198] M.H. Wright, The interior-point revolution in optimization: History, recent developments, and lasting consequences, *Bulletin of the American Mathematical Society* 42 (2004), 39–56.

[199] S.J. Wright, *Primal-Dual Interior-Point Methods*. Philadelphia: SIAM, 1997.

[200] Y. Ye, *Interior-point Algorithms: Theory and Analysis*. New York: John Wiley & Sons, 1997.

[201] W.H. Young, On classes of summable functions and their Fourier series, *Proceedings of the Royal Society London, Series A* 87 (1912), 225–229.

[202] P.B. Zwart, Nonlinear programming: Counterexamples to two global optimization algorithms, *Operations Research* 21 (1973), 1260–1266.

Index

Active constraint, 377, 435
Active Set
    Algorithm, 442
    application to (LIP), 441
Activity analysis problem, 10
Adding more activities, 110
Affine, 24
    hull, 51, 553
    minorant, 574
Altman condition, 308
Anti-cycling procedure, 100
Approximation
    piecewise linear, 259
Arc, 470
    differentiable, 471
    feasible, 470
    tangent to, 471
Artificial variables, 86
Asymptotic
    error constant, 277
    order of convergence, 277

Ball, 549
    closed, 549
    open, 549
Barrier function
    algorithm, 520
    logarithmic, 467
    methods, 467, 518–524
    primal, 525–529
    primal-dual, 529–533
    versus penalty function, 522
Basic

    feasible solution, 63, 188, 189
    solution, 54
    variables, 56, 460
Basis, 47, 52
    feasible and infeasible, 63
    in A, 53
    starting, 197
Benders Decomposition, 140, 220–224
Big-M Method, 91
Bisection Method, 279
Bisymmetric matrix, 464
Bland's Rule, 99
Block-angular structure, 208
Boundary point, 552
Bounded-variable
    linear program, 30, 187–196
        basic feasible solution, 188
        optimality condition, 188
    nonlinear program, 446
Box-constrained problems, 446
Bracken, J., 258
Breakpoints, 166
    lambda (λ), 167, 173
Broyden, C.G., 348

Canonical form of an LP, 66–67
Cardinality of a set, 51, 537
Carrying plane of a set, 553
Cartesian product of sets, 539
Cauchy-Schwartz inequality, 561
Changing
    model dimensions, 140, 141

© Springer Science+Business Media LLC 2017
R.W. Cottle and M.N. Thapa, *Linear and Nonlinear Optimization*,
International Series in Operations Research & Management Science 253,
DOI 10.1007/978-1-4939-7055-1

CPSIA information can be obtained
at www.ICGtesting.com
Printed in the USA
LVOW13*0601080618
580064LV00009B/59/P